Fourth Edition

MARVIN C. MEYER & O. WILFORD OLSEN'S

Essentials of PARASITOLOGY

BY GERALD D. SCHMIDT

MARVIN C. MEYER
University of Maine

O. WILFORD OLSEN
Colorado State University

GERALD D. SCHMIDT
University of Northern Colorado

wcb
Wm. C. Brown Publishers
Dubuque, Iowa

This book is dedicated to Dr. Marvin C. Meyer and to Dr. O. Wilford Olsen. They are mentors to us all.

Cover photo by Lauritz A. Jensen

Library of Congress Catalog Card Number: 87–70259

ISBN 0–697–04932–9

Printed in the United States of America by Wm. C. Brown Publishers
2460 Kerper Boulevard, Dubuque, IA 52001

10 9 8 7 6 5 4 3 2 1

Contents

Preface

When I entered the field of parasitology as a graduate student, both Professors Marvin C. Meyer and O. Wilford Olsen were internationally recognized as leaders in education and research in the discipline. I have always felt a sense of pride in having completed my master's and doctor's degrees under the stern, demanding leadership of Dr. Olsen, a man I later came to know as a compassionate friend. When I was asked to prepare the fourth edition of *Essentials of Parasitology,* I felt a close bond with these men and with the ideals that they have exemplified.

I have followed the basic format of previous editions but added and deleted as the forces of an ever-changing field have dictated. First, I have updated the classifications of various phyla, especially the Subkingdom Protozoa, which has changed so much in recent years. Because it is now very difficult to collect specimens of most wild birds and mammals, I deleted many exercises that depended on such hosts. Of those exercises that remain, I have indicated commercial sources of supply, so far as they exist. Of course, each instructor will have a personal collection upon which to draw for more esoteric exercises.

Of the life cycle exercises in previous editions, I have deleted some, added others, and placed them in Appendix 3. Appendix 1, Reagents and Solutions, will continue to be a valuable resource for students from their first course in parasitology throughout the rest of their careers.

New pedagogy includes end-of-chapter review questions; new terms identified by boldface type; space for student notes and sketches; and thoroughly updated references.

This manual is not considered to be a textbook, but rather a supplemental guide to laboratory procedures, both in the classroom and in the research laboratory. Several excellent textbooks are currently available, and this manual is designed to complement any of them. Its approach is classical, because we know that one cannot run before learning to walk, nor conduct research at the frontier of parasitological knowledge without a basic understanding of the parasites themselves.

I would like to thank the following people for reviewing the manuscript and offering many helpful comments throughout the revision: Omar M. Amin, University of Wisconsin–Parkside; Ralph P. Eckerlin, George Washington University; John P. Harley, Eastern Kentucky University; Antonio E. Harrises, Salem State College; Lauritz A. Jensen, University of Health Sciences; John S. Mackiewicz, State University of New York at Albany; Donald A. Munson, Washington College; John Murad, Louisiana Tech University.

Gerald D. Schmidt
Greeley, Colorado

SECTION 1 The Single-celled Animal Parasites

Chapter 1 Subkingdom Protozoa

This subkingdom includes many species, most of which are parasitic. They infect all kinds of animals (and some plants), including other protozoans and multicellular creatures as large as humans, elephants, and whales. In some cases they are among the most serious of parasites, being the cause of much morbidity and death.

In parasitology, as in other dynamic fields of zoology, the classification of this huge group of parasites fluctuates. The taxonomic scheme used here follows that of a prestigious international committee of experts. Undoubtedly, even this classification will change soon. An abbreviated classification in most current use is presented in table 1.1.

Table 1.1. The Classification of the Parasitic Protozoa.

Phylum	Subphylum	Class	Subclass	Order	Suborder	Family
SARCOMASTIGOPHORA	Mastigophora	Zoomastigophorea		Kinetoplastida	Trypanosomatina	Trypanosomatidae
				Retortamonadida		Retortamonadidae
				Diplomonadida		Hexamitidae
				Trichomonadida		Trichomonadidae
				Hypermastigida	Trichonymphina	
	Opalinata	Opalinatea		Opalinida		Opalinidae
	Sarcodina	Lobosia		Amoebida		Endamoebidae
				Schizopyrenida		Schizopyrenidae
APICOMPLEXA		Sporozoea	Gregarinia	Eugregarinida	Acephalina	Monocystidae
					Cephalina	Gregarinidae
			Coccidia	Eucoccida	Adeleina	Haemogregarinidae
					Eimeriina	Eimeriidae
						Sarcocystidae
					Haemosporina	Plasmodiidae
						Haemoproteidae
						Leucocytozoidae
			Piroplasmia			Babesiidae
MYXOZOA		Myxosporea		Bivalvulida		Myxosomatidae
				Multivalvulida		
MICROSPORA		Microsporea		Microsporida		Nosematidae
CILIOPHORA		Kinetofragminophorea	Vestibulifera	Trichostomatida		Balantidiidae
		Oligohymenophorea	Hymenostomatia	Hymenostomatida		Ophryoglenidae
			Peritrichia	Peritrichida		Urceolariidae
		Polymenophorea	Spirotrichia	Heterotrichida	Clevelandellina	Plagiotomidae

After Schmidt and Roberts (1985), with slight modification.

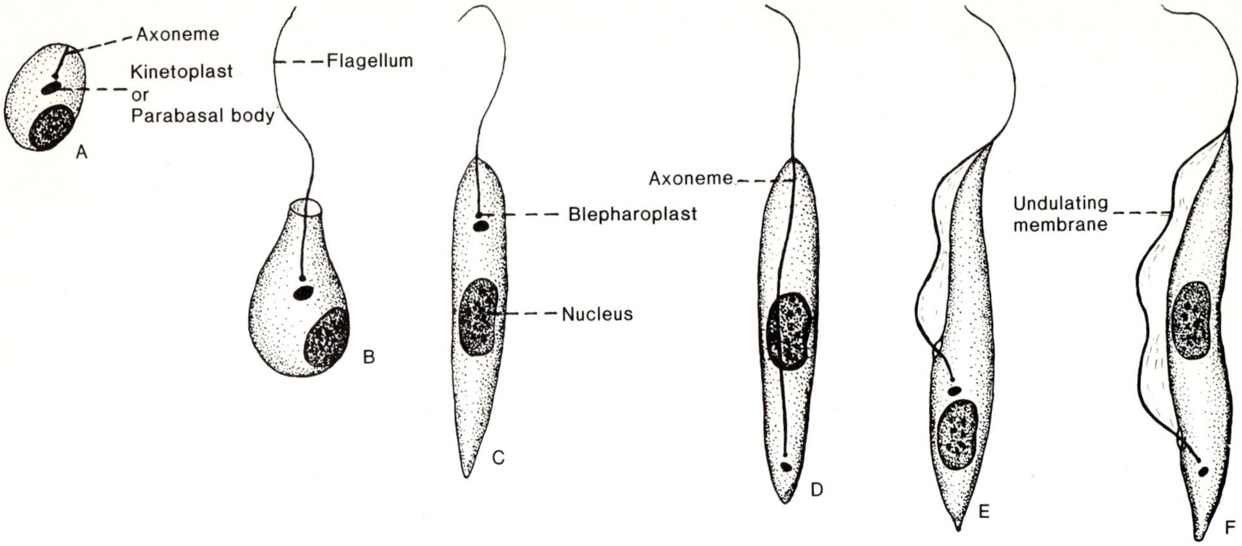

Fig. 1.1. Types of developmental stages of Trypanosomatidae. (A) Amastigote; (B) Choanomastigote; (C) Promastigote; (D) Opisthomastigote; (E) Epimastigote; and (F) Trypomastigote.

Phylum Sarcomastigophora, Subphylum Mastigophora, Class Zoomastigophorea

The forms included in this class bear one or more whiplike flagella and lack chloroplasts. The group contains many species parasitic in vertebrates, invertebrates, and plants, as well as free-living forms.

Order Kinetoplastida

Family Trypanosomatidae: The Hemoflagellates

Hemoflagellates are characterized by a spindle-shaped body containing a central nucleus, a **basal body,** or **kinetosome,** from which the flagellum arises, and a **kinetoplast,** a relatively large, deeply staining body situated near the base of the flagellum. Since the kinetoplast is a mass of DNA within the single mitochondrion, it stains similar to the nucleus and is seen with a light microscope. Some forms bear an **undulating membrane.** The group includes both pathogenic and nonpathogenic species. Most are **heteroxenous,** living in two hosts during their life cycle, while some are **monoxenous,** existing within a single host, such as an arthropod.

The members of this group are parasites in the alimentary canals of insects and other invertebrates, in milkweed latex, or in the blood or tissues of vertebrates. Life cycles may be extremely complicated, in some cases involving several different developmental stages, and will be illustrated in the laboratory exercises to follow.

Hoare and Wallace (1966) proposed a system of terminology based upon characteristics of the flagellum, its arrangement in the body as determined by its starting point (indicated by the position of the kinetoplast), and its course and point of emergence (fig. 1.1).

The word root "**-mastigote,**" combined with appropriate prefixes, forms the following descriptive terminology:

1. **Amastigote,** represented by rounded forms devoid of an external flagellum, as in the genus *Leishmania.*
2. **Promastigote,** represented by forms with an antenuclear kinetoplast, and a flagellum arising near it and emerging from the anterior end of the body, as in the genus *Leptomonas.* Promastigote forms are intermediate stages in the life cycle of many hemoflagellates.
3. **Opisthomastigote,** represented in the genus *Herpetomonas* by a form with a postnuclear kinetoplast, a flagellum arising near it and then passing through the body and emerging from its anterior end. This form is found in *Herpetomonas muscarum* in the intestine of houseflies.
4. **Epimastigote,** represented by forms with a juxtanuclear kinetoplast, a flagellum arising near it and emerging from the side of the body to run along a short undulating membrane; in genus *Blastocrithidia* and in

stages of *Trypanosoma* and *Herpetomonas. Blastocrithidia gerridis* is a common parasite of water striders and hence is an easily obtainable species for laboratory examination. **Be careful**—water striders can bite!

5. **Trypomastigote,** represented by forms with a postnuclear kinetoplast, a flagellum arising near it and emerging from the side of the body to run along a long undulating membrane, as in genus *Trypanosoma.* The trypomastigotes represent the final developmental stage in most species of the genera *Trypanosoma* and *Herpetomonas.* This form will be examined in preparations of *Trypanosoma cruzi* and *T. brucei* subspp.

6. **Choanomastigote,** usually with an antenuclear kinetoplast, a flagellum arising from a wide, funnel-shaped pocket and emerging at the anterior end of the body (typical of genus *Crithidia*).

Life Cycles of Hemoflagellates

Two types of life cycles occur. In one, transmission is generally by ingestion of infective stages passed in the feces of infected animals. These include species of *Leptomonas, Crithidia, Blastocrithidia,* and *Herpetomonas,* which occur in insects and arachnids. In the other, transmission is by sucking arthropods and leeches. Two means of entering the vertebrate host occur here. In one, the infective stage is injected directly during the bloodsucking activity of the infected invertebrate host. This is known as infection from the anterior station, i.e., from the mouth, and is inoculative in nature. In the other method, the infective form is voided in the feces while the arthropod sucks blood and is later ingested or rubbed into abrasions of the skin. This method is known as infection from the posterior station and is contaminative in nature (fig. 1.2).

Multiplication, involving two or more of the developmental stages, usually takes place in the invertebrate host. A few species have developed independence from the invertebrate host and are transmitted venereally, as in the case of *Trypanosoma equiperdum* of horses. In this case, all multiplication occurs in the vertebrate host.

Most species may be transmitted mechanically on the mouthparts of bloodsucking arthropods which, if interrupted while feeding on an infected animal, quickly resume feeding on an uninfected one. Surgical instruments, likewise, may serve as a means of mechanical transmission.

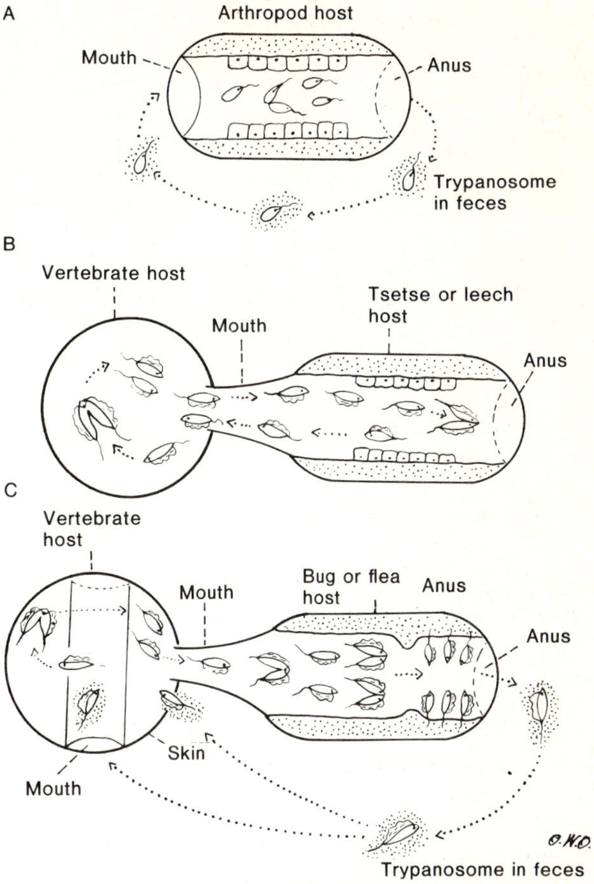

Fig. 1.2. Three basic life cycles of Trypanosomatidae.
A. *Direct infection. Leptomonas, Crithidia, Blastocrithidia,* and *Herpetomonas* voided in the feces of spiders and insects are infective directly to other hosts when ingested by them. This one host cycle is the monoxenous type.
B. *Indirect infection by inoculation. Trypanosoma brucei, vivax, congolense,* and other species are inoculated directly into the bloodstream of the vertebrate host by the mouthparts of the bloodsucking tsetses. Coming from the mouth, the infection is said to be from the anterior station. Trypanosomes are transmitted in this manner by leeches to fish and *Leishmania* to humans by sandflies. This and the following two-host cycle (C) are the heteroxenous type.
C. *Indirect infection by contamination.* Infection of rats with *Trypanosoma lewisi* and human with *T. cruzi* is by the contaminative method from the feces. Fleas for *lewisi* and bugs for *cruzi* defecate while feeding. Great numbers of infective trypanosomes are voided in the feces. When these are swallowed by the vertebrate host or contaminate its abraded skin, infection follows. With the infective stages appearing in the feces of the invertebrate host, infection is said to be from the posterior station.

Trypanosoma brucei gambiense and *T. b. rhodesiense*

These two subspecies are morphologically identical, so examination of either species will suffice for the other. Two forms of **African sleeping sickness—Gambian** and **Rhodesian**—are caused by these protozoans. Further, the parent subspecies, *Trypanosoma brucei brucei,* is responsible for the immensely important disease of livestock known as **nagana.** Their vector is any of several species of **tsetse fly,** *Glossina* (see p. 223).

Examine a slide of infected blood containing trypomastigotes. With the oil-immersion lens it should be possible to see the undulating membrane, flagellum, nucleus, and kinetoplast, which is rather small in this species.

Using a low-power lens, examine a tsetse fly mounted on a slide. How does it compare in size and shape with a housefly? Actually it is fairly closely related to the common housefly, but as you can see it has cutting-sponging mouthparts, rather than the wetting-sponging type found in houseflies.

Trypanosoma cruzi

This species is distributed throughout most of South, Central, and North America, where it causes **Chagas' disease** in humans. Although human cases are rare in the United States, the parasite has been found in a wide variety of reservoir mammalian hosts. Several species of **"kissing bugs,"** or **cone-nosed bugs** (Reduviidae, Triatominae) are alternate hosts and vectors. The flagellate is pleomorphic, changing its form according to the host, organ, and tissue it inhabits.

Search a stained blood film for the trypomastigote, which has a prominent kinetoplast and commonly assumes a question-mark shape. Typically the number of bloodstream forms is low, so considerable effort may be required to find one.

Stained sections of infected heart or intestine will reveal a few **pseudocysts,** containing clusters of amastigotes. Prepared smears of sliced, infected heart will allow more detailed examination of amastigote morphology, as well as intermediate stages between amastigote and trypomastigote.

When this parasite is grown in vitro, it reverts to the epimastigote form, typical of its appearance in the gut of its insect host. Examine a slide prepared from such a culture.

The most common triatomine genera that serve as vectors of *T. cruzi* are *Triatoma, Rhodnius* and *Panstrongylus* (see p. 227). Using a dissecting microscope, carefully examine one or more of these insects. Note that the head is narrow, with large eyes located midway or far back on the sides of the head. Two ocelli are usually present behind the eyes. The slender antennae have four segments. Bugs that are similar in appearance are common, and it requires some expertise to differentiate them. Triatomines are secretive, seldom being seen in daylight, and so are rarely caught by the casual collector. **Warning:** all bugs in this family are capable of inflicting a very painful bite and so must be handled with care.

Leishmania donovani

The amastigote form is dominant in vertebrates infected with this genus. Several mammalian species, including humans, suffer visceral infection, especially in the spleen and liver, causing the disease **kala-azar.** Bloodsucking **sandflies** of the genus *Phlebotomus* in the Eastern hemisphere and *Lutzomyia* in the Western hemisphere are the vectors (see p. 219).

Microscopic examination of a smear or section from an infected spleen will reveal abundant parasites. One must be careful to differentiate them from spleen and blood cells, which are much larger and do not contain the characteristic kinetoplast.

In vitro, the parasite changes to the promastigote form, corresponding to the insect-dwelling stage.

Trypanosoma lewisi

This species commonly occurs in wild rats. It can be reared easily in culture or in great numbers by injecting a small volume of infected blood or cultured flagellates into a young white rat.

Examine living trypanosomes in a drop of blood drawn from a heavily infected rat, viewing it with both the low and high powers of the microscope. Close the diaphragm to the point that it gives good contrast to see the otherwise transparent organisms. When available, a phase-contrast microscope is preferable.

Observe the activity of the trypanosomes. Red blood cells in the vicinity of each trypanosome are moving in response to its rapid lashing action. Introduce a drop of neutral red vital stain under the cover glass to color the trypanosomes and facilitate study of their actions.

Study stained smears of blood taken from experimentally infected rats shortly after infection and at the end of the first and second weeks postinfection.

Multiplication is by binary division which at first is incomplete, resulting in rosettes of small epimastigotes still attached by their posterior ends. These rosettes are most prevalent during the early period of infection when the trypanosomes are multiplying rapidly. Look for them.

Smears made at the end of the first week show smaller trypanosomes than those made two weeks or longer after infection.

Note the shape of the body, the nature of the undulating membrane, and the location of the internal organs, which should be identified.

Other trypanosomes that may be obtained for comparison both as living and stained specimens include *T. rotatorium* from frogs and *T. avium* from birds (transmitted by certain species of *Simulium* **blackflies** [Desser et al., 1975]). More *T. rotatorium* occur in the blood from kidneys and liver of infected frogs than in peripheral blood, and *T. avium* is more abundant in the bone marrow. A small amount of marrow from the femoral bone macerated with a toothpick in a drop of saline solution is more likely than blood smears to reveal the trypanosomes of infected birds. A drop of neutral red will render the hemoflagellates more visible.

Order Retortamonadida

Family Retortamonadidae

Two species of this family are found as harmless commensals of humans. Recognition of them is important so that they will not be mistaken for pathogenic forms.

Chilomastix mesnili

Examine prepared slides of **trophozoites** and **cysts.** The trophozoite body shape is pyriform, with the anterior end bluntly rounded, and the posterior long and drawn out (see fig. 1.3). The cytoplasm is reticulated and contains bacteria in food vacuoles. The spherical, vesicular nucleus is located near the anterior end of the body, has a distinct membrane, and a small eccentrically located endosome. Three flagella arise from the kinetoplast, situated slightly anterior to the nucleus; a fourth, shorter flagellum lies in the cytostomal cleft. Cysts are lemon shaped, with a nipplelike structure anteriorly, surrounded by thin walls except where thickened at the nipple end. The internal structure resembles that of the trophozoite.

The life cycle is direct, being transmitted only in the cystic stage, as trophozoites cannot survive in the acidic environment of the stomach.

Retortamonas intestinalis is basically similar to *C. mesnili* but is only 4 μm to 9 μm long in the trophozoite stage. It is not common in humans, being found mainly in monkeys and chimpanzees.

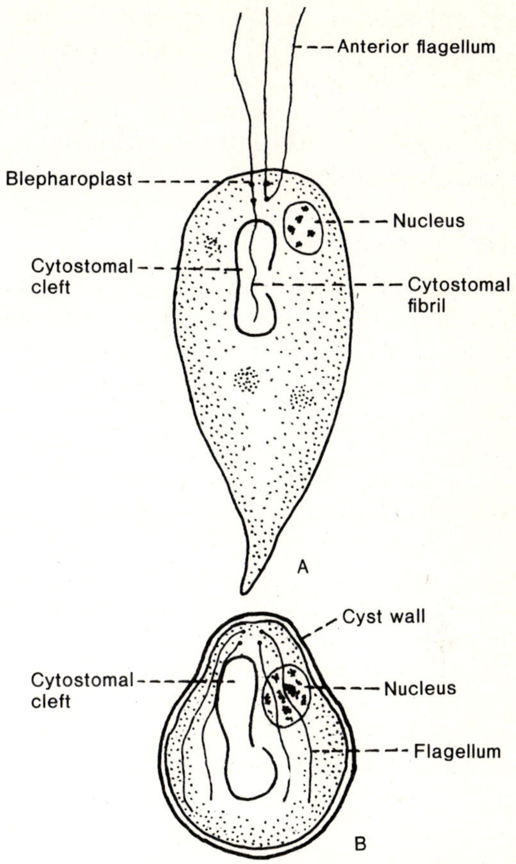

Fig. 1.3. *Chilomastix mesnili.* (A) Trophozoite; (B) cyst.

Order Diplomonadida

Family Hexamitidae: Flagellates of the Small Intestine

Organisms of this family are easily recognized because they have two equal nuclei lying side by side. Further, they are the only protozoa found in the small intestine of humans.

Giardia lamblia

This parasite (fig. 1.4) is common in humans and other mammals, especially in areas where drinking water is contaminated by feces. Wild and domestic animals, including beavers, sheep, and dogs, are **reservoir** hosts.

Trophozoites are passed in abundance during the early, diarrheic phase of infection. Examine a stained fecal smear for this quaint, tennis racquet-shaped stage. There are four pairs of flagella, but these usually cannot be seen except in special preparations. The nuclei each have a very large endosome. One or two dark staining **median bodies** lie across the posterior third of the body. Along the median line are two parallel rods, the **axonemes** of the ventral flagella (with associated microtubules).

Cysts of this species begin to appear in the stools later in the infection. Young cysts may have only two nuclei, but most will have four.

Find a cyst, with the high dry lens of your microscope, then advance to oil immersion. Note the rather elongate shape and an apparent space between the protoplasm and cyst wall. **Ghost cysts** are commonly encountered. These have only a small residual mass inside the otherwise normal cyst wall. Why they were not successful is not known.

Giardia muris, a species not infective to humans, can be found in species of wild mice, rats, and voles.

Order Trichomonadida

Family Trichomonadidae: Flagellates of the Alimentary and Reproductive Systems

Trichomonads are characterized by a tuft of flagella at the anterior end of the body, one of which is recurrent along a short undulating membrane. This flagellum may or may not continue as a free member. There is a single, large nucleus lacking a conspicuous nucleolus. A stout, median rod, the **axostyle,** extends the length of the body to emerge as a peglike appendage at the

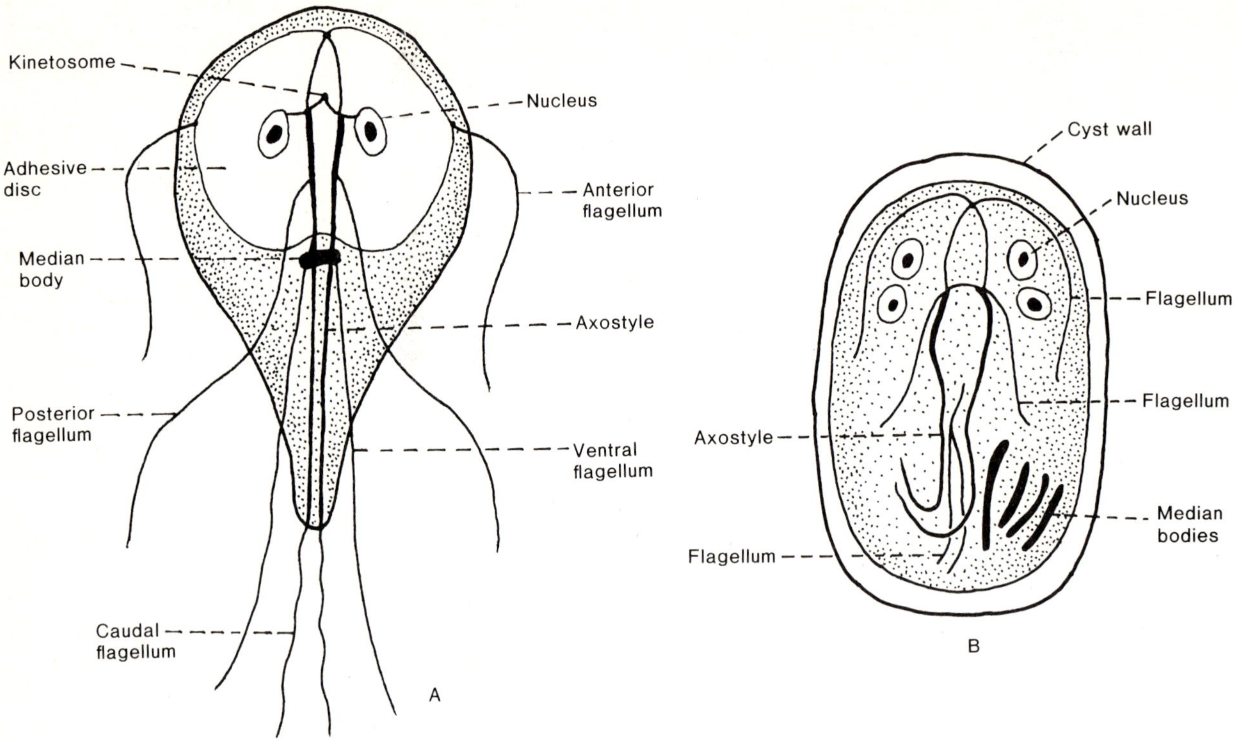

Fig. 1.4. *Giardia lamblia.* (A) Trophozoite, ventral view; 12–15 µm long. (B) Cyst; about 7–12 µm long.

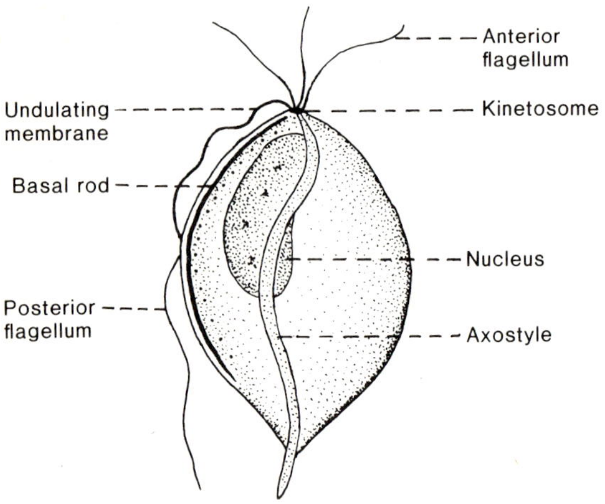

Fig. 1.5. *Trichomonas vaginalis.* The trophozoite is 7–32 µm long.

posterior end. The morphologies of all the genera and species in the family are similar. Examine whatever species may be available.

Trichomonas tenax

As in all species in this family, there is no cyst stage. A harmless commensal, *T. tenax* is found in the mouth of humans. It cannot survive outside the body, and its main mode of transmission is by kissing. Examine a prepared slide for the characteristics of the family. Using a blunt toothpick, carefully scrape a bit of debris from your teeth and place it in a drop of saliva on a slide. What can you see with your microscope? You will find a lot of living things. Is one of them *T. tenax?*

Trichomonas vaginalis

This species is an important pathogen of men and women, although women usually suffer from it more than men. It lives in the genital and urinary tracts of both sexes and is usually transmitted by sexual contact. Its morphology is nearly identical to that of *T. tenax,* being somewhat larger and also differing in minute morphological details (see fig. 1.5).

Examine a prepared slide that demonstrates this species.

Pentatrichomonas hominis

This third trichomonad of humans is a harmless commensal of the large intestine (fig. 1.6). Most trophozoites bear five anterior flagella, placing them in a different genus from *Trichomonas,* which usually has three or four. Laboratory material may be difficult to obtain, but a very common, morphologically identical species is always available in the form of *Pentatrichomonas muris* (fig. 1.7). Found in the cecum of white mice, it is a handy laboratory demonstration.

Open the cecum of a freshly killed mouse. Place a tiny amount of its contents into a drop of saline on a slide, and add a cover glass. It is desirable to study the

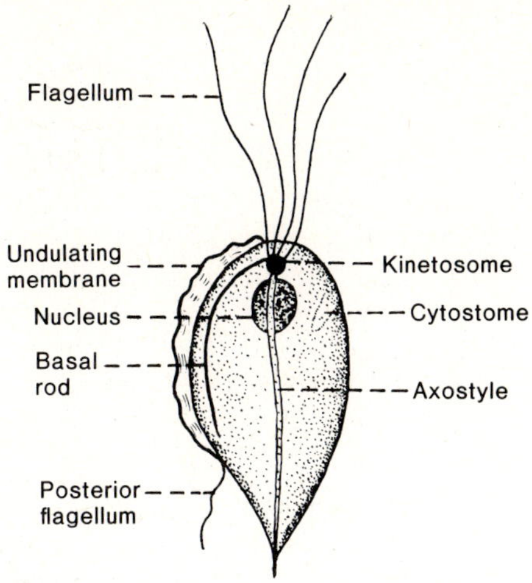

Fig. 1.6. *Pentatrichomonas hominis.* The trophozoites are 9–20 μm long.

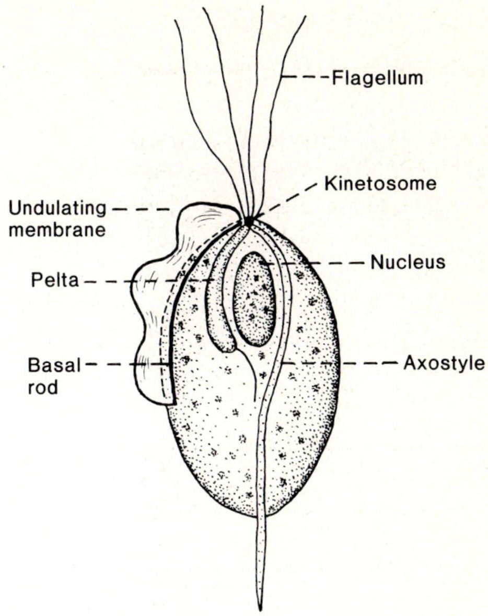

Fig. 1.7. *Pentatrichomonas muris.* The trophozoites are 8–20 μm long.

flagellates with the oil-immersion lens. This can be done only if the cover glass is stuck to the slide, so that it will not move up and down as the microscope is focused. Thus, before applying the cover glass, place a small amount of vaseline or other grease to the slide or underside of the cover glass.

You will see myriads of organisms moving about at random, even with a low-power lens. Spirillum and rod-shaped bacteria are hardly more abundant than are trichomonads. Increase magnification. As the flagellates cool down from mouse body temperature, they will slow their movements, eventually stopping altogether. At this time, with oil immersion, the undulating membrane and even flagella are clearly visible.

Dientamoeba fragilis

Even though this cosmopolitan commensal of humans has no free flagella, it contains ultrastructural rudiments of flagellar basal structures (fig. 1.8). The only species in the genus, it infects about 4% of the human population. It sometimes is associated with diarrhea.

Examine a stained population of *D. fragilis.* Only trophozoites are known; cysts apparently do not occur. They are 6 to 12 μm long. About 60% of them have two nuclei.

The mode of transmission of this organism seems to be within the egg of a pinworm, *Enterobius vermicularis* (see p. 159).

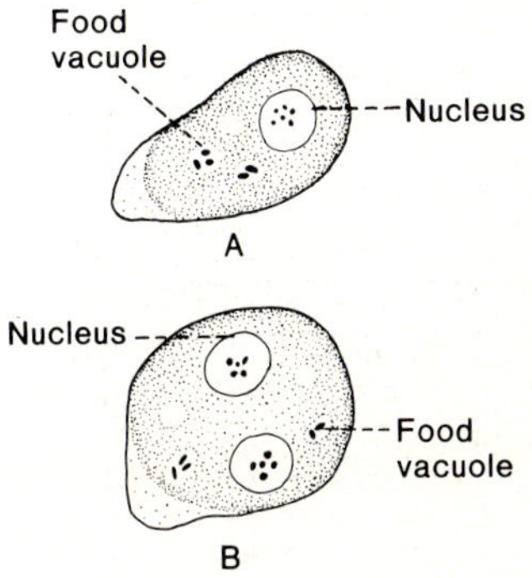

Fig. 1.8. *Dientamoeba fragilis.* (A) Mononucleate form; (B) binucleate form. They are 3.5–12 μm long.

Order Hypermastigida

These complex flagellates are mutualistic symbionts of termites and other insects, such as woodroaches. Because all termites are infected with them, they are convenient to demonstrate in a laboratory exercise.

Crush the head of a live termite or woodroach with forceps. Place the body in 0.4% saline, and tease it apart with needles. Several kinds of flagellates should be present, most conspicuously *Trichonympha.* These large, pear-shaped organisms are covered with hundreds of long flagella on their anterior half. They have a single nucleus and numerous food vacuoles.

Subphylum Opalinata

Opalinids are covered with many oblique rows of cilia, causing them to resemble ciliates. However, they have numerous nuclei of similar kind, whereas ciliates have two kinds of nuclei. They inhabit the rectum, urinary bladder, and cloaca of amphibians and occasionally fishes. Representative genera are *Opalina, Protoopalina, Zelleriella,* and *Cepedea.*

Opalina obtrigonoidea

This and other species of the genus have an oval, flattened body covered with rows of cilia and devoid of mouth, cytopyge, and contractile vacuoles. The body is uniformly covered with cilia arranged in rows. The body contains many nuclei of the same kind. Fully developed individuals exceed 300 μm in length and appear as opalescent bodies to the naked eye. Examine living specimens in saline solution and slides of stained ones.

Protoopalina mitotica

This species and others of the genus commonly occur in the colons of amphibians. The body is spindle shaped and has two large nuclei of equal size. Examine permanent mounts of fixed and stained specimens, as well as living individuals in saline solution to which some vital dye such as neutral red has been added.

Notes and Sketches

Subphylum Sarcodina, Class Lobosea

This class includes the typical amebas and a few species of interest to parasitologists. They have no tests, and their pseudopodia are lobose or filiform.

Order Amoebida

These are uninucleate forms with no flagellate stage.

Family Endamoebidae

Endamoebidae contains the genera *Entamoeba, Iodamoeba,* and *Endolimax,* among others, which occur in the alimentary canal of humans as well as in many other species of animals. Most of the species in these genera are harmless commensals, but some are serious parasites.

The species are differentiated primarily on the basis of the nuclear structure (table 1.2).

The life cycle generally consists of four stages: (1) ameboid **trophozoites,** (2) **precysts,** (3) **cysts,** and (4) **metacystic trophozoites,** also called **amebulae** (fig. 1.9).

Entamoeba gingivalis

This is an exception in which only a trophozoite stage occurs. It lives between the base of the teeth and gums in many persons, where ample material for study can be obtained. With the aid of a small mirror and the flat end of a clean wooden toothpick dipped in 95% ethyl alcohol and dried before use, material can be removed from around the base of the teeth for examination. If all students in the laboratory cooperate, someone may be found to be infected and able to provide material for all to see. The older the person, the more likely there will be an infection.

Both the slide and cover glass should be warmed and the material from around the teeth liquefied with a drop of clear saliva. Examine the preparation with high power of the microscope with the diaphragm closed down to give contrast. When available, a phase-contrast microscope is preferred. Epithelial tissue from the gums may be present and must be recognized. White blood cells also are easily misidentified as *E. gingivalis.* Close scrutiny of the nucleus quickly provides a clue to the cell's identity.

Table 1.2. Characteristics of Common Intestinal Amebae of Humans.

Characteristic	*Entamoeba histolytica*	*Entamoeba coli*	*Endolimax nana*	*Iodamoeba buetschlii*
TROPHOZOITES Size	20–30 μm	15–50 μm	6–15 μm	4–20 μm
Cytoplasm	Finely granular, nonvacuolated; ingested RBC's diagnostic	Coarsely granular; vaculated; bacteria, yeast ingested	Clear and finely granular; bacterial inclusions	Finely granular; bacterial inclusions
Nucleus (Stained)	Fine membrane, delicate beading; endosome central	Coarse membrane, coarse beading; endosome eccentric	No beading; large single, deeply stained endosome	No beading; large single, deeply stained endosome
CYSTS Size, shape	Roundish; 10–20 μm	Roundish; 10–33 μm	Ovoidal; 5–14 μm	Irregular; 6–15 μm
Nuclei number	Mature, 4	Mature, 8	Mature, 4	Mature, 1
Nuclei structure	As in trophozoites	As in trophozoites	As in trophozoites	Eccentric endosome beside granular mass
Chromatoidal bodies	Thick or slender bars; with rounded ends	Like glass splinters or irregular	No true chromatoids	No true chromatoids
Glycogen mass (in iodine)	Diffuse, brown. In young cysts	Large, deep brown. In young cysts	Occasionally present in young cysts	Large or small, sharply delimited, deep brown

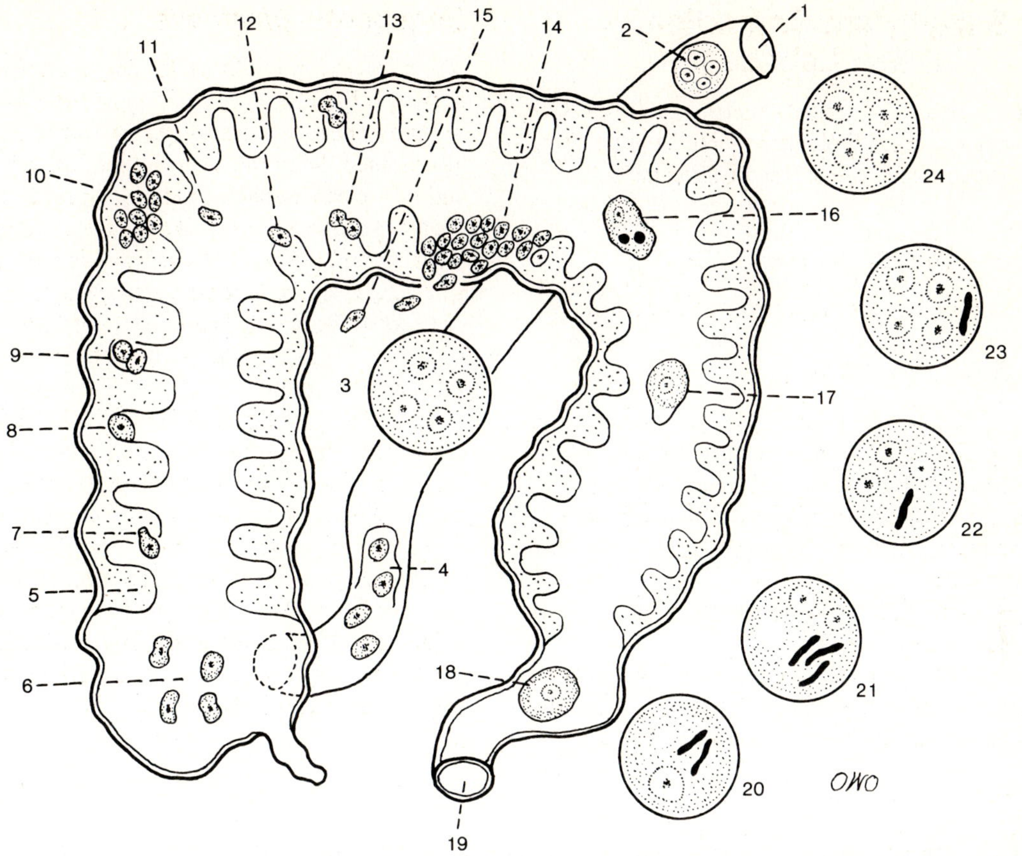

Fig. 1.9. Life cycle of *Entamoeba histolytica*
A. *In small intestine* (1–4). 1. Small intestine leading from mouth; 2. mature tetranucleate cyst having entered intestine via mouth with contaminated food or water; 3. ripe cyst; 4. excystation and freeing of infective entamebulae.
B. *Multiplication phase and pathology in large intestine* (5–15). 5. Intestinal mucosal folds or villi; 6. four amebulae free in lumen of colon; 7. amebula attacking mucosa at tip of villus; 8. amebula in crypt of mucosa; 9. multiplication of amebula by binary fission; 10. nest of amebulae in crypt invading mucosa; 11. amebula freed from crypts; 12. amebula attacking mucosa; 13. binary fission; 14. number of amebulae have destroyed large area of mucosa and intestinal wall; 15. some amebulae enter blood vessels of hepatic portal system and are carried to liver, lungs, and brain, where they colonize and multiply.
C. *Preparation in colon for encystment* (16–18). 16. Ameba containing two erythrocytes; 17. precystic stage, food material has been eliminated; 18. rounded precystic stage passing from sigmoid colon; 19. anus.
D. *Development of cyst to infectivity outside body of host* (20–24). 20. Uninucleate cyst containing chromatoidal bars and vacuole; 21. binucleate cyst; 22. trinucleate cyst in process of nuclear division; 23. tetranucleate cyst with remnant of chromatoidal bars, no vacuole; 24. mature infective cyst.

The living trophozoites measure up to 35 μm in diameter, averaging 10 to 20 μm. Active individuals show a band of clear ectoplasm surrounding the grayish granular endoplasm. The pseudopodia are constantly changing in shape and size as the organism moves around on the slide. Food vacuoles containing bacteria, leukocytes, and epthelial cells in various stages of digestion are present. See figure 1.10 for an example of *E. gingivalis*.

Entamoeba terrapinae

Living trophozoites usually are available in cultures from biological supply houses. Observe the live specimens, noting the clear ectoplasm and darker, more granular endoplasm whose food vacuoles contain bacteria and particles of media.

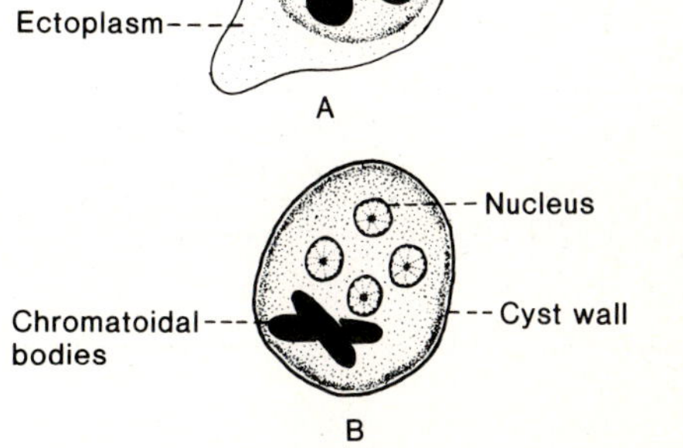

Fig. 1.10. *Entamoeba gingivalis.* The trophozoite is 12–30 μm wide.

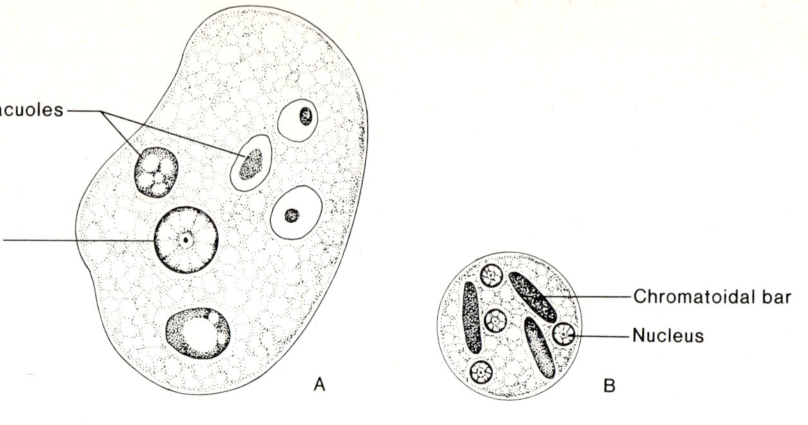

Fig. 1.11. *Entamoeba histolytica.* (A) Trophozoite; and (B) cyst. Trophozoites usually are 18–25 μm wide; the cysts are 10–20 μm wide. (From Schmidt, G. D., and L. S. Roberts. 1985. *Foundations of Parasitology,* 3d. ed. The C. V. Mosby Co., St. Louis.)

Entamoeba histolytica

This is the pathogenic species that invades the tissues of the human large intestine and elsewhere. A small form that is like histolytica but with slight anatomical differences is known as *E. hartmanni.* Other related amebae which may deserve consideration are *E. dispar* and *E. polecki.* For our purpose, efforts will be directed to recognition of *E. histolytica,* the pathogenic species (fig. 1.11).

Properly fixed trophozoites stained in iron hematoxylin are roundish in shape and have a bluish gray reticulate cytoplasm. It is differentiated into clearer ectoplasm and denser endoplasm. They usually measure 20 to 30 μm in diameter. Red blood cells are commonly ingested and undergo digestion in the food vacuoles. The nucleus is spherical, measuring 4 to 7 μm in diameter, and is highly vesicular. There is a small centrally located endosome or karyosome. Small, evenly distributed chromatin granules appear on the inner surface of the thin nuclear membrane.

The occasional presence of red blood cells, the delicate nuclear membrane lined with small chromatin particles, and minute, centrally located endosome are important diagnostic characteristics of the trophozoites of this species.

The cystic phase consists of precystic and encysted forms. The precystic entamebae expel the food material and assume a spherical shape. Since they resemble encysted forms, they are useful in diagnosing infections.

Cystic forms are spherical bodies 10 to 20 μm in diameter. The cyst wall is unstained and hyaline in appearance, the finely reticulated cytoplasm is bluish grey, and the nuclear membrane and granules black. The nucleus is similar in appearance to that of the trophozoite. In addition to the nucleus, the cytoplasm of young cysts contains one to several large black **chromatoidal bodies** with rounded, smooth ends. Round spaces appearing in the cytoplasm are glycogen vacuoles. As maturation of the cyst progresses, the nucleus divides once to form two small nuclei, each of which divides to produce a total of four in the fully developed cyst. Occasionally, additional divisions of the nuclei result in aberrant forms with more nuclei. In ripe cysts, the chromatoidal bodies are lacking, having been absorbed. After excystment the cell divides into four **amebulae,** each with one nucleus.

The nuclear membrane is lined with delicate chromatin granules, the minute, centrally located endosome, four small nuclei, and thick chromatoidal bodies with smooth, rounded ends are important diagnostic characteristics of cysts of this species. Size in itself is of little value since the cysts of other species are similar.

Formalinized specimens stained in Lugol's solution are used in routine diagnoses. The cytoplasm is lemon or greenish yellow in color and the nuclear material and chromatoidal bodies refractile. Glycogen vacuoles, when present, are yellowish brown. The cyst wall does not stain.

The life cycle of *Entamoeba histolytica* is shown in figure 1.9.

Entamoeba coli

This is a common nonpathogenic species occurring in the human colon. (See fig. 1.12.)

In most individuals, the body outline is somewhat rounded and may show division between ectoplasm and endoplasm. The food vacuoles contain bacteria and fecal debris but not red blood cells. The nuclear membrane is thicker than that of *E. histolytica.* The peripheral chromatin granules are coarser, often in block

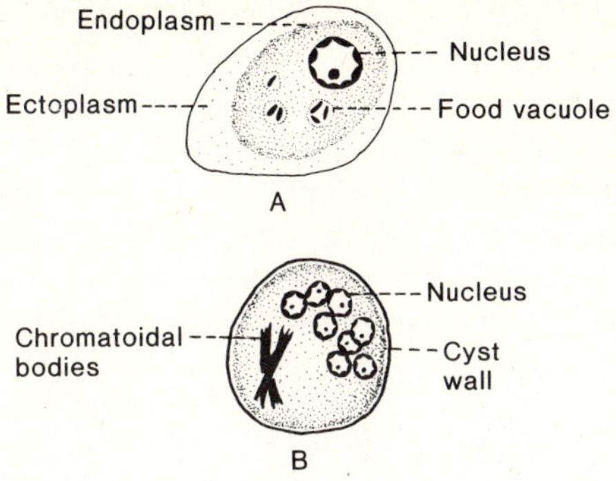

Fig. 1.12. *Entamoeba coli.* (A) Trophozoite, 20–30 μm wide; (B) cyst, 10–25 μm wide.

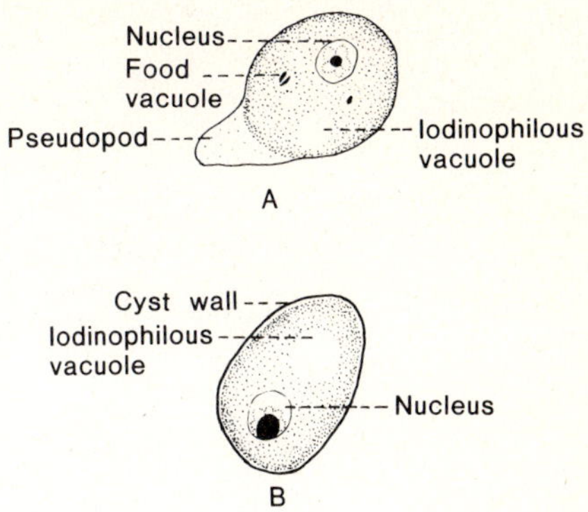

Fig. 1.13. *Iodamoeba buetschlii.* (A) Trophozoite, 10–12 μm wide; (B) cyst, 8–15 μm wide.

form. The endosome is much larger, about 1 μm in diameter, and usually located eccentrically. The nucleus is 5 to 8 μm in diameter.

The constant presence of bacteria in the food vacuoles, the thick nuclear membrane, coarseness of the chromatin on the nuclear membrane, and eccentric location of the endosome are the diagnostic characteristics of the trophozoites of this species.

In cysts, the cytoplasm is finely reticulated, and in young cysts with only one or two nuclei, a large well-defined glycogen vacuole may be present. The chromatoidal bodies are less abundant than in *E. histolytica,* and when present they are filamentous and irregularly shaped fragments with sharp, splintered ends. Usually there are eight nuclei, but there may be only one, two, or four, as in each of the second and third divisions the nuclei divide simultaneously. The cyst wall does not stain, appearing as a colorless, hyaline, doubly outlined capsule surrounding the cyst.

The double outline of the cyst wall, the presence of one to eight nuclei, the comparatively thick nuclear membrane, the large eccentrically located endosome, and the characteristic chromatoidal bodies are the important diagnostic characteristics of the cysts of this species.

Formalinized material stained with iodine shows the characteristic features described above in the same manner as in *E. histolytica.*

Iodamoeba buetschlii

The vesicular nucleus of the trophozoite measures about 3 to 4 μm in diameter. The cytoplasm is reticulated or alveolated, depending upon the extent of digestion of the bacteria. The endosome, which is about one-half the diameter of the nucleus, is typically surrounded by small spherules that do not take stain, so that the achromatic interspherule substance may appear as a reticulum. The endosome is found in various positions in the nucleus, and the well-defined nuclear membrane is free of chromatin granules.

The comparatively large nucleus, large endosome surrounded by the peculiar spherules, presence of bacteria in the cytoplasm, and absence of red blood cells are important diagnostic characteristics of the trophozoites of this species.

The cyst contents are reticulated and one (occasionally two) glycogen vacuole is always present. The single nucleus is usually situated close to the vacuole. The endosome is often attached to the nuclear membrane and may be crescent shaped. Specimens usually measure 6 to 15 μm in diameter.

Great variation in shape of the cyst, presence of only one nucleus, the peculiar structure of the nucleus, the comparatively large endosome, and very large **iodinophilous glycogen vacuole** are important diagnostic characteristics of cysts of this species.

Refer to figure 1.13.

Endolimax nana

The body is rounded; the cytoplasm is reticulated and contains bacteria. There is a vesicular nucleus, measuring about 1.5 to 3 μm in diameter. The nuclear membrane is delicate, without a distinct inner layer of chromatin. Usually the endosome is triangular, square, or irregularly angular in shape and may be in the center, off-center, or attached to the nuclear membrane. In the latter case, a strand may be seen connecting it with a smaller chromatin mass on the opposite side of the nucleus.

The comparative small size, peculiar variation in the endosome, and absence of red blood cells are important diagnostic characteristics of the trophozoites of this species.

The cytoplasm of the cyst is finely reticulated. The nuclei vary in number from one to four, and their structure varies as in that of the trophozoites but appears to be characterized by an angular endosome and its variable location within the nucleus. Specimens usually measure 6 to 15 μm in diameter.

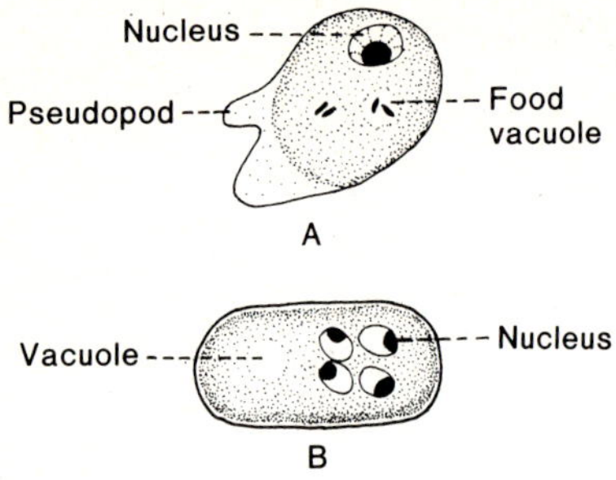

Fig. 1.14. *Endolimax nana.* (A) Trophozoite, 8–10 μm wide; (B) cyst, 5–14 μm long.

The oval shape of many of the cysts, the number of nuclei, the single or divided endosome of the nuclei, centrally situated or in contact with the nuclear membrane, and the absence of chromatoidal bodies are important characteristics of the cysts of this species.

See figure 1.14.

Naegleria fowleri

This is the cause of a rare but usually fatal disease called *primary amebic meningoencephalitis.* The organism is common as a free-living ameboid form in soil and as a biflagellated form without pseudopodia in water. When the flagellated stage is forced up the nose of a swimmer, it can migrate along the olfactory nerves into the brain. There it causes ulcers that are rapidly fatal. It is found in most countries, including the United States, and is a good example of a facultative parasite.

Examine a prepared slide of a section through a brain lesion. Notice the difference between healthy tissue and the cyst. Can you identify amebas in the preparation?

Review Questions

1. What is meant by monoxenous and heteroxenous?
2. Why does a kinetoplast stain darkly? What is its relationship to the centriole and mitochondrion?
3. Name and describe the body forms of Trypanosomatidae.
4. Which important family of flagellates has no cyst stages?
5. Which ameba of humans has no cyst stage?
6. Which ameba has cysts with splintery chromatoidal bars? In which species are they blunt?
7. What is the primary way that parasitic amebae are transmitted from host to host? Are there any exceptions?
8. What is a facultatively parasitic ameba that can cause a fatal disease?

Notes and Sketches

Phylum Apicomplexa

The phylum is characterized structurally by an **apical complex** in the **sporozoite** and **merozoite** revealed by the electron microscope. The apical complex consists of (1) the terminal **polar ring** and its associated structures, such as the **conoid** (a truncated cone of fibrils); (2) **rhopteries** (elongate electron-dense bodies attached to the polar ring); and (3) **micronemes,** which appear as slender twisted bodies that join a duct system with the rhopteries that opens anteriorly (fig. 1.15).

Only **microgametes** (slender active male cells) have flagellar locomotor organelles.

All Apicomplexa are parasitic, commonly histozoic, and infect members of all animal phyla. They include some of the most serious parasites of humans and domesticated animals. The life cycles are complex. Some have direct life cycles and others indirect in which an intermediate host is involved. Species with a direct life cycle have resistant spores or **oocysts** that bridge the gap in the external environment between hosts. Those with an indirect cycle remain in an internal environment and therefore have no need for the protective oocyst.

Class Sporozoea
Subclass Gregarinia

Gregarinia occur in the intestine, reproductive organs, celomic spaces, and other cavities of many kinds of aquatic and terrestrial invertebrates.

Although unimportant economically as parasites, their generalized life cycle makes them significant objects for study to aid in understanding the important sporozoan parasites of humans, domestic mammals, and birds.

Gregarines are numerous, with *Monocystis* and *Zygocystis* as common representatives in the seminal vesicles of earthworms.

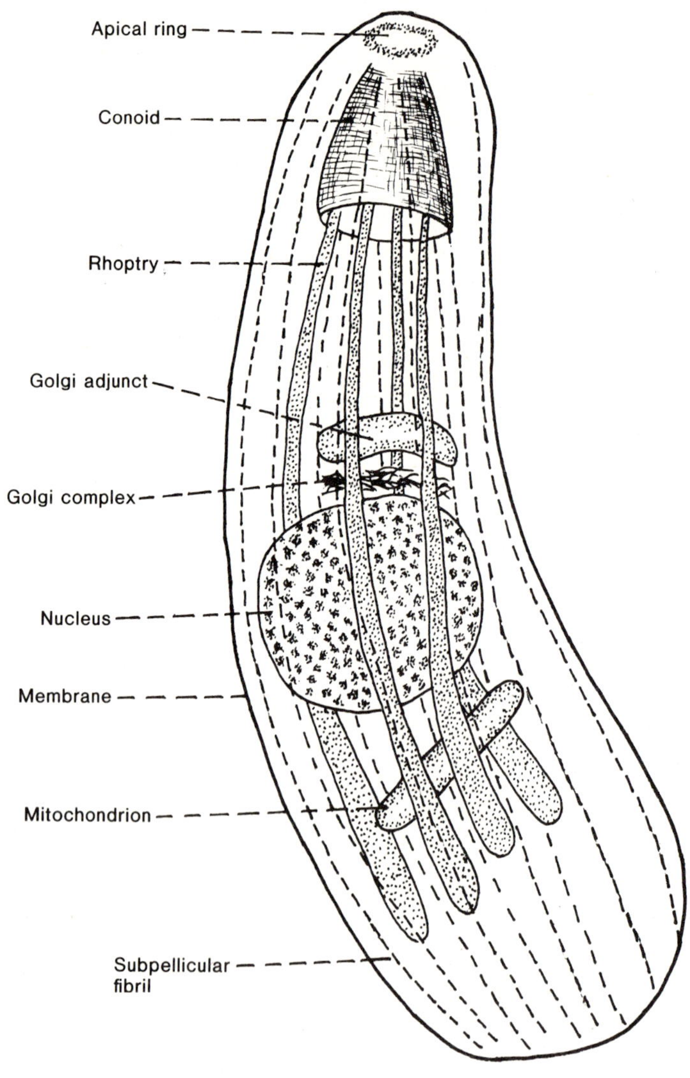

Fig. 1.15. Reconstruction of a zoite of *Toxoplasma gondii,* showing the elements of the apical complex. These can be seen only with the aid of an electron microscope.

Order Eugregarinida

Family Monocystidae

Monocystis lumbrici

Despite the fact that *M. lumbrici* is economically unimportant as an earthworm parasite, the generalized life cycle and the close relationship to the economically important avian and mammalian species of Coccidia make it a desirable species for study. (See fig. 1.16.)

To obtain living *Monocystis,* anesthetize an earthworm by placing it in 7% alcohol for about 30 minutes. After such treatment, make a cut through the dorsal body wall by inserting only the point of the scissors about midway between the head and the clitellum. Extend the cut to both sides. Beginning at the lateral points of the cut, make two parallel cuts through the body wall up to near the head. Then grasp the loose flap with forceps and gently pull it forward. This will expose the cream-colored seminal vesicles situated alongside the esophagus. Using forceps, transfer a small portion of seminal vesicle to a clean slide and add a drop of saline solution. Tease apart well with dissecting needles, so as to make a thin smear of the vesicle contents, add a cover glass and apply enough pressure to further spread the smear, and examine under a compound microscope.

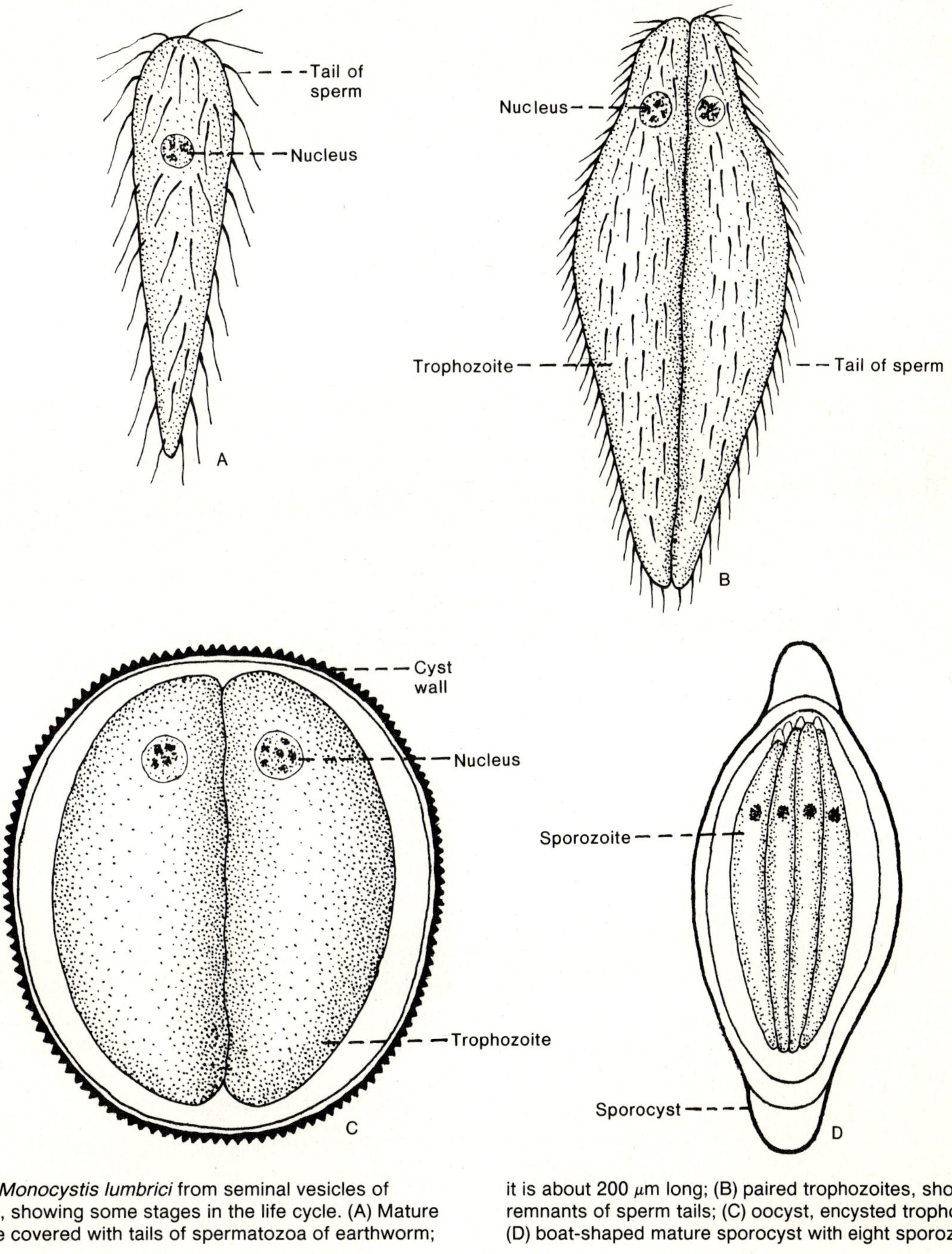

Fig. 1.16. *Monocystis lumbrici* from seminal vesicles of earthworm, showing some stages in the life cycle. (A) Mature trophozoite covered with tails of spermatozoa of earthworm; it is about 200 μm long; (B) paired trophozoites, showing remnants of sperm tails; (C) oocyst, encysted trophozoites; (D) boat-shaped mature sporocyst with eight sporozoites.

When sporocysts are ingested by a worm, the eight sporozoites are liberated by the digestive enzymes. The sporozoites then traverse the gut wall to reach the seminal vesicles, where they penetrate bundles of developing sperm. By this time each sporozoite has developed into a trophozoite. Since adjacent sperm cells are deprived of their nourishment, they slowly disintegrate, their tails adhering to the trophozoite, giving it the appearance of a ciliated structure. Trophozoites now pair off and become encysted, forming **oocysts,** after which each undergoes numerous divisions resulting in many gametes. After formation of the oocyst, the former trophozoites are called gametocytes by some workers. Membranes of the gametocytes disintegrate, and gametes, apparently from different trophozoites, unite to form zygotes. While still within the oocyst, a wall develops around each zygote to form an elongate structure known as a **sporocyst.** Within each sporocyst, three successive divisions occur, resulting in eight mononucleate sporozoites. The large forms commonly seen in seminal vesicle smears are trophozoites and sporocyst-filled oocysts.

Miles (1962) has shown experimentally that (1) the chief, if not the only, way worms are infected in nature is through ingesting mature sporocysts, and (2) sporocysts do not leave the host through openings in the body wall but are released after death and decay of the host. Earthworm-eating birds and mammals doubtless play an important role in the dissemination of sporocysts with their feces.

Family Gregarinidae

Gregarina spp.

Members of this genus represent common **cephaline gregarines.** They occur commonly in the digestive tract of cockroaches, grasshoppers, and mealworms.

Remove the intestine from a decapitated host, tease it apart in saline solution on a slide, add a cover glass, and examine under a microscope. Small, young stages of the parasite are inside the epithelial cells. Older and larger individuals are attached to the cells by the small anterior end of the body, with the large hind part hanging free in the lumen of the gut.

The body of the mature gregarine is divided into two major parts: the anterior **protomerite** and posterior **deutomerite.** The protomerite bears anteriorly a small, simple, knoblike **epimerite,** which is embedded in the cytoplasm of the epithelial cell.

The pendant deutomerite detaches to form the **sporadin.** Two or more sporadins unite in tandem to produce a condition known as **syzygy** (fig. 1.17). Find syzygyous forms.

Eventually two sporadins attach along the long axis of their bodies and secrete a cyst about both. In your dissection, find the conjugating sporadins known as a

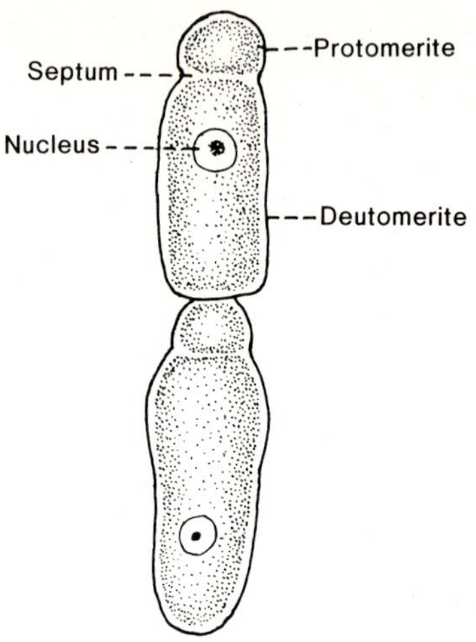

Fig. 1.17. *Gregarina blattarum,* a cephaline gregarine with two individuals in syzygy, showing the small anterior protomerite without the knoblike epimerite, and the large posterior deutomerite. Each individual is up to 350 μm long.

gametocyte. The nucleus of each gametocyte undergoes repeated division, and the particles migrate to the surface, where each becomes enclosed in a bit of cytoplasm. These are gametes. Union of two gametes produces a zygote in which develop the banana-shaped sporozoites. Find these various stages.

Subclass Coccidia
Order Eucoccida

Coccidia are intracellular (cytozoic) parasites occurring in many different kinds of cells of vertebrates and in the intestinal epithelium of invertebrates such as arthropods and molluscs.

Species of *Eimeria* (Eimeriidae) cause great economic loss in domestic and game animals. Serious outbreaks of infection known as **coccidiosis** occur in poultry, pigeons, rabbits, muskrats, foxes, mink, cattle, and sheep. *Toxoplasma* (Sarcocystidae) is of medical importance.

Life cycles are both direct and indirect. They consist of three phases, each with distinctive forms and functions. The phases are **schizogony** or **merogony, gametogony,** and **sporogony.** Schizogony is a form of multiple asexual reproduction inside the host cells. It results in numerous daughter cells all at once. Gametogony is the formation from the schizonts of sex cells—minute flagellated **microgametes** (♂) and large nonflagellated **macrogametes** (♀). Sporogony includes formation of the zygote and multiple division, or fission, of it to produce the sporozoites which are the infective stage.

Family Haemogregarinidae

These primitive Coccidia are parasitic in the erythrocytes of frogs and turtles and in the intestinal epithelium of leeches.

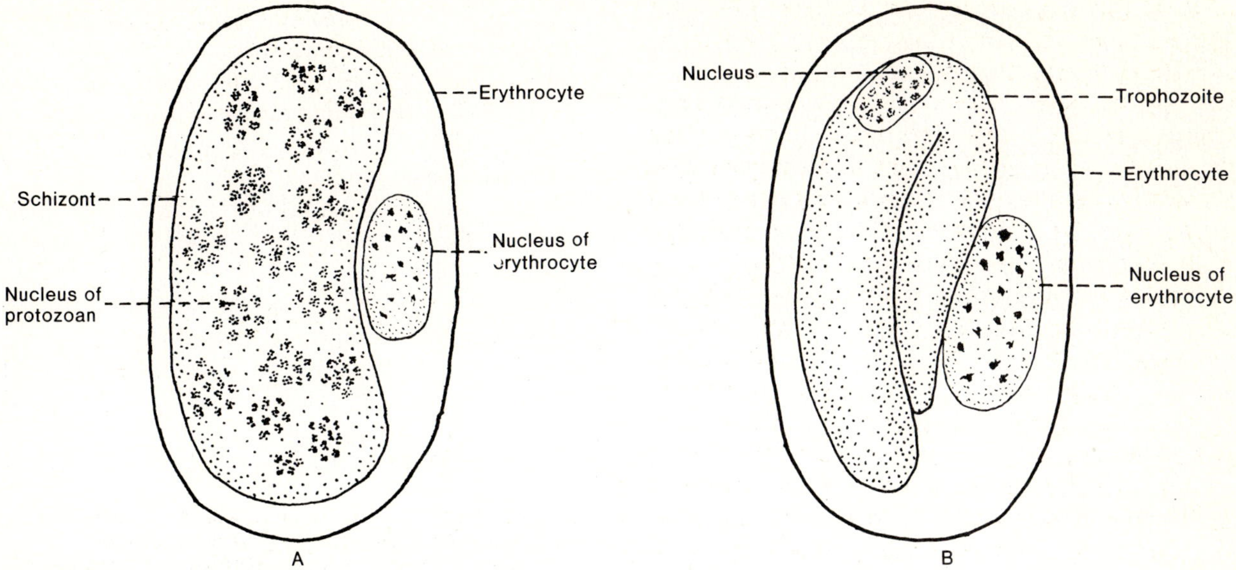

Fig. 1.18. *Haemogregarina stephanowi.* (A) Schizont in erythrocyte; (B) young U-shaped trophozoite in erythrocyte.

Haemogregarina stepanowi

Feeding leeches inject sporozoites that penetrate erythrocytes of the circulating blood. As they grow, the nucleus divides a number of times, forming **merozoites.** Examine a stained smear of circulating blood and locate a multinucleated **schizont** (fig. 1.18A) and a schizont containing formed merozoites. From slides of stained bone marrow, find the folded, somewhat U-shaped schizonts in erythrocytes (fig. 1.18B) that produce more merozoites, some of which continue schizogony in new blood cells and others that develop into **gametocytes,** which in turn divide to form micro- and macrogametes. Single large micro- and macrogametes occur in erythrocytes. When infected blood cells are ingested by leeches, the micro- and macrogametes are released in the gut and fuse to form zygotes. When injected by feeding leeches into turtles or frogs, the zygotes enter red blood cells and undergo gametogony, forming sporozoites.

Family Eimeriidae

The economically important genera of Eimeriidae are *Eimeria* and *Isospora.* They differ mainly in details of development within the **oocysts.** *Eimeria* oocysts have four **sporocysts** containing two sporozoites each (fig. 1.19), whereas *Isospora* oocysts have two sporocysts containing four sporozoites each. In both cases the ripe oocyst contains eight sporozoites.

The life cycle consists of three phases, each with its distinctive forms and functions. They are merogony (also known as schizogony), gametogony, and sporogony. Merogony and gametogony take place inside cells, whereas sporogony occurs outside the host in the soil or litter. Each of these phases will be considered in studying the Coccidia.

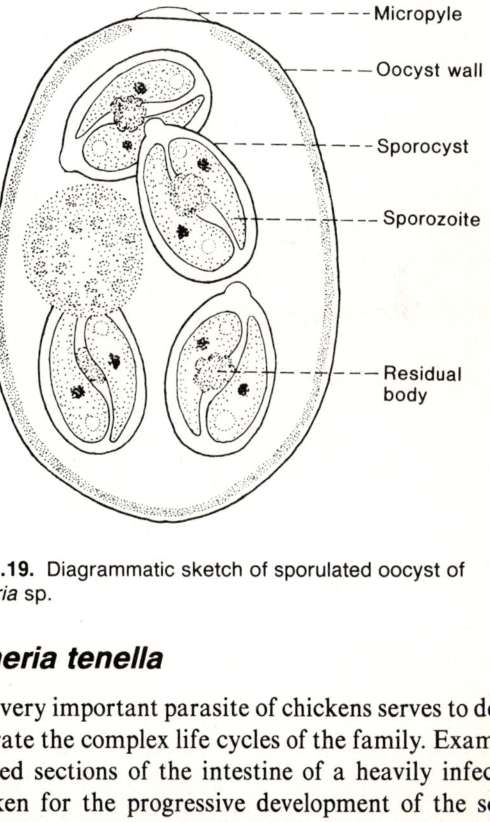

Fig. 1.19. Diagrammatic sketch of sporulated oocyst of *Eimeria* sp.

Eimeria tenella

This very important parasite of chickens serves to demonstrate the complex life cycles of the family. Examine stained sections of the intestine of a heavily infected chicken for the progressive development of the schizogonous and gametogonous stages.

Schizogonous Stage. This is the asexual phase of multiplication inside the epithelial cells of the intestine. Four different forms appear in succession.

1. **Trophozoite.** These are the small single-celled, uninuclear forms inside the epithelial cells. The small nuclei of the parasite are surrounded by a clear halo and are easily distinguished from the large nucleus of the epithelial cells.

2. **Schizont.** After a period of growth, the trophozoite matures, and the nucleus undergoes a series of divisions, resulting in 32 small ones. This process of nuclear division is known as schizogony and produces the schizonts. They are easily recognized by the many small nuclei scattered in the cytoplasm.

3. **Segmenter.** After division of the nucleus, a bit of cytoplasm surrounds each nuclear fragment, forming many tiny uninucleate individuals. The body containing these is known as a segmenter, since the original uninucleate cell has segmented to form many tiny individuals.

4. **Merozoites.** As individual bodies in the segmenter mature, they become spindle-shaped merozoites. The schizogenous stage terminates with the appearance of the merozoites. They are liberated into the lumen of the intestine, where other epithelial cells are invaded by the merozoites to begin a new schizogenous cycle or to initiate the gametogonous cycle. In the latter case, the individual merozoites may be considered as being of different sexes.

Find epithelial cells filled with numerous merozoites and cells that have ruptured, liberating the merozoites.

Gametogonous Stage. Some merozoites, upon entering epithelial cells, transform into gametocytes, or sexual cells, thereby initiating the gametogonous phase. Two types of cells are formed. They are macrogametocytes and microgametocytes.

1. **Macrogametocytes.** These are large cells located inside the epithelial cells, whose nucleus is crowded to

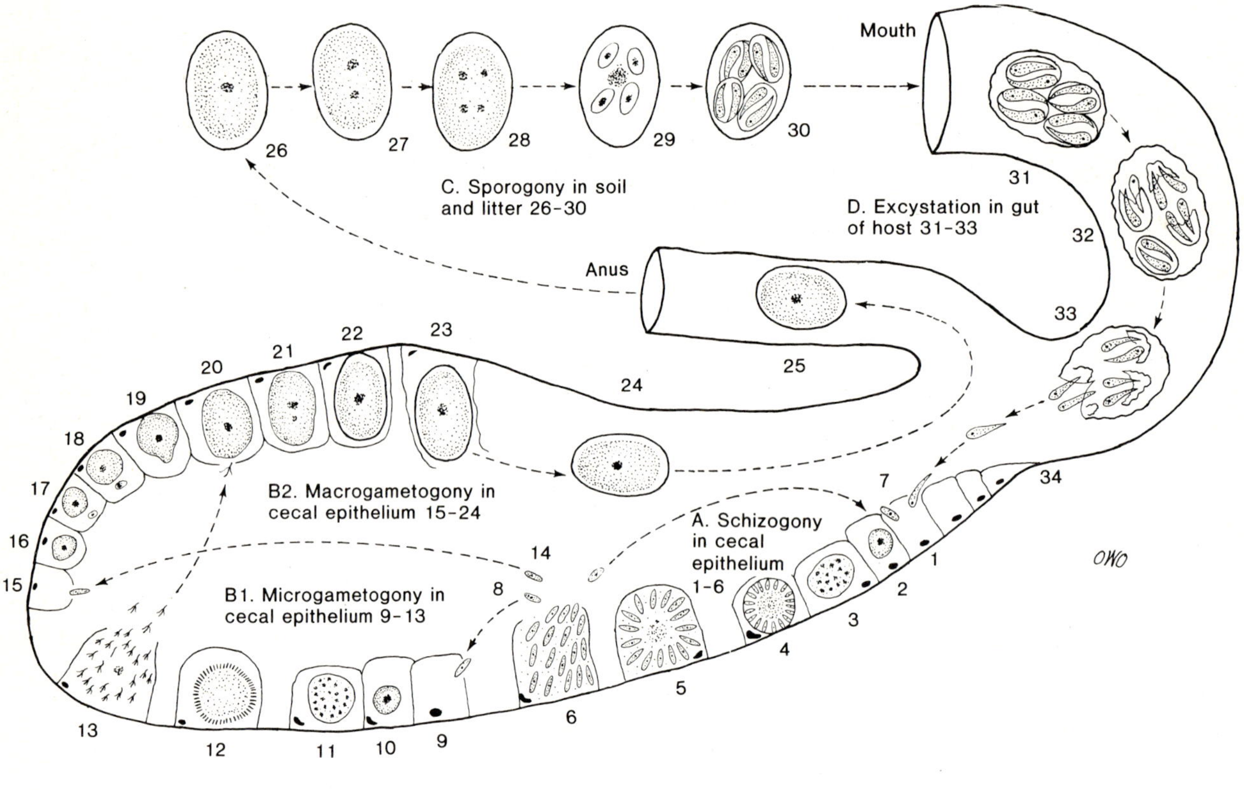

Fig. 1.20. Life cycle of *Eimeria tenella* of chickens.
A. *Schizogony in cecal epithelium* (1–7). 1. Sporozoite enters epithelial cell; 2. growing trophozoite; 3. nucleus undergoes multiple division to form schizont; 4. each nucleus together with a bit of cytoplasm forms a merozoite; 5. fully developed merozoites; 6. schizont and epithelial cell rupture, releasing merozoites into caecal lumen; 7. merozoites enter other caecal cells to repeat schizogonic cycle several times.
B. *Gametogony in caecal epithelium* (8–25).
Microgametogony. 8. Some schizogonic merozoites are destined to become microgametocytes; 9. merozoite penetrates caecal epithelial cell to begin cycle; 10. growth of microgametocyte; 11. multiple division of nucleus is beginning formation of microgametes; 12. formation and alignment of microgametes over surface of residual cellular cytoplasm; 13. caecal cell ruptures, liberating mature, biflagellate microgametes into caecel lumen.
Macrogametogony. 14. Schizogonic merozoites predetermined to initiate macrogametogenous cycle; 15. merozoite penetrates caecal epithelial cell; 16. formation of macrogametocyte; 17. meiotic division of nucleus with extrusion of half of chromosomes to form macrogamete; 18. haploid macrogamete; 19. mature macrogamete; 20. fertilization of macrogamete; 21. fusion of micro- and macrogamete to form zygote; 22. oocyst; 23. mature oocyst ruptures epithelium; 24. oocyst free in caecum; 25. oocyst in large intestine will be voided with feces.
C. *Sporogony in soil and litter* (26–30). 26. Mature oocyst has not begun development; 27. first nuclear division; 28. second and final nuclear division preparatory to forming sporoblasts; 29. formation of four sporoblasts; 30. infective or ripe oocyst containing four sporocysts each with two sporozoites.
D. *Infection of chicken host, excystation of spoprozoites in crop, gizzard, and intestine* (31–34). 31. Oocyst in crop where action of CO_2 and enzymes cause its wall to wrinkle and weaken; 32. oocyst in gizzard under mechanical and contained chemical action rupture sporocysts, releasing sporozoites inside; 33. sporozoites freed from oocyst in small intestine; 34. liberated sporozoites enter caecal lumen preparatory to penetrating epithelial cells.

one side. In sections stained with hematoxylin and eosin, the large cytoplasmic granules of the macrogametocytes stain bright pink. The cells have a thick hyaline wall and large centrally located nucleus. Upon reaching maturity, the numerous macrogametocytes become macrogametes ready for fertilization. Each one completely fills an epithelial cell. They usually appear in groups.

2. **Microgametocytes.** These differ from the macrogametocytes in that the nucleus breaks up into many small bits that migrate to the outer surface of the cytoplasm, where they align themselves in an orderly fashion. Each tiny nucleus is clothed separately in a bit of cytoplasm to form minute, slender, biflagellated microgametes. They are liberated from the cell into the intestinal lumen. The appearance of the male and female gametes terminates the two intracellular phases of gametogony.

Find the characteristic microgametocytes with the many small nuclei and those with fully developed microgametes.

Fertilization occurs in the intestine to form the zygote. It develops a thick cyst wall and is known as an oocyst that is released from the epithelial cell and voided with the feces for further development outside the host.

3. **Sporogony.** Sporogonous development is a process of asexual multiplication in which the sporozoites are formed.

The oocyst is the beginning of the sporogonous stage. When voided with the feces, the oocysts, with equally rounded ends, and without a micropyle, measure 24 to 30 × 14 to 20 μm. Each one contains a large spherical cytoplasmic body. Sporulation, the process of forming

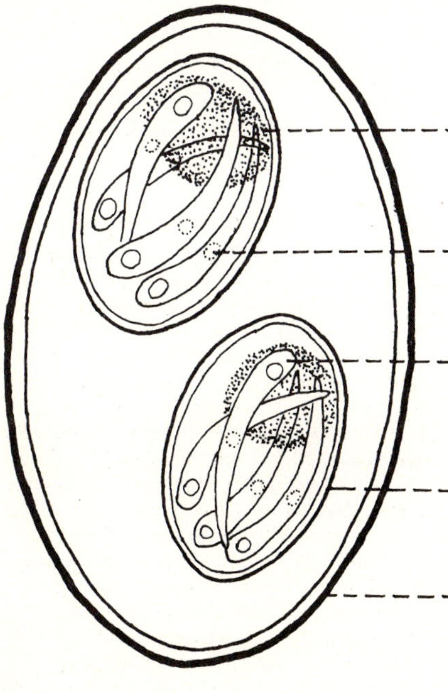

Fig. 1.21. Sporulated oocyst of *Toxoplasma gondii,* 10–12 μm long.

the sporozoites by a process of division, begins and is completed in 48 hours at 33 C. The cytoplasmic mass divides twice to form four sporoblasts. The next step is a single division of each sporoblast, with the formation of two sporozoites enclosed in a sporocyst. Thus each ripe oocyst contains four sporocysts, each of which contains two sporozoites (fig. 1.20). Sporozoites are the infective stage. When the ripe oocyst is swallowed by a suitable host, the sporozoites, liberated under the influence of digestive juices of the intestine, penetrate epithelial cells and begin the cycle anew.

Eimeria stiedae is a highly pathogenic species in the liver of rabbits. Compare a slide of this parasite with *E. tenella.*

Eimeria species are very common in mammals, especially ruminents and rodents. To demonstrate oocysts, obtain fresh feces from a potential host, such as a sheep, calf, or squirrel. Place it into a 0.5% solution of potassium dichromate in a shallow dish, such as a large petri dish. When passed in the feces, oocysts are unsporulated; seven to ten days at room temperature are required for the sporont to develop to the infective oocyst. Bacterial development is retarded by the KCr_2 during this time. Then, follow the procedure for a sugar or zinc sulfate flotation as described on page 255, to isolate oocysts.

Species of *Isospora* also are common, especially in carnivores and birds. In contrast to *Eimeria,* oocysts of *Isospora* contain two sporocysts, each with four sporozoites.

Family Sarcocystidae

This family is characterized by cysts or pseudocysts (a group of sporozoans not surrounded by a cyst wall) in tissues other than the gut lining. The cysts produce small, slow-moving merozoite-like bodies called **bradyzoites** that multiply by **endodyogeny** (formation of two daughter cells from the mother cell). *Toxoplasma* and *Sarcocystis* are common representatives.

Toxoplasma gondii

This species (see fig. 1.21) resembles *Isospora bigemina* of cats; in fact, what has been called the small race of *I. bigemina* is indeed *T. gondii.* Its life cycle differs from that of *Isospora* in that in addition to the sexual phase, there is a secondary host in which a special type of asexual proliferation known as endodyogeny occurs in cells of the striated muscles, retina, and central nervous system of birds and mammals (fig. 1.22).

Oocysts developing in the epithelial cells of the intestine of cats and passed in the feces contain two sporocysts, each with four sporozoites when mature. When these are swallowed by cats, the oocysts release the

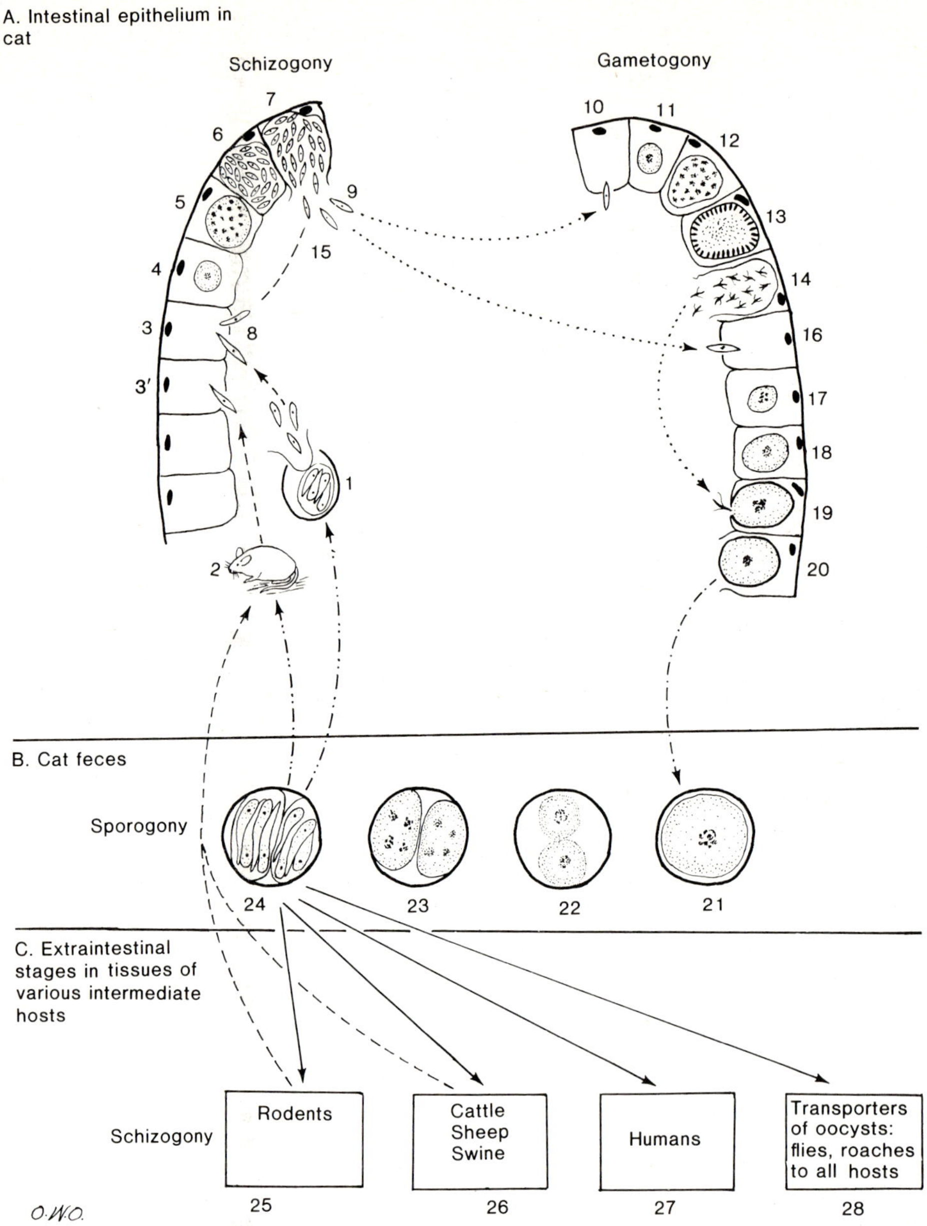

Fig. 1.22. Life cycle of *Toxoplasma gondii.*
A. *In intestinal epithelium of cat.* **Schizogony** (1–8).
1. Ruptured oocyst liberating sporozoites in intestine; 2. mouse infected with bradyzoites (merozoites) from ingesting sporulated oocysts; 3. bradyzoite entering cell; 3′. sporozoite entering cell; 4. trophozoite. 5. schizont; 6. segmenter; 7. merozoites (tachyzoites) escaping from ruptured cell; 8. merozoites infect other cells to reinitiate another schizogonous cycle. **Gametogony** (9–20). 9. Some merozoites initiate the gametogonous (sexual) cycle; 10. merozoite penetrates cell; 11. young microgametocyte (♂); 12. nucleus divides in early development of microgametes; 13. microgametes forming on surface of cytoplasm; 14. fully developed flagellated microgametes; 15–20 represent development of macrogamete and occyst (♀); 15. Merozoite initiates female gametogonous cycle; 16. merozoite enters cell; 17–18. developing macrogamete; 19. fertilization of mature macrogamete and formation of zygote; 20. zygote ruptures cell and enters lumen of intestine.
B. *In cat feces and soil.* **Sporogony** (21–24). 21. Undeveloped oocyst, 22. oocyst with two sporoblasts, 23. formation of two sporocysts, each with four nuclei, 24. infective (ripe) oocyst with two sporocysts each with four sporozoites.
C. *Extraintestinal stages in tissues of various intermediate hosts.* **Epidemiology** (25–29). 25. Rodents become infected by swallowing oocysts (24) or eating infected meat of various kinds (26), and become a source of infection to cats; 26. livestock are infected by swallowing oocysts; 27. humans acquire the infection from swallowing oocysts from cat feces and by eating inadequately cooked beef, mutton-lamb, and pork; 28. insects such as flies and cockroaches feeding on cat feces swallow oocysts and transport them to many places, including human food.

sporozoites in the small intestine. Some of them penetrate intestinal epithelial cells and undergo the sexual phase of development, as in the case of *Eimeria,* and produce more oocysts.

When other animals that serve only as intermediate host (mice, humans, cattle, sheep, and others) swallow mature oocysts, the sporozoites make their way into the blood and are distributed to all parts of the body. They enter various kinds of cells, mainly those of the striated muscles, retina, and brain. Following rapid asexual proliferation, the host cells rupture and liberate trophozoites called **tachyzoites** (*tachy-* refers to rapid development), which are free for a short time in the blood and abdominal serous exudate as small banana-shaped bodies. The trophozoites enter new host cells and initiate another asexual cycle.

As the number of trophozoites increases from repeated cycles of proliferation, the host reacts by producing specific antibodies that curtail proliferation. When this stage is reached, the infection passes into the chronic stage characterized by the formation of intracellular cysts packed with numerous trophozoites designated as **bradyzoites** (*brady-* refers to slow development). With retention of trophozoites in the cysts and absence of active reproduction in new cells, immunity declines and the loaded cysts deteriorate and eventually rupture, releasing the infective bradyzoites. Thus a succession of proliferative and cystic phases appear in infected intermediate hosts.

Trophozoites in cysts or free in the flesh of intermediaries are the principal source of infection to carnivorous animals, including humans. Ripe oocysts from fecal contamination of the environment by cats provide infection for grass-eating animals and to persons who may swallow them. In addition to the oral route of infection, congenital infection may occur when pregnancy and the proliferative phase of toxoplasmosis with free trophozoites in the circulation occur simultaneously.

Examine free trophozoites (tachyzoites) from smears of serous exudate of infected mice and in pseudocysts from sections of infected tissue. The latter are round to oval, intracellular bodies packed with trophozoites (bradyzoites).

Trophozoites are banana-shaped crescents 4 to 7 μm long by 2 to 4 μm in diameter. They possess an apical ring, conoid, micronemes, subpellicular fibrils, nucleus, Golgi apparatus, mitochondria, and endoplasmic reticulum (see fig. 1.15). Most of these are ultrastructures.

Oocysts are oval, surrounded by a smooth double wall, and measure about 10 by 12 μm. Sporulated oocysts contain two oval sporocysts about 6 to 8 μm long, each with four elongated, curved sporozoites 2 by 7 μm in size, and a residuum of cytoplasm. Instructions for the isolation of oocysts are included under Special Techniques and Further Notes, page 255.

Sarcocystis spp.

Elongated cysts known as **sarcocysts, Miescher's tubules,** and more recently as **zoitocysts** (a cystic stage in the late development of Sarcocystidae) occur commonly in striated muscles, both cardiac and voluntary. In mammals they occur in all these muscles but in birds usually only in the breast muscles. Cysts vary in size but generally are around 1 cm long by 1–2 mm in diameter. The outer surface of the rather thick cyst wall may be smooth or rough, with fibers (**cytophaneres**) that extend into the surrounding muscle tissue. Numerous banana-shaped merozoites develop inside the sarcocysts.

Only recently has the life cycle of *Sarcocystis* been solved. *Sarcocystis fusiformis* from dogs and cattle is an example. The intestinal phase, gametogony, occurs in the intestinal epithelium of dogs, where *Isospora*-like oocysts are produced. Sporogony takes place in the dog feces and in the soil. Cattle become infected when ripe oocysts are swallowed with the grass. It is thought the ubiquitious soil mites might ingest ripe oocysts and carry them in their intestines onto the grass, making them readily available to grazing cattle. In the bovine intestine, the sporozoites are freed. They penetrate the intestinal wall, enter the circulatory vessels, and are carried to the striated and cardiac muscles. Inside the muscle cells they undergo merogony, forming merozoites (tachyzoites) that may repeat the cycle in the cells. Finally the sarcocysts produce merozoites (bradyzoites) capable of forming micro- and macrogametocytes in the intestinal epithelium of the definitive hosts (fig. 1.23).

Observe slides showing whole sarcocysts. Sectioned slides show the wall, which is smooth externally in some hosts and spiny in others. Internally the sarcocyst is divided by septa and contains large numbers of merozoitelike bodies called bradyzoites.

Family Plasmodiidae

Plasmodiidae are parasites of the blood cells of vertebrates, where merogony occurs and gametogony begins, and of bloodsucking Diptera in whose bodies gametogony is completed and sporogony occurs. Malaria-producing *Plasmodium* is an example.

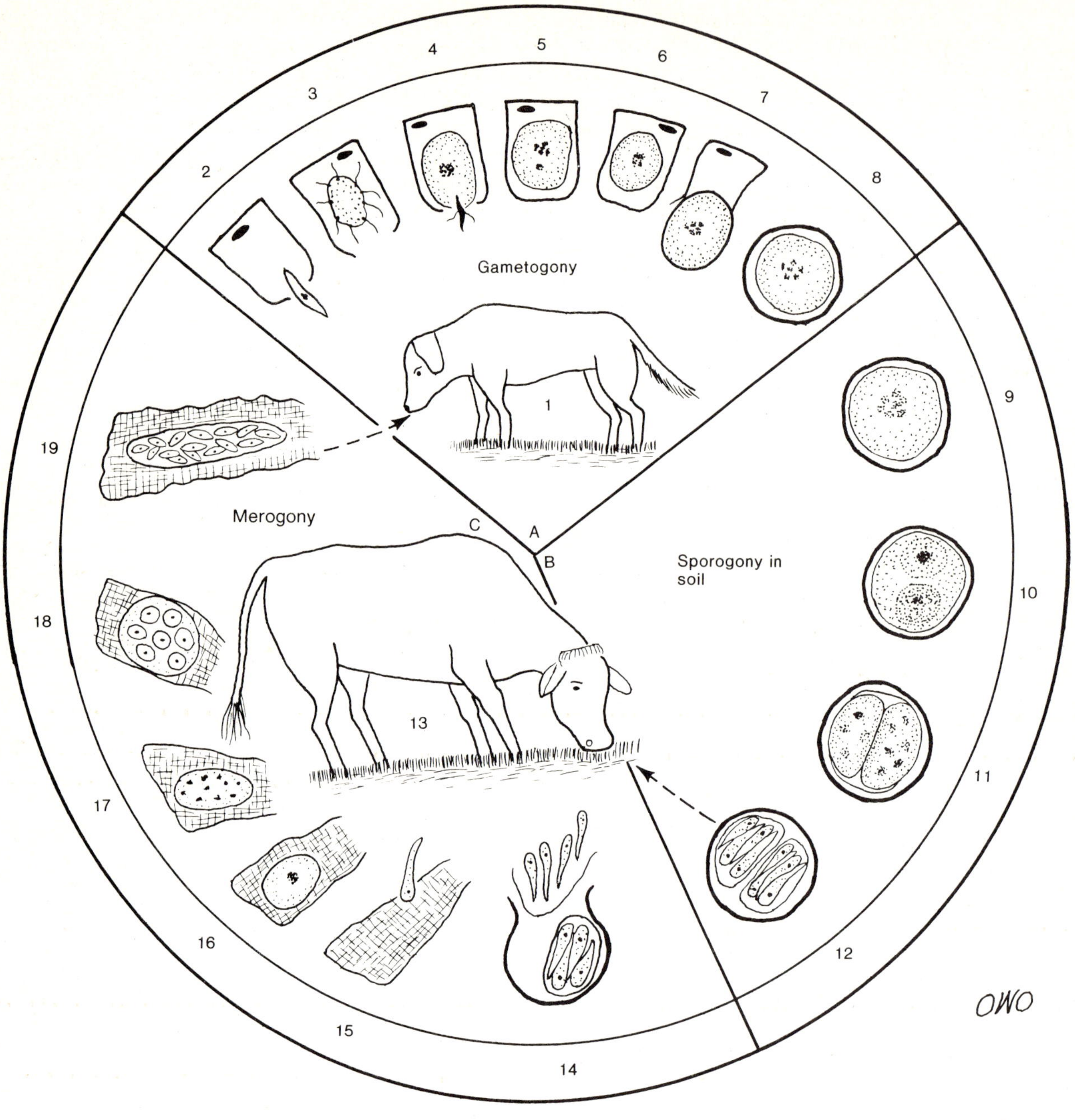

Fig. 1.23. Life cycle of *Sarcocystis fusiformis.*
A. *Gametogony in definitive hosts* (1–8). 1. Dog definitive host, along with coyotes, foxes, cats, raccoons, and humans, in whose intestinal epithelium gametogony occurs; 2. merozoite freed from sarcocyst in intestine of dog, etc., enters intestinal epithelial cell to begin gametogonous phase; 3. mature microgametocyte produces microgametes which escape from host cell; 4. mature macrogamete being fertilized by microgamete; 5. fertilized zygote; 6. mature oocyst; 7. oocyst escaping from host cell; 8. oocyst free in lumen of intestine.
B. *Sporogony in feces of dog and in soil* (9–12). 9. Mature oocyst; 10. oocyst with two sporoblasts; 11. oocyst with developing sporocysts, each with four nuclei; 12. ripe oocyst with two sporocysts, each containing four sporozoites.
C. *Schizogony in striated and cardiac muscles of intermediate host* (13–19). 13. Bovine intermediate host infected by swallowing ripe oocysts that are free or possibly ingested by soil mites and carried onto the grass; 14. oocyst ruptures in intestine, freeing sporozoites; 15. having entered circulatory system from the intestine, sporozoites distributed to striated and cardiac muscles which they enter; 16. sporozoite transforms into preschizont trophozoite; 17. early schizont with multiple nuclei; 18. segmenter schizont with numerous individual merozoites (tachyzoites) which may escape and form more schizonts; 19. sarcocyst (Miescher's tubule, zoitocyst) filled with merozoites (bradyzoites) infective to definite hosts.

The student should recognize the normal constituents of the blood of vertebrates before undertaking a study of *Plasmodium* and related forms. It is also urged that the student be familiar with the preparation and staining of blood smears. For instructions consult page 253.

Plasmodium vivax

This is the most common and widespread of the species of malaria infecting humans. In vivax malaria the asexual cycle is completed in a 48-hour period, and its effect on people is less severe than in the case of some other species. In the past, the disease caused by *P. vivax* was referred to as **benign tertian malaria.**

Like other Coccidia, *Plasmodium* has a merogonous, a gametogonous, and a sporogonous stage in its life cycle. The first phase of the merogonous stage occurs in the liver, while the second phase and part of the gametogonous stage occur in the red blood cells of lizards, birds, and mammals. The concluding part of the gametogonous and all of the sporogonous stages occur in mosquitoes.

Stages in the life cycle are found in liver sections and blood smears of vertebrates and in dissections of the alimentary canal and salivary glands of infected mosquitoes. Examine the appropriate preparations for the various stages described below. The life cycle is shown in figure 1.24.

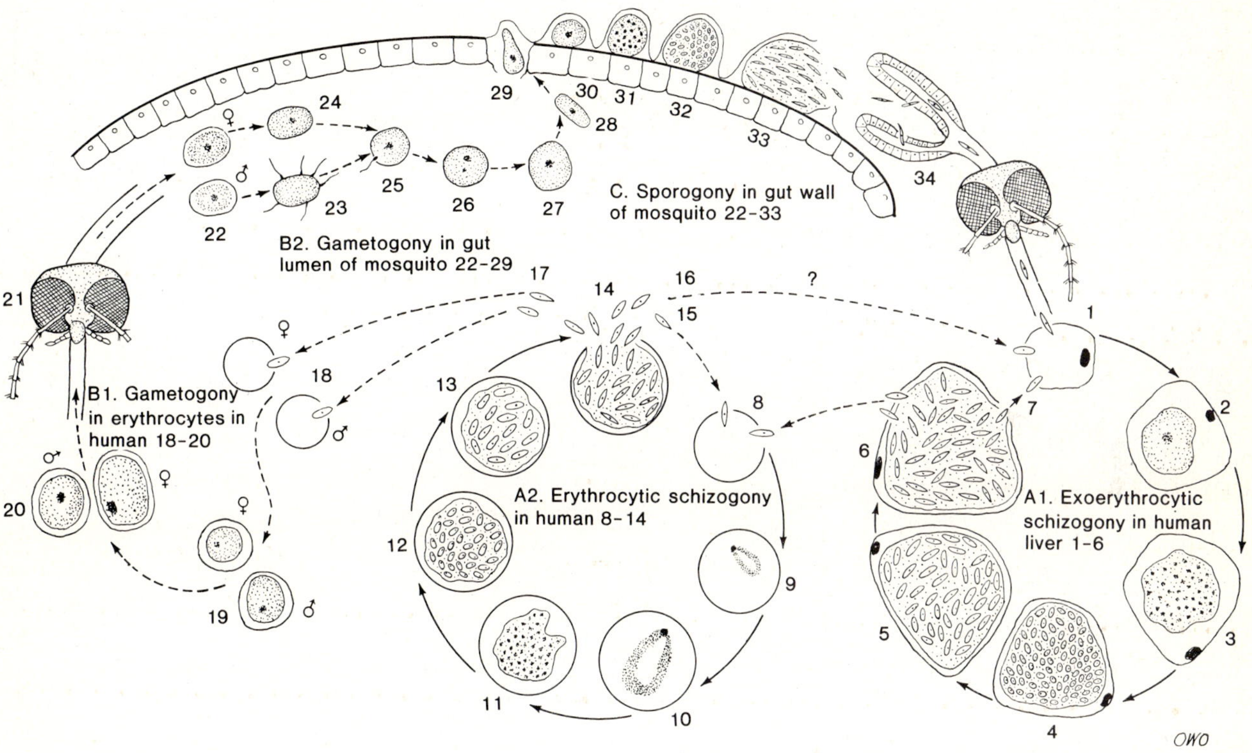

Fig. 1.24. Life cycle of *Plasmodium vivax*. A. *Schizogony* (1) Exoerythrocytic schizogony in liver of human, afebrile phase (1–7). 1. Sporozoite injected into bloodstream by feeding mosquito enters parenchymal cells of liver; 2. growing trophozoite; 3. multinucleate schizont; 4. segmenter containing immature merozoites; 5. segmenter with maturing merozoites; 6. mature schizont ruptures, releasing merozoites; 7. some merozoites enter other parenchymal cells to continue preerythrocytic cycle.
(2) Erythrocytic schizogony in red blood cells of human, febrile phase (8–16). 8. Merozoites from hepatic schizont enter bloodstream and penetrate erythrocytes; 9–10. young trophozoites in ring stage; 11. young schizont with multiplying nuclei; 12. young segmenter with developing merozoites; 13. mature segmenter with fully developed merozoites; 14. ruptured segmenter and erythrocyte release merozoites into bloodstream; 15. some merozoites enter erythrocytes to repeat the cycle at 48-hour intervals; 16. other merozoites enter liver to perpetuate that cycle. It now appears that steps 7 and 16 do not occur. Instead, dormant sporozoites called **hypnozoites** are believed to be the cause of relapse in *P. vivax, P. ovale* and *P. cynomolgi*.
B. *Gametogony.* (1) In erythrocytes of human (17–20). 17–18. Merozoites of erythrocytic schizogony destined to become gametocytes enter red blood cells to develop; 19. developing gametocytes; 20. mature macro- and microgametocytes. (2) In midgut to mosquito (21–28). 21. Infection of mosquitoes occurs when gametocytes are sucked up in red blood cells; 22. gametocytes still inside red blood cells; 23. microgametocyte freed and exflagellating to form slender microgametes; 24. freed macrogametocyte; 25. macrogamete being fertilized by microgamete; 26–27. zygotes; 28. ookinete (motile zygote).
C. *Sporogony in gut wall of mosquito* (29–34). 29. Ookinete penetrates wall of gut, remaining between epithelium and basement membrane; 30. young oocyst; 31. developing multinucleate oocyst; 32. oocyts containing young sporozoites; 33. ripe oocyst ruptures, releasing numerous sporozoites into hemocoel; 34. sporozoites enter lumen of salivary glands and are injected along with saliva into human by feeding mosquito.

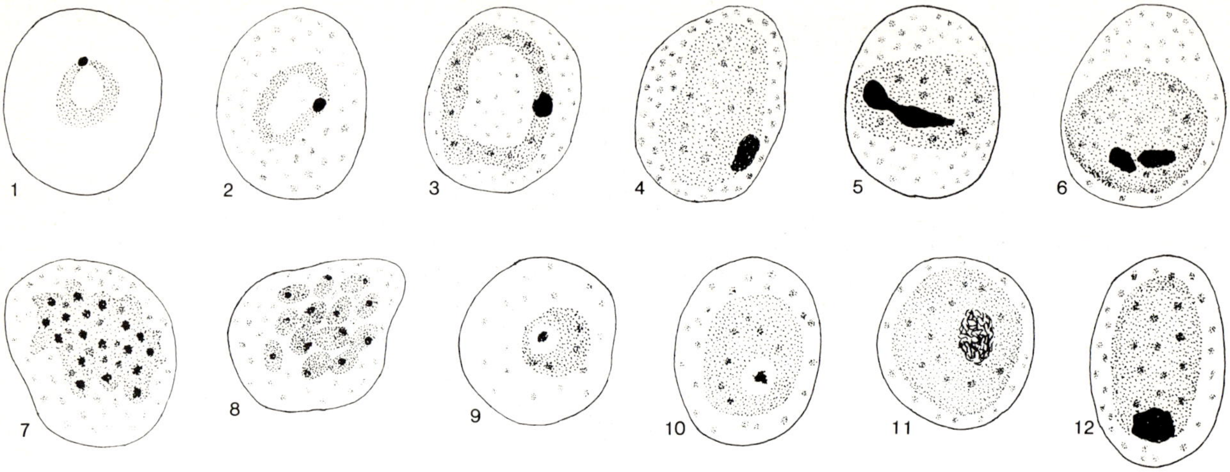

Fig. 1.25. Some representative stages in the development of *Plasmodium vivax* as shown in stained smears of human blood. Darkly stippled areas inside the erythrocytes show the parasite with its densely staining nuclear material. Small lightly stippled areas throughout the erythrocyte are Schüffner's dots. 1. Early ring-stage trophozoite in erythrocyte; 2. slightly older trophozoite in which Schüffner's dots have appeared throughout the cytoplasm of the erythrocyte; 3. still older trophozoite in which the ring is greatly enlarged; 4. mature trophozoite; 5. transformation of trophozoite to schizont by beginning division of the nucleus; 6. early schizont with two separate nuclei; 7. advanced schizont with many nuclei in cytoplasm of parasite; 8. segmenter where each nucleus has taken a bit of cytoplasm to form a merozoite; 9, 10. developing gametocytes, male and female; 11. microgametocyte; 12. macrogametocyte.

Merogonous Cycle. Merogonous stages consist of two parts: the **preerythrocytic stage** in the liver and the **erythrocytic stage** in the red blood cells.

The preerythrocytic phase is initiated when sporozoites are injected into the blood by infected mosquitoes while feeding. Sporozoites quickly enter liver cells, where merogonous multiplications take place, producing numerous merozoites known as **metacryptozoites.** (They are morphologically similar to other merozoites.) They enter the bloodstream, come in contact with the red blood cells, penetrate them, and initiate the erythrocytic merogonous phase of the life cycle.

Examine a section of liver for the large schizonts among the liver cells.

In *Plasmodium gallinaceum* of chickens, the preerythrocytic schizonts occur in the endothelial cells of the brain capillaries. Examine a demonstration slide that shows the large schizonts in the capillaries. Note the numerous nuclei that will form the metacryptozoic merozoites. These enter the bloodstream, penetrate the nucleated red blood cells, and initiate the erythrocytic phase.

The erythrocytic phase, with its different stages, will be studied from blood smears taken from patients infected with *P. vivax*. Representative stages are shown in figure 1.25.

1. **Trophozoites.** Upon entering the red corpuscles, metacryptozoites transform into ameboid trophozoites. A young trophozoite appears as a ring of blue cytoplasm with a single ruby-red nucleus on one side, giving the appearance of a signet ring—hence the designation **ring stage.** As the ameboid trophozoites grow, they appear in various shapes, causing the infected corpuscles to increase in size and vary in shape.

The membrane of cells infected with some species become stippled with fine pinkish spots or granules known as **Schüffner's dots.** A brownish pigment known as **hemozoin** is separated from the iron-bearing hemoglobin by the parasites. It appears in the older trophozoites and subsequent stages.

Find the unicellular trophozoites in various stages of growth, together with the pigment and Schüffner's dots.

2. **Schizont.** Upon completion of the first division of the nucleus of the trophozoite, it becomes a schizont. Division of the nucleus continues until 12 to 24 minute nuclei appear.

Locate a schizont with many nuclei scattered throughout the uniform cytoplasm.

3. **Segmenter.** The cytoplasm of the schizont divides into small masses with each bit surrounding a nucleus to form a clump of small bodies. This is a segmenter.

Find segmenters in the red blood cells.

4. **Merozoites.** Each nucleus of the segmenter with its particle of cytoplasm becomes a merozoite. The mature segmenter ruptures the parasitized blood corpuscle, releasing the spindle-shaped merozoites into the bloodstream with resultant chills and fever in the host. In mature segmenters, the pigment usually appears as a single mass and remains in the residuum of cytoplasm. The merozoites then parasitize red blood cells

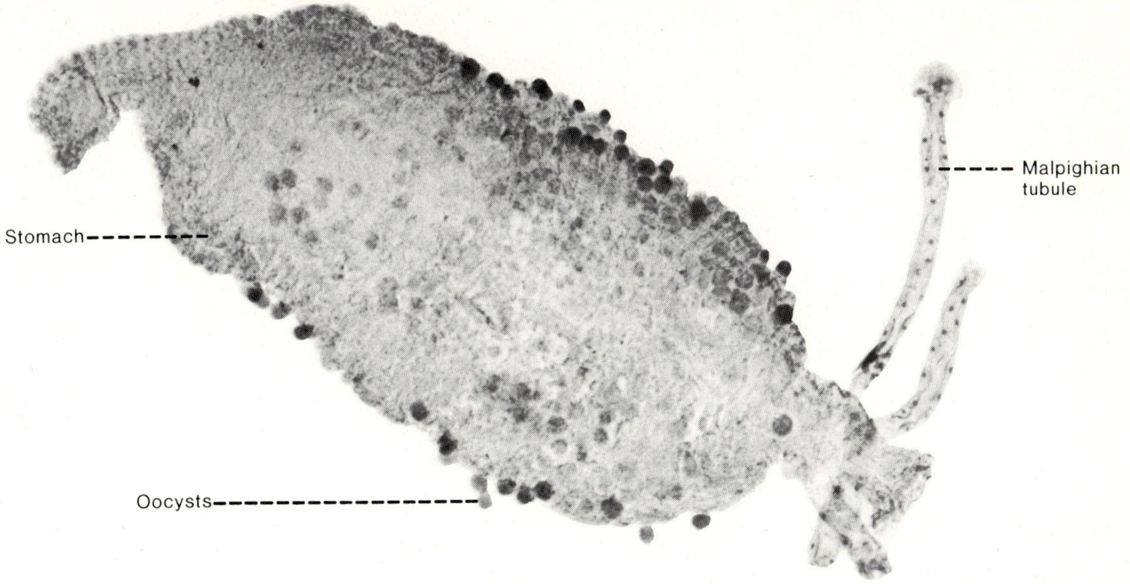

Fig. 1.26. Oocysts of *Plasmodium* sp. on the midgut of a mosquito.

and initiate a new schizogonous cycle leading to the gametogonous stage. Merozoites forming gametocytes are either male or female. The formation of merozoites occurs at intervals of 48 hours. In *P. vivax,* the prepatent period, the interval between the introduction of an organism and the demonstration of its presence in the host, is usually ten to 12 days.

Observe merozoites in the red blood cells. Merozoites are difficult to find in most smears. Nevertheless, one should watch for the free, minute, spindle-shaped bodies.

Gametogonous Cycle. The gametogonous cycle is the beginning of the sexual stages. Large, uninucleated parasites with smooth contour that fill the red corpuscles are mature gametocytes. They consist of the male microgametocytes and the female macrogametocytes.

1. **Microgametocyte.** The microgametocyte has pale blue or pinkish blue cytoplasm and a large nucleus with irregularly distributed granules. It fills the red blood cell, often distorting it.

2. **Macrogametocyte.** This form has dense, blue cytoplasm containing a small, compact nucleus that may have a dark red nucleolus.

The gametocytes do not develop beyond maturity in the blood cells. Development continues and is completed in the stomach of mosquitoes.

Find representatives of both sexes of gametocytes in the red blood cells.

3. **Microgamete.** Upon entering the stomach of mosquitoes, microgametocytes produce six to eight long, slender, filamentous microgametes by a process of exflagellation. They are attached to the residual mass of cytoplasm for a short time, after which they become free to seek and penetrate a macrogamete. The macrogametes are difficult to distinguish from gametocytes.

Observe the demonstration slide showing exflagellation of a microgametocyte in the stomach of a mosquito.

4. **Macrogamete.** The macrogamete is similar to the macrogametocyte in appearance.

5. **Zygote.** Fertilization takes place in the mosquito's stomach when a microgamete enters a macrogamete to form a zygote. In the various species of *Plasmodium,* the zygote is motile and is designated as an **ookinete.** Formation of the zygote terminates the gametogonous cycle.

Sporogonous Cycle. The sporogonous phase commences when the ookinete begins to develop.

1. **Oocyst.** When the migrating ookinete comes to rest between the epithelial cells and basement membrane of the mosquito's stomach, it is known as a young oocyst. Nuclear division commences and continues to the point where there are vast numbers of minute nuclear particles, known as **sporoblasts.**

As growth continues, the oocyst increases greatly in size. Bits of cytoplasm surround each nuclear particle to form myriads of sporozoites. The enlarged oocyst ruptures, liberating the sporozoites in the hemocoel.

See a slide of mosquito stomach, showing oocysts of various sizes attached to the wall (fig. 1.26).

2. **Sporozoite.** When the ripe oocyst ruptures, up to 200,000 sporozoites are liberated into the hemocoel of the mosquito. They travel to the salivary glands and

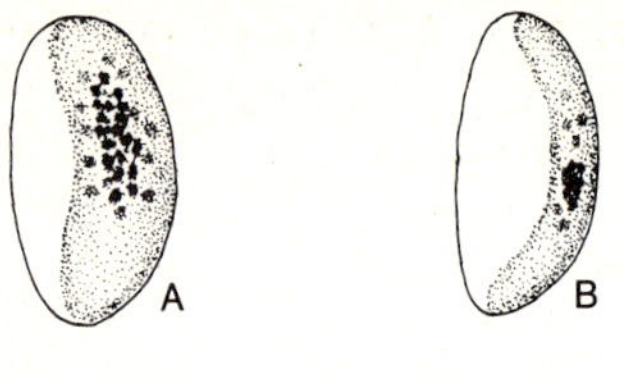

Fig. 1.27. Gametocytes of *Plasmodium falciparum* in human erythrocytes distorted by the characteristic shape of the microgametocyte (A) and the macrogametocyte (B).

migrate through the cells into the lumen of the glands. They are injected into the host with the contents of the salivary glands when the mosquitoes feed. Entrance of sporozoites into the body of the vertebrate host initiates the preerythrocytic phase of the schizogonous cycle.

See demonstration slide of sporozoites.

Three other species of plasmodia causing malaria in humans are generally accepted. They are (1) *Plasmodium falciparum,* (fig. 1.27), the cause of **falciparum malaria;** (2) *P. malariae,* (fig. 1.28), the cause of **malariae** infection; and (3) *P. ovale,* the most recently accepted species, the cause of **ovale malaria.** (See table 1.3.)

In falciparum malaria, schizogony is completed in 36 to 48 hours, usually requiring the longer period. In this species, only the ringlike trophozoites and elongated, somewhat crescent-shaped macrogametocytes and bean-shaped microgametocytes appear in the circulating blood. Multiple ring stages in a single host cell are common. They often lie against the host cell membrane, and are then called appliqué forms. Schizonts and segmenters are in capillaries and blood sinuses of the internal organs and in the bone marrow.

This species differs conspicuously from the others in humans in its crescent- or bean-shaped gametocytes and in the absence of schizonts and segmenters in the circulating blood cells.

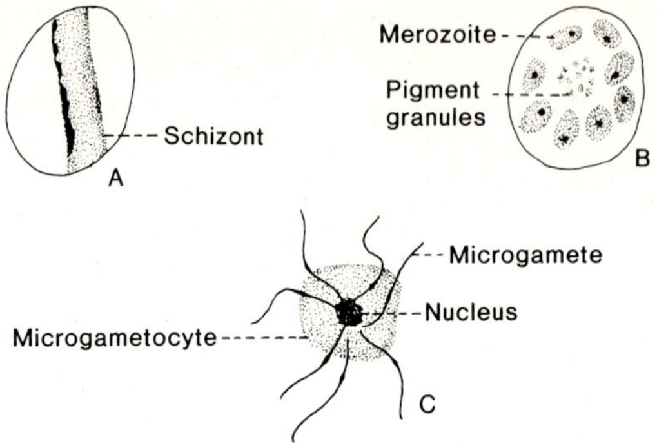

Fig. 1.28. *Plasmodium malariae.* (A) Band-shaped schizont in erythrocyte; (B) segmenter with characteristically small number of merozoites; lightly stippled areas represent pigment granules; (C) exflagellating microgametocyte forming elongated, uninucleate microgametes from nucleus.

Observe smears of human blood showing the distinctive gametocytes in the greatly distorted red corpuscles.

In *Plasmodium malariae* infections schizogony occurs at 72-hour intervals. Schizonts often appear as bandlike organisms, extending across the red blood cell. Segmenters contain six to 12, usually eight to 10, robust merozoites instead of the 12 to 18 smaller ones in *P. vivax* and *P. falciparum.* Gametocytes are oval and similar to those of *P. vivax;* being smaller, they do not expand and distort the blood corpuscles.

Observe smears containing blood infected with *P. malariae.* Note the characteristic morphology of the schizonts, segmenters, and gametocytes of this species.

Family Haemoproteidae

Members of this family are parasitic in reptiles, birds, and mammals. Birds are common hosts. Large pigmented crescent- or C-shaped gametocytes occur in the erythrocytes, particularly surrounding the nucleus in

Table 1.3. Criteria for Differential Diagnosis of *Plasmodium* spp. in Humans.

P. vivax	*P. falciparum*	*P. ovale*	*P. malariae*
1. Trophozoites ameboid 2. Segmenters form about 16 merozoites 3. Host cell enlarged, decolorized, frequently with Schüffner's dots 4. Parasites relatively large 5. Gametocytes rounded	1. Ring stages small, often with two chromatin dots 2. Segmenters usually form about 16 merozoites 3. Appliqué forms frequent; Maurer's clefts present 4. Larger trophozoites and schizonts not usually in peripheral blood 5. Gametocytes crescent shaped	1. Trophozoites not ameboid 2. Segmenters usually form eight merozoites 3. Host cells somewhat enlarged, sometimes decolorized with oval distortion. Schüffner's dots heavy 4. Parasites large 5. Gametocytes rounded	1. Trophozoites often band form 2. Segmenters usually form eight merozoites 3. Host cells not enlarged, decolorized, or stippled; hemozoin granules large, abundant 4. Parasites large 5. Gametocytes rounded

Modified from Schmidt and Roberts (1985).

blood cells of birds and reptiles. Schizonts are in endothelial (lining) cells of small blood vessels in the liver and lungs. Bloodsucking flies other than mosquitoes serve as vectors in which gametogony takes place.

Haemoproteus columbae occurs in the erythrocytes of pigeons only as gametocytes. The crescent-shaped gametocyte surrounds much of the nucleus of the red blood cell (fig. 1.29). The life cycle is depicted in figure 1.30. Schizonts are found in the endothelial cells of the

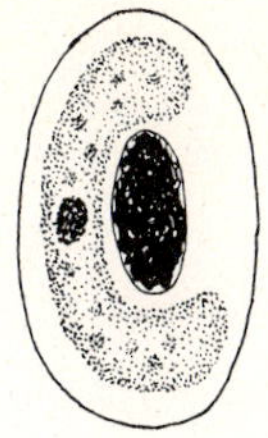

Fig. 1.29. Mature C-shaped gametocyte of *Haemoproteus sp.* partially surrounds dark nucleus of avian erythrocyte.

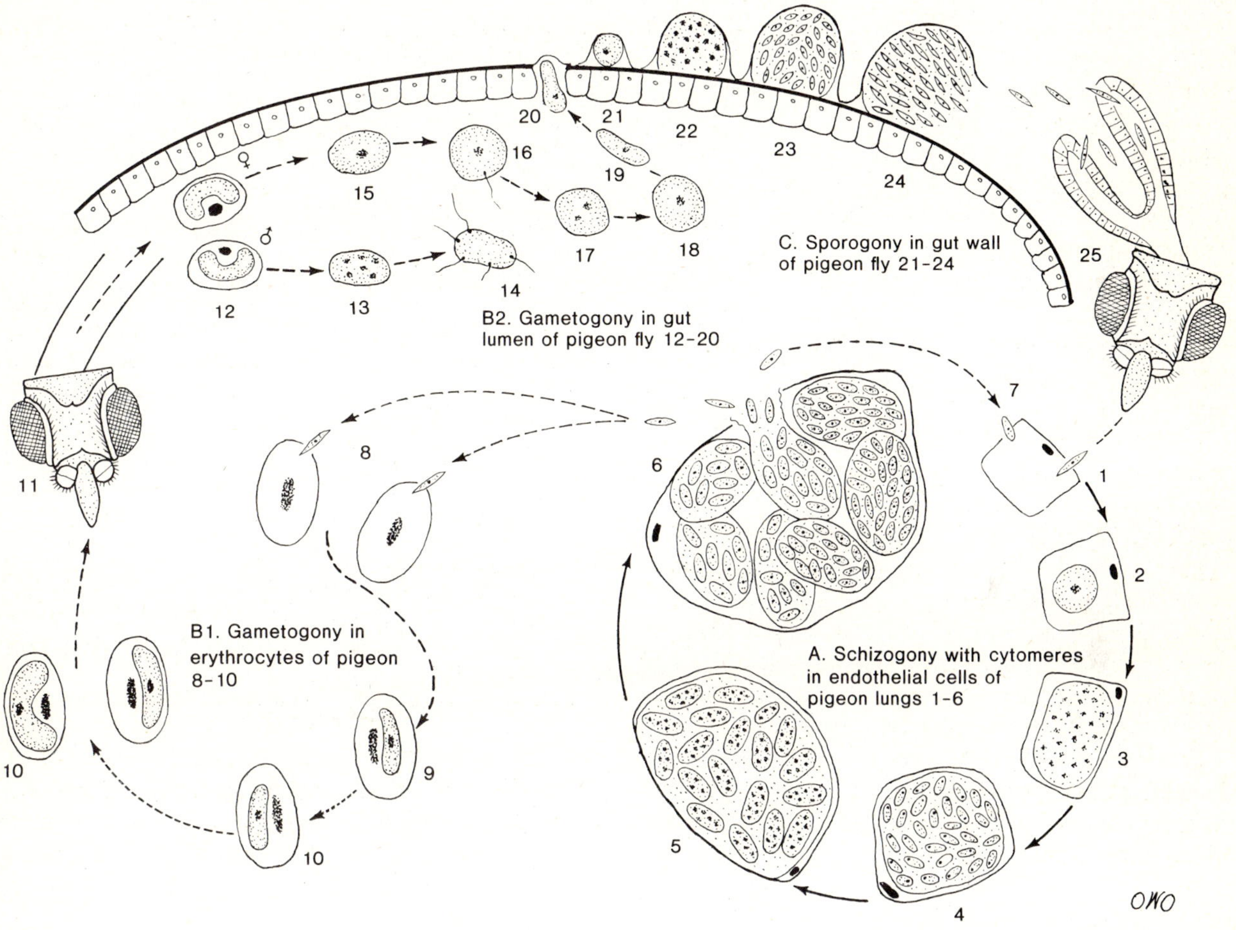

Fig. 1.30. Life cycle of *Haemoproteus columbae.*
A. *Schizogony in endothelial cells of lungs of pigeon* (1–7).
1. Sporozoite from bloodsucking pigeon louse fly injected into bloodstream enters endothelial cells of lung capillaries;
2. trophozoites; 3. multinucleate schizont; 4. uninucleate cytomeres in schizont; 5. multinucleate cytomeres; 6. mature megaloschizont with cytomeres ruptures, releasing fully developed cytomeric merozoites; 7. some merozoites enter endothelial cells to continue megaloschizogonic cycle.
B. *Gametogony.* (1) In erythrocytes of pigeon (8–10).
8. Megaloschizogonic merozoites destined to become gametocytes enter red blood cells; 9. developing gametocytes; 10. mature macro- and microgametocytes.
(2) In lumen of gut of pigeon louse fly (11–19). 11. Pigeon louse fly infected by swallowing gametocytes;
12. gametocytes inside erythrocytes; 13. free multinucleate microgametocyte; 14. exflagellation of microgametocyte forms slender microgametes; 15. freed macrogametocyte;
16. macrogamete being fertilized by microgamete; 17–18. zygote; 19. ookinete (motile zygote).
C. *Sporogony in gut wall of pigeon louse fly* (20–25).
20. Ookinete penetrates gut wall, coming to rest between epithelium and basement membrane; 21. young oocyst;
22. multinulceate oocyst; 23. young sporozoites form from nuclei and cytoplasm; 24. ripe oocyst ruptures, releasing numerous sporozoites into hemocoel; 25. sporozoites enter lumen of salivary glands are injected into bloodstream of pigeon by feeding fly.

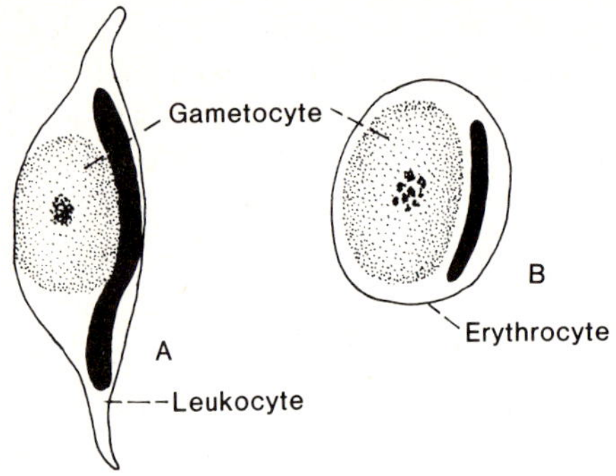

Fig. 1.31. Gametocytes of *Leucocytozoon* in blood cells of avian host. (A) Gametocyte developed from merozoites originating in megaloschizonts enter monocytes and lymphocytes, causing them to elongate; (B) merozoites from hepatic schizonts penetrate erythrocytes, producing round gametocytes.

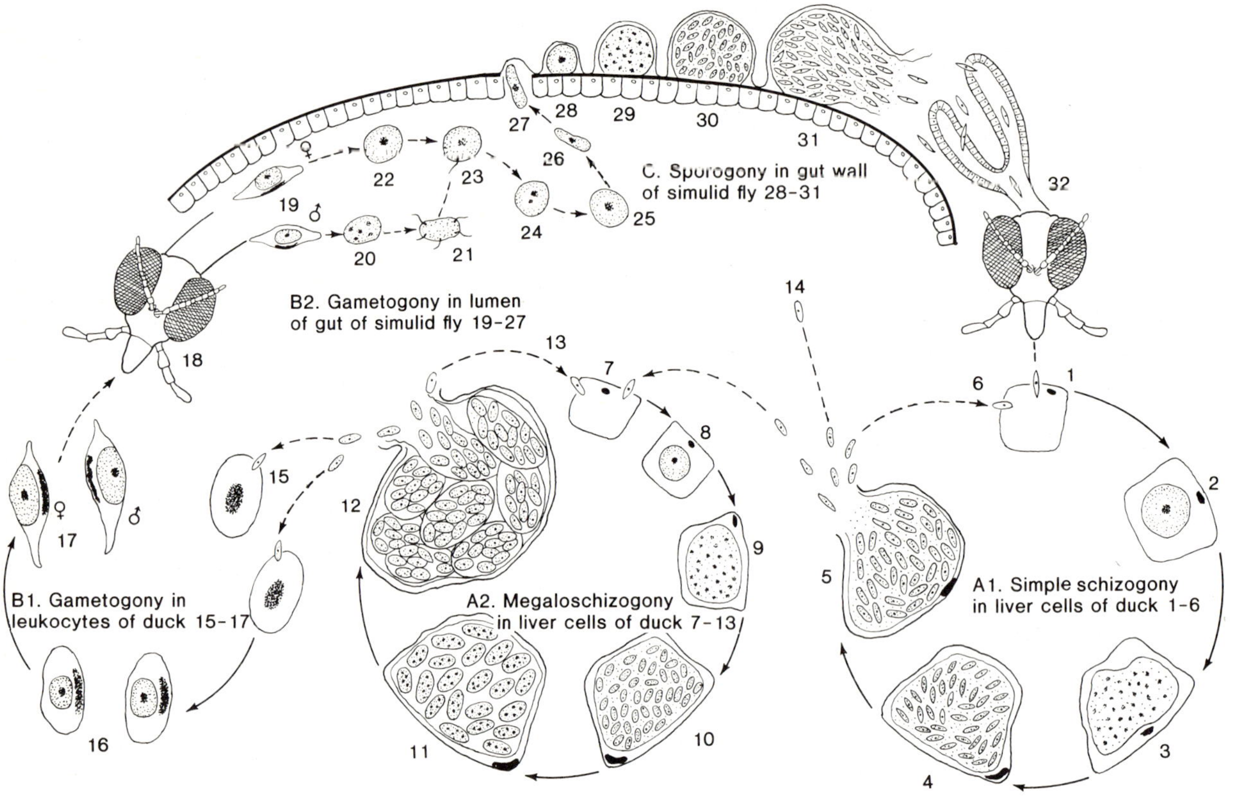

Fig. 1.32. Life cycle of *Leucocytozoon simondi.*

A. *Schizogony.* (1) In liver, simple schizogony (1–6).

1. Sporozoites from blackfly enter macrophages of liver to initiate simple schizogony; 2. trophozoite; 3. schizont with multiple nuclei; 4. developing merozoites in segmenter; 5. ruptured segmenter releasing mature merozoites; 6. merozoites enter other macrophages to continue this type of schizogony.

(2) In liver, megaloschizogony (7–13). 7. Merozoites from simple schizogony enter other macrophages of liver to begin the cycle of megaloschizogony; 8. trophozoite; 9. schizont with multiple nuclei; 10. each nucleus takes a bit of cytoplasm to form a cytomere; 11. nucleus of each cytomere divides many times, greatly increasing progeny; 12. mature cytomeric merozoites released; 13. some merozoites reenter macrophages to begin another cycle of cytomeric schizogony.

(3) In other tissues (14). 14. Merozoites enter macrophages of heart, lungs, intestine to initiate both simple and megaloschizogony.

B. *Gametogony.* (1) In leukocytes of duck (15–17).

15. Megaloschizogonic merozoites enter leukocytes to form gametocytes; 16. young micro- and macrogametocytes in leukocytes; 17. elongate gametocytes (round gametocytes originating from merozoites of simple schizogony develop in young erythrocytes).

(2) In lumen of gut of blackfly (18–26). 18. Feeding blackfly becomes infected by ingesting gametocytes; 19. gametocytes inside leukocytes; 20. free microgametocyte; 21. exflagellation of microgametocyte to form slender microgametes; 22. free macrogametocyte; 23. fertilization of macrogamete by microgamete; 24–25. zygote; 26. ookinete (motile zygote).

C. *Sporogony in wall of midgut of blackfly* (27–32).

27. Ookinete penetrates gut wall, coming to rest between epithelium and basement membrane; 28. young oocyst; 29. multinucleate oocyst; 30. young sporozoites developing from separate nuclei; 31. ripe oocyst ruptures, releasing mature sporozoites into hemocoel of blackfly; 32. sporozoites enter salivary glands of blackfly and are injected into the ducks by feeding flies.

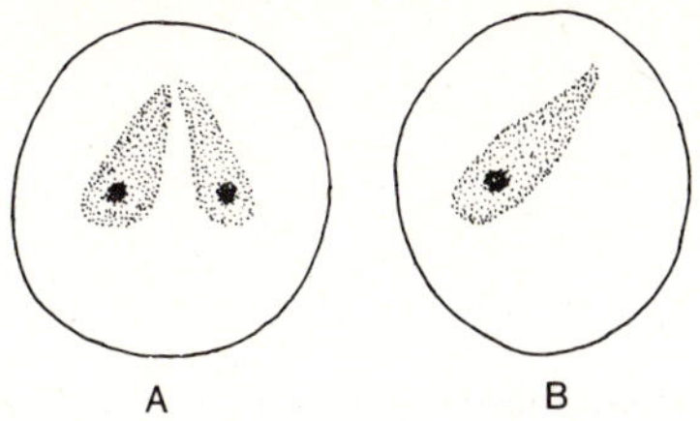

Fig. 1.33. *Babesia bigemina* in erythrocytes of bovine host. (A) Cell with two pyriform-shaped parasites, a typical arrangement; (B) cell with a single elongated parasite.

capillaries of the lungs. Gametogony and sporogony occur in the stomach of **louse flies** (*Lynchia* and related genera) similar to *Plasmodium* in mosquitoes. *Parahaemoproteus nettionis,* found in ducks and geese, is transmitted by **biting midges** (*Culicoides* sp.).

Examine stained smears of pigeon blood for infected erythrocytes containing the large, pigmented, crescent-shaped gametocytes partially surrounding the nucleus of the host cell. Two gametocytes may be present in a single erythrocyte. Immature, noncrescent gametocytes appear in the cytoplasm of the erythrocytes.

Other common species of *Haemoproteus* are *H. lophortyx* of quail, *Parahaemoproteus nettionis* of ducks, and *P. fringillae* and *P. garnhami* from various species of sparrows. Midges of the genus *Culicoides* serve as vectors for these species.

Family Leucocytozoidae

Members of this family are very similar to those of the Plasmodiidae and Haemoproteidae except that pigment granules and schizogony are not found in host erythrocytes. The asexual stages remain in the internal organs, such as the spleen and lungs. All are parasites of birds.

Leucocytozoon simondi

This common parasite can be found as ovoid or fusiform gametocytes in red and white blood cells of domestic and wild ducks and geese (see fig. 1.31). Mortality is high among domestic flocks, especially young birds. It is transmitted by **blackflies** of the genus *Simulium,* and in North America *S. rugglesi* is the main vector. Sporozoites in the anseriform host develop into small hepatic schizonts, and their progeny, the merozoites, penetrate erythrocytes and develop into round gametocytes. Syncytia from hepatic schizonts are phagocytized by cells of the lymphoid macrophage system and develop into large **megaloschizonts** in many organs of the body. Merozoites from the latter penetrate monocytes and lymphocytes and develop into gametocytes that are fusiform. The life cycle (fig. 1.32) differs from *Plasmodium* in that there is no erythrocytic schizogony. Other species of *Leucocytozoon* occur in turkeys, grouse, and song birds.

Subclass Piroplasmea

Piroplasmea are minute parasites in the erythrocytes of mammals and are transmitted by ticks. Gliding and flexion of the body are the only known means of locomotion. Pigments are not produced by intraerythrocytic forms. An apical complex is present though reduced. Sexual reproduction, though suspected, has not been proved.

Family Babesiidae

Babesia bigemina

This parasite appears as small, paired pyriform, round or oval, occasionally single bodies in the erythrocytes of cattle (fig. 1.33). The pyriform shape is the most common. The pyroplasms multiply by binary fission in the erythrocytes, ultimately destroying them.

Boophilus annulatus, a one-host tick, is the vector in the United States and Mexico. Piroplasms ingested by feeding ticks multiply asexually in the intestinal epithelium, liberating forms that enter the Malpighian tubules for another round of multiplication. Progeny from the Malpighian tubules enter the developing ovaries and eggs and undergo further multiple fission, producing great numbers of **vermicules.** They eventually enter the body cavity of the developing ticks and come to rest in the salivary glands of the young ticks. Feeding ticks inject the parasites with the salivary secretion. The parasites enter the erythrocytes. This is an example of transovarian transmission.

The life cycle is shown in figure 1.34.

Southern cattle with their ticks carried *Babesia bigemina* to nonimmune northern cattle during the great drives northward in the past, resulting in numerous deaths. The disease was eliminated from the United States by the systematic dipping of all southern cattle, which eventually eradicated the *Boophilus* tick hosts.

Find the characteristic, paired pyriform *Babesia bigemina* in stained blood smears of infected cattle.

A related species is *Babesia microti.* Normally a parasite of small mammals, such as voles, it has been recognized as the agent of a malarialike disease in humans, sometimes called **Nantucket Island fever.** Its vector is the tick *Ixodes dammini.* Morphologically, this parasite is very similar to *Babesia bigemina.*

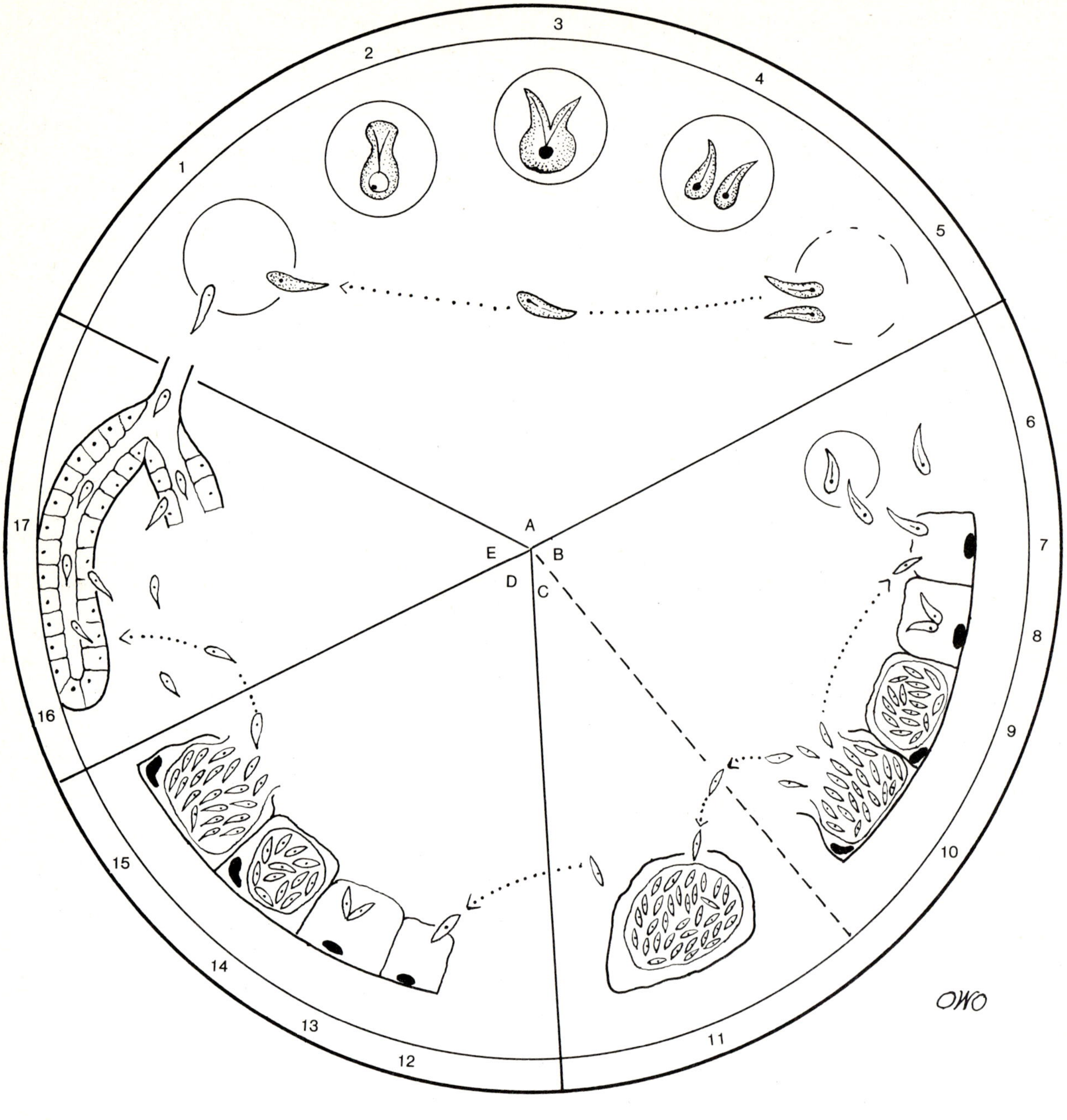

Fig. 1.34. Life Cycle of *Babesia bigemina.*
A. *Bovine host, in erythrocytes* (1–5). 1. Erythrocyte infected by vermicules injected by nymphal ticks to form a trophozoite; 2. trophozoite preparing to divide by binary fission; 3. fission underway; 4. fission completed with two daughter trophozoites; 5. trophozoites destroy erythrocytes, become free in bloodstream, and infect other red blood cells (1).
B. *Tick host (adult), in intestine and Malpighian tubules* (6–10). 6. Infected erythrocytes or free trophozoites ingested by female tick (*Boophilus spp.*); 7. liberated trophozoite enters intestinal epithelial cell of intestine; 8. division begins to form vermicules; 9. mass of vermicules resulting from repeated binary fission; 10. epithelial cell ruptures from pressure and vermicules enter other cells, including those of Malpighian tubules; others escape into hemocoel.
C. *Tick host (adult female), in ovary and eggs* (11).
11. Vermicules enter ovary and eggs where they increase greatly in number.
D. *Larval tick, in intestinal epithelium* (12–15).
12. Vermicules from eggs enter intestinal epithelial cells of embryonic ticks; 13. multiplication by binary fission begins; 14. cells fill with vermicules; 15. loaded cells rupture, liberating vermicules into hemocoel of nymphal tick.
E. *Nymphal tick, in hemocoel and salivary gland* (16–17).
16. Vermicules in hemocoel; 17. vermicules penetrate cells of salivary glands, enter ducts, and are injected with saliva by feeding nymphs.

Review Questions

1. What organelles give the phylum its name?
2. Define syzygy.
3. Name the body regions of a cephaline gregarine.
4. Define and compare sporogony, schizogony, and gametogony.
5. Compare oocysts of *Eimeria* and *Isospora*.
6. What are tachyzoites and bradyzoites?
7. What is the name of the motile zygote of *Plasmodium?*
8. Which genus of malarialike organisms has no pigment?
9. What are the vectors of *Babesia?*
10. What vectors are involved in the life cycles of *Plasmodium, Leucocytozoon,* and *Haemoproteus?*

Notes and Sketches

Phylum Myxozoa

In this phylum the spores are of multicellular origin and are surrounded by two, three, or sometimes more **valves** of various shapes. They have one or more **polar capsules** and are mostly parasites of fishes. From one to four polar capsules can be found at one end of the spore, except in the suborder Bipolarina, where one capsule is located at each end of the spore. More than 700 species are described in this phylum. Most are host and tissue specific.

Family Myxosomatidae

The unusual spores with the coiled polar filament(s) inside the polar capsules characterize this group structurally. They are common parasites within the cells of kidneys, liver, spleen, gills, and integument, as well as in the lumen of the gall and urinary bladders.

Myxosoma sp.

The spore varies from pyriform, circular, or ovoid in shape. There are two pear-shaped capsules. The sporoplasm has no iodinophilous vacuole. (See fig. 1.35.) *Myxosoma catostomi* is prevalent in the muscles and connective tissue of the common sucker. *Myxosoma cerebralis* occurs in the cartilage and perichondrium of trout, causing twist or **whirling disease.** The spores are 10 to 11 μm in length.

The life cycle is complex, involving two hosts. Spores reach the environment when infected fish die and decompose or are eaten by scavengers. Alternatively, when infected fish are cannibalized or eaten by a predator, viable cysts can be released in the feces. The cyst is not infective to trout. First it must be eaten by a tubificid annelid worm, in which it enters the intestinal epithelium and grows into a sporocyst containing eight organisms known as **triactinomyxons.** This form has three taillike filaments, a slender stalk, or **style,** capped with an **epispore** containing 32 to 50 spherical sporozoites. There are three polar capsules at the apex of the epispore. (The change from two capsules in the spore to three capsules in the epispore raises some doubt in the mind of the author as to the accuracy of this life cycle, but these are the facts as are known at this writing.)

Trout become infected when they eat the tubificid. Also, the triactinomyxon may escape the annelid and, when in contact with a fish's gill, penetrate into the capillaries to be disseminated throughout the body. The entire life cycle requires from one to one and one-half months at 12° C.

Observe slides containing spores of some of the Myxosomatidae. Note the pustulelike lesions on the skin of fish. Open a pustule, transfer the contents to a slide, and observe them under the high power of a microscope. Add a drop of iodine solution to determine whether there is an iodinophilous vacuole.

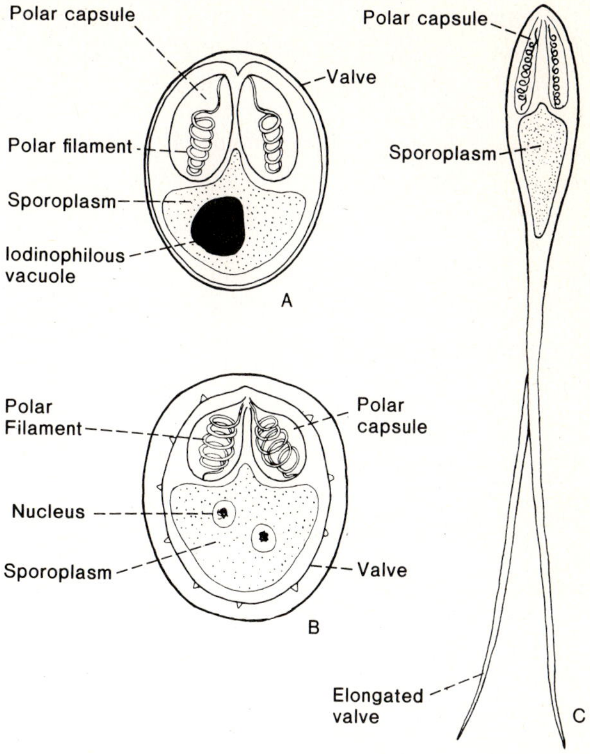

Fig. 1.35. Some representative Myxosoma. (A) Frontal view of *Myxobolus,* showing iodinophilous vacuole in sporoplasm; (B) frontal view of *Myxosome* which has no iodinophilous vacuole; (C) frontal view of *Henneguya.*

Myxobolus sp.

The spores are somewhat similar to those of *Myxosoma* except that there is an iodinophilous vacuole in the sporoplasm and there may be one or two polar capsules. *Myxobolus orbiculatus* occurs in the muscles of shiners and *M. intestinalis* in the intestinal wall of sunfish. Spores are small, being 9 to 13 μm in size and somewhat circular or oval in shape.

Observe smears containing spores that have been stained with iodine and sections of gut showing the lesions filled with developing forms and spores.

Henneguya sp.

Typical spores are somewhat spindle shaped in frontal view. Each of the two valves of the cyst continues posteriorly as a long, slender extension. There are two polar capsules and a posterior iodinophilous vacuole. *H. exilis* occurs in the gills and skin of catfish and *H. microspora* in bass and sunfish. Spores proper are 18 to 20 μm long but with the valvular extensions attain lengths of 60 to 70 μm.

Observe a demonstration slide of *Henneguya,* noting the spindle-shaped spore and the long extensions of the valves. Collect and examine catfish, bass, and sunfish for infections. Stain living spores with Lugol's iodine solution to demonstrate the iodinophilous vacuole.

Phylum Microspora

The spore of this group consists of a single valve containing one long, tubular polar filament through which the sporoplasm escapes. Microspora are intracellular parasites, mostly of invertebrates and primarily of arthropods.

Family Nosematidae

Nosematidae is a large family representive of the Microspora.

Nosema apis, a cosmopolitan and prevalent parasite of honeybees, causes great damage in apiaries. When swallowed by bees, the spores in the midgut extrude the polar filament, which pierces an epithelial cell. The sporoplasm leaves the valve by passing through the long tubular polar filament and into the interior of the cell. Multiplication (sporogony) is rapid, destroying the infected epithelial cells and killing the bees. Masses of spores discharged in the feces are the source of infection of the other bees in the hive.

If specimens are available, observe the single-valved spores (4–6 μm long by 2–4 μm wide) containing the extremely long, slender, coiled polar filament.

Another famous microsporidian is *Nosema bombycis* of the larvae of silkworm moths. As in bees, the crowded conditions under which the caterpillars are reared favor rapid spread of the parasite.

Review Questions

1. Which species causes whirling disease in trout?
2. What is the principal difference between a myxozoan and a microsporan?
3. Is there an insect vector in the life cycle of either?
4. What is contained within the polar capsules?
5. What is the host of *Nosema apis?*
6. How do fish become infected with whirling disease?

Notes and Sketches

Phylum Ciliophora

The body of Ciliophora is covered with simple cilia or ciliary organelles. Most species are free-living, some are commensals in or on organisms, and a few of them are parasitic.

Class Kinetofragminophorea

Family Balantidiidae

Balantidium coli

This common species appears to be a commensal of the caecum and large intestine of swine. In humans it is a pathogenic parasite that invades the intestinal mucosa in a manner similar to that of *Entamoeba histolytica*. Examine stained slides of trophozoites and cysts of this organism.

The large, egg-shaped **trophozoite,** measuring 50 to 100 × 40 to 60 μm, is surrounded by a pellicle from which arise cilia arranged in rows. One end is more pointed than the other, and at this end there is a cleft, known as the peristome, from which leads the cytostome. At the more rounded end is the cytopyge. Within is a large, slightly curved macronucleus, usually situated near the middle of the body. Lying within the concavity of the macronucleus, and in contact with it, is the much smaller micronucleus. The two contractile vacuoles, one in each end, appear as clear areas, and the food particles stain variously.

Cysts stain poorly, the cyst wall remaining unstained, and the macronucleus appears as the most conspicuous structure.

Life Cycle

Trophozoites (fig. 1.36A) in the intestine multiply by binary fission and feed on bacteria and particulate material. When carried into the rectum, the trophozoites round up and secrete a double wall about themselves before being voided with the feces. Infection results from swallowing the encysted stage (fig. 1.36B) with contaminated food or water.

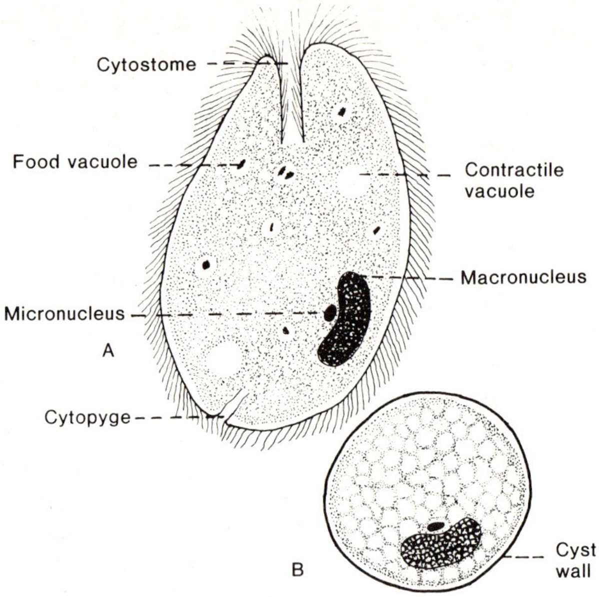

Fig. 1.36. *Balantidium coli.* (A) Mature trophozoite; (B) encysted stage.

Family Ophryoglenidae

Ichthyophthirius multifiliis, commonly known as "ick," occurs on the skin, gills, and sometimes the cornea of freshwater fish. It is especially prevalent on pond, aquarium, and hatchery fish, being embedded in whitish pustules. It may prove fatal by making the skin of the fish host susceptible to attack by fungi.

Examine a section of infected fish skin. (Live material may be available from the local pet store.) Fully developed ciliates are oval in shape and up to 1 mm in length. The body is covered by numerous rows of cilia and bears a circular, unciliated cytostome at one end. The large macronucleus is horseshoe shaped, and the small, spherical micronucleus lies in the concavity of the former. Numerous vacuoles occur in the cytoplasm.

If slides are available, study an infected fish gill (fig. 1.37).

Family Plagiotomidae

Most species in this family are quite large, robust parasites of the intestines of vertebrates and invertebrates. The entire body has tiny cilia in longitudinal rows. Masses of cilia, fused at their bases and thus forming an undulating membrane, extend from the anterior end to deep within the cytopharynx.

A common genus is *Nyctotherus* (fig. 1.38), easily obtainable from the colon of toads and frogs, the gut of cockroaches, and the intestine of earthworms.

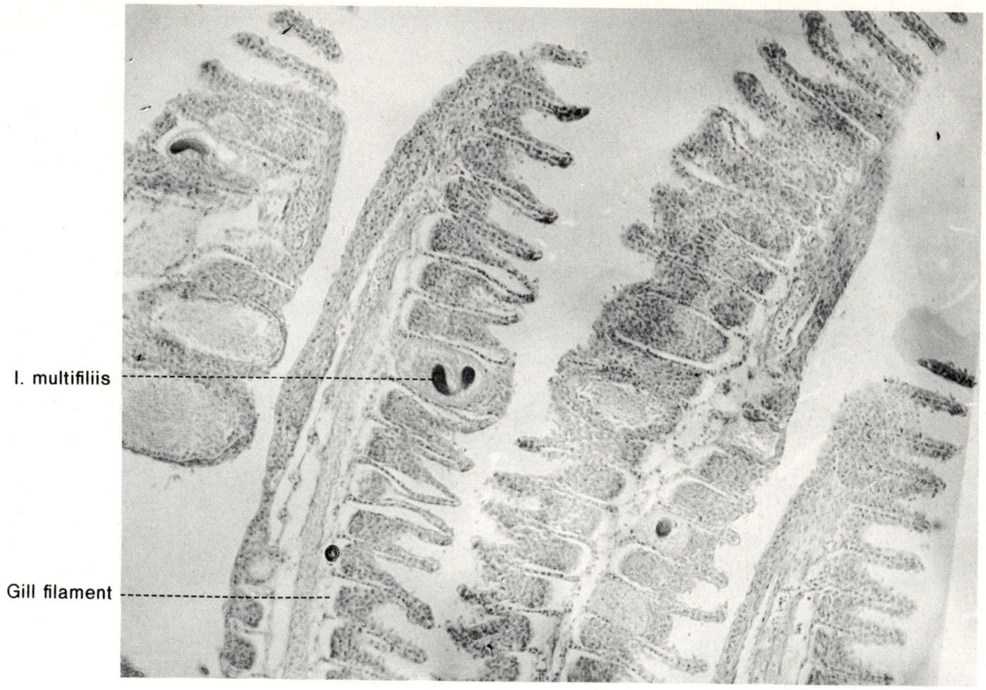

Fig. 1.37. *Ichthyophthirius multifiliis,* photomicrograph of section of infected fish gill.

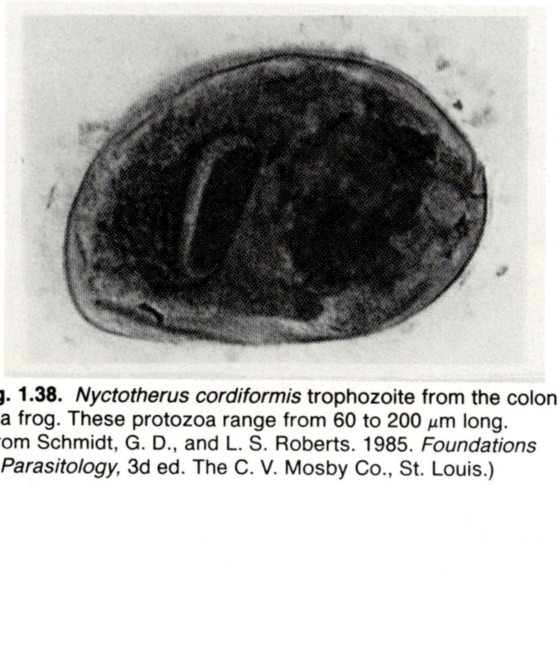

Fig. 1.38. *Nyctotherus cordiformis* trophozoite from the colon of a frog. These protozoa range from 60 to 200 μm long. (From Schmidt, G. D., and L. S. Roberts. 1985. *Foundations of Parasitology,* 3d ed. The C. V. Mosby Co., St. Louis.)

Review Questions

1. Which ciliate can invade the liver of humans?
2. What is the cause of "ick"?
3. What are swarmers?
4. What are the functions of the macronucleus and micronucleus?
5. What is the most common nonhuman reservoir of *Balantidium coli?*
6. Name the oral and anal locations of a ciliate.

Notes and Sketches

References

General

Aikawa, M., and Sterling, C. H. 1974. *Intracellular Parasitic Protozoa.* Academic Press, New York, 76 pp.

Baker, J. R. 1969. *Parasitic Protozoa.* Hutchinson University Library, London, 174 pp.

Farmer, J. N. 1980. *The Protozoa: Introduction to Protozoology.* The C. V. Mosby Co., St. Louis.

Kudo, R. R. 1966. *Protozoology.* 5th ed. Charles C. Thomas, Springfield, 1174 pp.

Levine, N. D. 1973. *Protozoan Parasites of Domestic Animals and of Man.* 2nd ed. Burgess Publ. Co., Minneapolis, 406 pp.

Levine, N. D. et al. 1980. A newly revised classification of the Protozoa. *J. Protozool.* 27:37–58.

Sarcomastigophora

Adler, S. 1964. Leishmania. In *Advances in Parasitology,* ed. B. Dawes. Academic Press, New York, vol. 2, pp. 35–96.

Albach, R. A., and Booden, T. 1978. Amoebae. In *Parasitic Protozoa,* ed. J. P. Kreier. Academic Press, New York, vol. 2.

Camp, R. R.; Mattern, C. F. T.; and Honigberg, B. M. 1974. Study of *Dientamoeba fragilis* Jepps and Dobell. I. Electron microscopic observations of binucleate stages. II. Taxonomic position and revision of the genus. *J. Protozool.* 21:69–82.

Desser, S. S.; McIver, S. B.; and Jez, D. 1975. Observations on the role of simuliids and culicids in the transmission of avian and anuran trypanosomes. *Int. J. Parasitol.* 5:507–9.

Elsdon-Dew, R. 1968. The Epidemiology of Amoebiasis. In *Advances in Parasitology,* ed. B. Dawes. Academic Press, New York, vol. 6, pp. 1–62.

Garnham, P. C. C. 1965. The leishmanias, with special reference to the role of animal reservoirs. *Amer. Zool.* 5: 141–51.

Hoare, C. A. 1967. Evolutionary Trends in Mammalian Trypanosomes. In *Advances in Parasitology,* ed. B. Dawes. Academic Press, New York, vol. 5, pp. 47–91.

Hoare, C. A., and Wallace, F. G. 1966. Developmental stages of trypanosomatid flagellates: a new terminology. *Nature* 212:1385–86.

Honigberg, B. M. 1978. Trichomonads of Importance in Human Medicine. In *Parasitic Protozoa,* ed. J. P. Kreier. Academic Press, New York, vol. 3.

Jirovec, O., and Petrou, M. 1968. *Trichomonas vaginalis* and Trichomoniasis. In *Advances in Parasitology,* ed. B. Dawes. Academic Press, New York, vol. 6, pp. 117–88.

Köberle, F. 1968. Chagas' Disease and Chagas' Syndromes: The Pathology of American Trypanosomiasis. In *Advances in Parasitology,* ed. B. Dawes. Academic Press, New York, vol. 6, pp. 63–116.

Lumsden, W. H. R. 1965 and 1970. Biological Aspects of Trypanosomiasis Research. In *Advances in Parasitology,* ed. B. Dawes. Academic Press, New York, vol. 3, pp. 1–57; vol. 8, pp. 227–49.

Meyer, E. A., and Radulescu, S. 1979. *Giardia* and Giardiasis. In *Advances in Parasitology,* ed. W. H. R. Lumsden. Academic Press, New York, vol. 17.

Wessenberg, H. 1978. Opalinata. In *Parasitic Protozoa,* ed. J. P. Kreier. Academic Press, New York, vol. 3.

Zuckerman, A., and Lainson, R. 1977. Leishmania. In *Parasitic Protozoa,* ed. J. P. Kreier. Academic Press, New York, vol. 1.

Apicomplexa

Bruce-Chwatt, L. 1980. *Essential Malariology.* William Heinemann Medical Books, London.

Coatney, G. R. 1976. Relapse in malaria—an enigma. *J. Parasitol.* 62:3–9.

Fallis, A. M. 1965. Protozoan life cycles. *Amer. Zool.* 5:85–94.

Frenkel, J. K. 1973. *Toxoplasma* in and around us. *BioScience* 23:343–51.

———. 1977. *Besnoitia wallacei* of cats and rodents: with a reclassification of other cyst-forming isosporoid Coccidia. *J. Parasitol.* 63:611–28.

Garnham, P. C. C. 1966. *Malaria Parasites and Other Haemosporidia.* Blackwell Sci. Publ., Oxford, 1114 pp.

———. 1967 and 1973. Malaria in Mammals Excluding Man. In *Advances in Parasitology.* ed. B. Dawes. Academic Press, New York, vol. 5, pp. 139–204; vol. 11, pp. 603–30.

Hewitt, R. 1940. Bird Malaria. *Amer. J. Hyg. Monog. Ser.* No. 15, 228 pp.

Horton-Smith, C., and Long, P. L. 1963 and 1968. Coccidia and Coccidiosis in the Domestic Fowl and Turkey. In *Advances in Parasitology,* ed. B. Dawes. Academic Press, New York, vol. 1, pp. 67–107; vol. 6, pp. 313–25.

Huff, C. G. 1963 and 1968. Experimental Research in Avian Malaria. In *Advances in Parasitology,* ed. B. Dawes. Academic Press, New York, vol. 1, pp. 1–65; vol. 6, pp. 293–311.

Kreier, J. P., ed. 1980. *Malaria.* Academic Press, New York, vols. 1, 2, and 3.

Lotze, J. C., and Leek, R. G. 1961. A practical method of culturing coccidial oocysts in tap water. *J. Parasitol.* 47:588–90.

———, and ———. 1968. Excystation of sporozoites of *Eimeria tenella* in apparently unbroken oocysts in the chicken. *J. Protozool.* 15:693–97.

Maegraith, B. 1968. Liver Involvement in Acute Mammalian Malaria with Special Reference to *Plasmodium knowlesi* Malaria. In *Advances in Parasitology,* ed. B. Dawes. Academic Press, New York, vol. 6, pp. 189–231.

Miles, H. B. 1962. The mode of the transmission of the acephaline gregarine parasites of earthworms. *J. Protozool.* 9:303–6.

Walker, A. J. 1968. Manual for the Microscopic Diagnosis of Malaria. 3rd ed. Pan American Health Organization, Washington, D.C., *Sci. Publ.* No. 161. 117 pp.

Wilcox, Aimee. 1960. Manual for the Microscopial Diagnosis of Malaria in Man, U.S. Dept. Health, Educ., Welfare. *Public Health Service Publ.* No. 796, 80 pp.

Myxozoa

Hoffman, G. L. 1975. Whirling disease (*Myxosoma cerebralis*); control with ultraviolet irradiation and effect on fish. *J. Wildlife Dis.* 11:505–7.

Myers, T. U.; Scola, J.; and Sunnibs, E. 1970. Modes of transmission of whirling disease of trout. *Nature* (London) 227:622–23.

Noble, E. R. 1944. Life cycles in the Myxosporidia. *Q. Rev. Biol.* 19:213–35.

Sprague, V. 1982. Myxozoa. In *Synopsis and Classification of Living Organisms,* ed. S. P. Parker. McGraw-Hill, New York, vol. 1, pp. 595–97.

Wolf, K., and Markiw, M. E. 1984. Biology contravenes taxonomy in the Myxozoa: new discoveries show alternation of invertebrate and vertebrate hosts. *Science* 225:1449–52.

Microspora

Sprague, V. 1982. Microspora. In *Synopsis and Classification of Living Organisms,* ed. S. P. Parker. McGraw-Hill, New York, vol. 1, pp. 589–94.

Ciliophora

Corliss, J. O. 1979. *The Ciliated Protozoa. Characterization, Classification, and Guide to the Literature.* Pergamon Press, Oxford.

Davis, H. S. 1946. *Care and Diseases of Trout.* U.S. Dept. Interior, Fish and Wildlife Service. Res. Rept. No. 12, 98 pp. (*Chilodon* [now *Chilodonella*], *Trichodina,* and *Ichthyophthirius,* pp. 37–45.)

Sources of Study Materials for Haemoflagellates*

Slides

Glossina sp.: T, W
Leishmania donovani, spleen smear: T, W
Leishmania donovani, spleen section: T, W
Leishmania donovani, in vitro culture: T, W, Tr, C
Leishmania donovani, liver section: W
Leishmania tropica, ulcer smear: W
Rhodnius prolixus, "kissing bug": W
Trypanosoma brucei gambiense, blood smear: W, Tr, C
Trypanosoma brucei rhodesiense, blood smear: W
Trypanosoma brucei rhodesiense, brain section: W, Tr
Trypanosoma cruzi, blood smear: T, W
Trypanosoma cruzi, heart smear: W
Trypanosoma cruzi, section of heart muscle or brain: T, W, C
Trypanosoma cruzi, in vitro culture: T, W, C
Trypanosoma equiperdum: W, C
Trypanosoma lewisi, blood smear: T, W, Tr, C
Trypanosoma musculi, blood smear: W

*Code letters stand for sources listed on page 45.

Live Material

Crithidia fasciculata: W, ATCC
Herpetomonas muscarum: W, ATCC
Leishmania tarentalae: W, ATCC
Leptomonas pessoai: W
Trypanosoma lewisi: W, ATCC
Trypanosoma ranarum: W, ATCC

Sources of Study Materials for Intestinal and Genital Flagellates

Slides

Chilomastix mesnili, trophozoite: T, W, C
Dientamoeba fragilis, smear: T, W, C
Giardia lamblia, cysts: T, Tr, W, C
Giardia lamblia, trophozoite: T, Tr, W, C
Histomonas meleagridis, liver: W
Opalina sp., section of infected frog intestine: T
Opalina sp., trophozoites: Tr, T, W, C
Trichomonas vaginalis, smear: T, Tr, C
Trichonympha, smear from termite: T, C, W
Tritrichomonas (=*Pentatrichomonas*) *muris,* smear: T, C

Live Material

Trichomonas vaginalis: W, ATCC
Trichonympha sp.: W, C
Tritrichomonas augusta: W, ATCC

Sources of Study Materials for Sarcodina

Slides

Endolimax nana: T, W, C
Entamoeba coli, cysts: Tr, T, W, C
Entamoeba coli, trophozoite: Tr, T, C, W
Entamoeba histolytica, cysts: Tr, T, W, C
Entamoeba histolytica, section of infected intestine: T, W, C
Entamoeba histolytica, trophozoite: Tr, T, W, C
Iodamoeba buetschlii, trophozoites: T, W, C
Naegleria fowleri, section of brain lesion; culture forms: W

Live Material

Entamoeba invadens, trophozoites from reptiles: ATCC
Entamoeba terrapinae, trophozoites from reptiles: ATCC

Sources of Study Material for Sporozoa

Slides

Eimeria stiedae, liver section: W, C
Eimeria tenella, section of infected chicken intestine: T, W
Gregarina, trophozoites from insect: T, W, C
Haemogregarina, gametocytes in frog blood: Tr, W, T, C
Haemoproteus columbae, gametocytes in pigeon blood: T, W
Leucocytozoon, gametocytes in bird blood: T
Monocystis, in section or smear of earthworm seminal vesicle: Tr, T, W, C
Plasmodium cynomolgi, in monkey blood: T, W
Plasmodium falciparum, gametocytes: T, W, C
Plasmodium falciparum, section of infected brain: T, W
Plasmodium falciparum, trophozoites in blood: T, Tr, W, C
Plasmodium gallinaceum, in bird blood: T
Plasmodium malariae, blood smear: W
Sarcocystis, sections of zoitocysts in muscle: Tr, T, W, C
Toxoplasma gondii, section of infected cat intestine: W
Toxoplasma gondii, section of infected liver, brain: W
Toxoplasma gondii, zoites in tissue smear: T, W, C

Live Material

Plasmodium berghi: ATCC

Sources of Study Material for Myxozoa

Slides

Leptotheca ohlmacheri, in frog kidney: T
Myxosoma, section of infected fish gill with cysts: C

Sources of Study Materials for *Microspora*

Live Material

Nosema lacusta: ATCC
Nosema necatrix: ATCC

Sources of Study Materials for Ciliophora

Slides

Balantidium coli, cysts: Tr, T, W, C
Balantidium coli, section of infected intestine: T, W, C
Balantidium coli, trophozoites: Tr, T, W, C
Ichthyophthirius, section of infected fish skin: T
Nyctotherus, section of infected earthworm intestine: T

Addresses of Sources for Study Materials

ATCC—American Type Culture Collection. c/o Order Department, 12301 Parklawn Drive, Rockville, Maryland 20852 USA (Telephone 800–638–6597)

C—Carolina Biological Supply Company. Powell Laboratories Division, Gladstone, Oregon 97027 USA (Telephone 503–636–1641)

T—Turtox, Inc. 5000 West 128th Place, Alsip, Illinois 60658 USA (Telephone 312–371–5500)

Tr—Triarch Inc., P.O. Box 98, Ripon, Wisconsin 54971 USA (Telephone 414–748–5125).

W—Ward's Natural Science Establishment, Inc., Rochester, New York 14692 USA (Telephone 800–962–2660).

SECTION 2 Helminths

Chapter 2 Phylum Platyhelminthes

The phylum Platyhelminthes consists of four classes. They are the primarily free-living **Turbellaria** and the parasitic **Monogenea, Trematoda,** and **Cestoidea.** The basic scheme of classification follows that of Schmidt and Roberts (1985).

Parasitic Platyhelminthes are characterized by: (1) body covered with a **syncytial tegument;** (2) organs of attachment are **suckers** and **hooks;** (3) an incomplete digestive tract without anus in flukes and completely absent in tapeworms; (4) **acelomate** but the space between the internal organs filled with loose **parenchyma;** (5) muscle layers strongly developed; (6) circulatory, respiratory, and skeletal systems absent; (7) excretory system consists of numerous **flame cells** connected by tubules which, in the flukes, empty into an excretory bladder but open posteriorly directly from the excretory canals in tapeworms; (8) nervous system a pair of anterior ganglia connecting with up to three pairs of longitudinal nerve trunks; (9) usually monoecious, i.e., both sexes present in the same individual, with fertilization internal; and (10) life cycles complex with intermediate hosts in digenetic flukes and cestodes.

Class Turbellaria

All biology students are familiar with *Dugesia,* the planarian. This is an example of a free-living, predaceous turbellarian. Indeed, most turbellarians are free-living, but at least 27 families among the 12 orders in this class have symbiotic species. Most are commensals of echinoderms, crustaceans, sipunculids, arthropods, molluscs, and other invertebrates. A few are truly parasitic. Because Platyhelminthes is a predominantly parasitic phylum, the symbiotic turbellarians offer tantalizing clues to the evolution of parasitism.

Most such species are unavailable to biology students, but two genera can usually be obtained.

Bdelloura candida

This worm belongs to the **order Tricladida,** the same order that contains *Dugesia.* It lives on the book gills of horseshoe crabs and is very common. It has well-developed eyespots near its anterior end and a large adhesive disc on its posterior end. It feeds on particles of food torn apart by the gnathobases of its host and lays its eggs in capsules on the gills of its host. Apparently it is transmitted from one host to another during copulation of the horseshoe crabs.

Examine a prepared slide of *Bdelloura.* Notice particularly its resemblance to the free-living *Dugesia.* (See also fig. 2.1.)

Syndesmis sp.

These curious worms are in the family Umagillidae, **order Neorhabdocoela.** They live in the intestine of sea urchins, but almost nothing is known about their biology and physiology. They do appear to feed on ciliates, which also are symbiotic in the urchin's gut. Live or preserved sea urchins obtained from a biological supply house are likely to be infected and so are sources of material.

Dissect a sea urchin in search of *Syndesmis* or a related genus. Examine a prepared slide and identify the organs labeled in figure 2.2.

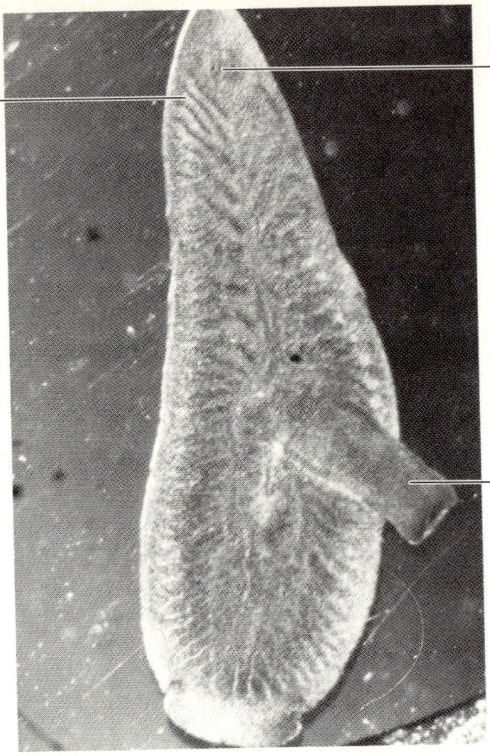

Fig. 2.1. *Bdelloura candida,* a triclad turbellarian from the gills of a horseshoe crab. Note the eyespots and the huge midventral pharynx. (From Schmidt, G. D., and L. S. Roberts. 1985. *Foundations of Parasitology,* 3d ed. The C. V. Mosby Co., St. Louis.)

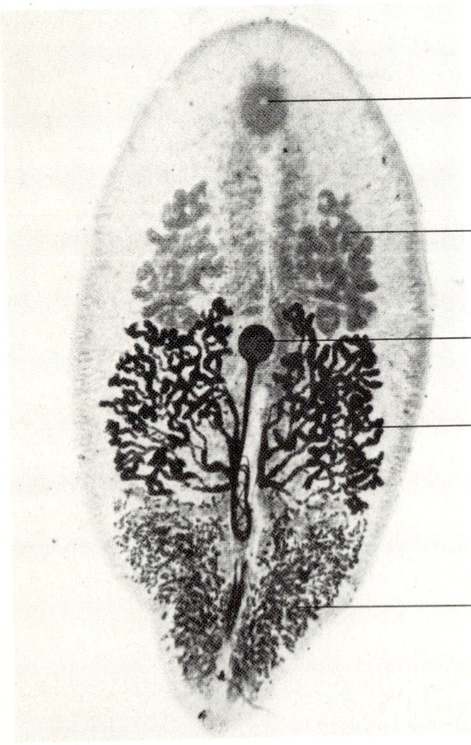

Fig. 2.2. *Syndesmis* sp., a neorhabdocoel turbellarian from the intestine of a sea urchin. (From Schmidt, G. D., and L. S. Roberts. 1985. *Foundations of Parasitology,* 3d ed. The C. V. Mosby Co., St. Louis.)

Class Monogenea

These are external parasites on the skin and gills of aquatic vertebrates, primarily fish. Some occur in the uteri of fish. A few species live in the mouth and urinary bladder of amphibians and reptiles. Adhesive organs are known as haptors; **prohaptor** at the anterior end of the body and **opisthaptor** at the rear. The life cycle is direct; the egg hatches into an **oncomiracidium** which is ciliated and bears numerous hooks (see fig. 2.4B), so that the larva is well adapted both for swimming and for attachment. In some species, eggs hatch in the uterus (ovoviviparity), and the newly hatched larvae somewhat resemble the adults.

Prepared slides of monogeneans are not always available for sale, so this laboratory exercise will vary according to whatever specimens the instructor has. The following descriptions illustrate the major characteristics of typical groups but can be modified to fit available study material.

Subclass Monopisthocotylea

Characterized by the undivided **opisthaptor,** which may have internal septa. There are one to three pairs of large hooks and 12 to 16 **marginal hooklets.** The **prohaptor** consists of glandular areas of suckers or **pseudosuckers** located outside the buccal cavity. Intestinal caeca may

Table 2.1. Classification of the Flukes (Abbreviated).

Class	Subclass	Order	Superfamily	Family
MONOGENEA	Monopisthocotylea		Acanthocotyloidea	Acanthocotylidae
			Tetraoncoidea	Tetraoncidae
			Gyrodactyloidea	Gyrodactylidae
			Dactylogyroidea	Dactylogyridae
				Calceostomatidae
			Udonelloidea	Udonellidae
			Capsaloidea	Capsalidae
				Loimoidae
				Microbothriidae
				Monocotylidae
	Polyopisthocotylea		Megaloncoidea	Megaloncidae
			Avielloidea	Aviellidae
			Polystomatoidea	Polystomatidae
				Hexabothriidae
				Sphyranuridae
			Chimaericoloidea	Chimaericolidae
			Microcotyloidea	Microcotylidae
				Axinidae
			Diplozooidea	Diplozooidae
			Diclidophoroidea	Diclidophoridae
				Dactylocotylidae
				Mazocraeidae
				Discocotylidae
TREMATODA	Aspidogastrea			Aspidogastridae
	Didymozoidea			Didymozoidae
	Digenea	Strigeata	Strigeoidea	Strigeidae
				Diplostomatidae
				Cyathocotylidae
			Clinostomatoidea	Clinostomatidae
			Schistosomatoidea	Schistosomatidae
				Sanguinicolidae
				Spirorchiidae
			Azygioidea	Azygiidae
			Cyclocoeloidea	Cyclocoelidae
			Brachylaemoidea	Brachylaemidae
			Fellodistomatoidea	Fellodistomatidae
			Bucephaloidea	Bucephalidae
		Echinostomata	Echinostomatoidea	Echinostomatidae
				Cathaemasiidae
				Philophthalmidae
				Psilostomatidae
				Fasciolidae
			Paramphistomatoidea	Paramphistomatidae
			Notocotyloidea	Notocotylidae
		Plagiorchiata	Plagiorchioidea	Plagiorchiidae
				Dicrocoeliidae
				Prosthogonimidae
				Microphallidae
				Eucotylidae
				Lecithodendriidae
			Allocreadioidea	Allocreadiidae
				Gorgoderidae
				Troglotrematidae
		Opisthorchiata	Opisthorchioidea	Opisthorchiidae
				Heterophyidae
			Hemiuroidea	Hemiuridae
				Halipegidae

or may not be branched or united posteriorly. Testes one, two, three, or many, usually postovarian. The cirrus consists of a simple or complex cuticularized structure. A **genito-intestinal canal** extending from the oviduct to the intestinal caeca is lacking. **Oviparous** or **viviparous;** eggs usually have a polar filament at one or both ends.

Family Gyrodactylidae

These are small flukes with poorly developed vitellaria, no eyespots or vagina, and are viviparous with young in the uterus. They are parasites of fish and amphibians. There are many species.

Gyrodactylus spp.

Species of this genus are parasitic on the gills of freshwater and marine bony fishes. (See fig. 2.3.)

Description

The small body is elongate. The opisthaptor is without divisions; it bears a pair of large **anchors** connected by a ventral and dorsal **bar;** there are 16 **marginal hooklets.** There is a single pair of lobelike **head glands.** The mouth is subterminal and followed by a globular pharynx; the short esophagus divides into a pair of intestinal caeca, terminating near the posterior end of the body.

The testis is a small median body lying between or behind the intestinal caeca; the cirrus is armed with spines at its opening. There is a complex sclerotized accessory piece associated with the copulatory organ. The genital pore is submedian behind the pharynx. A fully developed embryo is in the uterus; the ovary is posttesticular, median; vitellaria form two lobes surrounding the ends of the caeca.

When born, the precocious larvae appear much like their parents. They attach to the gills of their hosts and grow directly without metamorphic changes into adults. When fully developed, the larva contains a less-developed larva in its uterus; before birth a second larva appears in the uterus of the first, a third inside the second, and even a fourth inside the third.

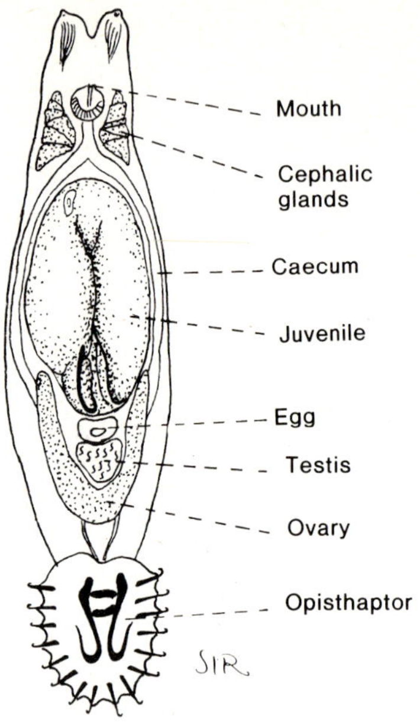

Fig. 2.3. *Gyrodactylus medius.*

Family Dactylogyridae

These are small, oviparous flukes with eyes and vagina. The vitellaria are well developed. They are parasites on the gills of freshwater and marine fishes.

Dactylogyrus vastator

These flukes occur commonly on the gills of carp. They are especially detrimental to young fish in crowded ponds. (See fig. 2.4.)

Description

Adults are up to 1.2 mm long by 0.3 mm wide. There are two prominent head lobes, each with glands. Two pairs of eyespots are located anterior to the pharynx. The opisthaptor bears one pair of large anchors with bifurcate roots united by a single rod-shaped bar; there are seven pairs of small marginal hooklets. There is a single, oval testis located slightly posterior to the ovary. The small oval ovary is situated near the midbody. The vaginal opening is toward the right margin of the body, slightly preequatorial.

Life Cycle

The life cycle is direct, meaning no intermediate host is involved in its completion (fig. 2.5). Adult worms lay their eggs on the gills of the carp host. There are two kinds of eggs, summer and winter eggs. During summer when the water is warm, summer eggs are laid. They develop quickly and hatch on the fish, freeing an **oncomiracidium** which develops directly, metamorphosing into an adult fluke. Several generations of worms appear each summer. Toward the end of the summer when the water cools, winter eggs are laid. They fall off the fish, dropping to the bottom of the ponds where they remain dormant during the winter. With the onset of spring and warming of the water, development begins and soon hatching takes place. Newly hatched oncomiracidia swim in search of a carp host. Upon reaching one, they migrate to the gills, where final development is completed. Laying of summer eggs then begins.

Family Capsalidae

Characterized by a flat, oval body with a well-developed muscular disc-shaped opisthaptor bearing three pairs, occasionally two pairs, of anchors; marginal hooklets may or may not be present. The prohaptor consists of paired suckerlike structures. There are two pairs of eyes. Parasitic on marine fish.

Neobenedenia melleni

This species occurs on the skin and gills, in nasal cavities, and around the eyes of a wide variety of marine teleosts. More than 50 species representing 18 families serve as hosts. (See fig. 2.6.)

Description

Adults are up to 5 mm long. An ophisthaptor 1.2 mm in diameter is attached to the body by a slender stalk; there are three pairs of anchors with slightly forked

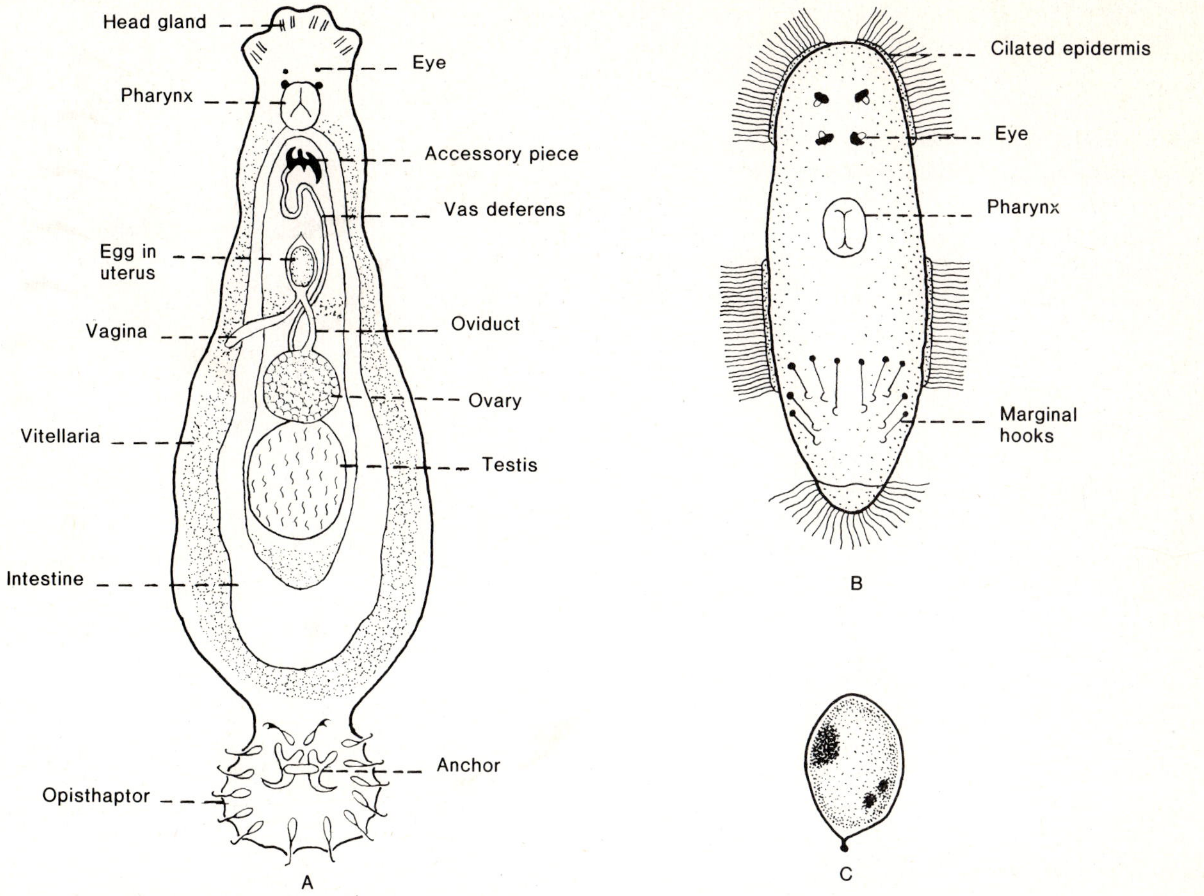

Fig. 2.4. *Dactylogyrus vastator.* (A) Adult; (B) oncomiracidium; (C) egg with characteristic terminal knob.

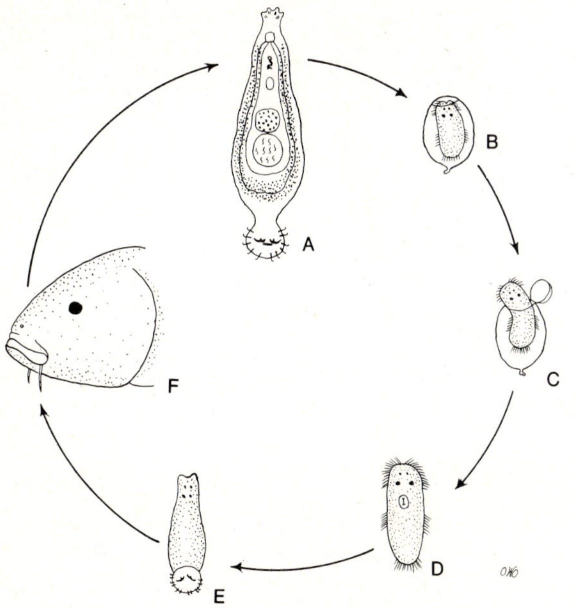

Fig. 2.5. Life cycle of *Dactylogyrus vastator.* (A) Adult fluke on gills of carp lays unembryonated eggs; (B) eggs fall to bottom of pond, embryonate; (C) eggs hatch; (D) oncomiracidium swims in search of fish; (E) oncomiracidium attaches to carp, crawls to gills, tranforms to larva; (F) larva grows directly in adult fluke on gills of fish.

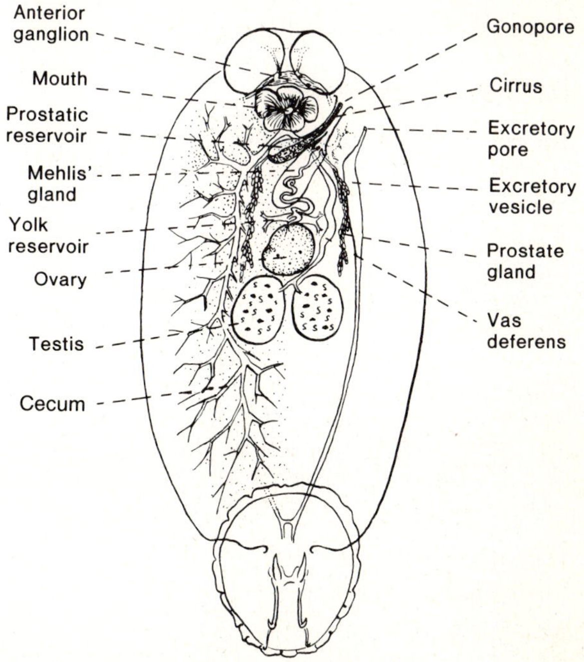

Fig. 2.6. *Neobendenia melleni.*

bases, and 14 small accessory branched hooks arranged radially around the margin of the sucker. The prohaptor, a pair of large unarmed suckerlike organs, lies at the anterior extremity of the body. Oral sucker is lobed; the intestine bifurcates at the mouth to form two branched caeca that extend to near the posterior end of the body.

There are two roundish testes arranged transversely near the middle of the body and a voluminous, slender, pear-shaped seminal vesicle with a pair of long external follicular prostate glands. The common genital pore opens near the left side of the mouth. The ovary is located medianly near the anterior margin of the testes; there is a well-developed seminal receptacle, and a sinuous duct that terminates in a voluminous shell gland lying parallel to the cirrus pouch and opening in the common genital atrium. The vitellaria occupy most of the space between the organs. The eggs are tetrahedral in shape, usually with one very long and two much shorter filaments with hooklike tips.

Subclass Polyopisthocotylea

Posterior holdfast (opisthaptor) consists of suckers, or clamps, or anchors. Oral sucker present and with a pair of internal suckers; eyes generally absent; genito-intestinal canal usually present; intestine usually with two branches, sometimes united caudally; testes many.

Characterized by an opisthaptor consisting of two or more suckers or clamps. The prohaptor is usually devoid of adhesive glands and the mouth surrounded by an oral sucker. These worms are parasitic on the skin or gills, or in the oral cavity or urinary bladder, of fishes, amphibians, and reptiles.

Family Polystomatidae

The opisthaptor bears three pairs of cup-shaped, muscular suckers, with a number of marginal hooklets, and with or without haptoral anchors. The mouth is terminal or subterminal and surrounded by a muscular sucker. Polystomatids are parasitic in the mouth, esophagus, or urinary bladder of amphibians and reptiles.

Polystomoides oris

This and related species occur in the mouth of the painted turtle (*Chrysemys picta*) and other species of North American turtles. (See fig. 2.7.)

Description

The fusiform body is up to 4.2 mm long by 1.6 mm wide. The opisthaptor bears the usual six suckers. There are two pairs of large hooks between the posterior pair of suckers, and 16 larval hooks: six between the anterior pair of suckers, four between the posterior pair of suckers and one in each sucker. A vagina opens ventrally near each lateral margin of the body near the union of the first and second thirds of the body. The oral sucker is large and subterminal; the pharynx is as large as or slightly larger than the oral sucker; the esophagus is short, and the intestinal caeca long and diverticulate anteriorly. The spined genital pore is median and directly behind the intestinal bifurcation.

There is a median, single, large testis a short distance behind the intestinal bifurcation. The cirrus bears 24 to 27 spines. A small, comma-shaped ovary is anterior to the testis and to the left of the midventral line of the body. The vitelline follicles surround the lateral and ventral sides of the intestinal caeca from the pharynx to the opisthaptor. A uterus is absent; the ootype contains a single egg 250 × 180 μm.

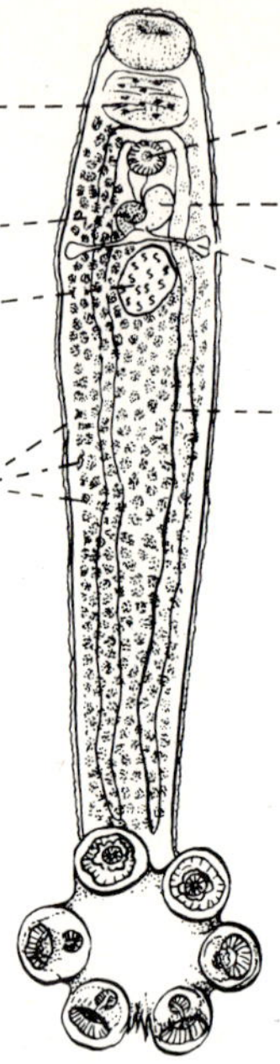

Fig. 2.7. *Polystomoides oris.*

Polystoma nearcticum

This species, originally regarded as a subspecies of *P. integerrimum,* the well-known European species, occurs in the urinary bladder of adult tree frogs (*Hyla versicolor* and *H. cinerea*), and on the gills of their tadpoles. There are two adult forms: the branchial form on the gills and the bladder form in the urinary bladder.

Description

Branchial forms are up to 5 mm long by 0.8 mm wide. The opisthaptor has six pedunculate suckers. Hooks are rudimentary or absent. The testis is spherical and the ovary elongate. Bladder forms are up to 4.5 mm long by 1.5 mm wide. The opisthaptor has six muscular suckers and one pair of large hooks. The testis is multilobate and the ovary is comma shaped. Eggs of both forms average 300 × 150 μm.

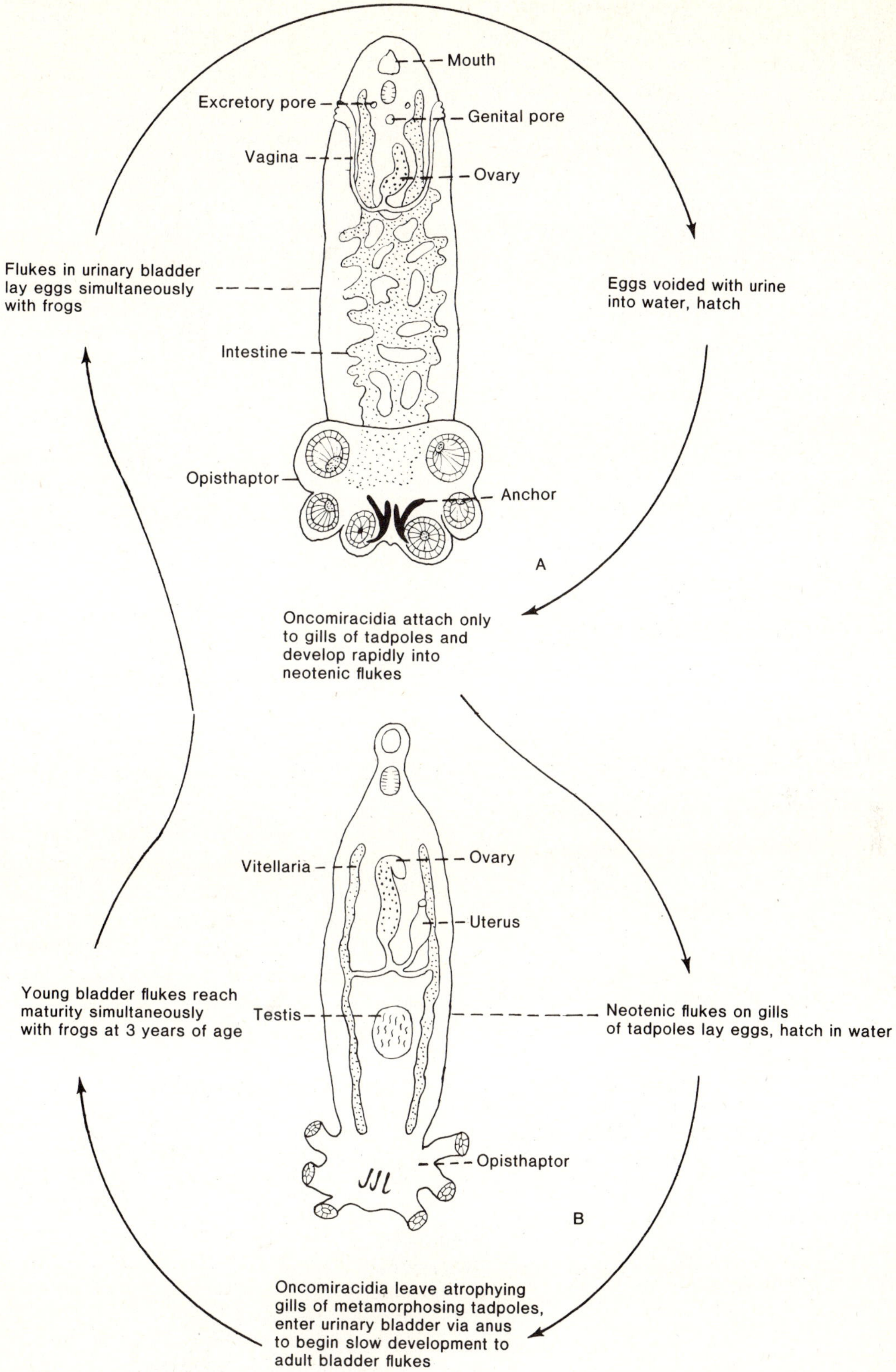

Fig. 2.8. Life cycle of *Polystoma nearcticum.* (A) Bladder phase of cycle; (B) gill phase of cycle.

Life Cycle

As a result of one form of adult worms occurring in the urinary bladder of adult tree frogs and another form of adults living on the external gills of the tadpoles, the life cycle is somewhat complicated (fig. 2.8). Eggs from the bladder form produce oncomiracidia which, when attaching to the gills of young tadpoles, undergo accelerated development of the reproductive organs. But when the larvae hatched from eggs produced by the neotenic gill form, or those from eggs of the bladder form, encounter an older tadpole undergoing metamorphosis, they enter the urinary bladder through the cloacal opening and attain maturity in the spring of their third year of life, at which time the frog also reproduces for the first time. Under the influence of hormones, which appear in the host's urine at the breeding season in spring, the bladder form lays eggs that pass out when the frog enters the water to breed.

Family Sphyranuridae

These are readily recognized by the bilobed opisthaptor bearing only two large cuplike, muscular suckers. They are parasites on the skin and gills of amphibians.

Sphyranura oligorchis

This polystome from the gills of the perennibranchiate mudpuppy (*Necturus maculatus*) is characterized by an opisthaptor with two large, cuplike, muscular suckers and two large hooks.

Description

Adults are up to 4 mm long by 0.7 mm wide. The bilobed opisthaptor with two large, muscular suckers is wider than the body; each sucker has a single hooklet inside; there is a large anchor near the posterior margin of each sucker and 14 hooklets on the margin of the opisthaptor. The terminal funnel-shaped mouth is surrounded by an oral sucker followed by a conspicuous muscular pharynx, a short esophagus, and intestinal caeca that are confluent posteriorly.

About six testes are arranged linearly in the posterior half of the body; a vas deferens extends anteriorly to the common genital pore just posterior to the bifurcation of the esophagus. The cirrus is armed with spines. A small pyriform ovary lies to the left of the midline and anterior to the testes; the uterus is largely anterior to the ovary; the vagina is double but does not open to the outside. A genito-intestinal canal opens into the left caecum near the level of the ovary. The vitellaria are follicular and extend along the intestinal caeca from the ovary posteriorly. Usually only a single egg, 28 × 41 μm in size, is present in the uterus. (See fig. 2.9.)

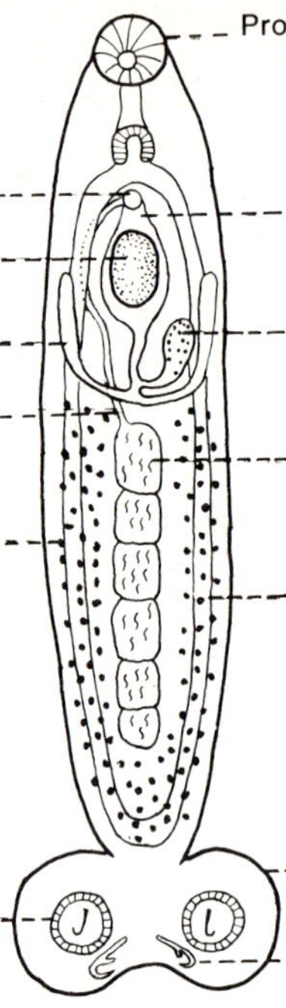

Fig. 2.9. *Sphyranura oligorchis.*

Life Cycle

Adults on the gills of the host lay their eggs, which settle to the bottom of the pond. Development is slow at room temperature, as hatching takes about a month. If the newly hatched larvae succeed in making contact with a host, they promptly attach and migrate to the gills. Maturity is completed in about two months.

Family Discocotylidae

The sclerotized clamps are symmetrically arranged with four in each lateral row. A pair of small suckers opens into the mouth cavity, which is followed by a muscular pharynx, esophagus, and intestinal caeca with numerous diverticula. Parasitic on fish.

Octomacrum lanceatum

This species is parasitic on the gills of the common sucker and chub sucker. It is a good example of clamp-bearing members of the Polyopisthocotylea. (See fig. 2.10.)

Description

Adults are up to 6 mm long by 2 mm wide. The opisthaptor is rectangular, and set off from the body by a slight, broad constriction. There are two rows of four suckers with edges in contact on each side of the opisthaptor; each sucker is supported by a complicated sclerotized clamp consisting of a central arched piece and two others, each beginning as a single part near

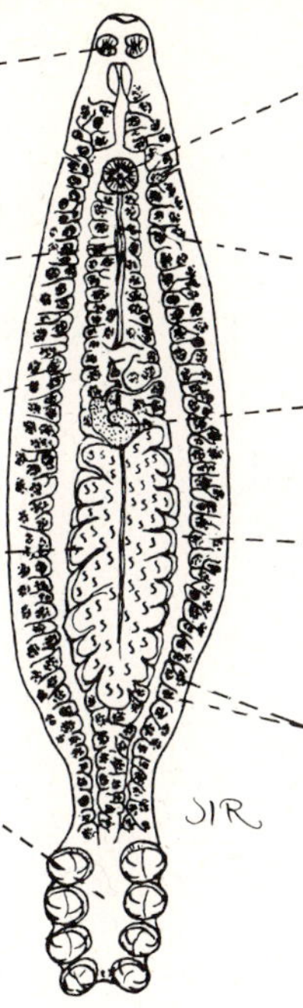

Fig. 2.10. *Octomacrum lanceatum.*

the middle of the anterior margin of the sucker and extending around the end, where it divides into two branches, one on the posterior margin of the sucker and the other in between.

The mouth, at the anterior tip of the body, has two lateral suckers inside; it is followed by a muscular pharynx, an esophagus, and two long single, diverticulate intestinal caeca.

The male reproductive system consists of a multilobate testis and a vas deferens that ascends directly anteriorly to the genital pore, opening through a muscular, hookless genital sucker at the level of the intestinal bifurcation. The coiled ovary lies at the anterior end of the testis. A short distance from the ovary, the oviduct divides with one branch forming the uterus and going to the common genital pore and the other to the genito-intestinal canal. The vitellaria fill all space not occupied by the other organs.

Key to Superfamilies of the Class Monogenea

Capital letters in parentheses refer to hosts: F–Fish, A–Amphibia, R–Reptilia, M–Mammalia, C–Crustacea, and O–Cephalopods. Small letters following each capital letter refer to the type of host within the taxonomic group: f–freshwater fish, h–holocephalians, m–marine teleosts, and e–elasmobranchs. The generic names of figures 2.3 through 2.18 designate the relationship to the superfamilies appearing in the key.

1 Opisthaptor a single sucker or disclike organ with or without septa; with one to three pairs of anchors; prohaptor as head glands; mouth not surrounded by a sucker; genito-intestinal canal usually absent. Subclass MONOPISTHOCOTYLEA 2
Opisthaptor with two or more suckers or rows of cuticularized clamps; prohaptor a sucker or paired suckers; head glands generally absent; genito-intestinal canal present. Subclass POLYOPISTHOCOTYLEA 7

2(1) Opisthaptor with minute suckerlike structure on posterior margin (fig. 2.11) (Fm) .... **Acanthocotyloidea**
No small sucker on opisthaptor; marginal hooklets usually present 3

3(2) Opisthaptor one- or two-lobed, thin; one or two pairs of anchors supported by cross bars; marginal hooklets present; cirrus cuticularized, usually with accessory piece; anterior end with well-developed glands 4
Opisthaptor a muscular disc, with or without one to three pairs of anchors not supported by cross bars; marginal hooklets present or absent; intestine single or bifurcated; cirrus without accessory piece; anterior end with preoral suckers and/or glandular areas; genito-intestinal canal absent 6

4(3) Intestine single (fig. 2.12) (Ffm) .... **Tetraoncoidea**
Intestine bifurcate; opisthaptoral hooks large, with or without cross bars 5

5(4) Opisthaptoral anchors large, with cross bars; vitellaria near posterior end of body; vagina absent; ovoviviparous (fig. 2.3) (Ffm, A, C, O) .... **Gyrodactyloidea**
Opisthaptoral anchors present, occasionally without cross bars; vitellaria, lateral along full length of intestinal caeca; vagina present or absent; oviparous (fig. 2.4) (Ffm) .... **Dactylogyroidea**

6(3) Intestine single; on copepods infesting fish (Fme, C) .... **Udonelloidea**
Intestine bifurcate; on fish (fig. 2.13) (Feh) .... **Capsaloidea**

7(1) Opisthaptor with three pairs of anchor complexes alone or in combination with clamps; prohaptor consists of paired intrabuccal suckers (fig. 2.14) (Fm) .... **Megaloncoidea**
Opisthaptor with either muscular suckers or clamps but not both; prohaptor variable 8

8(7) Opisthaptor with six muscular suckers and four large anchors on slender stalk; head with glands opening on each side (fig. 2.15) (Ff) .... **Avielloidea**
Opisthaptor not on long slender stalk 9

9(8) Opisthaptor sessile, with six muscular suckers (two in Sphyranidae), appendixlike prolongation present or absent; with or without one to

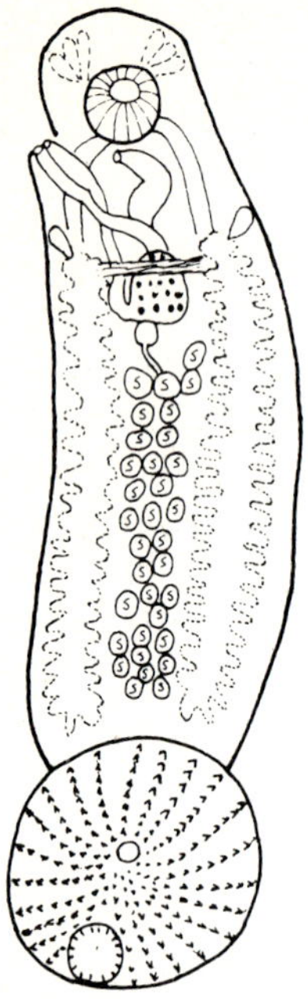

Fig. 2.11. *Acanthocotyle williamsi.*

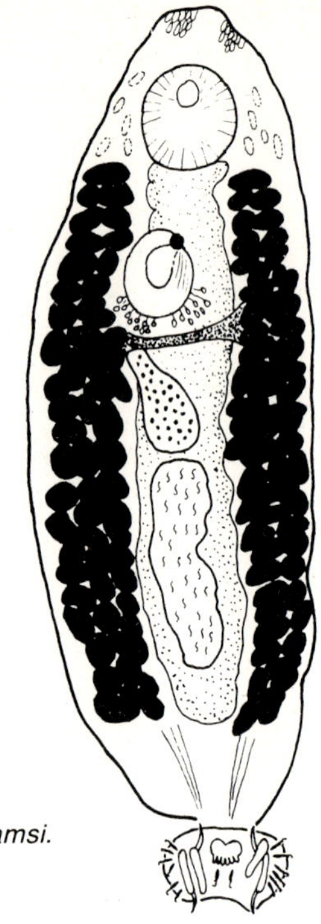

Fig. 2.12. *Tetraonchus alaskensis.*

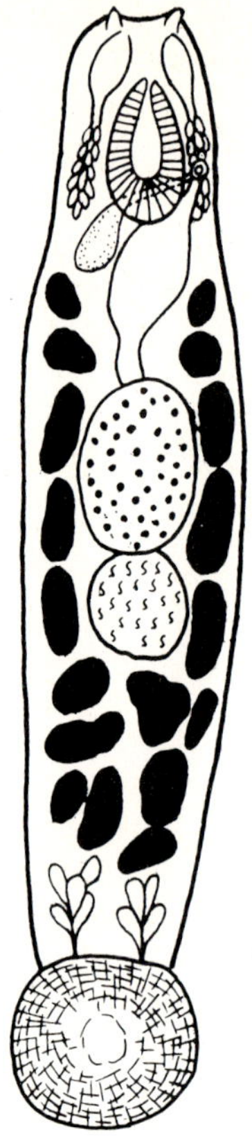

Fig. 2.13. *Udonella caligorum.*

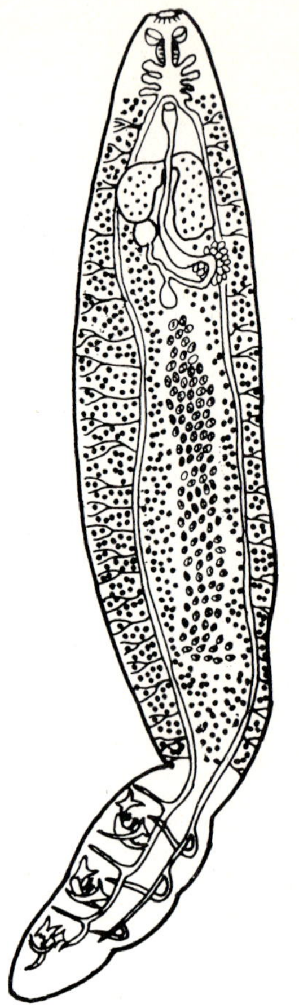

Fig. 2.14. *Megaloncus arelisci.*

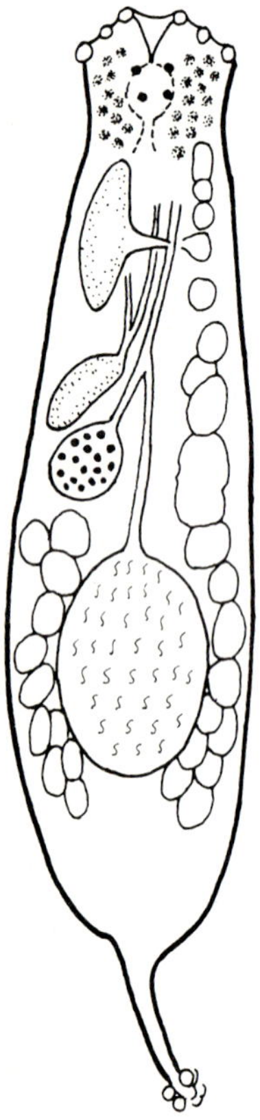

Fig. 2.15. *Aviella baikalensis.*

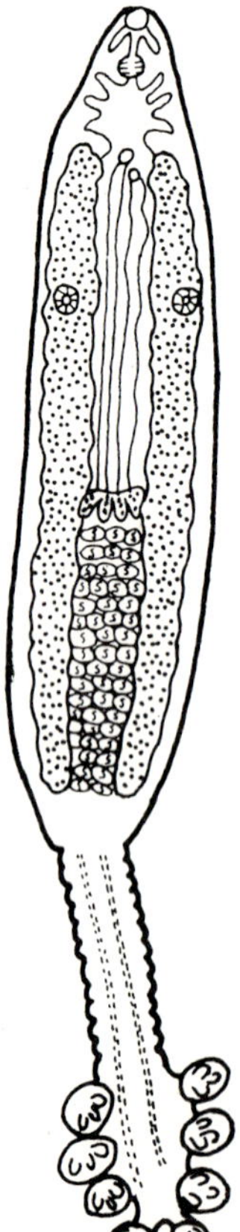

Fig. 2.16. *Chimaericola leptogaster.*

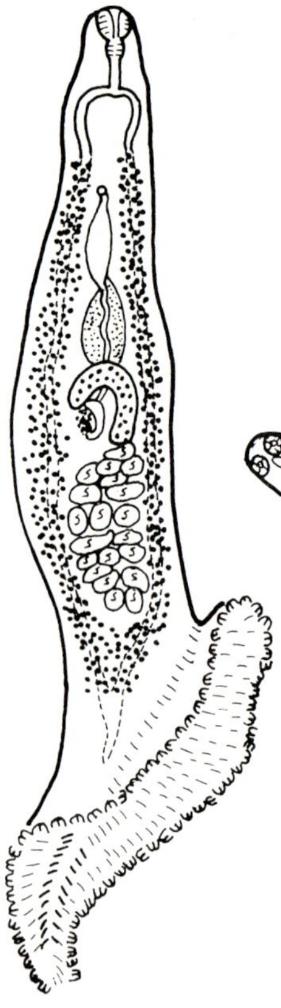

Fig. 2.17. *Microcotyle donavani.*

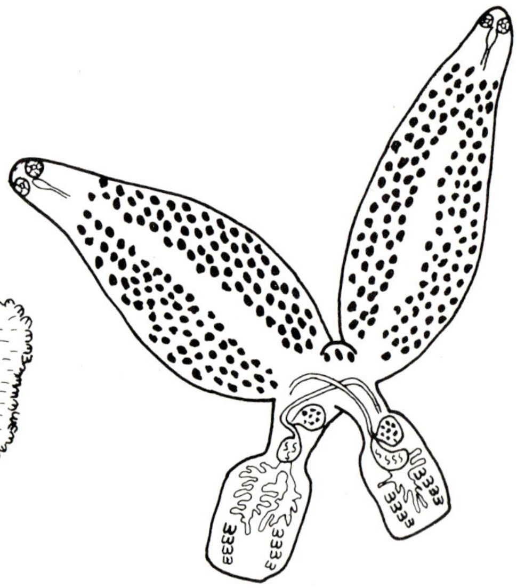

Fig. 2.18. *Diplozoon paradoxum.*

three pairs of posterior anchors; prohaptor a muscular oral sucker (figs. 2.7, 2.8, 2.9) (A, R, M, Fm) .... **Polystomatoidea**

Prohaptor not a well-developed muscular sucker surrounding mouth 10

10(9) Uterus forming many vertical or transverse loops in broader anterior portion of body, posterior part of body tapering; opisthaptor with two rows of simple clamps; oral sucker weak or lacking (fig. 2.16) (Fh) .... **Chimaericoloidea**

Uterus not forming such loops..................... 11

11(10) Clamps of opisthaptor numerous, in two rows on the symmetrical or asymmetrical posterior end of body; prohaptor paired suckers opening in mouth cavity (fig. 2.17) (Fm) .... **Microcotyloidea**

Clamps of opisthaptor not more than four per side 12

12(11) Adults fused in the form of an X; intestine single with lateral branches (fig. 2.18) (Ff) .... **Diplozooidea**

Adults not fused but independent; intestine bifurcate (fig. 2.10) (Ffm) .... **Diclidophoroidea**

Notes and Sketches

Review Questions

1. What are two names for the excretory cells of Platyhelminthes?
2. What is the host and site of infection of *Bdelloura candida?*
3. Is *Syndesmis* monoecious or dioecious?
4. What is a possible function of the coiled filament on the egg of *Syndesmis?*
5. Name the adhesive organs of the Monogenea. What specializations have you observed?
6. Do all monogeneans lay eggs? What is the functional significance of viviparity in *Gyrodactylus?*
7. What is a genito-intestinal canal?
8. Review the life cycle of *Polystoma nearcticum.* What influences the morphological types in this pleomorphic species?
9. Name the sclerotized portions of the opisthaptor.
10. What are the two types of prohaptors?
11. Name the ciliated larva of most monogeneans.
12. How do fish become infected with monogeneans?

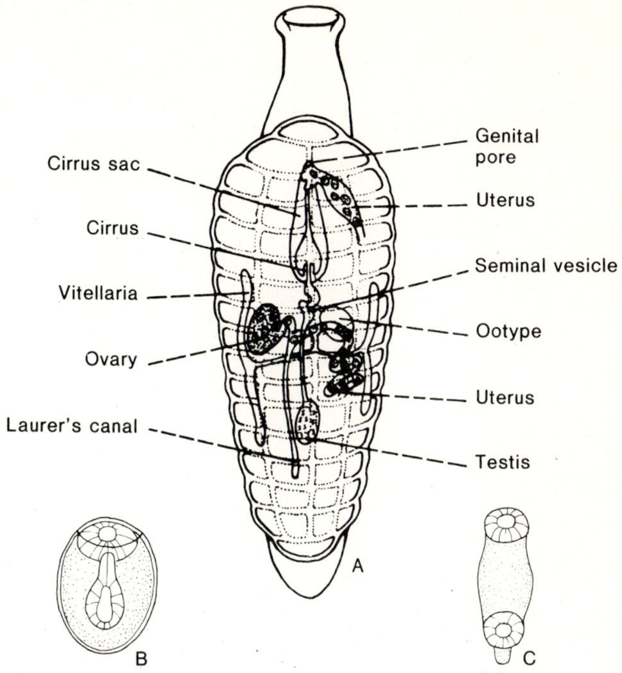

Fig. 2.19. *Aspidogaster conchicola.* (A) Adult; (B) embryonated egg, showing oral and ventral suckers of larva; (C) newly hatched larva.

Class Trematoda

The class Trematoda consists of those Platyhelminthes usually with a muscular oral sucker surrounding the mouth at the anterior end of the body and one or more suckers on the ventral surface of the body. The digestive tract is ⋀ shaped. They are adult endoparasites in vertebrates, with a few exceptions, and usually develop to some extent in molluscs.

Subclass Aspidogastrea

The Aspidogastrea (also called Aspidobothrea, or Aspidocotylea) are considered by some workers as a transitional group intermediate between the class Monogenea and the subclass Digenea. Anatomically, they differ from the Monogenea in lacking the opisthaptor with its sclerotized accessories and in having a **simple unbranched intestine,** and they differ biologically from the Digenea in the absence of an alternation of generations in the life cycle.

This small group of trematodes is characterized by the large compartmentalized **adhesive organ** called *Baer's Disc,* that encompasses almost the entire ventral surface of the body. They are endoparasites of molluscs, fish, and turtles.

Family Aspidogasteridae

The ventral sucker is large in size, oval, or elongated in shape, and composed of numerous **alveoli,** or compartments, arranged in one or several longitudinal rows. The oral sucker is rather poorly developed; a pharynx is present, and the intestine is a single, simple sac. Testes are single or double. The common genital pore is median and anterior to the ventral sucker. The ovary is prestesticular; the vitellaria are paired.

Aspidogaster conchicola

See figure 2.19. This cosmopolitan fluke infects the kidney and pericardium of freshwater clams. It may occur accidentally in gastropods and some fish. Adults are up to 2.7 mm long by 1.2 mm wide. The large areolated ventral sucker occupies almost the entire ventral surface of the body; it consists of four rows of quadrangular sucking grooves. The narrow anterior neck, consisting of about one-fifth of the total body length, bears a terminal mouth.

The single testis lies in the posterior third of the body. A large cirrus sac opens into a median common genital pore on the anterior end of the ventral adhesive organ. An oval ovary lies to the right of the median line, near the equator of the sucker. Vitellaria are two tubular, lateral structures lying in the middle third of the ventral sucker. Operculate eggs, embryonated when laid, are 128–130 $\times$ 48–50 μm. Newly hatched larvae show the two well-developed suckers of about equal size. As growth continues, the ventral sucker will increase in size and occupy the entire ventral surface of the body.

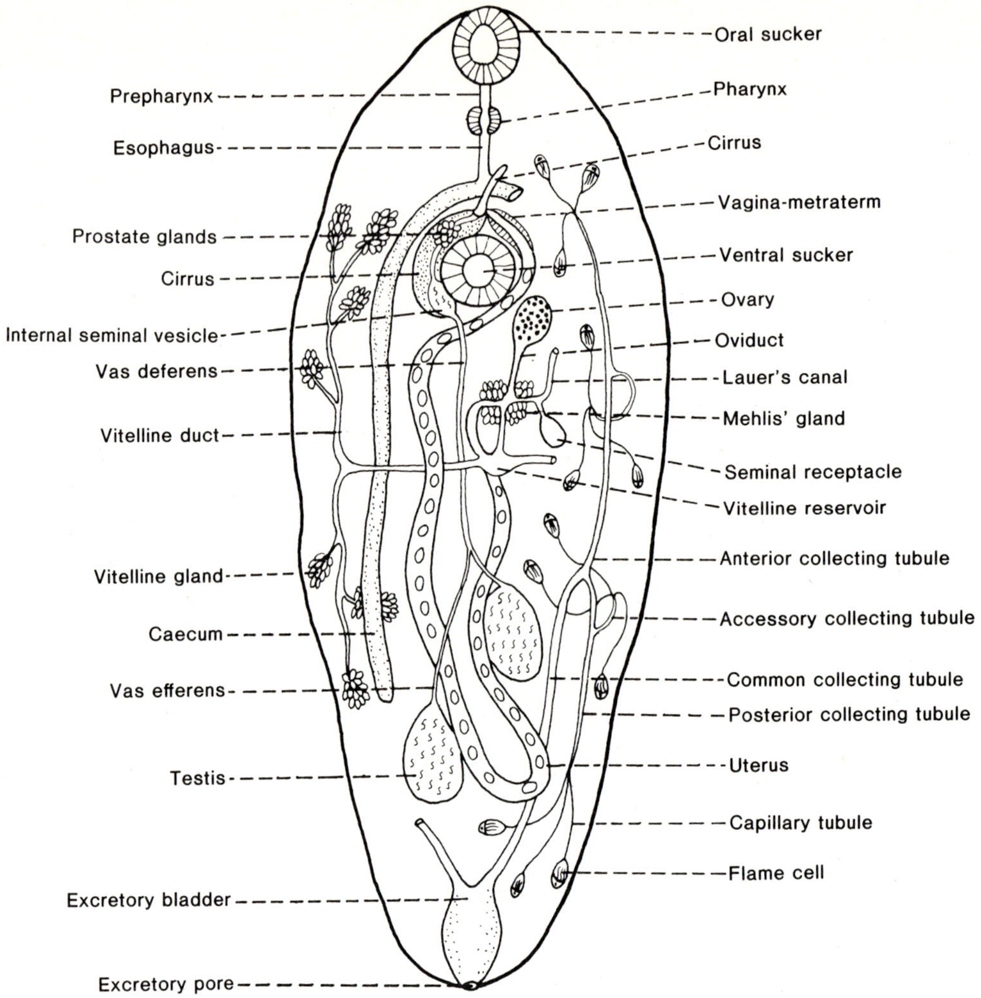

Fig. 2.20. Diagrammatic figure of adult digenetic trematode, showing principal parts of anatomy.

Life Cycle

How infection of new clam hosts occurs has not been determined. It has been suggested that the eggs hatch in the clams and the larvae develop in situ. Considering the prevalence of infection of clams, it seems more likely that the eggs are voided by the host, hatch in water, and larvae are drawn into the gill chambers and enter the host by way of the nephridiopore. The life cycle is direct, and growth of the larvae is one of gradual transformation through four arbitrary developmental stages. For life cycle details consult Olsen (1974) and Huehner and Etges (1977).

Subclass Digenea

Adults are endoparasites, occurring in all classes of vertebrates, and, in a few instances, in invertebrates. The larval stages have free-living phases necessary for getting from one host to the next. There are at least three parasitic stages that develop in the molluscan intermediate host. The following section includes an account of the body types of adults and a basic description of each of the other generative stages.

Adults. Principal organ-systems generally considered in the classification are: (1) the holdfast organs; (2) digestive; (3) reproductive; and (4) excretory systems. Anatomy of a generalized fluke is given in figure 2.20, showing the relationship between the various organ systems. There are seven distinct body types (fig. 2.21), which are: (A) **distomes** (the largest group), in which the ventral sucker is close to the oral sucker in the anterior region of the body; (B) **monostomes** with or without an oral sucker at the anterior end of the body and always without a ventral sucker (see fig. 2.55) (all others, except gasterostomes, have two suckers); (C) **gasterostomes** with only the oral sucker, which is located midventrally (in all others the oral sucker is at the anterior end of the body); (D) **amphistomes** with

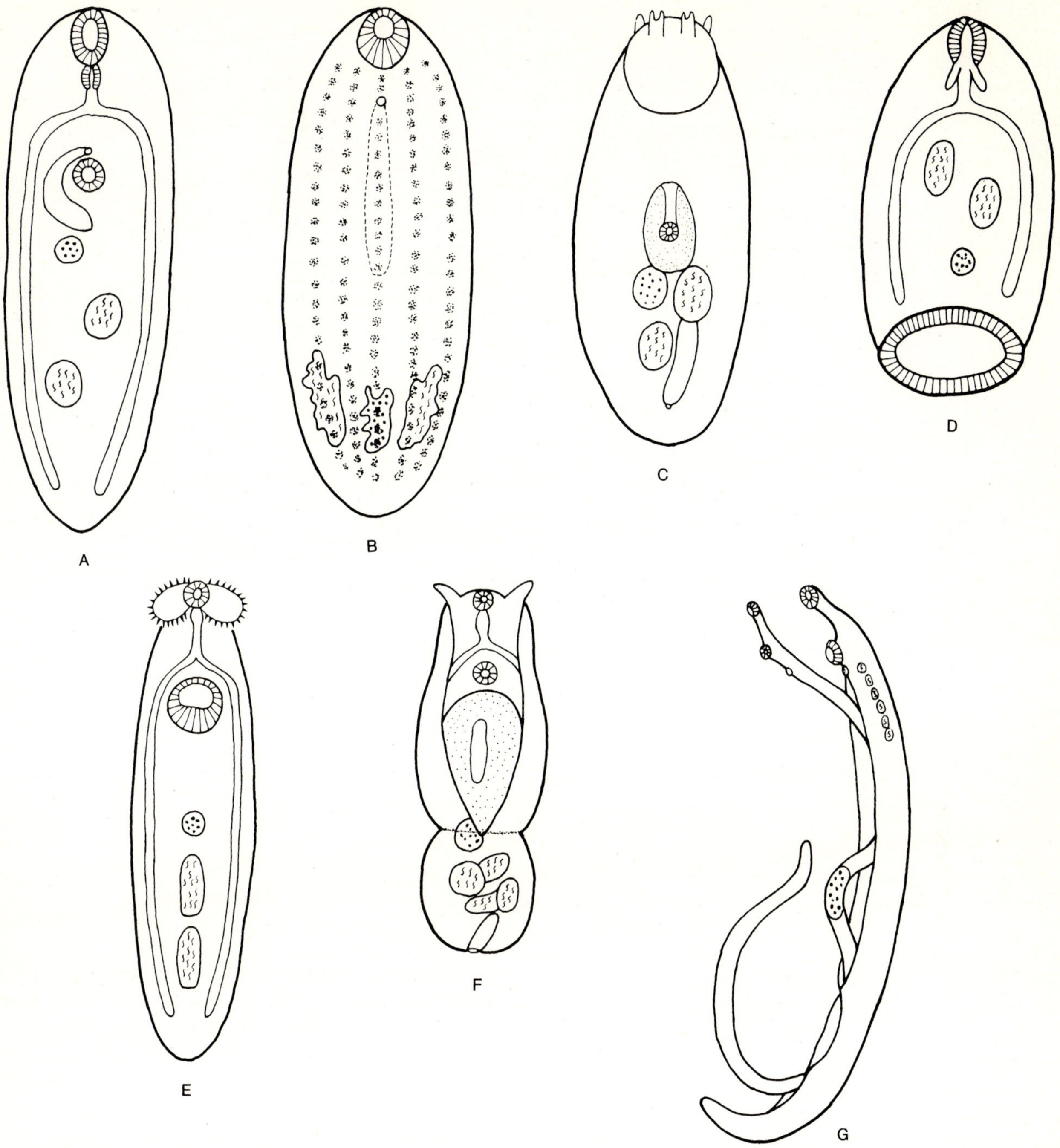

Fig. 2.21. Some basic body types of common adult digenetic trematodes. (A) Distome; (B) monostome; (C) gasterostome; (D) amphistome; (E) echinostome; (F) strigeoid; and (G) schistosome.

the oral sucker and ventral sucker at the anterior and posterior extremities, respectively, of the body; (E) **echinostomes** with a collar of spines around the oral sucker; (F) **strigeoids** with a transverse equatorial constriction that divides the body into a forebody with a holdfast organ and a hindbody containing the reproductive organs; and (G) **schistosomes,** which live in the blood vessels of the final host; some are monoecious and others dioecious, and in the latter case the males have a **gynecophoric canal** where the female is held. Instructions for the recovery of parasites and preparation of permanent mounts for microscopical examination are found on page 243.

Eggs. Eggs of most species are operculate, blood flukes being a notable exception in having nonoperculate eggs. The shells are smooth and thin. The eggs of some species are undeveloped when laid but fully embryonated in others. The eggs of some species hatch in the water, whereas others hatch in the intestine of the molluscan first intermediary.

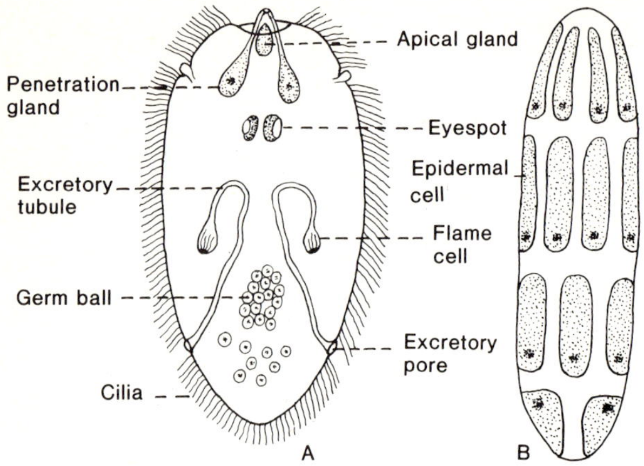

Fig. 2.22. Miracidium. (A) Shows basic internal anatomy; (B) shows tiers of epidermal plates, which vary in number and arrangement in different species.

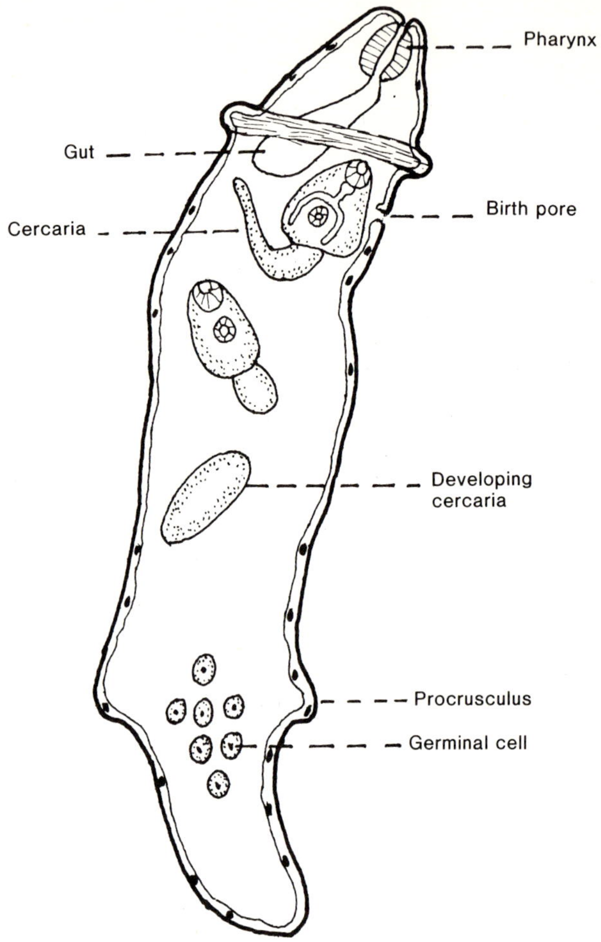

Fig. 2.24. Daughter redia.

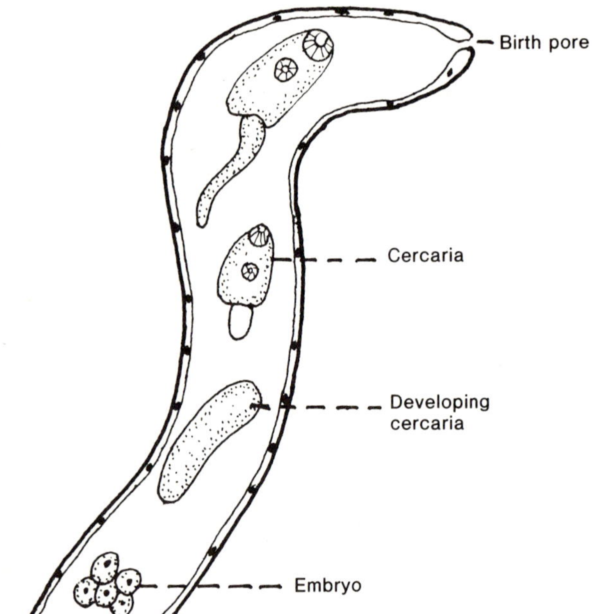

Fig. 2.23. Daughter sporocyst.

Miracidium. Miracidia are remarkably uniform throughout the trematodes. The body is more or less fusiform and ciliated. At the anterior end is a median apical gland and one to several bilaterally arranged pairs of penetration glands. A mass of undifferentiated cellular material known as the germ ball lies inside the miracidium (fig. 2.22). The surface of the miracidium is covered by tiers of epidermal plates that encircle the body and bear the cilia. The number of tiers and plates is characteristic of the species of trematode.

Mother sporocyst. The miracidium sheds its ciliated covering inside the mollusc and transforms directly into an elongated mother sporocyst. The germ ball begins to differentiate to form either daughter sporocysts or rediae, depending on the species of flukes, but not both in the same species.

Daughter sporocysts. Sporocysts produced by the mother sporocysts have a tubular body, usually simple, but branched in some species. There is no mouth or intestine. Reproduction in the molluscan host is asexual from germ balls, producing another generation of sporocysts or one of cercariae. Offspring escape through a birth pore into the tissues of the mollusc (fig. 2.23).

Redia. The mother sporocysts of some species produce rediae instead of daughter sporocysts. Anatomically rediae differ from sporocysts in having a well-defined muscular pharynx and short, saclike gut. A second generation of rediae may be produced. The end product of asexual development of the rediae, like that of the sporocysts, is cercariae. Progeny leaving the rediae through a birth pore are located in the tissues of the molluscan host (fig. 2.24).

Cercaria. Cercariae are the end product of asexual reproduction in the molluscs and are tailed, immature,

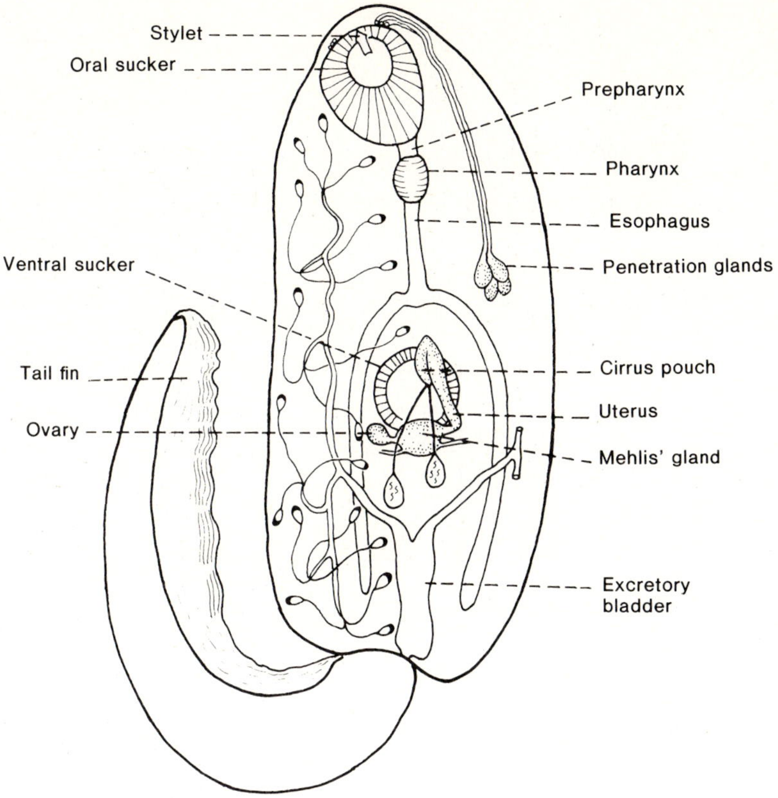

Fig. 2.25. Cercaria; for labeling of excretory system see figure 2.20.

sexual forms, resembling adults in general body conformation. In general, they emerge from the molluscs into the water. They alternate periods of swimming by violently lashing the tail with short times of resting. The basic organ systems generally are outlined to the extent that the parts are recognizable. Some species have a set of penetration glands in each side of the anterior part of the body; their ducts open at the anterior margin of the oral sucker.

The excretory system consists of flame cells, their individual ducts, collecting ducts, all bilaterally arranged, and a medial excretory bladder, which opens externally through the excretory pore at or near the posterior extremity of the body. Extending from each side of the anterior end of the excretory bladder is a common collecting tube, which divides into an anterior and a posterior collecting tube. From each of these, there extends a definite number of accessory tubules, which in turn branch into capillaries, each terminating in a flame cell.

The flame cells of each species are distributed in a definite pattern, which may be expressed by several types of simple formulae. One type is given below for the flame cell pattern shown in figure 2.20 as

$$2[(3 + 3) + (3 + 3)] = 24,$$

and again in figure 2.25 as

$$2[(3 + 3 + 3) + (3 + 3 + 3)] = 36.$$

The numeral 2 indicates that the flame cells are distributed bilaterally in the body, and the brackets refer to one side. The first set of parentheses relates to the anterior collecting tube and the second to the posterior one. The three digits within the parentheses refer to the number of accessory tubules; the numerical value of each digit states the number of flame cells that empty through their capillaries into each accessory tubule. The second set of parentheses refers to the posterior collecting tube and its components. Thus, this diagrammatic excretory system consists of a total of 36 flame cells arranged in groups of three on each of 12 accessory tubules, with three anteriorly and three posteriorly on each side.

Cercariae, when fully developed, are of different body types useful in identification. The basic types presented (fig. 2.26) are: (A) **gasterostome** with oral sucker on midventral surface; (B) **amphistome** with a sucker at each end of the body; (C) **monostome** with only an oral sucker that is at the anterior extremity of the body; (D) **gymnocephalous** without spines around or a stylet in the oral sucker and with a long naked tail;

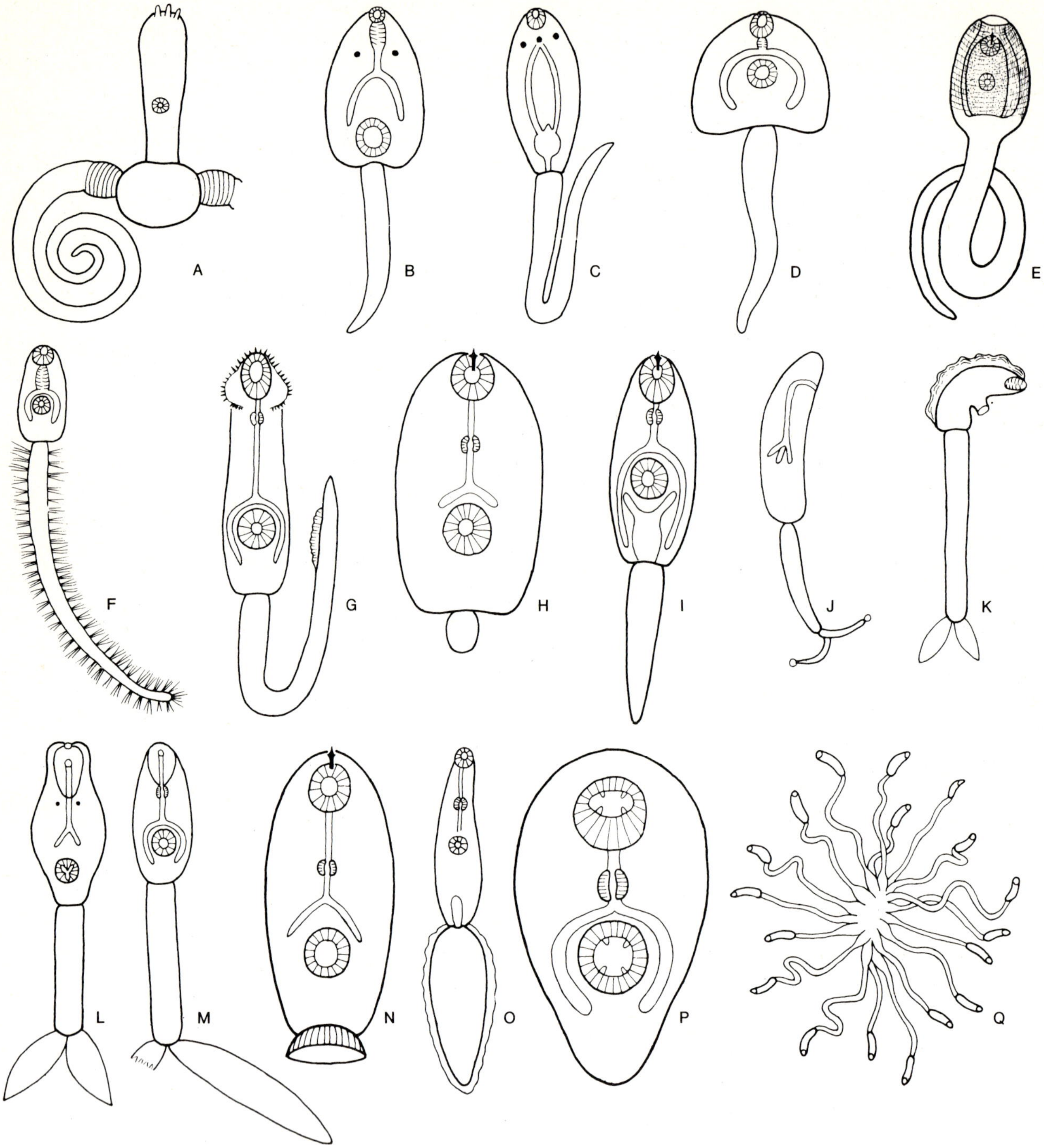

Fig. 2.26. Body types of some common cercariae. (A) Gasterostome; (B) amphistome; (C) monostome; (D) gymnocephalus; (E) cystophorous; (F) trichocercous; (G) echinostome; (H) microcercous; (I) xiphidiocercous; (J) sanguinicolid; (K) lophocercous; (L) aphryngeate furcocercous; (M) pharyngeate furcocercous; (N) cotylocercous; (O) rhopalocercous; (P) cercariaea; and (Q) rat-king.

(E) **cystocercous** with a chamber in the anterior part of the massive tail into which the cercaria may be withdrawn; (F) **trichocercous** with long slender tail bearing numerous bristles; (G) **echinostome** with a spine-bearing collar around the oral sucker; (H) **microcercous** with short, stumpy tail and a stylet in the oral sucker; (I) **xiphidiocercous** with a spine in the dorsal part of the oral sucker and a long, slender tail; (J to M) **furcocercous** types; (J) **sanguinicolid** with forked tail and no suckers; (K) **lophocercous** with forked tail and a fin over the dorsal surface of body; (L) **apharyngeate** schistosome with forked tail; (M) **pharyngeate** strigeoid with forked tail; (N) **cotylocercous** with short, cuplike tail; (O) **rhopalocercous** with tail as broad or greater than the width of the body; (P) **cercariae** without tail; and (Q) **rat-king,** consisting of a mass of cercariae attached together by the distal ends of the large tails.

In general, cercariae attach to objects in water or burrow into the bodies of aquatic animals (second intermediate hosts), drop the tail, and secrete a protective cyst about themselves from material produced by cystogenous glands. Cercariae of blood flukes do not encyst but burrow into the final host and develop directly into adults. Encysted cercariae are known as **metacercariae.**

Key to Basic Types of Cercariae

- I. No suckers
  - A. **Sanguinicolid** cercaria: fork-tailed, mouth opens subterminally (fig. 2.26J)
- II. Single sucker
  - A. **Gasterostome:** oral sucker on midventral side of body, gut single (fig. 2.26A)
  - B. **Monostome:** oral sucker at anterior end of body; gut bifurcate (fig. 2.26C)
- III. Two suckers
  - A. Tail simple
    1. **Amphistome:** ventral sucker at posterior extremity of body, in all others ventral sucker located otherwise (fig. 2.26B)
    2. **Echinostome:** oral sucker surrounded by crown of spines, in all others oral sucker without crown of spines (fig. 2.26G)
    3. **Gymnocephalous:** oral sucker without stylet; tail long, slender, simple (fig. 2.26D)
    4. **Xiphidiocercous:** oral sucker armed with a single stylet; tail long, body with or without eyespots (fig. 2.26I)
    5. **Microcercous:** oral sucker armed with stylet; tail stumpy (fig. 2.26H)
    6. **Cystophorous:** tail very large with cuplike anterior chamber containing small cercaria; oral sucker with stylet (fig. 2.26E)
    7. **Cotylocercous:** tail short, cuplike; oral sucker with stylet (fig. 2.26N)
    8. **Trichocercous:** tail long, slender, and with lateral hairlike tufts; oral sucker without stylet (fig. 2.26F)
    9. **Rhopalocercous:** tail flat, broader than body; oral sucker without stylet (fig. 2.26O)
    10. **Rattenkönig** or rat-king: long tails attached at tips, forming clumps of cercariae; oral sucker without stylet (fig. 2.26Q)
  - B. Forked tail
    1. **Lophocercous:** with longitudinal fin over entire length of dorsal side of body (fig. 2.26K) or on tail
    2. **Apharyngeate** furcocerous cercaria (fig. 2.26L)
    3. **Pharyngeate** furcocerous cercaria (fig. 2.26M)
  - C. Tailless
    1. **Cercariaea:** large body (fig. 2.26P)

Metacercaria. Upon encystment, metacercariae develop to the infective stage. When swallowed by the final host, they are released from the cyst by the digestive processes, and the young flukes go to the respective parts of the body where development to adulthood is completed. Usually, the metacercariae are prototypes of the adults into which they develop (fig. 2.27). The

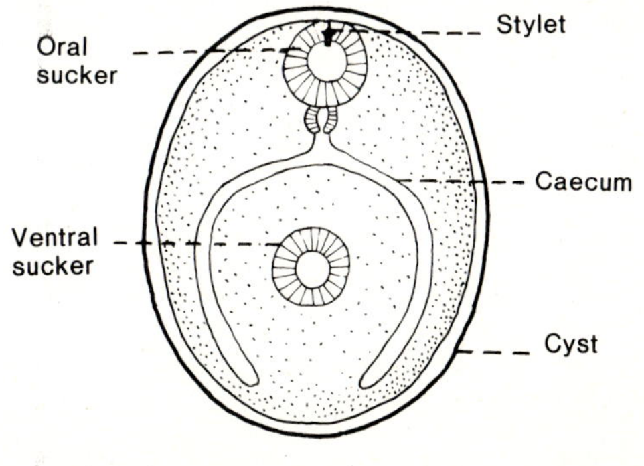

Fig. 2.27. Metacercaria.

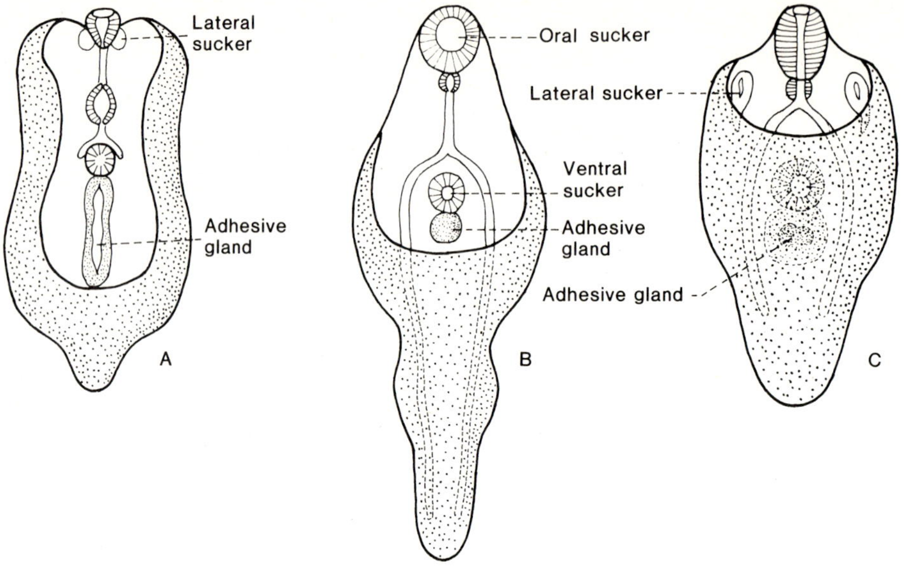

Fig. 2.28. Three typical unencysted metacercariae of strigeoid flukes. (A) Tetracotyle; (B) neascus; and (C) diplostomulum.

metacercariae of the strigeoid flukes are known as the **tetracotyle, neascus,** and **diplostomulum** types, depending on the structure of the respective adults (fig. 2.28).

Each of the above larval stages, miracidium through metacercariae, should be studied in the living state whenever possible, with the aid of neutral red, Nile blue sulphate, or other suitable vital stain. Subsequent studies should be made on permanent mounts of fixed and stained specimens.

Parasitology laboratories should maintain several balanced aquaria in which representatives of the local species of aquatic snails are breeding. Likewise, terraria should be kept for the common species of land snails. With such breeding populations, parasitologists have readily available uninfected snails to use in life cycle studies whenever adult flukes and their eggs are available.

Basic Life Cycles

Life cycles of digenetic trematodes follow a basic pattern. It includes a molluscan first intermediate host in which a definite series of sequential stages develop from the miracidium. These stages include **mother sporocyst, daughter sporocyst,** and **cercaria,** or **mother sporocyst, redia,** and **cercaria.** In another group, it consists of mother rediae and cercariae. Cercariae transform into metacercariae as the infective stage outside the molluscan intermediary.

An example of a typical life cycle is given in figure 2.29. In this cycle, the miracidium develops into a mother sporocyst, which produces mother rediae in the first intermediary. These rediae produce a second generation, daughter rediae, which produce cercariae. The mature cercariae escape from the rediae and snail into the water. Upon contacting a fish, the cercariae attach to the skin, enter, and develop into metacercariae. Infection of the final host occurs when the second intermediary harboring metacercariae is eaten. Metacercariae excyst and develop to adulthood in the site specific for the species. In this example, it is the mouth of certain fish-eating birds.

Following the example of a typical life cycle, it is important to understand the major variations that occur. These are shown in figure 2.30. Upon entering the molluscan intermediary, development of the miracidium follows either of two major patterns, according to the genetic code of each. In one line, all miracidia transform directly into mother sporocysts (1, 2, 3, 4, 5, 6), while in the other line all miracidia transform into mother rediae (7, 8). In the mother-sporocyst lineage (1–6), the mother sporocysts produce sporocysts which in turn produce cercariae (1–3), or they produce mother rediae with cercarial progeny (4–6). In the miracidium-redia lineage, all miracidia transform directly into mother rediae (7–8), with some lines producing daughter rediae and cercariae (7) and others cercariae directly (8).

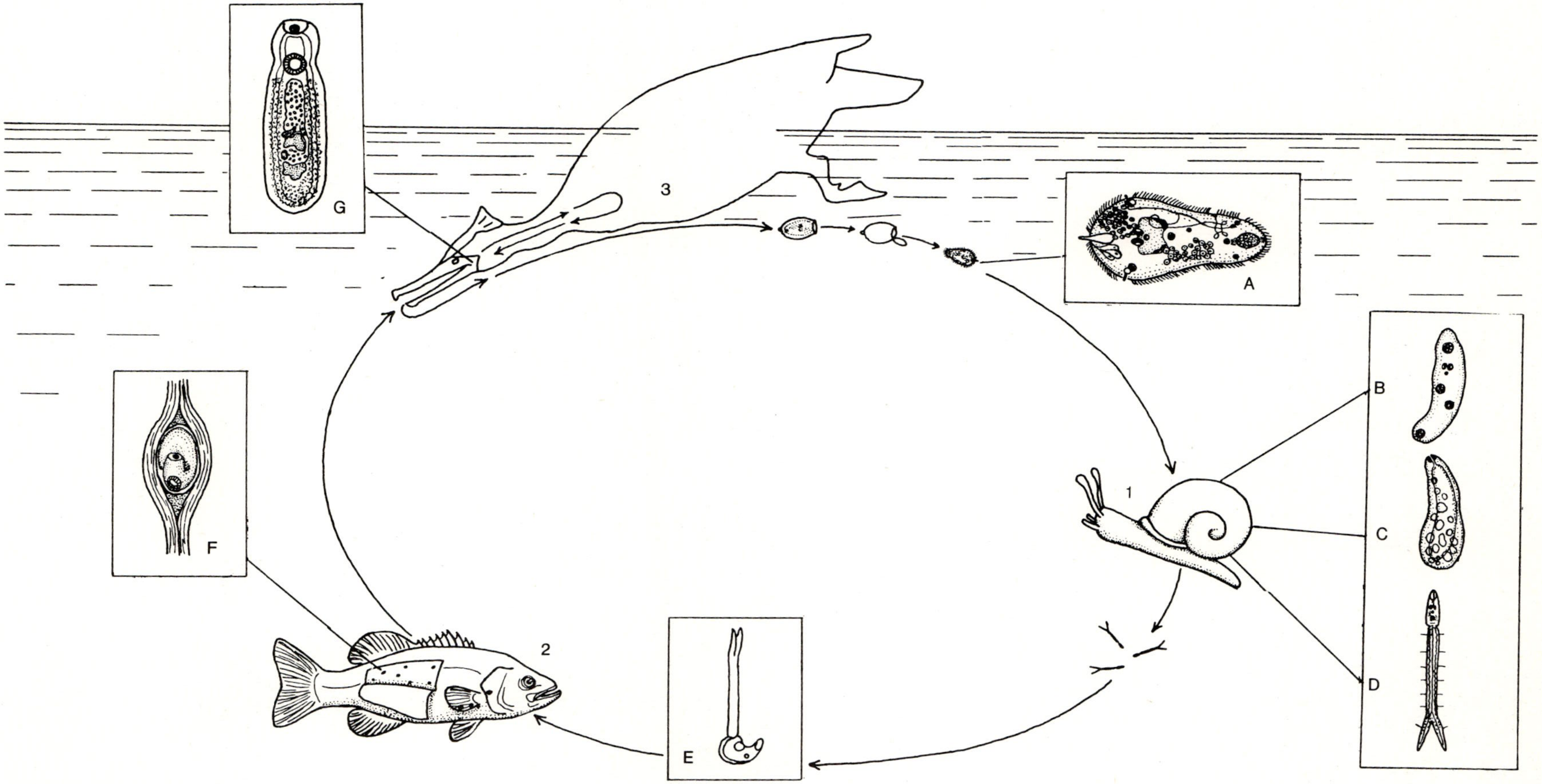

Fig. 2.29. Life cycle of *Clinostomum complanatum* is an example in which there are two intermediate hosts, the second containing the metacercariae. Infection of the definitive host takes place when animals harboring metacercariae are eaten. There are many examples of this type of life cycle. (A) Free-swimming miracidium; (B) mother sporocyst; (C) daughter redia, containing developing cercariae; (D–E) cercariae; (F) metacercaria in muscles of fish second intermediate host; (G) adult fluke in mouth and gullet of cormorant final host. (1) Snail, *Helisoma,* first intermediary; (2) fish second intermediary; (3) cormorant final host.

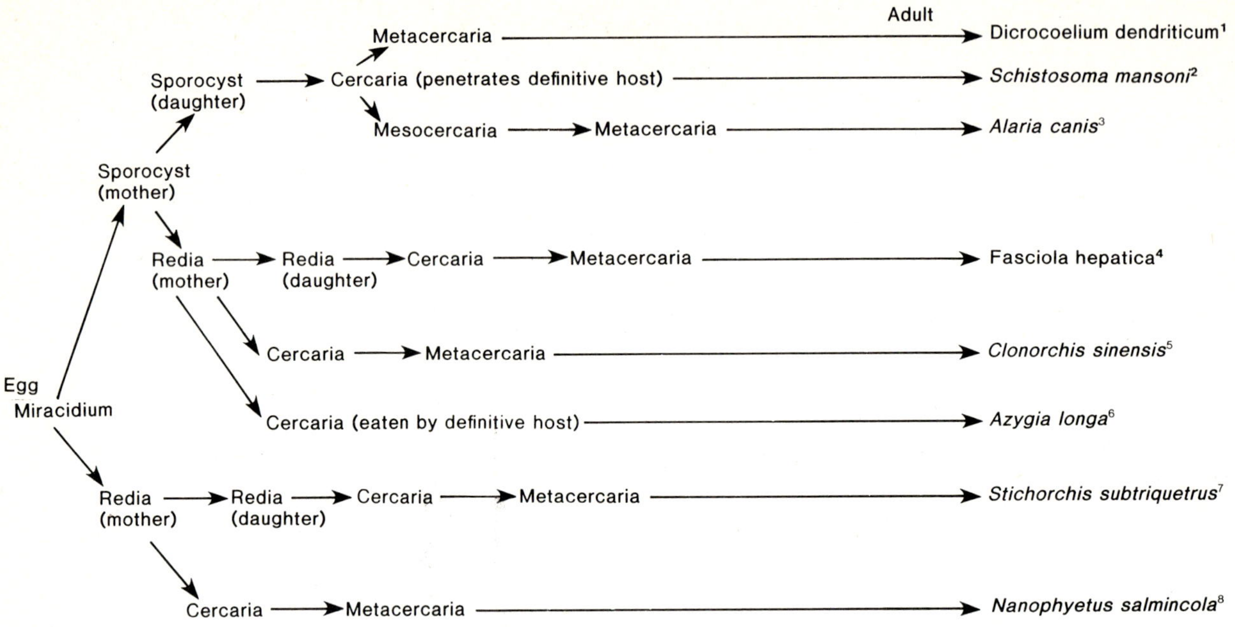

Fig. 2.30. Variations in life cycles of digenetic trematodes. Modified from Schell, 1970.

Following the above plan, the type of life cycle of each major species discussed in the manual is designated by a parenthesized supra number corresponding to that in the figure. For example, *Uvulifer amblo-plitis*[(1)] refers to (1) in the figure, showing the life cycle has the following larval stages: (a) miracidium, (b) mother sporocyst, (c) daughter sporocyst, (d) cercaria, and (e) metacercaria.

Four principal types of life cycles, based on the presence or absence of the metacercarial stage and how infection of the definitive host occurs, may be recognized.

The metacercarial stage is absent in some species. In the case of blood flukes of the family Schistosomatidae, the cercariae penetrate the skin of the definitive host, enter directly into the bloodstream, and develop to maturity. With members of the Azygidae, the large furcocysticercous cercariae swimming in the water are eaten by the fish definitive hosts and develop directly to maturity in the alimentary canal.

A metacercarial stage is present but varies in its location. In the Notocotylidae, Amphistomatidae, and Fasciolidae, as examples, it is on vegetation or other objects in the water. The definitive host becomes infected by swallowing them with vegetation. In many other families, cercariae penetrate a second intermediate host, which may be an arthropod or vertebrate, and transform into metacercariae. Infection of the definitive hosts occurs when animals harboring the metacercarial states are ingested.

Notes and Sketches

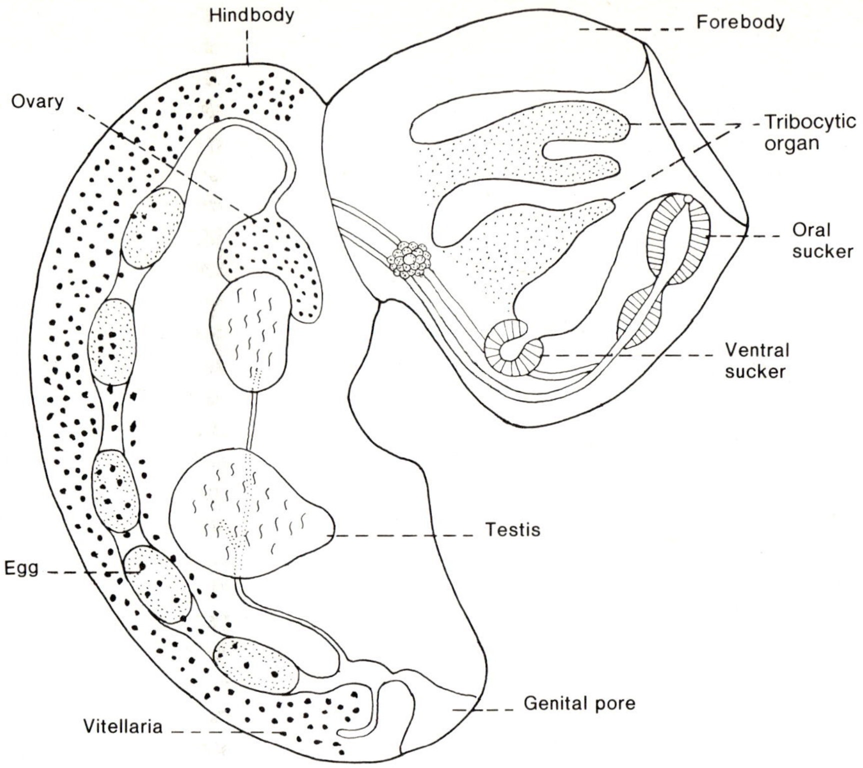

Fig. 2.31. *Cotylurus flabelliformis.*

Review Questions

1. What is the name of the ventral adhesive organ on an aspidogastrean? What are its individual compartments called?
2. Is there an intermediate host in an aspidogastrean life cycle?
3. What are the seven types of adult body forms in the Digenea?
4. What is an operculum? Which flukes lack it?
5. Discuss the "germ ball" theory of asexual development.
6. What are the morphological differences between a sporocyst and a redia?
7. What is meant by a "flame-cell formula"?
8. Describe the different forms of cercariae.
9. What are the types of strigeoid metacercariae? Give an example of each.
10. Which of the life cycle stages of the Digenea may be free-living, and which are always parasitic?

Examples of Representative Orders of Digenetic Flukes

Order Strigeata

Cercariae with forked tail; usually with pharynx and two suckers. Adults occur in birds and mammals.

Family Strigeidae

This is a large family characterized by a somewhat cup shaped forebody containing a two-lobed **tribocytic organ** and a hindbody that is cylindrical.

Cotylurus flabelliformis[1]

This species is very common in the small intestine of ducks in North America (fig. 2.31). Pharyngeate, furcocercous cercariae (see fig. 2.26M) penetrate snails and develop to infective metacercariae of the tetracotyle type (see fig. 2.28A).

Description

The body is divided by a deep transverse constriction into a cup-shaped forebody and a cylindrical hindbody about twice the length of the former.

The body is strongly flexed dorsally at the junction of the two regions. The tribocytic organ consists of two transverse, partly cleft lips which often protrude from the forebody. The oral sucker is located dorsally at the anterior margin of the forebody; the oval ventral sucker is behind the oral sucker, and in the bottom of the cup-shaped forebody. The digestive tract consists of a well-developed oval pharynx, an esophagus of about equal length, and intestinal caeca that terminate somewhat short of the posterior end of the hindbody.

Testes occupy the second and third quarters of the hindbody. From the dorsal view they are bean shaped, and from the side heart shaped. The vasa efferentia unite to form a tubular seminal vesicle that enlarges into a globular ejaculatory pouch located at the posterior margin of the hind testis. The common genital pore is large, funnel shaped, and located subdorsally near the posterior end of the hindbody.

The oval ovary lies on the dorsal side of the anterior testis. The oviduct, after giving off Laurer's canal which opens dorsally, extends caudad to Mehlis' gland. From here the anterior limb of the uterus extends forward to the ovary, bends ventrad, and continues posteriorly to the common genital pore. Vitellaria are in a single field in the ventral half of the hindbody, extending the full length of it. Eggs operculate, 100–112 × 68–76μm.

"White grub," the metacercaria of *Posthodiplostomum minimum,* is common in the heart, liver, and mesenteries of bluegill sunfish. Examination of such fish constitutes an interesting laboratory exercise.

Life Cycle

Characteristic of this life cycle is the tetracotyle type of metacercariae (fig. 2.28A) found only in the Strigeidae. The unembryonated eggs hatch in water, and the miracidia attack and enter several species of snails, among which are *Physa sayi, P. parkeri, Lymnaea emarginata, L. stagnalis,* and *L. perampla.* Miracidia transform into mother sporocysts, which in turn produce daughter sporocysts. These give birth to cercariae. After leaving the first snail, the cercariae enter other snails (*L. emarginata, L. stagnicola, L. reflexa, L. palustris,* and *Fossaria obrussa*) and develop into the type of metacercariae known as tetracotyles. Final host becomes infected upon eating the snails containing the tetracotyles.

Family Schistosomatidae

These flukes occur in the blood vessels of birds and mammals and are known collectively—together with the Spirorchiidae, of turtles, and Sanguinicolidae, of fish—as blood flukes. Schistosomatidae are dioecious, with males being the larger. Usually, the male has a ventral groove (**gynecophoric canal**) in which the long, slender female is carried. Apharyngeate, fork-tailed cercariae (fig. 2.26L) penetrate directly into the final host and develop to maturity, thereby eliminating the metacercarial stage. In this respect, this family of blood flukes is biologically unique.

Schistosoma mansoni[2]

This species occurs commonly in the mesenteric venules of the lower part of the ileum and large intestine of humans across equatorial Africa, parts of South America, including the coastal areas of Venezuela and Brazil, and in Puerto Rico.

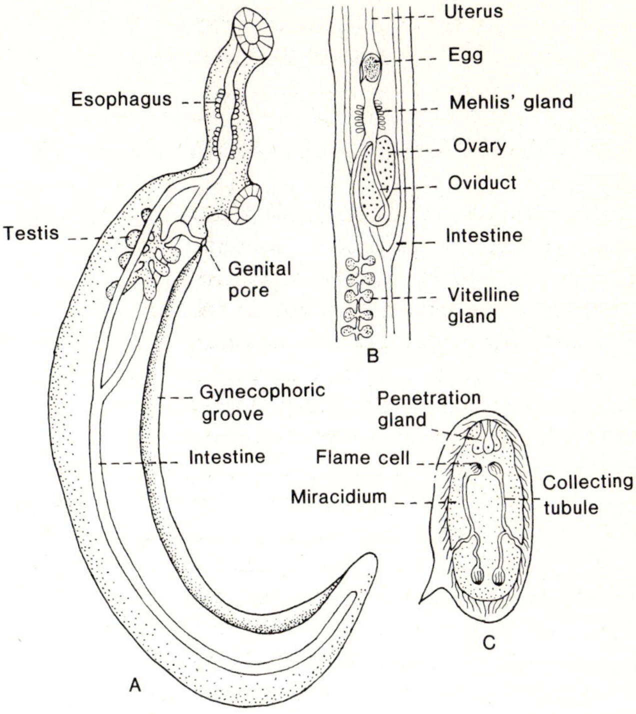

Fig. 2.32. *Schistosoma mansoni.* (A) Adult male; (B) segment of adult female, showing reproductive organs; (C) embryonated egg containing fully developed miracidium.

Description

The characteristic of prime significance is the occurrence of separate male and female individuals. (See fig. 2.32.)

Male. The body is up to 12 mm long with many conspicuous **tuberculations** over the entire surface caudad from the ventral sucker. Lateral margins of the body extend as flaps to form a ventral gynecophoric canal in which the female is held. The oral sucker is terminal, with the ventral sucker a short distance posterior. The gynecophoric canal extends from the ventral sucker to the posterior end of the body. The digestive tract consists of the apharyngeate esophagus and the caeca, which bifurcate at the level of the ventral sucker, reuniting preequatorially and continuing to the caudal tip of the body. A cluster of seven (three to thirteen) small testes just behind the anterior bifurcation of the intestine opens by means of a short duct into the genital pore in the anterior part of the gynecophoric canal.

Female. The slender body is up to 17 mm long. The oral sucker is terminal and the ventral sucker only a short distance caudad. The intestine is similar to that of the male, dividing at the level of the ventral sucker and reuniting preequatorially. The elongate ovary lies between the caeca of the intestine; there is a small crooked seminal receptacle near its posterior end; the oviduct extends forward, soon enlarges to form the ootype, which is surrounded by Mehlis' glands, and then merges with the uterus, which continues as a rather

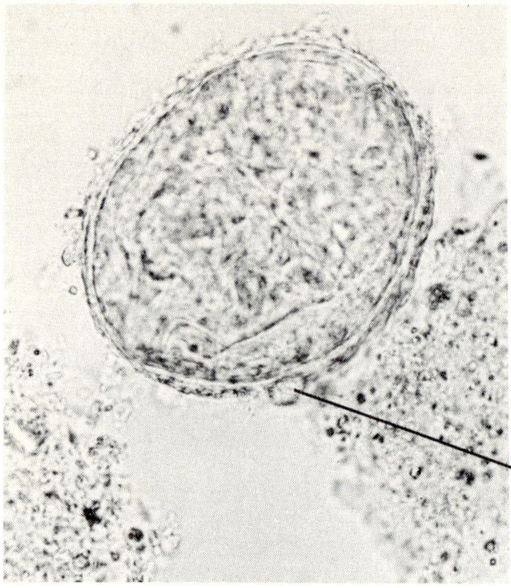

Fig. 2.33. *Schistosoma mansoni.* Oval at both ends and provided with a sharp lateral spine; size 114–175 × 45–68 μm.

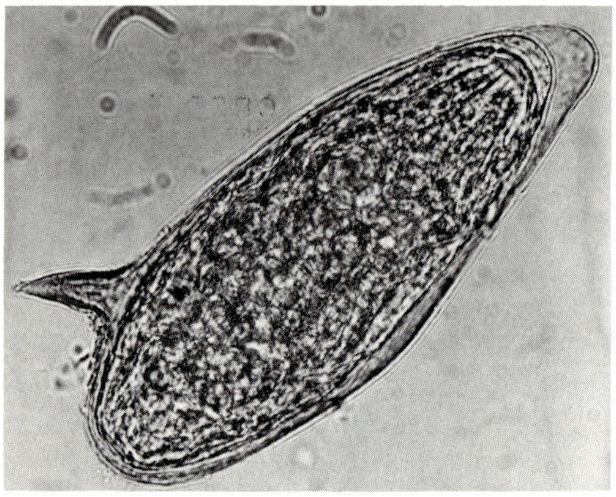

Fig. 2.34. *Schistosoma japonicum.* Ovoidal, except for a shallow depression near one end, from which extends a short knoblike spine; pale yellow in color; size 65–90 × 50–65 μm.

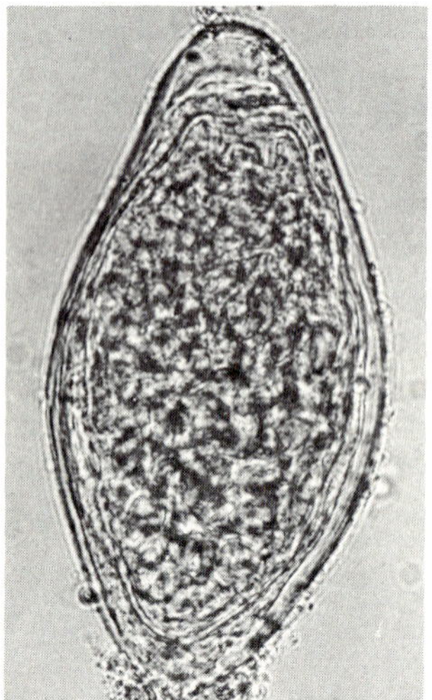

Fig. 2.35. *Schistosoma haematobium.* Oval at one end and conical at the other, tapering to a distinct spine; size 112–170 × 40–70 μm.

straight tube to the genital pore immediately posterior to the ventral sucker. Usually one to four eggs (fig. 2.33) appear at a time in the short uterus. The vitelline gland extends as a cylindrical structure with lateral extensions to the posterior end of the body.

Other important human blood flukes include *S. japonicum* from the Orient, particularly China, Taiwan, Japan, and the Philippines, and *S. haematobium,* which is widespread throughout Africa and on Malagasy Republic (formerly Madagascar), as well as in some parts of the Near East. Males of *S. japonicum* have seven (six to eight) testes arranged in a line. The intestinal caeca unite postequatorially. The body covering is smooth. Males of *S. haematobium* have a cluster of four to five large testes. The integument is finely tuberculate, and the intestinal caeca unite postequatorially. The eggs of these three species, which are nonoperculate and contain fully embryonated miracidia, are illustrated in figures 2.34 and 2.35.

Life Cycle

As stated above, the life cycle of the Schistosomatidae and related families of blood flukes is unique in that the cercaria penetrate the final host and develop directly into the adult without going through the metacercarial stage.

Eggs deposited in the venules of the intestinal wall work their way into the lumen and are voided with the feces. Once in the water, the miracidia escape through a rent in the shell and swim away. In the Western hemisphere, the important snail intermediate hosts are the discoidal snails *Biomphalaria glabratus* and species of *Tropicorbis.*

Miracidia penetrate the snails, shedding the ciliated epithelium in the process. Inside the snail, they transform into mother sporocysts that produce numerous elongate daughter sporocysts. Periodically, these give rise to vast numbers of apharyngeate, fork-tailed cercariae (fig. 2.26L). Upon coming in contact with the human skin, the cercariae quickly burrow with the aid of the penetration glands and body movements through the skin. Inside the body, they migrate by way of the blood stream through the heart, lungs, back to the heart, and out to the mesenteric vessels and intestinal wall and liver. Prepatency takes five to seven weeks.

The life cycles of *S. japonicum* and *S. haematobium* are basically similar. Snail hosts of *S. japonicum* are species of the dextrally spired *Oncomelania,* whereas those of *S. haematobium* are species of the sinistrally spired *Bulinus* and *Physopsis.*

Family Spirorchiidae

These are parasites of the blood vessels of the small intestine of turtles. These flukes are monoecious, lack a ventral sucker and gynecophoric canal, and possess one or more testes. The esophagus is surrounded by gland cells.

Spirorchis parvus[2]

This species occurs in the arteries of the wall of the pyloric stomach and small intestine of the painted turtle, *Chrysemys picta.*

Description

Adults are up to 2mm long, thin, translucent, and lack a pharynx. There are four to five testes, and a seminal vesicle that opens through a laterally placed common genital pore caudad from the testes. The ovary is slightly anterior to the genital pore. Vitellaria fill the body from the bifurcation of the intestinal caeca to near the posterior end of the body (fig. 2.36). Eggs nonoperculate, average 54 × 38 μm.

Life Cycle

When the eggs reach water, the miracidia become active, and hatching occurs within four to six days. Miracidia attack and enter ram's horn snails (*Helisoma trivolvis, H. campanulata*); only young snails are susceptible to infection. Miracidia develop into mother sporocysts. Development is completed in about 18 days, at which time daughter sporocysts appear. These escape and migrate to the digestive gland, where apharyngeate, furcocercous cercariae develop and escape. Upon coming in contact with turtles, especially young ones, they penetrate the soft membranes and enter.

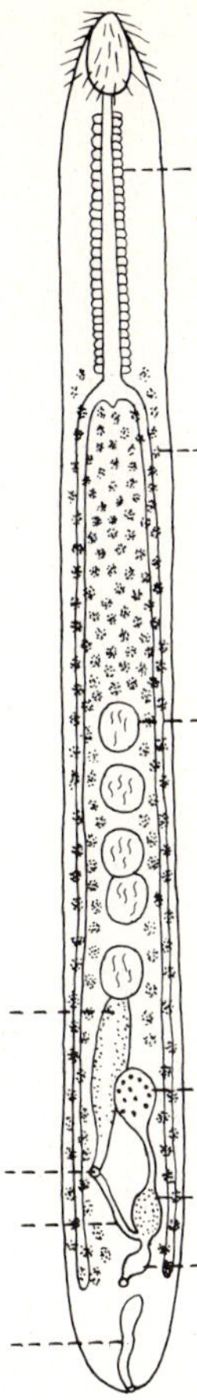

Fig. 2.36. *Spirorchis parvus.*

Order Echinostomata

Cercariae have large bodies, strong simple tails, and numerous cystogenous glands in the epidermis.

Family Echinostomatidae

Members of the family, common in reptiles, birds, and mammals, are characterized by a **head collar** armed with a single or double row of spines.

Echinostoma revolutum[4]

This is a cosmopolitan parasite of ducks and occasionally mammals, especially muskrats, and sometimes people (fig. 2.37). It shows great biological versatility in being able to parasitize so many species of animals in the course of development. Miracidia penetrate and develop in species of four genera of snails from three families. Cercariae encyst in many species of pulmonate snails, often those in which they developed. They also encyst and develop in fingernail clams, tadpoles, bullheads, and many other vertebrates and invertebrates.

Description

The body is about 15 mm long and with parallel sides, having a ratio of length to width of 15:1 in fully relaxed specimens. The size varies greatly, depending on the final host. The prominent **head collar** bears 37 spines, arranged in three groups, two of which are paired. They include two **corner groups** of five spines each, two lateral groups of six each, and a dorsal group containing 15 alternating spines. There are transverse rows of cuticular spines variable in number and size that cover the body posteriorly to the ventral sucker. The ventral sucker is larger than the oral sucker and varies in position from the anterior end of the body. The muscular pharynx is close to the oral sucker, followed by a short esophagus that branches at the anterior margin of the ventral sucker to form two slender, simple intestinal caeca extending to the posterior end of the body. The excretory bladder is a long, slender, saclike vesicle reaching to the posterior margin of the hind testis.

Gonads are arranged one behind the other, with the paired testes hindmost. Between the anterior testis and the ovary, which is foremost, is the Mehlis' gland. The oviduct extends posteriorly for a short distance and then turns forward as a long, convoluted uterus that opens through the common genital pore. A short cirrus pouch lies dorsal to the ventral sucker. Vitellaria extend laterally from the oral sucker to the posterior end of the body, being confluent behind the testes. Eggs operculate, 92–145 × 66–83 μm.

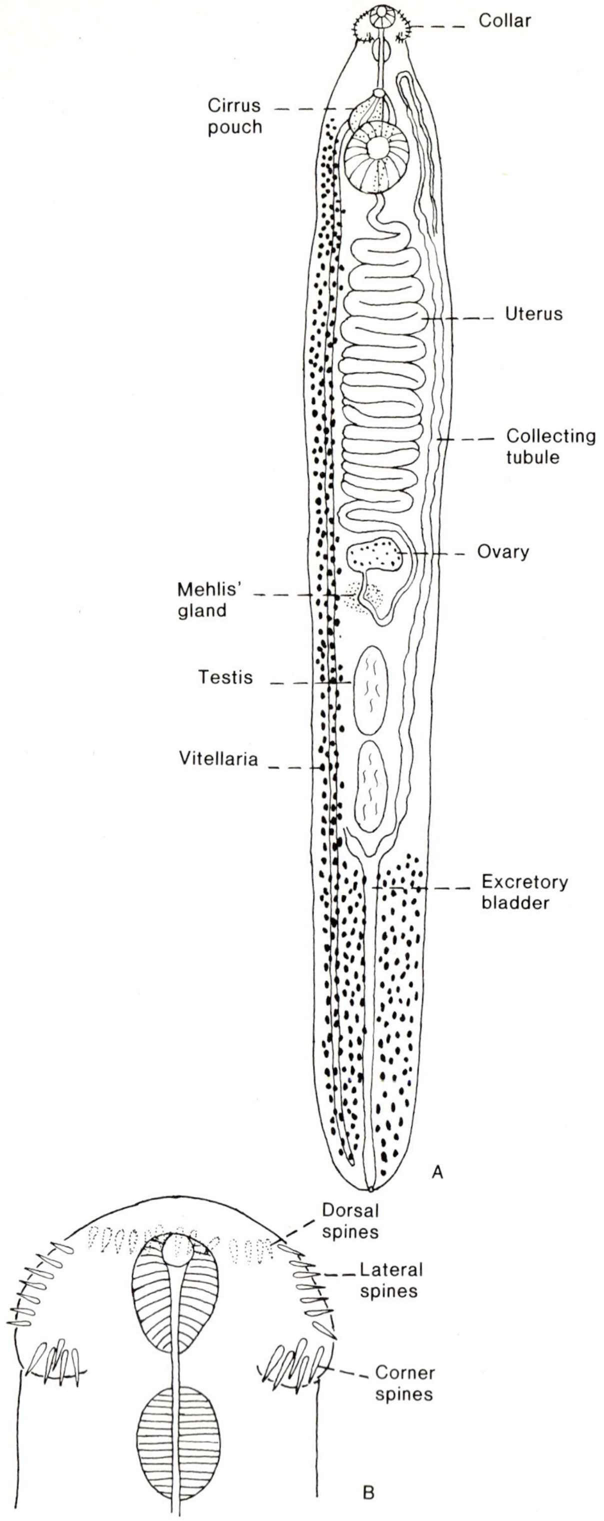

Fig. 2.37. *Echinostoma revolutum.* (A) Adult; (B) details of head spines.

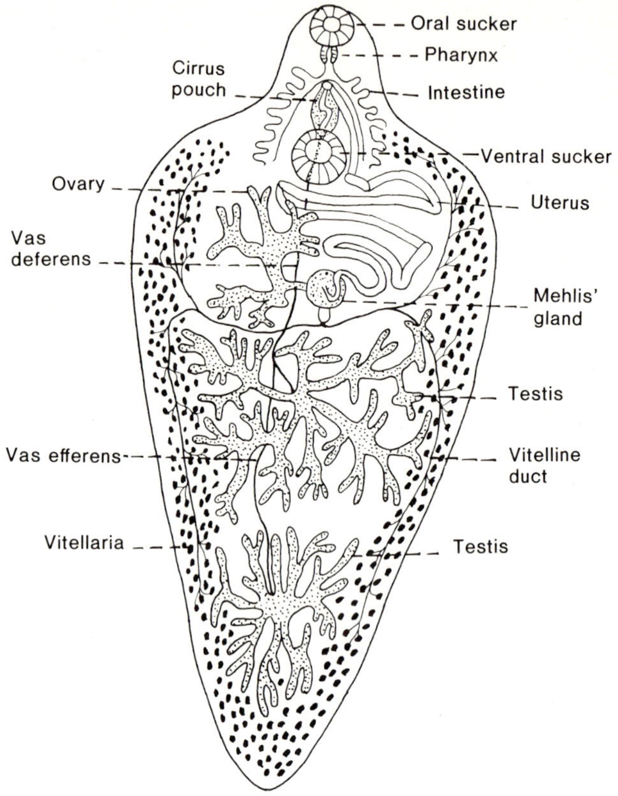

Fig. 2.38. *Fasciola hepatica.*

Life Cycle

Eggs hatch after about three weeks at room temperature, and the miracidia enter suitable snail hosts (*Helisoma trivolvis, H. antrosa,* and *Physa occidentalis*). Miracidia transform into mother sporocysts that give rise to two generations of rediae, the second of which produces echinostome cercariae in nine to ten weeks after infection of the snails. Cercariae enter and encyst in several species of pulmonate snails and tadpoles of frogs. *Physa gyrina, P. occidentalis, Helisoma trivolvis, Fossaria modicella,* and *Pseudosuccinea columella* are common molluscan second intermediaries. The final host becomes infected by ingesting the hosts with metacercariae.

Family Fasciolidae

These are parasites of the liver and intestine of mammals. They are large, flat distomes, with suckers close to each other. Intestinal caeca are highly branched or unbranched. Testes usually branched but may be smooth or lobed; ovary branched or smooth. The uterus has few coils and is anterior to the ovary. Vitellaria are profusely developed and widely distributed.

Fasciola hepatica[4]

This is the common liver fluke of sheep, cattle, goats, and rabbits. It is widespread throughout the world and

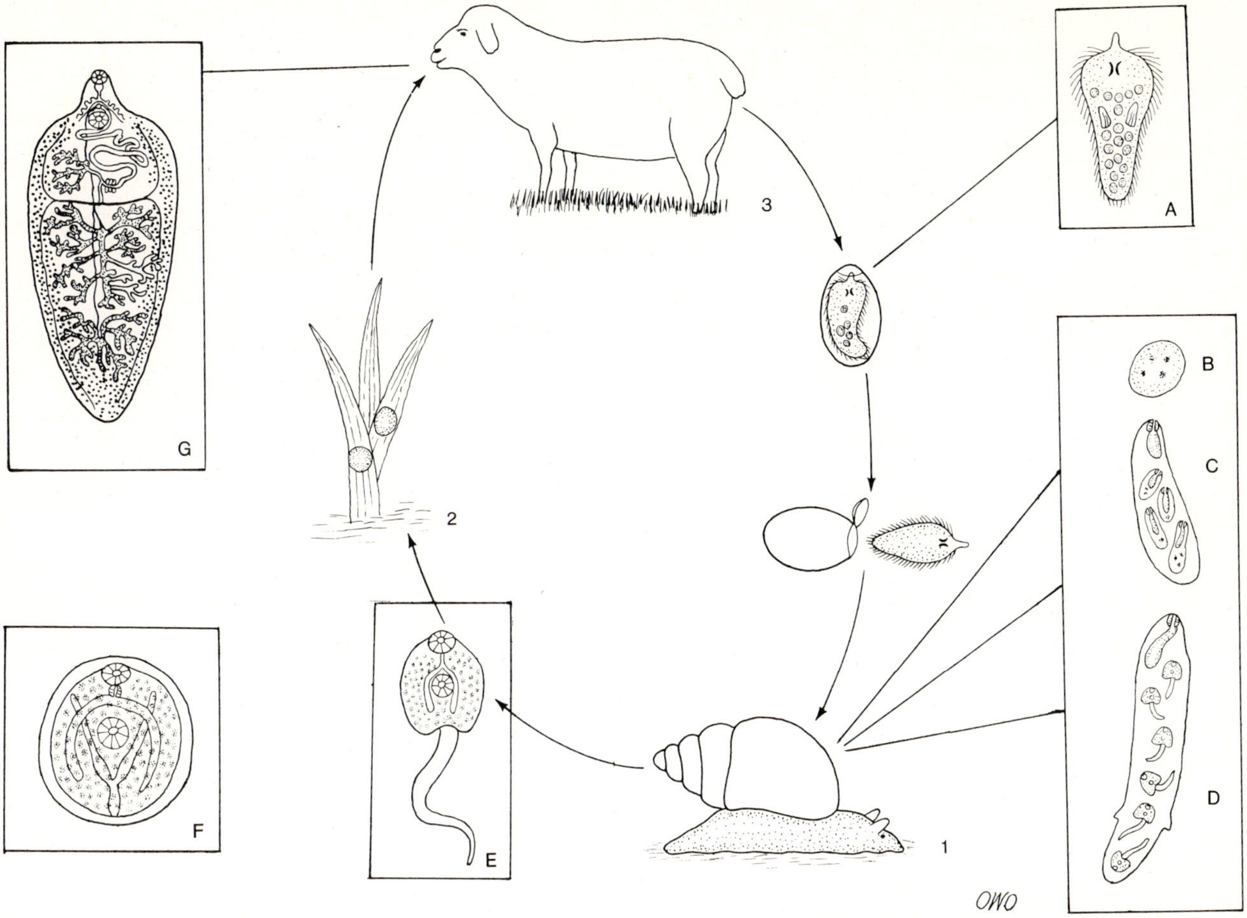

Fig. 2.39. Life cycle of *Fasciola hepatica* is an example in which there is only a molluscan intermediate host and the metacercariae are on vegetation. Infection of the definitive host occurs when it swallows the metacercariae with the forage. Other representatives of this type of life cycle include Paramphistomatidae and Notocotyliidae. (A) Free-swimming miracidium; (B) mother sporocyst; (C) mother redia; (D) daughter redia; (E) cercaria; (F) metacercaria; (G) adult fluke in liver of definitive host. (1) Snail first intermediary (*Stagincola bulimoides, Fossaria modicella,* and others); (2) metacercaria on grass; (3) sheep and cattle definitive hosts.

is of great economic importance to the livestock industry. It was the first digenetic trematode to have its life cycle elucidated. Simultaneously and independently in 1881, Leuckart in Germany and Thomas in Great Britain discovered the life cycle and described for the first time the larval development of digenetic trematodes. *Fasciola hepatica* has a life cycle differing from the intestinal form of the family in that the metacercariae migrate through the wall of the small intestine into the coelom into the liver. The liver capsule is penetrated by the young flukes, which burrow through the parenchyma and eventually reach the bile ducts, where they mature. Live specimens are easily obtained from the local abattoir.

Description

Fully grown specimens are up to 30 mm long by 13 mm wide. They are flat and leaf shaped, with a small cephalic cone on a broad anterior part. The anterior portion of the body has scalelike spines. The oral sucker is smaller than the nearby ventral sucker. There is a well-developed pharynx; the highly dendritic intestinal caeca occupy most of the body.

Two profusely branched testes occupy the posterior half of the body; vasa efferentia unite anteriorly to form the vas deferens, extending to the cirrus pouch at the anterior margin of the ventral sucker. The common genital pore is median and opens at the level of the bifurcation of the intestinal caeca.

The dendritic ovary lies to the right of the body midline, anterior to the testes. A short oviduct enlarges to form the ootype surrounded by Mehlis' gland. The uterus loops anteriorly to the common genital pore. The vitellaria are profuse, extending the full length of the body. (See fig. 2.38.)

Life Cycle

The life cycle of *F. hepatica* is shown in figure 2.39. Eggs (fig. 2.40) are unembryonated when expelled in the feces. They require nine to ten days to mature at

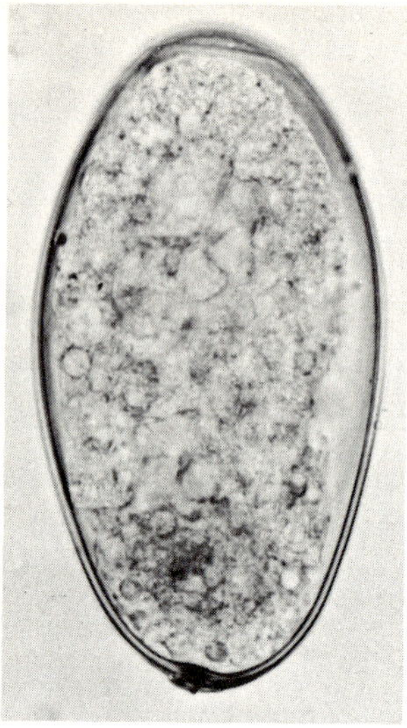

Fig. 2.40. *Fasciola hepatica*, egg. Operculate; delicate light brown in color; size 130–150 × 63–90 μm.

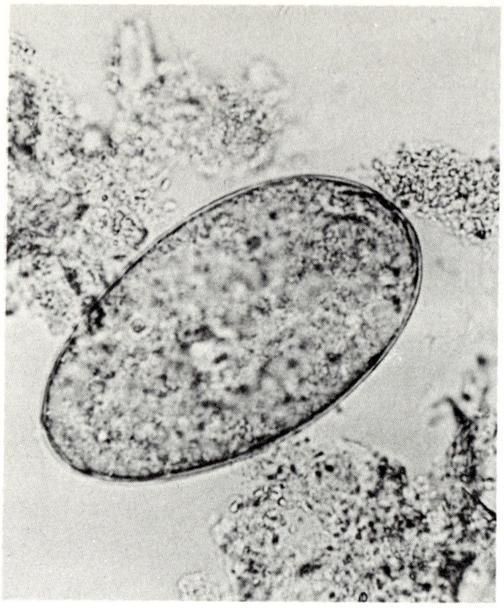

Fig. 2.42. *Fasciolopsis buski*, egg. Predominately oval; clear, yellowish-tinged shell with a small and slightly convex operculum; size 130–140 × 80–85 μm. Eggs are practically identical with those of *Fasciola hepatica*.

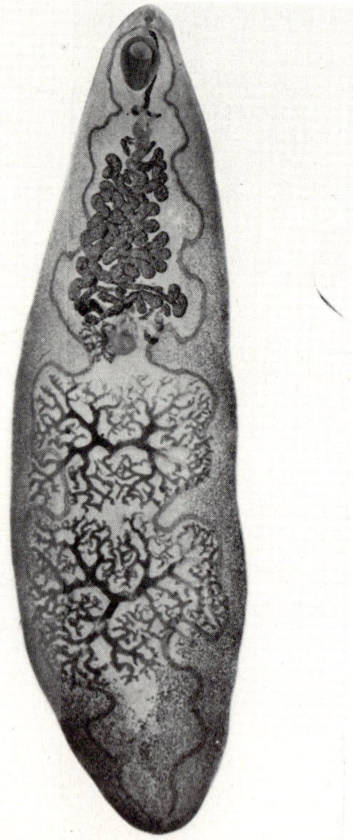

Fig. 2.41. *Fasciolopsis buski.* Note the dendritic testes and the unbranched caeca. The adult worm is up to 75 mm long and 20 mm wide. (From Schmidt, G. D., and L. S. Roberts. 1985. *Foundations of Parasitology,* 3d. ed. The C. V. Mosby Co., St. Louis.)

summer temperature. Upon hatching, the miracidia invade a suitable snail host and transform into sporocysts. Within five to seven weeks, two generations of rediae and the **gymnocephalus cercariae** have been produced. Common snail intermediate hosts in the United States are *Stagnicola bulimoides,* including its various subspecies, *Fossaria modicella,* and *Pseudosuccinea columella.* Elsewhere, other species of dextral lymnaeids serve as intermediaries. *Limnaea truncatula* is the common host in Europe and *Lymnaea tomentosa* in Australia. Cercariae encyst upon moist or submerged vegetation, where they remain until eaten by the final host. When viable cysts ingested by suitable hosts reach the duodenum, they excyst, and the metacercariae migrate through the intestinal wall into the coelom and to the liver, which they enter.

Fasciolopsis buski

This fluke (fig. 2.41) is peculiar among the Fasciolidae in that it inhabits the small intestine of its host, rather than the liver. It is a common parasite of humans and pigs in the Orient, where it may infect ten million or more people.

A giant among digenetic flukes, it is elongate-oval, with a length of up to 75 mm and a width of up to 20 mm. The anterior end is rounded, lacking the oral cone found on *Fasciola.* The acetabulum is larger than the oral sucker and is located close to it. The caeca are wavy but unbranched. The testes are tandem in the posterior half of the worm and are highly branched. Vitellaria are extensive, filling the lateral parenchyma to the posterior end. The uterus is short, with an ascending limb only. The eggs (fig. 2.42) are almost identical to those of *F. hepatica.*

Life Cycle

The life cycle of *F. buski* is similar to that of *F. hepatica*. One worm produces about 25,000 eggs per day. The eggs mature and hatch in about seven weeks. Snails of the genera *Segmentina* and *Hippeutis* serve as intermediate hosts. Cercariae encyst on underwater vegetation, including plants that are important foods for people, such as water chestnut, lotus, and water caltrop. Metacercariae are swallowed when these plants are eaten raw or peeled with the teeth. It takes about three months for the worms to mature.

Family Paramphistomatidae

These are parasites of the cloaca of amphibians and birds, and rumen of ruminants. The group is characterized by the ventral sucker being at the posterior end of the body, and by the ovary being posterior to the testes, which is uncommon.

Megalodiscus temperatus[4]

Prevalence of this species in frogs and the ease with which the life cycle can be studied make it an ideal model for class use. Specimens occur in the cloaca of various species of adult *Rana* and their tadpoles. Morphologies of species from mammals, such as *Paramphistomum cervi,* are similar.

Description

The body is cone shaped, with the ventral side somewhat flattened. They attain a length of 6 mm, a width of 2.5 mm, and a thickness of 2 mm. The oral sucker is well developed, with large, buccal pouches at its posterior end. A large ventral sucker that may have papillae in the center is located at the posterior end of the body. A short prepharynx is followed by a globular pharynx. Intestinal caeca begin at the pharynx and extend along the sides of the body as simple tubes to near the posterior end of the body. A large excretory bladder just anterior and dorsal to the ventral sucker opens on the dorsal side of the body near the caudal end.

Testes are arranged transversely near the posterior part of the anterior half of the body. A short vas deferens with a muscular end extends directly to the genital pore located at the beginning of the caeca. There is no cirrus pouch. The spherical ovary is located dorsally in the body on the median line near the ventral sucker. The oviduct extends caudad a short distance, enlarges to form the ootype surrounded by Mehlis' gland, and extends forward as broad uterine loops to the common genital pore. Laurer's canal extends from the oviduct to the dorsal side of the body above the ovary. Sparse vitellaria consist of two groups on each side of the body between the testes and ventral sucker. (See fig. 2.43.) Eggs operculate, 120–128 × 60–70 μm.

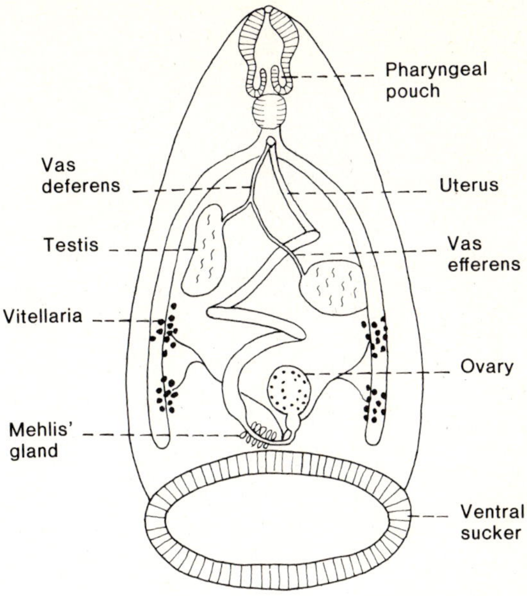

Fig. 2.43. *Megalodiscus temperatus.*

Life Cycle

Eggs hatch promptly upon entering the water, and miracidia penetrate young snails of *Helisoma* spp., where they transform into mother sporocysts. Three generations of rediae are produced. Within the final generation of redia, amphistome cercariae develop; they emerge and encyst on the surface of various species of frogs and their tadpoles. Ingestion by adult frogs of their own stratum corneum, on which the cercariae have encysted, results in infection. Tadpoles become infected by eating metacercariae in pond ooze, or by taking in cercariae during respiration. Prepatency may be as short as a month, but it is usually longer.

Notes and Sketches

Order Plagiorchiata

Cercariae lack caudal excretory vessels in all stages of development; oral sucker with or lacking **stylet.** Adults occur in all classes of vertebrates.

Family Haematoloechidae

Haematoloechus medioplexus(1)

This is one of the common lung flukes of adult *Rana pipiens* and *Bufo americanus* in North America.

Description

They are generally elongate worms, up to 8 mm long by 1.2 mm wide. The tegument is densely spined. Ventral sucker small, in first third of body, anterior to ovary. Ovary oval to round, situated submedially anterior to testes. Seminal receptacle posterior to and slightly overlying the ovary. Rounded testes posterior to ovary and seminal receptacle, diagonally situated in third quarter of body. Vitellaria consist of about ten groups of six or more follicles, on either side, extending along the caeca from the level of the posterior testis to near the anterior end. Uterus continues posteriorly in short transverse folds between testes, filling the posttesticular region; ascending limb fills anterior intercecal region with short, transverse folds. Genital pore situated slightly posterior to pharynx. Eggs operculate, 22–29 × 13–17 μm. Live worms should be placed in tap water so they will shed eggs, thereby uncovering the other internal structures. (See fig. 2.44.)

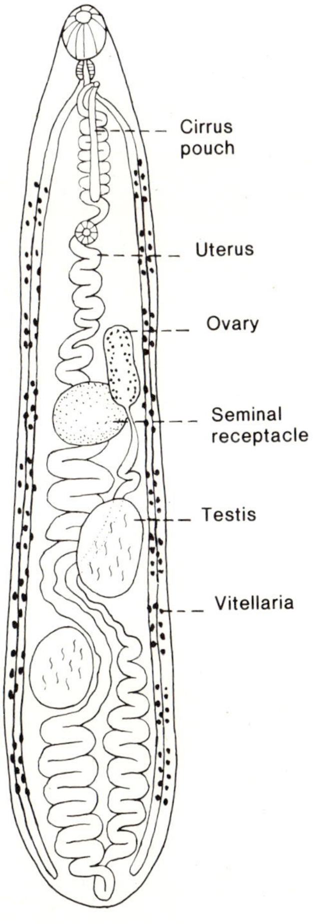

Fig. 2.44. *Haematoloechus medioplexus.*

Life Cycle

Eggs pass from the lungs into the mouth of the frog, are swallowed, and escape with the feces. When voided, the miracidia are mature, but unless the eggs are eaten by the flat snail *Planorbula armigera,* hatching does not occur. Daughter sporocyst gives rise to xiphidiocercariae, which remain suspended in the water. As a result of this aimless swimming, they are passively drawn by respiratory currents into the branchial basket of *Sympetrum* dragonfly naiads. Once in the second intermediate host, the cercariae enter and encyst in the gill lamellae. As the naiad metamorphoses, metacercariae collect in the posterior part of the body, where further development is contingent upon the adult dragonfly being eaten by a frog or a toad. Metacercarial excystment occurs in the stomach of the final host, after which the worms ascend the esophagus and enter by way of the glottis to the lungs.

Family Dicrocoeliidae

Parasites of the liver, bile ducts, gallbladder, and pancreas of reptiles, birds, and mammals. In addition to the location in the final host, the family is characterized by the posttesticular position of the ovary and small, brown, embryonated eggs whose operculum sets in a rimlike thickening.

Dicrocoelium dendriticum(1)

This species, the lancet fluke, occurs in the liver and gallbladder of sheep, cattle, deer, rabbits, and marmots. It will also develop in guinea pigs.

Description

The translucent, dorsoventrally flattened body measures up to 12 mm long by 2.5 mm wide. The oral sucker is subterminal; the slightly larger ventral one is in the anterior quarter of the body. The small, round pharynx lies next to the oral sucker; it is followed by the slender esophagus that extends about midway between the pharynx and ventral sucker, and bifurcates, sending two long, slender intestinal caeca to the beginning of the last quarter of the body. The common genital pore is just posterior to the intestinal bifurcation. The slender excretory bladder reaches anterior to the ovary. Slightly lobed testes are close together, obliquely arranged in the body with the anterior one near the posterior margin of the ventral sucker. The vasa efferentia unite at the level of the ventral sucker to form a short vas deferens

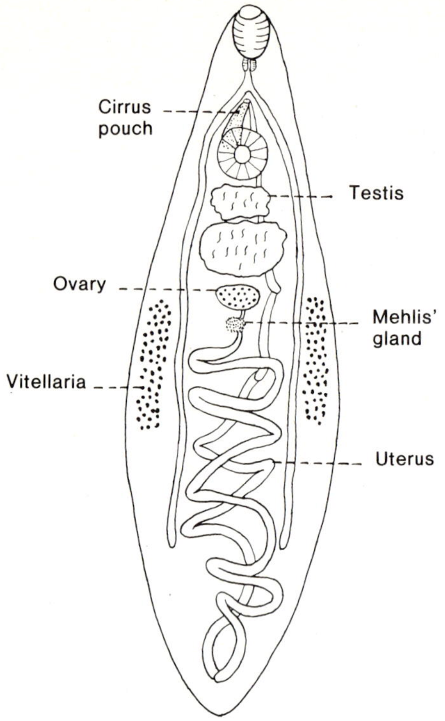

Fig. 2.45. *Dicrocoelium dendriticum.*

that enters the cirrus pouch, forming a sinuous seminal vesicle. The cirrus pouch reaches almost to the ventral sucker.

The somewhat oval ovary is located slightly posterior to the hind testis, a rather unusual position, as it is commonly anterior to them. A small seminal receptacle and Mehlis' gland are close behind. The uterus consists of a descending portion with many loops and an ascending one also with many loops, which cross each other below the gonads. The vitellaria consist of numerous, small follicles located extracaecally and extend from the posterior margin of the hind testis to or slightly caudad from the middle fifth of the body. Eggs operculate, 38–45 × 22–30 μm. (See fig. 2.45.)

Life Cycle

The land snail, *Cionella lubrica*, becomes infected by eating embryonated eggs containing miracidia. Sporocysts in the digestive gland of the snail become distended with the cercariae. The cercariae escape from the sporocyst periodically and accumulate in masses of mucus called **slimeballs.** These are sticky and adhere to vegetation and debris. Slimeballs are actually packages of numerous cercariae, in a covering which protects them against desiccation. They are carried away by the ant *Formica fusca*, which uses them as food and becomes infected by ingesting the cercariae. Encysted metacercariae accumulate in the abdomen and head of ants and await a possible transfer to one of the several final hosts. The final host becomes parasitized by accidentally ingesting the infected ants while grazing. Behavior of the ant usually is changed to increase the chance of its being eaten by the definitive host. For instance, when the temperature drops at evening, infected ants will grasp objects with their mandibles, holding on until being warmed the next morning. If they are attached to grass or other plants this makes them available to grazing animals.

Family Prosthogonimidae

These mature in the oviducts and bursa Fabricii of birds. They are small, clear flukes with the anterior end tapering and the posterior end broadly rounded. The genital pore is near the oral sucker. The vitellaria form grapelike clusters laterally. The excretory bladder is Y-shaped.

Prosthogonimus macrorchis[1]

This is the oviduct fluke of ducks, chickens, crows, English sparrows, ruffed grouse, and other birds.

Description

The body is thin, translucent, and generally oval in shape, with the anterior end somewhat narrow and the posterior end broadly rounded. Size varies, depending on the species of host, ranging from 2.5 to 2.8 mm long by 1.4 to 2 mm wide in English sparrows up to 6.3 to 8.5 mm long by 5.1 to 6 mm wide in chickens. The oral sucker is roughly half the size of the ventral one, with the latter located near the posterior end of the first third of the body. The muscular pharynx is followed by a short esophagus of about the same length that divides into two long slender intestinal caeca, which extend to near the posterior end of the body. The common genital pore opens on the anterior right margin of the oral sucker.

Large roundish testes lie side by side near the anterior part of the second half of the body. Slender vasa efferentia unite near the posterior margin of the ventral sucker to form the vas deferens, which continues directly to the cirrus sac. The latter extends posteriorly to near the end of the esophagus and opens through the common genital pore; it contains a convoluted seminal vesicle in the posterior end.

The multilobed ovary lies immediately behind the ventral sucker. Attached to the short oviduct directly behind the ovary is a small seminal receptacle. Highly convoluted, the uterus descends between the testes, fills the posttesticular part of the body, and ascends between the testes, filling most of the space between the vitellaria, the testes, and ventral sucker. It continues as a thin tube to the genital pore. Vitellaria, numbering eight to nine lateral clusters, begin at the level of the ventral sucker and end near the posterior margin of the testes. Eggs are operculate, with an abopercular protuberance, average 28 × 16 μm. (See fig. 2.46.)

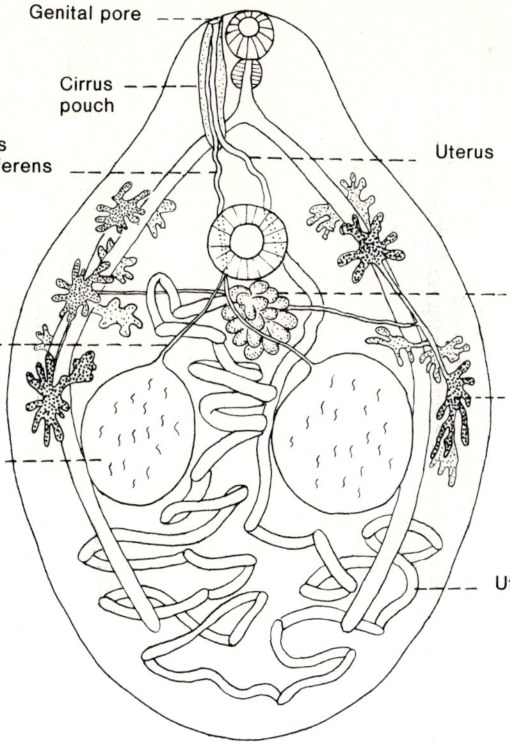

Fig. 2.46. *Prosthogonimus macrorchis.*

Life Cycle

Eggs, unembryonated when voided with the feces, develop in the water and hatch in the intestine of the snail intermediate host *Amnicola limosa porata.* Freed xiphidiocercariae that developed in sporocysts are drawn into the rectum of dragonfly naiads by the respiratory currents, burrow through the wall, and encyst in the abdomen. Species of *Leucorrhinia, Tetragoneuria, Epicordulia,* and *Mesothemis* dragonflies serve as second intermediate hosts. Birds become infected when they eat dragonflies, either as naiads or adults, containing metacercariae.

Family Troglotrematidae

Flukes with a plump or rounded body, the ventral sucker near the midbody, and the tegument spined. They are parasites of the kidney and skin of birds and the intestine, frontal sinuses, and lungs of mustelids. The family is best known for *Paragonimus westermani,* the Oriental lung fluke, which infects humans.

Paragonimus westermani[5]

Adults occur in cysts either singly or in pairs in the lungs. Common animal hosts belong to the Felidae (including tigers, lions, leopards, and other wild cats), swine, and dogs. Important endemic areas are in Korea, Japan, Taiwan, central China, and the Philippines.

Description

Adults are plump, spinose worms resembling a coffee bean in size and shape. Often they are quite brown due to the profuse brownish vitellaria and loops of the uterus filled with brown eggs. Worms are up to 12 mm long, 6 mm wide, and 5 mm thick. The oral sucker is slightly larger than the ventral sucker, which is located slightly anterior to the midbody. A muscular pharynx close to or in contact with the oral sucker is followed by an extremely short esophagus. Simple, sinuous intestinal caeca extend to the posterior end of the body. The excretory bladder extends forward as a narrow, irregular sac to the level of the posterior margin of the pharynx.

Highly lobed testes are arranged transversely about midway between the ventral sucker and hind end of the body. There is no cirrus or cirrus pouch. The common genital pore is near the posterior margin of the ventral sucker. The ovary is a large irregularly shaped organ to the left and slightly behind the ventral sucker, but anterior to the testes. A looped uterus lies opposite the ovary on the other side of the body. The vitellaria consist of extensively branched, small follicles filling the lateral fields of the body and overlapping the intestinal caeca the full length of the body. (See fig. 2.47.)

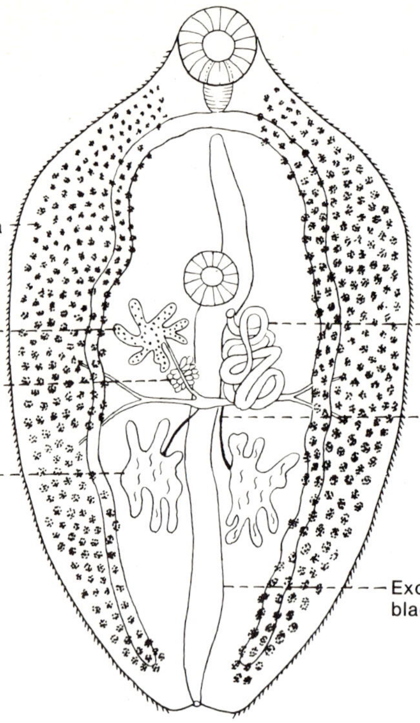

Fig. 2.47. *Paragonimus westermani.*

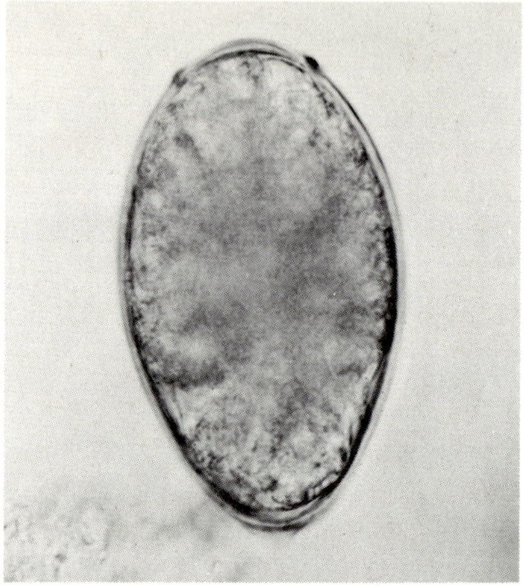

Fig. 2.48. *Paragonimus westermani,* egg. Broadly ovoidal, golden brown in color, somewhat flattened operculum; size 80–118 × 48–60 μm. As in eggs of *Clonorchis sinensis,* some eggs of *P. westermani* have a nipplelike protuberance at the abopercular end.

Life Cycle

Eggs (fig. 2.48) require 16 days or more to develop and hatch. The miracidia penetrate Pleuroceridae and Thiaridae gastropods. Mother sporocysts produce rediae which in turn spawn characteristic cercaria. They are small and of the microcercous type, with a small tail and bearing a well-developed stylet in the anterior part of the oral sucker (fig. 2.26H). Cercariae penetrate crayfish or crabs, enter the viscera and muscles, and encyst. The final host becomes infected by eating crustaceans containing living metacercariae. Upon being released from the cyst in the intestine, the young flukes burrow through the gut wall into the coelom, migrate anteriorly, passing through the diaphragm and into the pleural cavities, where they burrow into the lungs.

Paragonimus kellicotti

This is the North American representative of lung fluke. It appears to be a normal parasite of mink but occurs in cats, dogs, muskrats, swine, goats, and even humans. In surveys, it has been found in 15% of the mink examined in Michigan and 7% in Minnesota, and in 12% of the muskrats in Michigan. It is widespread in the United States.

Similarity between *Paragonimus westermani* and *P. kellicotti* has led some workers to consider them as synonyms. Because of the similarity of the two species, the basic anatomical description of *P. westermani* given above will suffice for *P. kellicotti.* The life cycle is basically similar to that of *P. westermani.* It is a long one and not well suited for a short term study.

Order Opisthorchiata

Cercariae with caudal excretory vessels; oral sucker lacking stylet.

Family Opisthorchiidae

Members of this family occur in the bile ducts and gallbladder (rarely in the intestine) of fish, reptiles, birds, and mammals. They are characterized by slender, flattened, transparent bodies. There is a well-developed digestive system and a ventral sucker in the anterior part of the undivided body. The ovary and testes are in the posterior half of the body, and the vitellaria are lateral to the intestinal caeca and pretesticular.

Clonorchis sinensis[5] (*Opisthorchis sinensis* of some authors)

An important fluke in the biliary passages and occasionally the pancreatic duct of humans and fish-eating mammals of the Orient, chiefly Japan, Korea, China, Taiwan, and Indo-China. (See fig. 2.49.)

Description

Adults are flat, thin, transparent, and somewhat spatulate in shape, with both ends attenuated. They are up to 25 mm long by 5 mm wide. The tegument is smooth. The oral sucker is somewhat larger than the ventral sucker or acetabulum, which is near the junction of the first and second quarters of the body.

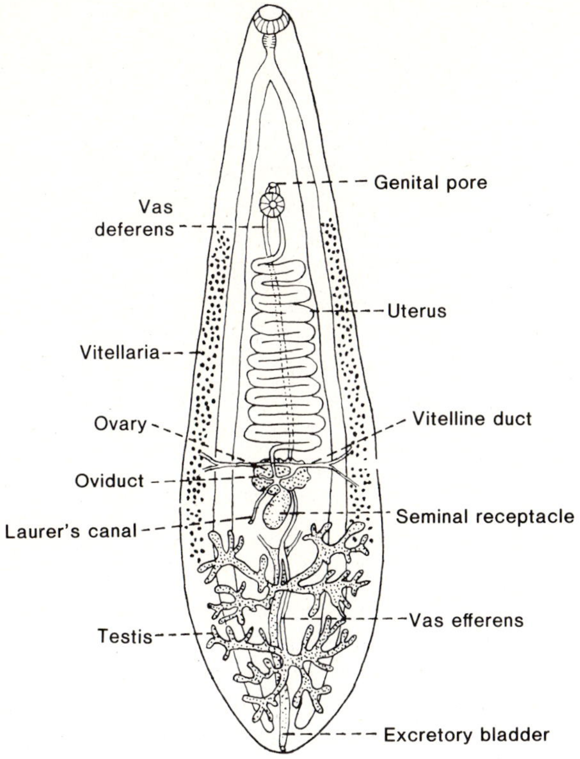

Fig. 2.49. *Clonorchis sinensis.*

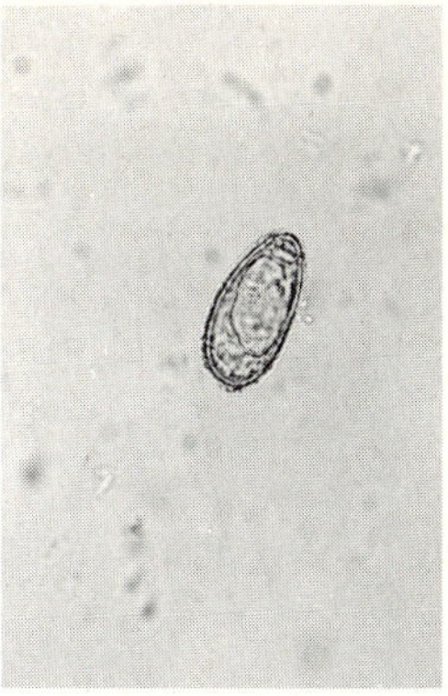

Fig. 2.50. *Clonorchis sinensis,* egg. Operculate; yellowish-brown color; shape of an old-fashioned light bulb; abopercular end sometimes with a nipplelike protuberance; size 26–30 × 15–17 μm.

The mouth opens at the bottom of the oral sucker and connects posteriorly with the small muscular pharynx; a very short esophagus continues and joins the two epithelium-lined intestinal caeca, which extend posteriorly to the end of the body, where they terminate blindly.

Large testes are branched, with one behind the other in the posterior third of the body. A slender vas efferens extends anteriorly from each testis to near the middle of the body, where they join to form the enlarged vas deferens that continues almost directly, dorsal to the uterus, to the male genital opening at the anterior margin of the ventral sucker. There is no cirrus, cirrus pouch, or prostate gland.

The small, slightly lobed and oval ovary lies a short distance anteriorly from the forward testis. Two tubules open into the short oviduct. The first one is from the large, oval seminal receptacle located directly posterior to the ovary. A slender duct, the Laurer's canal, which arises from the oviduct, is conspicuous. Slightly beyond the opening of the seminal receptacle, the common vitelline duct enters the oviduct. A short distance from this point, the oviduct enlarges to form the ootype which is surrounded by a lightly stained glandular mass known as Mehlis' gland. The uterus arises from the ootype and continues forward in compact, intercaecal coils to the genital atrium, where it opens beside the male duct. Viewed ventrally, the uterus opens to the right of the vas deferens.

Vitellaria, composed of delicate follicles, are extracecal, extending from near the posterior margin of the ventral sucker to the anterior side of the front testis. A common vitelline duct conveys yolk granules from the glands to the small yolk reservoir, which empties into the oviduct.

By focusing on the ventral sucker, determine if your specimen is mounted ventral or dorsal surface up. Another way to tell which surface is up is by the arrangement of the female reproductive organs. If the Mehlis' gland is on the right and the seminal receptacle to the left, with the ovary between the two, the specimen is mounted ventral side up; conversely, if the relationship of these organs is reversed, the dorsal surface is up.

Ova are produced in the ovary and passed into the oviduct, where they are fertilized by sperm from the seminal receptacle and provided with yolk and shell-forming material from the vitellaria. As they pass through Mehlis' gland, secretions involved in several aspects of production of the egg shell are supplied.

Life Cycle

The life cycle includes two intermediate hosts: Amnicolidae snails and fish, first and second intermediaries, respectively. The eggs (fig. 2.50) hatch only when eaten by a snail. The mother sporocysts produce rediae in which gymnocephalous cercariae with a longitudinal tail fin develop. Upon coming into contact with numerous species of freshwater fish (mostly Cyprinidae) of the Orient, the cercariae drop their tails, penetrate, and encyst in the muscles. When viable metacercariae are ingested with raw fish, they excyst in the intestine and migrate to the liver by way of the common bile duct (Sun et al., 1968).

Other often available Opisthorchiidae with their hosts include *Opisthorchis tonkae,* muskrats; *Amphimerus elongatus,* ducks; *A. pseudofelineus,* cats, dogs; *Metorchis conjunctus,* cats, dogs, fox, mink, raccoons, and occasionally humans in Canada and the northern United States.

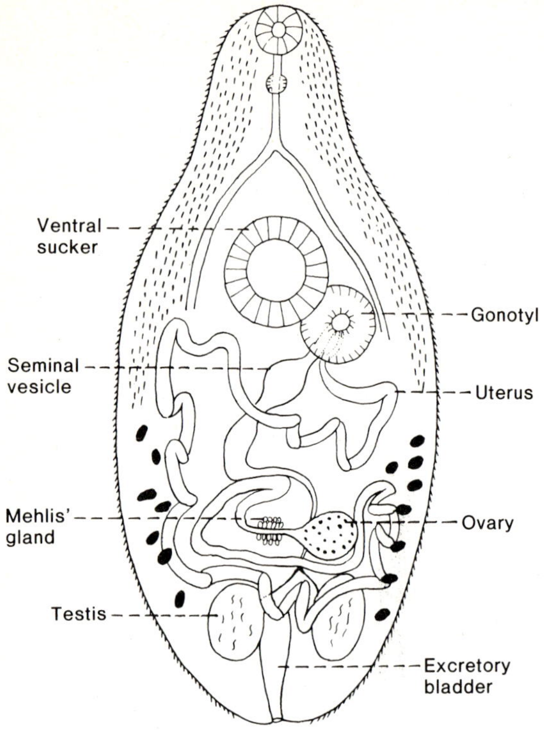

Fig. 2.51. *Heterophyes heterophyes.*

Family Heterophyidae

Members of this family are small to tiny flukes covered with scalelike spines, common in the intestine of birds and mammals, including humans, throughout the world. The ventral sucker is usually well developed and may or may not be enclosed by a large suckerlike genital atrium.

Heterophyes heterophyes[4]

This small, scaly fluke (see fig. 2.51) occurs frequently in humans in the Nile Delta and the Orient. The second intermediate host is a food fish, and infection is acquired when raw fish containing metacercariae are eaten. The flukes also occur in dogs, cats, foxes, and other piscivorous mammals, thus intensifying the infection in the endemic areas.

Description

Adult flukes measure up to 1.7 mm long by 0.4 mm wide. The entire body is covered with scales that have lateral projections and are more numerous anteriorly. The oral sucker is small, followed by a short prepharynx, tiny pharynx, and short esophagus that bifurcates into slender intestinal caeca, reaching to the posterior end of the first half of the body. A large genital sucker, or **gonotyl,** on the left margin of the acetabulum, bears numerous multidigitate spines. The excretory bladder is elongate and saclike, reaching almost to the ovary.

Small, oval testes, located side by side intercaecally, are in the posterior end of the body. The vasa efferentia unite slightly anterior to the ovary to form a voluminous U- or retort-shaped seminal vesicle that opens into the genital sucker through a muscular ejaculatory duct without benefit of a cirrus pouch or cirrus.

A small oval or round ovary lies medially a short distance anterior to the testes, with the Mehlis' glands and seminal receptacle close by. The uterus loops posteriorly on the right side of the body, traverses to the left, goes posteriorly between the testes, then runs anteriorly along the right side to the level of the ventral sucker and coils across the body to the left side, emptying through a metraterm into the genital atrium. About 14 coarse vitelline follicles lie extracaecally on each side in the posterior third of the body.

Life Cycle

Eggs hatch in the intestine of the brackish water snail *Pirenella conica* in Egypt and *Cerithidia cingulata* in the Orient. The intramolluscan stages include the mother sporocyst and two generations of rediae. Cercariae are of the lophocercous type (see fig. 2.26K) i.e., those with a long tail bearing a finlike structure and an oral sucker without armature such as a stylet. The second intermediate host is a fish, commonly a brackish water mullet, *Mugil* sp. Infection of the final hosts occurs when viable cysts are ingested with raw or incompletely cooked fish.

Key to Common Families of Digenea

Capital letters in parentheses refer to hosts: F–Fish, A–Amphibia, R–Reptiles, B–Birds, M–Mammals.

1 Mouth near midventral surface of body (fig. 2.52) .... (F) **Bucephalidae**
Mouth near anterior end of body 2

2(1) In blood vessels 3
Not in blood vessels 4

3(2) Monoecious (fig. 2.53) (F) **Sanguinicolidae;** (R) **Spirorchiidae**
Dioecious; male usually with gynecophoric canal or flaps to embrace female .... (B, M) **Schistosomatidae**

4(2) Encysted in tissues 5
In locations other than blood vessels or cysts in tissues 6

5(4) Body slender anteriorly, sometimes expanded posteriorly (fig. 2.54) .... (F) **Didymozoidae**
Body oval, fleshy, spiny (R, B, M) **Troglotrematidae**

6(4) Ventral sucker absent 7
Ventral sucker present 9

7(6) Testes median to caeca, which unite posteriorly to form ringlike intestine; may lack oral sucker (fig. 2.55) .... (B) **Cyclocoelidae**
Testes lateral to caeca 8

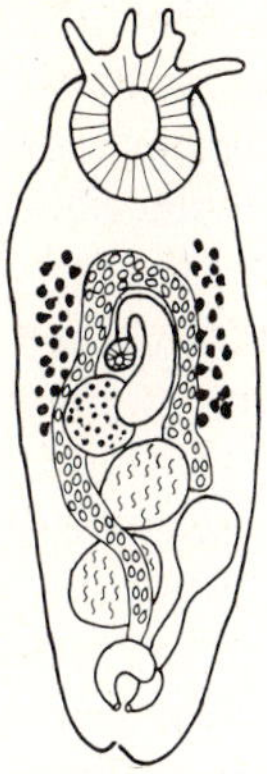

Fig. 2.52. *Bucephalus elegans.*

Fig. 2.53. *Sanguinicola occidentalis.*

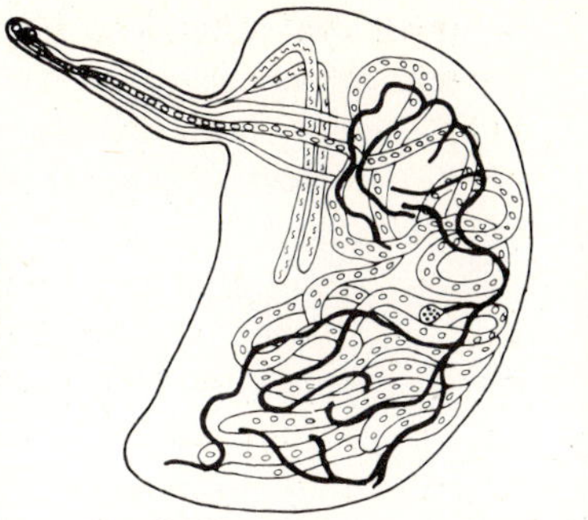

Fig. 2.54. *Didymocystis coatesi.*

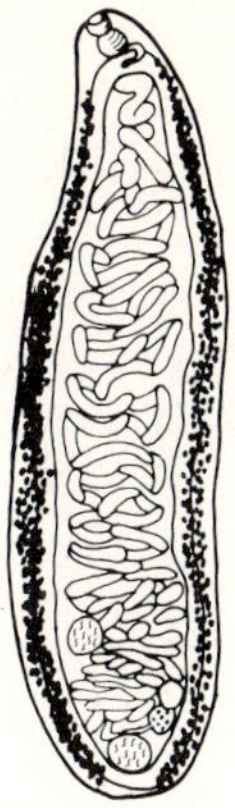

Fig. 2.55. *Cyclocoelium leidyi.*

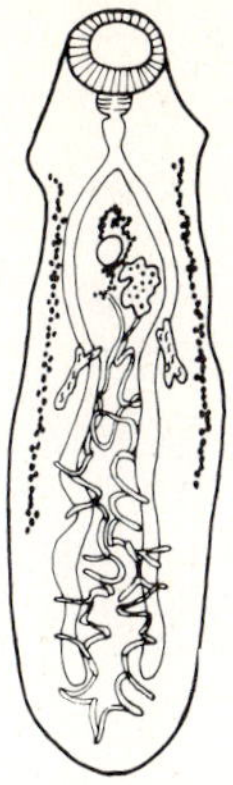

Fig. 2.56. *Eucotyle hassalli.*

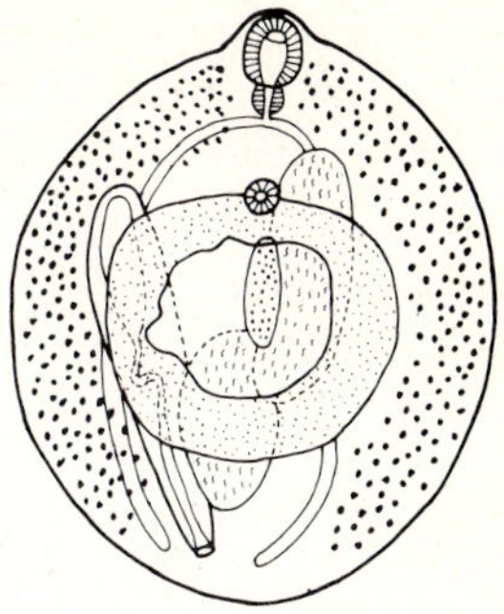

Fig. 2.57. *Cyathocotyle prussica.*

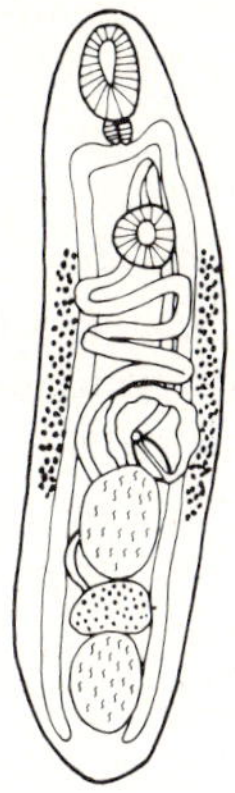

Fig. 2.58. *Brachylaema virginianus.*

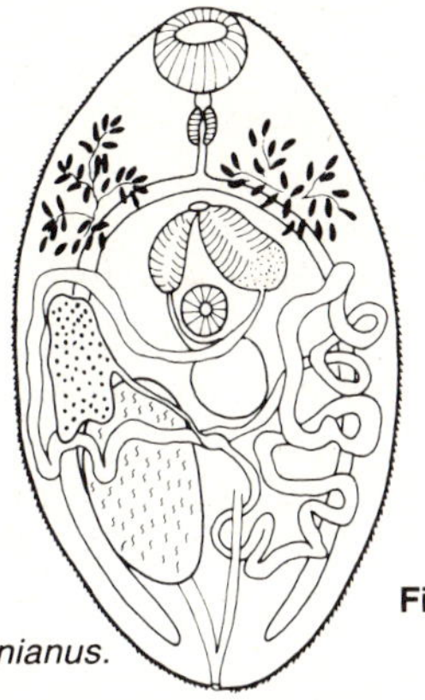

Fig. 2.59. *Monorchis monorchis.*

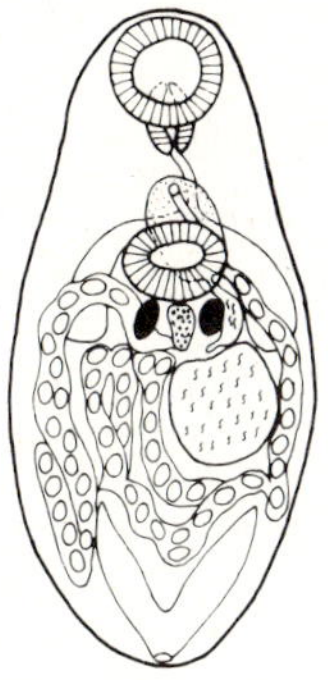

Fig. 2.60. *Haploporus benedeni.*

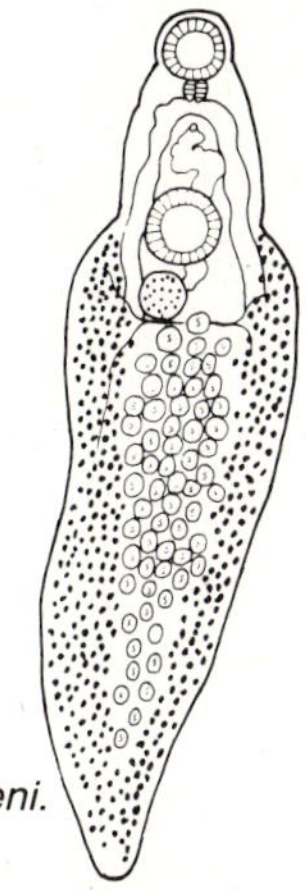

Fig. 2.61. *Orchipedium tracheicola.*

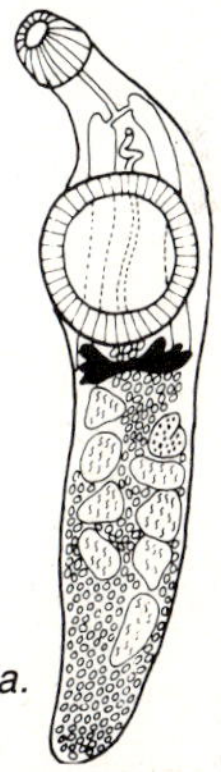

Fig. 2.62. *Gorgodera amplicava.*

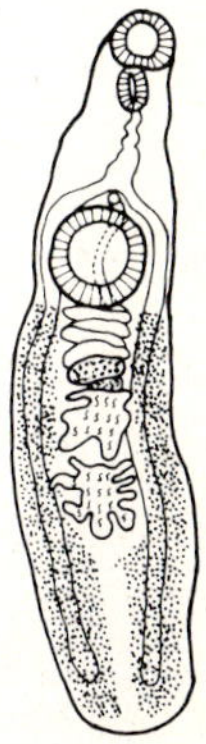

Fig. 2.63. *Cathaemasia nycticoracis.*

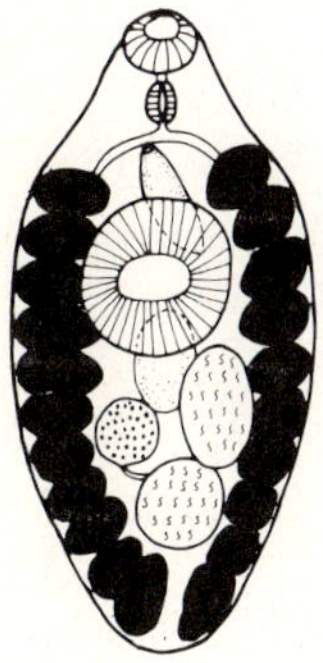

Fig. 2.64. *Psilostoma marilae.*

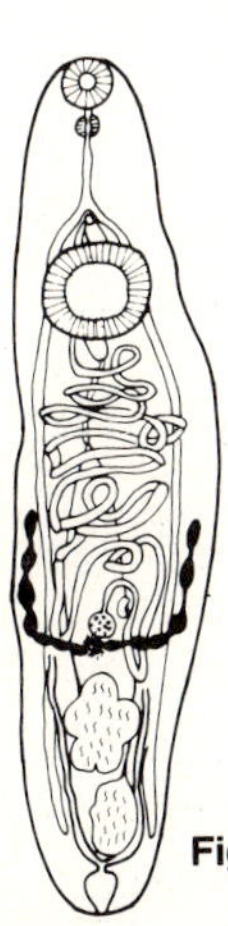

Fig. 2.65. *Philophthalmus lucipetus.*

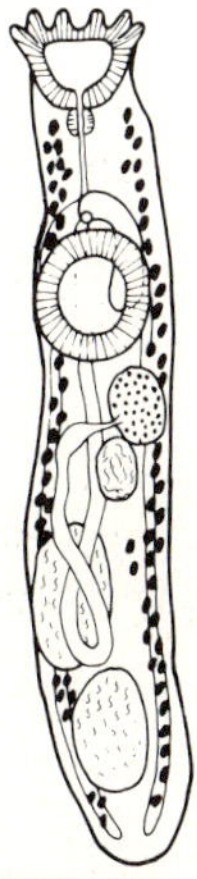

Fig. 2.66. *Bunodera leucopercae.*

Fig. 2.67. *Gorgoderina attenuata.*

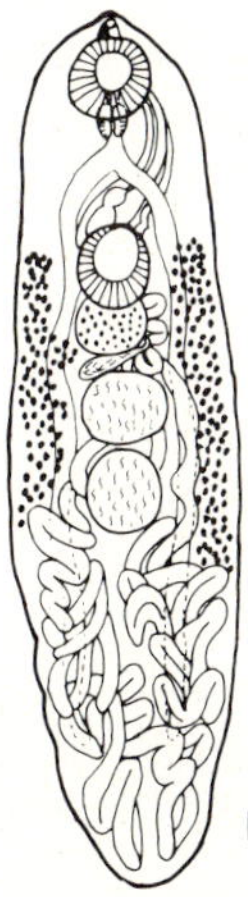

Fig. 2.68. *Cephalogonimus amphiumae.*

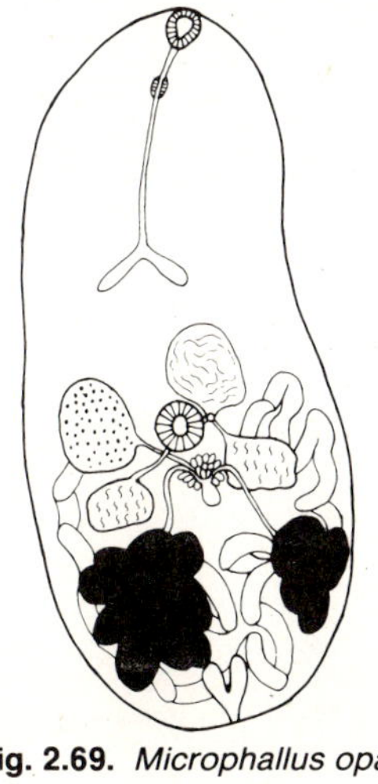

Fig. 2.69. *Microphallus opaca.*

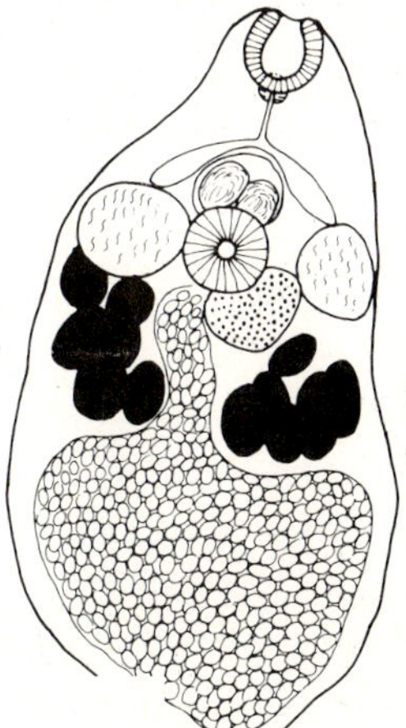

Fig. 2.70. *Lecithodendrium breckenridgei.*

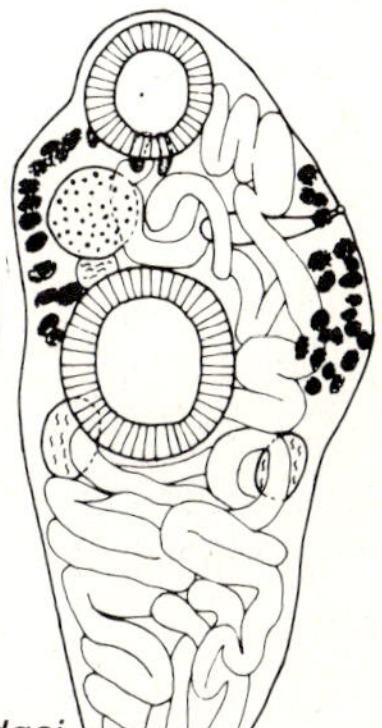

Fig. 2.71. *Laterotrema americana.*

8(7) Testes near midbody; cirrus pouch absent; ventral surface of body without rows of glands (fig. 2.56) .... (B) **Eucotylidae**
Testes near posterior extremity of body; cirrus pouch very long; ventral surface of body with rows of glands .... (B, M) **Notocotylidae**

9(6) Ventral sucker at posterior extremity of body; fleshy, somewhat cone-shaped flukes ....(F, A, R, B, M) **Paramphistomatidae**
Ventral sucker not at posterior extremity of body 10

10(9) Genital pore at posterior extremity of body; adhesive body posterior to ventral sucker 11
Genital pore not at posterior extremity of body; no adhesive body 13

11(10) Ovary between testes; body oval in outline, not divided into anterior and posterior parts; cirrus pouch large (fig. 2.57) .... (B) **Cyathocotylidae**
Ovary anterior to testes; cirrus pouch absent; body divided into anterior and posterior parts 12

12(11) Anterior part of body cup shaped, posterior part ovoid or cylindrical .... (R, B, M) **Strigeidae**
Anterior part of body flattened, posterior region cone shaped .... (R, B, M) **Diplostomatidae**

13(10) With gonotyl (genital sucker) near ventral sucker, latter often modified as a genital sinus; body scaly .... (B, M) **Heterophyidae**
Without gonotyl or modification of ventral sucker 14

14(13) Ovary behind testes* 15
Ovary not behind testes 16

15(14) Elongate, cylindrical body; gonads near posterior end of body; uterus pregonadal; vitellaria with few follicles at posterior end of body; suckers widely separated .... (A) **Halipegidae**
Elongated, flattened body; gonads in anterior half of body; uterus almost entirely postgonadal; vitellaria with many follicles near middle third of body; suckers close .... (A, R, B, M) **Dicrocoeliidae**

16(14) Ovary between testes; genital pore behind ventral sucker 17
Ovary anterior to testes; genital pore not behind ventral sucker 18

17(16) Uterus between anterior testis and fork of intestine; vitellaria lateral, extending back to anterior testis, not intercaecal posteriorly; gonads in posterior third of body (fig. 2.58) .... (B, M) **Brachylaemidae**
Uterus between testes and ventral sucker; vitellaria lateral, extending caudad from gonads, intercaecal posteriorly; gonads near middle of body ....(R, B, M) **Clinostomatidae**

18(16) One testis 19
Two or more testes 20

19(18) Pharynx small; vitellaria follicular but feebly developed (fig. 2.59) .... (F) **Monorchiidae**
Pharynx large; vitellaria compact, small (fig. 2.60) .... (F) **Haploporidae**

20(18) More than two testes 21
Two testes 22

21(20) Testes numerous; vitellaria follicular, extending in extracaecal spaces behind ventral sucker; uterus pregonadal (fig. 2.61) .... (B) **Orchipedidae**
Testes (actually two, divided into several irregular bodies) arranged in two longitudinal rows of four to six each; vitellaria compact, small, lateral; uterus descends to posterior end of body (*Gorgodera*) (fig. 2.62) ....(A) **Gorgoderidae**

22(20) Uterus with ascending limb only 23
Uterus with both ascending and descending limbs 30

23(22) Ovary and/or testes highly branched 24
Neither ovary nor testes branched 25

24(23) Ovary and testes profusely branched; large spiny flukes .... (M) **Fasciolidae**
Only testes branched; medium-size flukes with cuticle spined ventrally (fig. 2.63) .... (B) **Cathaemasiidae**

25(23) Head collar with single or double crown of stout spines around margin; suckers close .... (B, M) **Echinostomatidae**
Head collar absent 26

26(25) Prepharynx long; vitelline follicles small, abundant, lateral and extend to posterior extremity of body, intermingling medially; excretory bladder I-shaped ....(F) **Allocreadiidae**
Prepharynx short or absent; vitelline follicles fewer, do not reach posterior extremity of body 27

27(26) Cirrus pouch absent; seminal vesicle coiled; excretory bladder Y-shaped with long stem and short arms which do not extend anteriorly to ovary .... (R, B, M) **Opisthorchiidae**
Cirrus pouch present 28

*See couplet 34 where some genera of Lecithodendriidae have the testes anterior to ovary. In this case, they are in front of the ventral sucker, whereas in couplet 14 both testes and ovary are behind the ventral sucker.

28(27) Body wall thick, with wrinkled tegument; excretory bladder with stem reaching almost to posterior testis and arms to anterior end of body; cirrus pouch small and anterior to ventral sucker .... (F) **Azygiidae**
Body wall thin, not wrinkled; cirrus pouch large, extending far behind anterior margin of ventral sucker 29

29(28) Vitellaria numerous large follicles, filling lateral space behind ventral sucker; small flukes (fig. 2.64) .... (B) **Psilostomatidae**
Vitellaria tubular with six to seven follicles near middle of body; muscular medium-size flukes (fig. 2.65) .... (B) **Philophthalmidae**

30(22) Oral sucker with six anterior muscular processes; ventral sucker as large or larger than oral, slightly anterior to middle of body; vitellaria lateral, extend from pharynx to posterior end of body (fig. 2.66) .... (F) **Bunoderidae**
Oral sucker without such muscular processes 31

31(30) Vitellaria two small compact masses near middle of body; cirrus pouch absent (fig. 2.67) (*Gorgoderina*) .... (F, A, R) **Gorgoderidae**
Vitellaria follicular; cirrus pouch present 32

32(31) Genital pore at level of oral sucker; long cirrus pouch extends back to ventral sucker (fig. 2.68) .... (A, R, B) **Cephalogonimidae**
Genital pore much posterior to oral sucker 33

33(32) Caeca very short or barely reaching beyond ventral sucker 34
Caeca long, extending well beyond ventral sucker or to posterior extremity of body 35

34(33) Minute, usually pyriform flukes; testes postacetabular and postovarian; seminal receptacle absent (fig. 2.69) .... (B, M) **Microphallidae**
Small flukes, neither compact nor extended; testes periacetabular and preovarian; seminal receptacle present (fig. 2.70) .... (A, R, B, M) **Lecithodendriidae**

35(33) Testes and ovary near ventral sucker; cirrus pouch horizontal, genital pore on left lateral margin of body between suckers; uterus surrounds ventral sucker; caeca extend to posterior extremity of body (fig. 2.71) .... (B) **Stomylotrematidae**
Testes and ovary all distinctly postacetabular; cirrus pouch vertical, genital pore median; descending and ascending limbs of uterus pass between testes .... (A, R, B, M) **Plagiorchiidae**

Notes and Sketches

Review Questions

1. Some furcocercous cercariae are pharyngeate and others are apharyngeate. What is the medical implication of knowing the difference?
2. What is a tribocytic organ? What is its function?
3. Review the shapes of the eggs of the three most important species of *Schistosoma.* How do these eggs differ from those of other Trematodes?
4. What is the number and arrangement of testes in the species of *Schistosoma?*
5. How many circles of collar spines are found on *Echinostoma revolutum?*
6. Compare the locations of the acetabula of *E. revolutum* and *Fasciola hepatica.*
7. Which member of Fasciolidae has an unbranched intestine?
8. Name two families of digeneans in which the ovary is posterior to the testes.
9. Which type of cercaria is found in the Paramphistomidae?
10. What important fluke uses an ant as a second intermediate host?
11. Which digenetic trematode might a person find inside a hen's egg?
12. Where does a crab fit into the life cycle of *Paragonimus westermani?*
13. Does *P. westermani* have a cirrus and cirrus pouch?
14. What type of cercaria does *P. westermani* have?
15. In *Clonorchis,* which is the larger—the ovary or the seminal receptacle?
16. What is a gonotyl? What is its possible function?
17. How do humans become infected with *Schistosoma? Paragonimus? Clonorchis? Fasciola?*

References

Monogenea

Bychowsky, B. E. 1957. Monogenetic Trematodes. Their Systematics and Phylogeny. *Akad. Nauk SSSR,* Moscow. (1961. Amer. Inst. Biol. Sci., Washington, D.C., 627 pp.)

Chubb, J. C. 1977. Seasonal Occurrence of Helminths in Freshwater Fishes. Part. I. Monogenea. In *Advances in Parasitology,* ed. B. Dawes. Academic Press, New York, vol. 15, pp. 133–99.

Kearn, G. C. 1971. The Physiology and Behaviour of the Monogenean Skin Parasite *Entobdella soleae* in Relation to its Host (*Solea solea*). In *Ecology and Physiology of Parasites,* ed. A. M. Fallis. University of Toronto Press, Toronto, pp. 161–87.

Llewellyn, J. 1963 and 1968. Larvae and Larval Development of Monogeneans. In *Advances in Parasitology,* ed. B. Dawes. Academic Press, New York, vol. 1, pp. 287–326; vol. 6, pp. 373–83.

———. 1970. Monogenea. *J. Parasitol.* 56 (4, sect. 2, pt. 3); 493–504.

Schell, S. C. 1982. Trematoda. In *Synopsis and Classification of Living Organisms,* ed. S. P. Parker. McGraw-Hill, New York, vol. 1, pp. 740–807.

Yamaguti, S. 1963. Systema Helminthum. *Monogenea and Aspidocotylea,* vol. 4. Interscience Publishers Inc., New York, 699 pp.

Aspidogastrea

Rohde, K. 1972. The Aspidogastrea, Especially *Multicotyle purvisi* Dawes, 1941. In *Advances in Parsitology,* ed. B. Dawes. Academic Press, New York, vol. 10, pp. 77–151.

Huehner, M. K., and Etges, F. J. 1977. The life cycle and development of *Aspidogaster conchicola* in the snails, *Vivipotus malleatus* and *Goniobasis liviscens. J. Parasitol.* 63:669–74.

Digenea

LaRue, G. R. 1957. The classification of digenetic Trematoda: A review and a new system. *Exptl. Parasitol.* 6: 306–49.

Pearson, J. C. 1972. A Phylogeny of Life-cycle Patterns of the Digenea. In *Advances in Parasitology,* ed. B. Dawes. Academic Press, New York, vol. 10, pp. 153–89.

Schell, S. C. 1985. *Trematodes of North America North of Mexico.* Univ. Press of Idaho, Moscow, 263 pp.

Skrjabin, K. I., et al. 1947–1962. Keys to the Trematodes of Animals and Man. *Akad. Nauk SSSR,* Moscow. (1964. Univ. Illinois Press, Urbana, 351 pp.).

———. 1960. Trematodes of Animals and Man. *Akad. Nauk SSSR,* Moscow, vol. 17, 444 pp.; vol. 18, 532 pp. (Israel Program for Scientific Translations, 1964; 1965).

Yamaguti, S. 1958. Systema Helminthum. *Digenetic Trematodes,* vol. 1, Pts. 1 and 2. Interscience Publishers, Inc., New York, 1575 pp.

———. 1975. *Synoptical Review of Life Histories of Digenetic Trematodes of Vertebrates.* Keigaku Publishing Co., Ltd., Tokyo, 1100 pp.

Strigeata

Campbell, R. A. 1973. Studies on the biology of the life cycle of *Cotylurus flabelliformis* (Trematoda: Strigeidae). *Trans. Amer. Micr. Soc.* 92:629–40.

Hoffman, G. L. 1960. Synopsis of Strigeoidea (Trematoda) of fishes and their life cycles. *Bur. Sport Fish. Wildlife, Fishery Bull.* 175, 60:437–69.

Jordan, P., and Webbe, G. 1970. *Human Schistosomiasis.* Charles C. Thomas, Publishers, Springfield, Ill., 212 pp.

Thomas, J. D. 1973. Schistosomiasis and the Control of Molluscan Hosts of Human Schistosomes with Particular Reference to Possible Self-regulatory Mechanisms. In *Advances in Parasitology,* ed. B. Dawes, Academic Press, New York, vol. 11, pp. 307–94.

Ulmer, M. J. 1957. Notes on the development of *Cotylurus flabelliformis* tetracotyles in the second intermediate host (Trematoda: Strigeidae). *Trans. Amer. Micr. Soc.* 76:321–27.

Warren, K. S. 1973. The pathology of schistosome infections. *Helminthol. Abstr., Series A,* 42:591–633.

Echinostomata

Dawes, B., and Hughes, D. L. 1964 and 1970. Fascioliasis: the Invasive Stages of *Fasciola hepatica* in Mammalian Hosts. In *Advances in Parasitology,* ed. B. Dawes. Academic Press, New York, vol. 2, pp. 97–168; vol. 8, pp. 259–74.

Kendall, S. B. 1965 and 1970. Relationships between the Species of *Fasciola* and Their Molluscan Hosts. In *Advances in Parasitology,* ed. B. Dawes. Academic Press, New York, vol. 3, pp. 59–98; vol. 8, pp. 251–58.

Pantelouris, E. M. 1965. *The Common Liver Fluke* Fasciola hepatica. Pergamon Press, New York, 259 pp.

Plagiorchiata

Ameel, D. J. 1934. *Paragonimus,* its life history and distribution in North America and its taxonomy (Trematoda: Troglotrematidae). *Amer. J. Hyg.* 19:279–317.

Kingston, N. 1965. On the life cycle of *Brachylecithum orfi* Kingston and Freeman, 1959 (Trematoda: Dicrocoeliidae), from the liver of the ruffed grouse, *Bonasa umbellus* L. Infections in the vertebrate and molluscan hosts. *Canad. J. Zool.* 43:745–64.

———. 1965. On the morphology and life cycle of *Tanaisia zarudnyi* (Skrjabin, 1924) Byrd and Denton, 1950, from the ruffed grouse, *Bonasa umbellus* L. *Canad. J. Zool.* 43:953–69.

Millemann, R. E., and Knapp, S. E. 1970. Biology of *Nanophyetus salmincola* and "Salmon Poisoning" Disease. In *Advances in Parasitology,* ed. B. Dawes. Academic Press, New York, vol. 8, pp. 1–41.

Philip, C. B. 1955. There's always something new under the "parasitological" sun (the unique story of helminth-borne salmon poisoning disease). *J. Parasitol.* 41:125–48.

Yokogawa, M. 1965 and 1969. *Paragonimus* and Paragonimiasis. In *Advances in Parasitology,* ed. B. Dawes. Academic Press, New York, vol. 3, pp. 99–158; vol. 7, pp. 375–87.

Opisthorchiata

Komiya, Y. 1966. *Clonorchis* and Clonorchiasis. In *Advances in Parasitology,* ed. B. Dawes. Academic Press, New York, vol. 4, pp. 53–106.

Macy, R. W.; Cook, W. A.; and DeMott, W. R. 1960. Studies on the life cycle of *Halipegus occidualis* Stafford, 1905 (Trematoda: Hemiuridae). *Northwest Sci.* 34:1–17.

Sun, T.; Chou, S. T.; and Gibson, J. B. 1968. Route of the entry of *Clonorchis sinensis* to the mammalian liver. *Exptl. Parasitol.* 22:346–51.

Sources of Study Material for Turbellaria*

Bdelloura sp.: W
Syndesmis sp.: T

Sources of Study Material for Monogenea

Gyrodactylus sp.: C
Urocleidus sp.: C
Miscellaneous species: C

Sources of Study Material for Aspidogastrea

Aspidogaster sp.: C

Sources of Study Material for Digenea

Clonorchis sinensis, adult: W,C,T,Tr
Clonorchis sinensis, eggs: W,C,T,Tr
Clonorchis sinensis, metacercaria: C
Clonorchis sinensis, in liver; section: W,T
Clonorchis sinensis, sections: W,C,T,Tr

Dicrocoelium dendriticum, adult: C,T

Echinostoma revolutum, adult: W,C,T
Echinostoma revolutum, cercaria: W,T
Echinostoma revolutum, eggs: C
Echinostoma revolutum, redia: W,T

Fasciola gigantica, adult: T

Fasciola hepatica, adult: W,C,T,Tr
Fasciola hepatica, cercaria: W,T,Tr
Fasciola hepatica, eggs: W,C,T,Tr
Fasciola hepatica, metacercaria: W,C,Tr
Fasciola hepatica, miracidia: W,C,T,Tr
Fasciola hepatica, redia: W,C,T,Tr
Fasciola hepatica, sections: W,C,T,Tr
Fasciola hepatica, section of infected liver: W,T
Fasciola hepatica, section of snail: W,T
Fasciolopsis buski, adult: C,T,Tr
Fasciolopsis buski, cercaria: C
Fasciolopsis buski, eggs: C,T
Fasciolopsis buski, miracidia: C
Fasciolopsis buski, redia: C

Heterophyes heterophyes: C,T

Notocotylus sp.: C

Paragonimus westermani, adult: T
Paragonimus westermani, cercaria: W,C
Paragonimus westermani, eggs: W,C
Paragonimus westermani, metacercaria: W,C

*The addresses of the sources are listed on p. 45.

Paragonimus westermani, miracidia: W,C
Paragonimus westermani, redia: W,C
Paragonimus westermani, section of lung: T

Prosthogonimus macrorchis, adult: C,T

Schistosoma haematobium, adult: W
Schistosoma haematobium, eggs: W,T,Tr

Schistosoma japonicum, adult: W,C,T
Schistosoma japonicum, cercaria: C,T
Schistosoma japonicum, eggs: W,C,T
Schistosoma japonicum, miracidium: T
Schistosoma japonicum, section of intestine with eggs: W
Schistosoma japonicum, section of liver lesion: W
Schistosoma japonicum, section of prostate with eggs: T
Schistosoma mansoni, adults: W,C,T,Tr
Schistosoma mansoni, cercariae: W,C,T,Tr
Schistosoma mansoni, eggs: W,C,T,Tr
Schistosoma mansoni, miracidia: W,C,T,Tr
Schistosoma mansoni, sporocysts: W,C
Schistosoma mansoni, section of intestine: C,T
Schistosoma mansoni, section of liver lesion: W,C,T,Tr

Miscellaneous Larval Stages

Cercaria: C,W
Metacercaria: C
Redia: C
Sporocysts: C

Chapter 3 Class Cestoidea (Cestodaria and Eucestoda)

The class Cestoidea includes the **tapeworms.** As adults, they occur in the stomach, intestine, and bile ducts of mammals, the intestine of other classes of vertebrates, and in the coelom of some fishes, freshwater turtles, and freshwater oligochaetes.

In a few primitive taxa, the body of the adult consists of a single unit. These are **monozoic** forms since there is no chain of **proglottids.** Taxa, in which the **strobila** or body comprise a chain of proglottids, are **polyzoic.**

Typically an adult polyzoic tapeworm consists of a series of distinct body parts. Beginning anteriorly, they are: (1) the **scolex** for attachment and locomotion; (2) the **neck,** which, if present, is a zone of tissue proliferation; and (3) a **strobila** consisting of a chain of proglottids that originate by budding from the neck or scolex if a neck is not present. Proglottids immediately behind the neck are indistinct, being undeveloped, and the internal organs are undifferentiated. As the proglottids are pushed back by the formation of new ones at the neck, organogenesis occurs, so that a single mature strobila shows a complete developmental series, including immature, mature, and gravid proglottids.

With few exceptions tapeworms are monoecious and **protandrous,** i.e., the male reproductive organs develop prior to the female organs. The gonads are usually situated in the medullary zone, which is between the sheets of muscle fibers of the proglottids. Although there are *many exceptions,* the testes tend to be dorsal and anterior in the proglottid; the ovary and Mehlis' gland ventral and posterior. The vitellaria may be follicular and widely dispersed, or compact and located in lateral bands or behind the bilobed ovary. Genital pores are located laterally in some groups and ventrally in others.

The larval and **metacestode** stages of development between the **oncosphere** and the sexually mature adult occur in a variety of invertebrate and vertebrate hosts. Eggs are passed in the feces of the definitive host. Those laid singly are unembryonated and require a period in water for development; those contained in the uterus of voided gravid proglottids are already embryonated and infective. A very few species do not require an intermediate host, being able to complete the life cycle in a single host.

The class Cestoidea is divided into two subclasses: Cestodaria with two orders, all from fish or turtles, and Cestoda (or Eucestoda) with thirteen orders (Schmidt, 1986). Commonly encountered Cestoda include members of the orders Caryophyllidea in freshwater fish and aquatic oligochaetes, Pseudophyllidea in all classes of vertebrates, Proteocephalidea mostly in fish, but occasionally in amphibians and reptiles, and Cyclophyllidea, the largest order, primarily in birds and mammals but also in amphibians and reptiles. Only the Pseudophyllidea and Cyclophyllidea contain species occurring in humans and domesticated animals. Most of the remaining orders occur in elasmobranchs and contain relatively few species (table 3.1).

Subclass Cestodaria

Adult cestodarians resemble trematodes, except for the lack of digestive organs. There is no scolex or chain of proglottids, and there is only one set of reproductive organs, the cirrus, vagina, and uterine pore opening independently. These monozoic tapeworms differ from the Eucestoda in that their oncospheres have ten rather than six hooks and, accordingly, are known as **decacanths,** or **lycophores.** They are parasitic in the coelom

Table 3.1 Abbreviated Classification of Class Cestoidea, after Schmidt (1986).

Subclass	Order	Family
CESTODARIA	Amphilinidea	Amphilinidae
	Gyrocotylidea	Gyrocotylidae
EUCESTODA	Caryophyllidea	Caryophyllaeidae
		Balanotaeniidae
		Lytocestidae
		Capingentidae
	Trypanorhyncha	Tentaculariidae
		Gymnorhynchidae
		Otobothriidae
		Dasyrhynchidae
		several other families
	Pseudophyllidea	Diphyllobothriidae
		Amphicotylidae
		Triaenophoridae
		Bothriocephalidae
		several other families
	Lecanicephalidea	Lecanicephalidae
		Disculicipitidae
		others
	Tetraphyllidea	Phyllobothriidae
		Oncobothriidae
		others
	Proteocephalidea	Proteocephalidae
		Monticellidae
	Cyclophyllidea	Taeniidae
		Dilepididae
		Hymenolepididae
		Anoplocephalidae
		Davaineidae
		Mesocestoididae
		Dioecocestidae
		others
	Spathibothriidea	Spathibothriidae
		Cyathocephalidae
		Bothrimonidae
	Aporidea	Nematoparataeniidae
	Diphyllidea	Echinobothriidae
		Ditrachybothriidae
	Litobothridea	Litobothridae
	Nippotaeniidea	Nippotaeniidae
	Dioecotaeniidea	Dioecotaeniidae

and intestine of lower fishes and coelom of freshwater turtles in Australia. Some examples with their hosts include: *Gyrocotyle urna* from ratfish (*Hydrolagus colliei*) and *Amphilina bipunctata* from sturgeon. (See fig. 3.1.)

Subclass Eucestoda

These are the polyzoic tapeworms that have a scolex followed by a succession of internal and/or external proglottids (except in the order Caryophyllidea), each containing a set of male and female reproductive organs. An **oncosphere** with three pairs of hooks (**hexacanth**) hatches from the egg. A wide variety of both invertebrates and vertebrates act as intermediate hosts.

Order Caryophyllidea

The scolex is not distinct from the body and may show faint grooves or depressions (**loculi**), but true suckers or bothria are lacking. These are small monozoic forms whose genital pores open on the ventral surface of the body. Operculate eggs are unembryonated when laid. The life cycle may be direct or involve an intermediate host; the sexual stages are often considered progenetic procercoid or pleurocercoid larvae of an extinct ancestor. Parasitic in the intestine of Catostomidae, Cyprinidae and other fishes. In *Archigetes* (fig. 3.2) the sexual stages occur in oligochaete annelids as well as in fish.

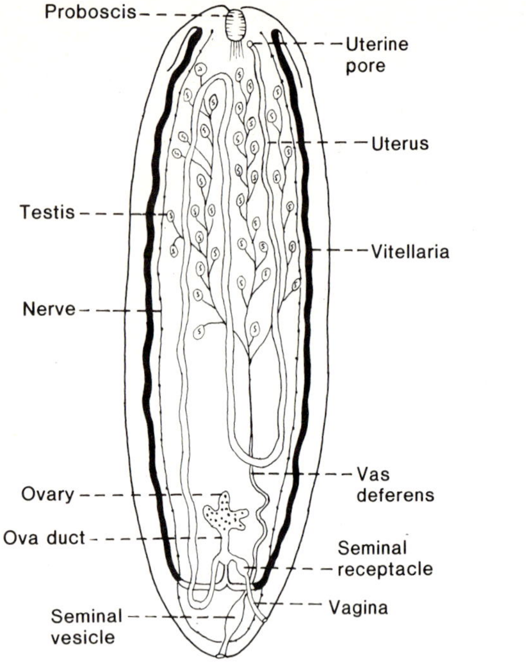

Fig. 3.1. *Amphilina foliacea.*

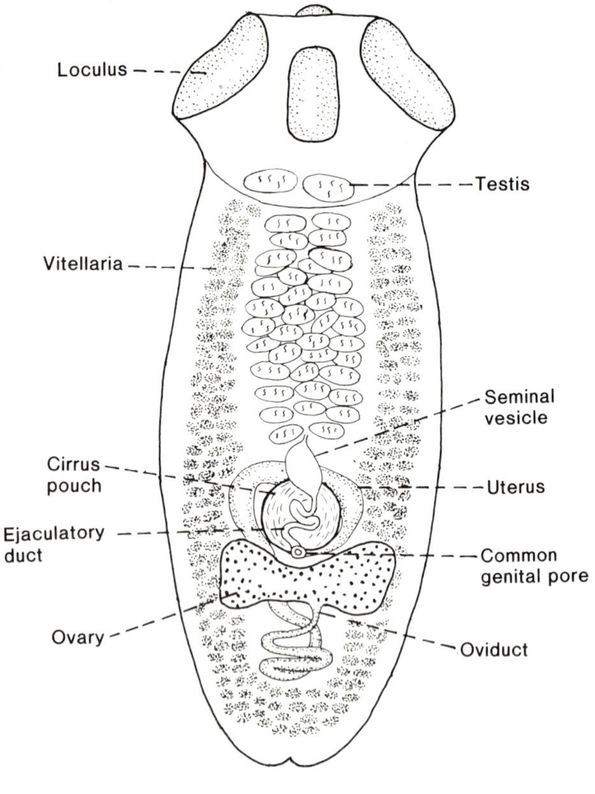

Fig. 3.2. *Archigetes iowensis.*

Family Caryophyllaeidae

Glaridacris catostomi

In North America, *G. catostomi* is a common parasite of the white sucker (*Catostomus commersoni*) and related fish. Other caryophyllaeid species that might be available to the student are similar to the following description.

Description

Adults measure up to 25 mm long, 1 mm in width, and are somewhat cylindrical in cross section. The scolex is short, with three elongate depressions on each side. Testes, 150 to 160, are arranged longitudinally and median. The ovary is dumbbell shaped, with the wings short, compact, and rounded. The preovarian vitellaria surround the testes; posterior to the ovary the vitellaria are numerous, forming a V-shaped cluster. The uterus, which opens slightly posterior to the cirrus pouch, is both post- and preovarian. Eggs operculate, with an abopercular boss, 37–48 × 20–31 μm.

Order Pseudophyllidea

Scolex typically with a dorsal and ventral slit or shallow depression (**bothrium**); majority of proglottids in similar stage of development, shedding eggs from a uterine pore; genital pores usually on the midventral surface although variations abound. Testes and vitelline follicles are numerous, small bodies scattered throughout the proglottid except for the region of the ovary and uterus. The life cycle normally involves three developmental stages and two intermediate hosts. Eggs are undeveloped when laid, usually operculate, and give rise to ciliated oncospheres known as **coracidia.** The next two stages are a **procercoid** in a copepod crustacean and a **plerocercoid** in all classes of vertebrates except birds. Adults are mainly in fish, but also in amphibians, reptiles, birds, and mammals.

Family Diphyllobothriidae

Diphyllobothrium latum

The fish tapeworm, *D. latum* (fig. 3.3), occurs in humans, bears, cats, dogs, and possibly other fish-eating carnivores. In North America, it is known from the Great Lakes region (Michigan, Minnesota, and parts of Canada). There are other North American records, but the evidence is not conclusive. *D. latum* is a monster among tapeworms, commonly reaching a length of more than 9 m, with a width of nearly 2.5 cm, and consisting of more than 4,000 proglottids. Östling (1961) reported recovering 16 scolices and 330 m of worm from an adult man in Finland. If the total strobilar length were associated with the 16 scolices, each worm averaged nearly 21 m.

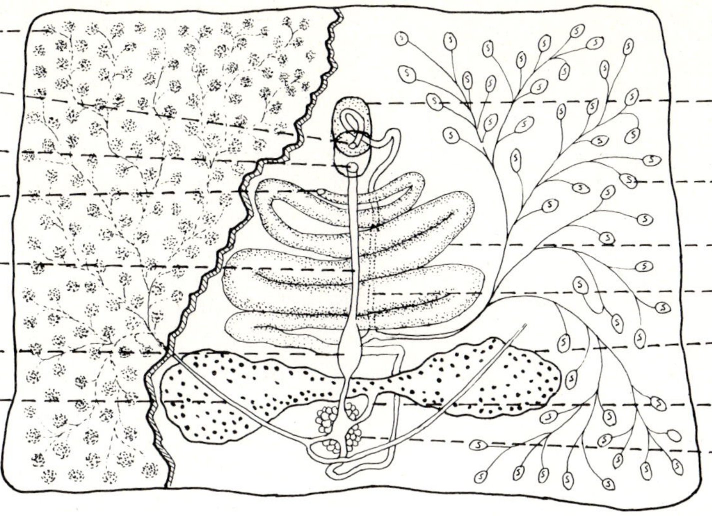

Fig. 3.3. *Diphyllobothrium latum.*

Description

The almond-shaped scolex has a pair of suctorial grooves, one dorsal and the other ventral, known as **bothria.** A short neck region separates the scolex from the proglottids. Anterior proglottids are broader than long, but posteriorly they are approximately square. The female system consists of (1) a bilobed ovary, situated posteriorly in the proglottid, (2) a vagina extending in a straight line forward from the ootype and opening in the genital atrium posterior to the cirrus pouch on the midventral surface of the proglottid, (3) follicular vitelline glands situated in layers, dorsal and ventral to the testes, and (4) the rosette-shaped uterus winding forward, in the form of lateral coils, from the ootype and opening through the uterine pore a short distance behind, and often slightly lateral to, the genital pore. The male system consists of (1) many testes and their ducts disposed in a single layer, largely obscured by the vitelline glands, and (2) the prominent cirrus pouch, with its cirrus, which opens in the genital pore anterior to the vagina.

Life Cycle

Eggs (fig. 3.4) develop in water and coracidia are formed, each of which consists of a ciliated embryophore enclosing an oncosphere. When the free-swimming coracidium is eaten by copepods (species of *Diaptomus* serve best), the ciliated embryophore is shed, and the oncosphere migrates to the hemocoel, where it develops into a procercoid. The procercoid is liberated when the copepod is eaten by certain freshwater fish and it, in turn, bores through the intestine to reach the body wall or viscera of the second intermediate host, where it develops into a plerocercoid. In North America, pike and walleyes (*Esox* and *Stizostedion*) are the most important hosts, but elsewhere other fish are involved. The plerocercoid is liberated when the fish host is eaten, and it develops into the adult in the intestine of the final host. This is the three-host cycle, determined experimentally, and undoubtedly occurs in nature. However, in the case of the larger fish, which are not apt to feed directly on copepods, it is more likely that in nature a smaller forage fish serves as a means of transfer between the copepod and the large carnivorous pike and walleye. Plerocercoids are able to reestablish themselves in fish after fish until they are ingested by a suitable final host in which maturity can be attained.

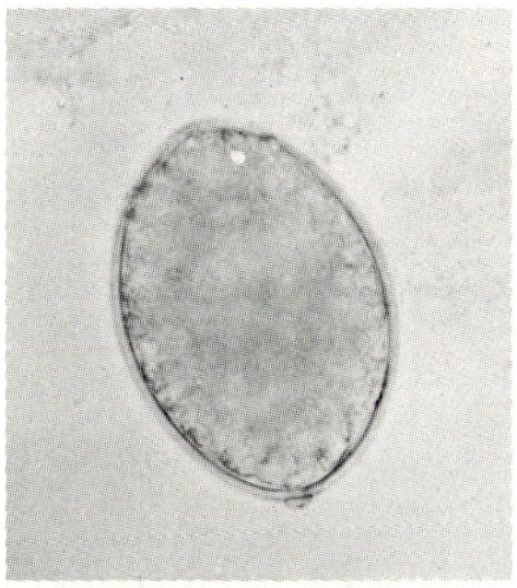

Fig. 3.4. *Diphyllobothrium latum,* egg. Operculate; broadly ovoidal, evenly rounded at both ends, abopercular end sometimes with a nipplelike protuberance; contents granular, mulberrylike; shell thin and light straw colored; size 62–68 × 40–88 μm.

Sparganosis

Sparganosis is an infectious disease of humans and other vertebrates caused by migrating plerocercoids or spargana larvae of tapeworms belonging to the genus *Diphyllobothrium* (=*Spirometra*). While there is some confusion about the species of *Diphyllobothrium,* in the Orient it is usually regarded as *D. erinacei,* and in the United States, where sparganosis is distributed over a broad area of the Atlantic and Gulf States, it is due to *D. mansonoides.*

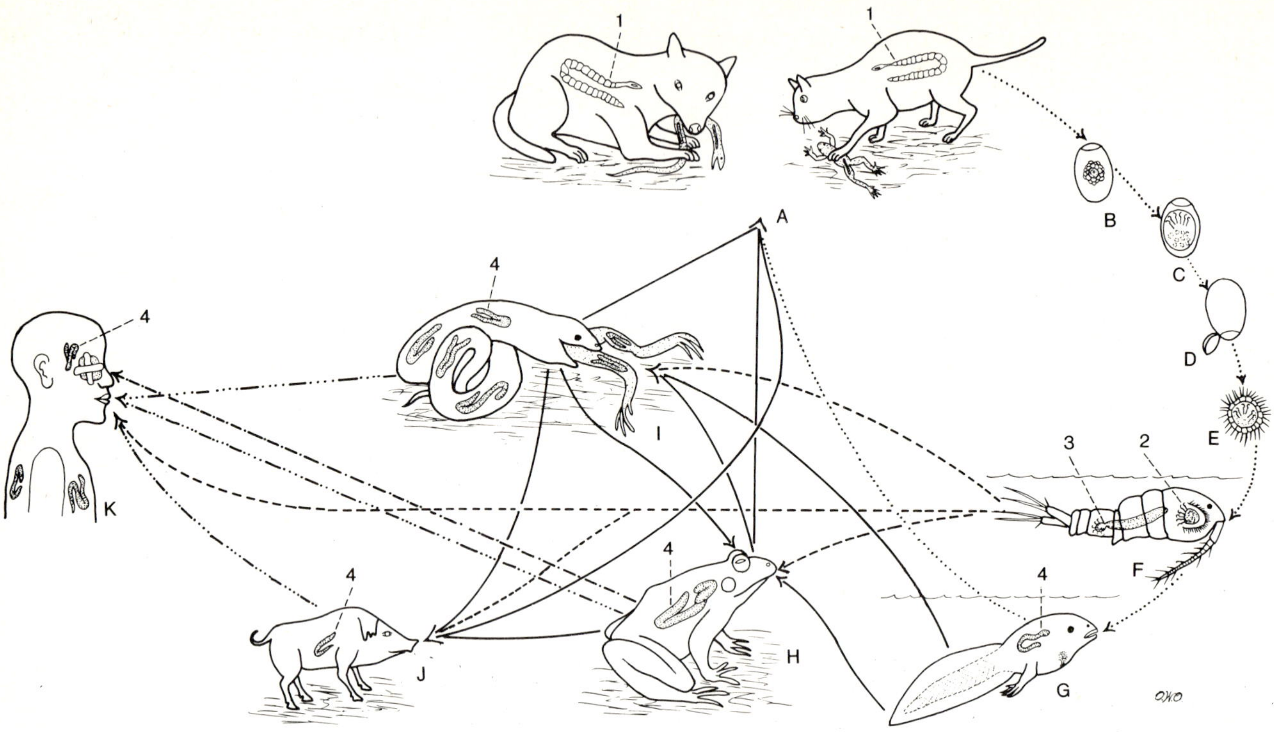

Fig. 3.5. Life cycle and pattern of circulation of *Diphyllobothrium*(*-Spirometra*). (A) Dog and cat definitive hosts; (B) undeveloped eggs passed in feces of dogs and cats; (C) embryonated egg in water; (D) empty shell of hatched egg; (E) ciliated coracidium containing the six-hooked hexacanth; (F) *Cyclops* first intermediate host; (G) tadpole second intermediary; (H–K) paratenic, or collector hosts; (H) frog; (I) snake; (J) feral swine; (K) human. (1) Adult cestode in small intestine of definitive host; (2) hexacanth being freed from ciliated covering in intestine; (3)procercoid in hemocoel; (4) plerocercoid or sparganum.
. . . Simplest type of life cycles consisting of definitive host, egg, larvae, and first intermediate host. ________________, Paratenic phase of life cycle that occurs when animals, other than definitive hosts, harboring plerocercoids, prey upon each other, thereby collecting plerocercoids (H,I,J). . . . , Paratenic phase of the life cycle that arises from hosts swallowing infected *Cyclops* while drinking water or pursuing food in the water (F, H, I, K); the procercoids develop into plerocercoids. -.-.-.-.-.-, Zoonotic phase of the life cycle that occurs when plerocercoids transfer from raw frog meat used as poultices on the human body. - . . . - . . . . - . . . . - . . . , Zoonotic phase of life cycle that results when meat of frogs, snakes, feral swine containing plerocercoids (H, I, J) is eaten by humans (K).

When the procercoid-infected copepods (species of *Cyclops* serve best), are eaten by certain amphibians, reptiles, and mammals, the plerocercoid stage develops in their tissues. Infection of the final host, cats and related mammals, comes from eating a second intermediate host containing plerocercoids. As a result of their resistance to both procercoids and plerocercoids, fish are unsuitable intermediate hosts.

When the plerocercoids of *Diphyllobothrium,* (for which the name *Sparganum* was used before their adult stage was known) get into an unsuitable vertebrate host (frogs, reptiles, and some mammals), they migrate through the intestinal wall and reestablish themselves in the subcutaneous tissues and muscles, where they grow until a host is reached in which maturity can be attained in the intestine. This is of much more than academic interest, because humans can serve as intermediate hosts for some species and definitive hosts for others. (See fig. 3.5.)

Sparganosis as a zoonosis results from: (1) swallowing procercoid-infected *Cyclops* in drinking water; (2) eating amphibians, reptiles, and mammals containing plerocercoids; and (3) applying plerocercoid-infected flesh of frogs, snakes, and possibly mammals to wounds or inflamed eyes as a poultice.

While the use of meat-poultices is a common practice in the Orient, where the spargana of *D. erinacei* are involved, they can be ruled out for America. Until recently it was presumed that zoonotic sparganosis in the United States came from drinking water containing the procercoid-infected copepods. But the finding of spargana in swine in Florida, and the discovery of natural infections of *D. mansonoides* in two amphibian, eight reptilian, and three mammalian species in Louisiana by Corkum (1966), who showed experimentally that swine are susceptible to spargana, suggest that human infection may occasionally result from consuming wild animals or pork containing the living spargana.

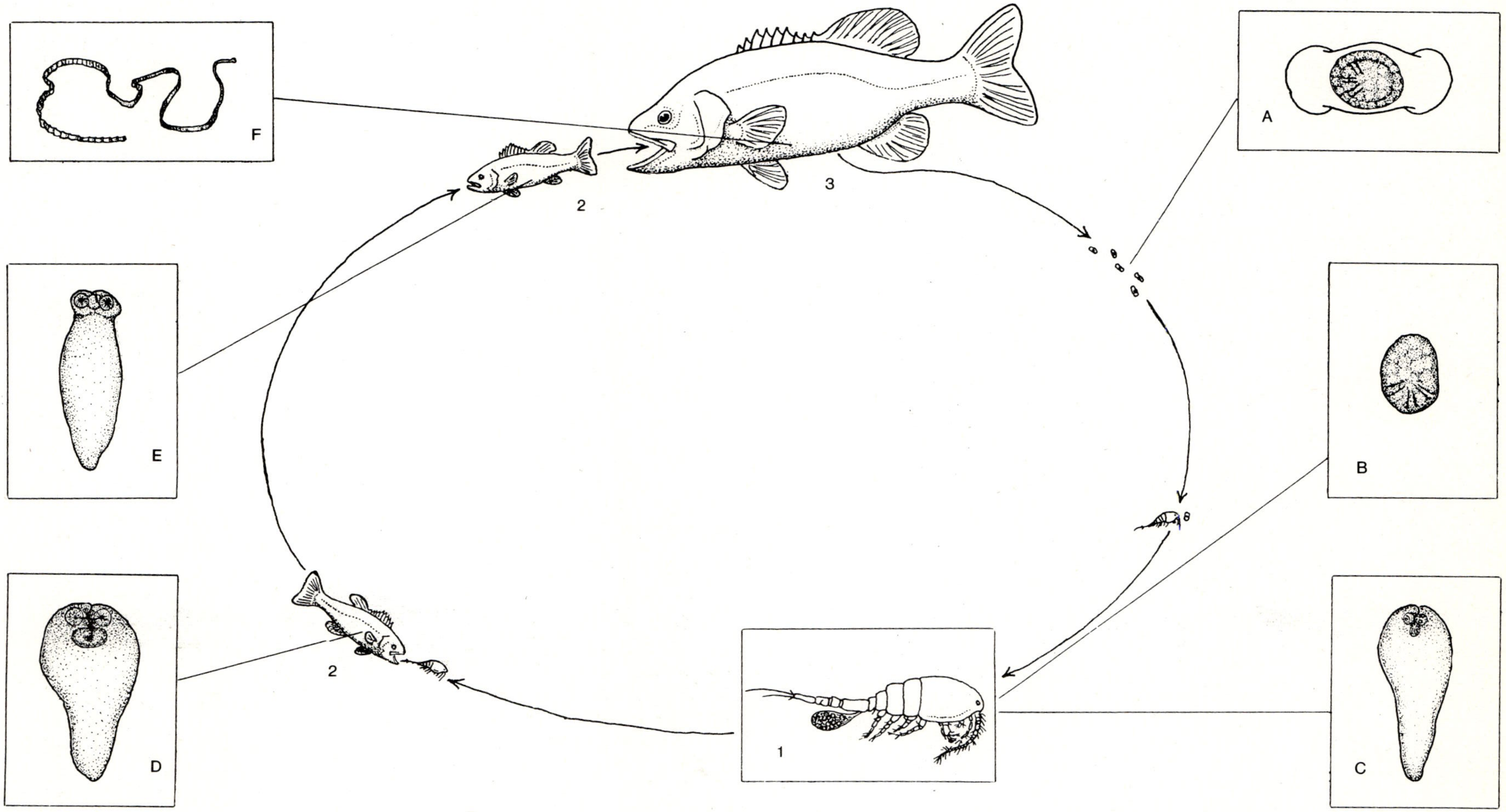

Fig. 3.6. Life cycle of *Proteocephalus ambloplitis.* (A) Egg containing oncosphere; (B) oncosphere in gut of copepod; (C) plerocercoid 1 in hemocoel of copepod; (D) plerocercoid 11 in parenteral cavity of fish; (E) later stage of same; (F) adult in enteral cavity of fish. Numbers 1, 2, and 3 indicate first, second, and third or final hosts, respectively. (After Hunter and Hunter [1929] from Meyer [1954].)

Order Proteocephalidea

Scolex with four muscular cuplike suckers, and sometimes a fifth terminal apical organ; genital pores lateral; vitellaria in lateral bands; gravid uterus with numerous lateral outgrowths and one or more median ventral openings, made by breaks or clefts in the body wall, through which eggs escape. The life cycle (fig. 3.6) involves a copepod, in which the oncospheres develop into infective plerocercoids. When the infected copepod is eaten by the suitable final host, fish, amphibian, or reptile, the adult worm develops. These common final hosts may also serve as paratenic hosts when they fall prey to larger fish, amphibians, or reptiles. A second intermediate host is apparently not obligatory.

Family Proteocephalidae

Proteocephalus ambloplitis

This species, known as the bass tapeworm, is widespread throughout much of North America. The extensive range is largely due to stocking infected fish, primarily black bass, in uninfected waters. In addition to bass harboring adult worms in the intestine, plerocercoids occur in the coelom and viscera, especially the gonads and liver. The parenteral plerocercoids in bass are much more damaging than are adults in the intestine. Plerocercoids cause extensive adhesions of the viscera and may injure the gonads, especially in the females, inhibiting spawning. (See fig. 3.7.)

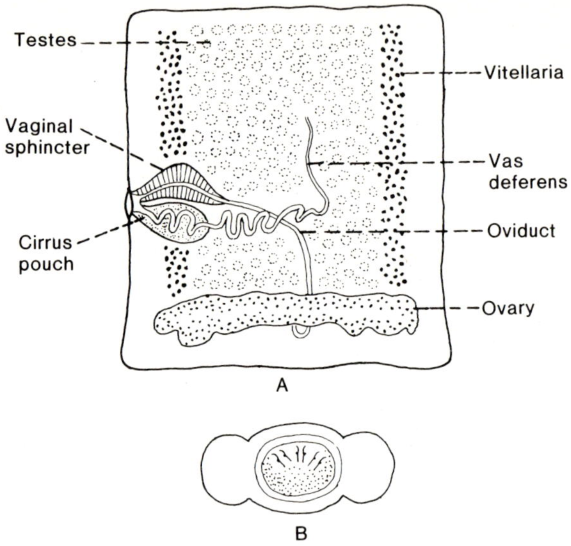

Fig. 3.7. *Proteocephalus ambloplitis.* (A) Mature proglottid; (B) egg.

Description

The scolex is globose, with four simple suckers and a vestigial apical organ. Vitellaria are arranged in lateral bands along the length of the proglottid. Numerous testes are bounded laterally by the vitellaria and posteriorly by the ovary. The ovary is in the form of a transverse band in the rear of the proglottid. Genital pore is lateral, alternating irregularly; the vagina opens in the genital atrium anterior to the cirrus pouch. Eggs are ellipsoidal, with three membranes, 43 × 36 μm.

Notes and Sketches

Order Cyclophyllidea

Scolex with four deeply cupped muscular suckers and usually a rostellum, which may be armed or unarmed with hooks; proglottids in all stages of development, gravid ones only near posterior end of strobila; uterus varies in form, but does not open through a special pore; genital pores usually lateral; vitellarium a single post-ovarian mass. The life cycle as a rule involves only one intermediate host, usually an arthropod or vertebrate. Adults are parasitic in amphibians (though rarely), reptiles, birds, and mammals.

Family Taeniidae

The family includes the majority of the more important adult tapeworms of humans, canids, and felids. This, combined with the occurrence of the metacestode stages also in mammals, including humans, makes them unrivaled in medical and economic importance. Eggs (fig. 3.8) of the various species of Taeniidae with their thick, striated shells and hexacanths are characteristic of the family but cannot be used to determine species.

The life cycles of the Taeniidae include a single intermediate host that is always a mammal. A variety of larval forms (metacestodes) are found in this family (fig. 3.9).

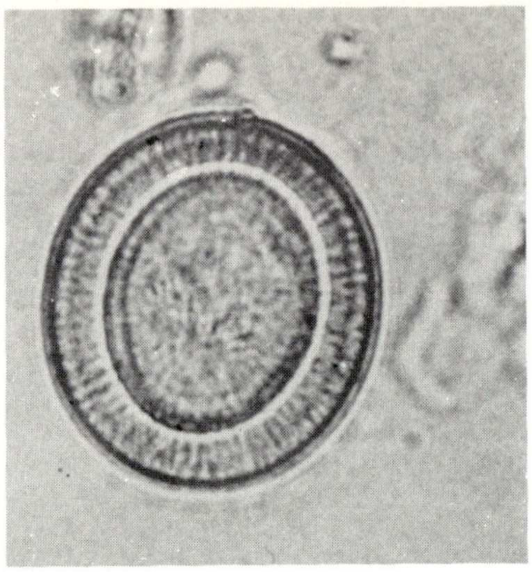

Fig. 3.8. *Taenia* spp., egg. Round or slightly ovoidal; embryophore commonly characterized by many radial striations; size 30–40 × 20–30 μm.

Taenia pisiformis

This is one of the most common tapeworms of dogs and related wild canids; it seldom occurs in cats.

Description

Mature worms up to 100 cm long and consisting of about 400 proglottids. Rostellum large and powerful, and armed with a double crown of 34 to 48 hooks of two sizes. The longer hooks are 200 to 269 μm (mean 240 μm), and the smaller ones are 114 to 172 μm (mean 140 μm) long. Vitelline gland often triangular with broad apex extending into interovarian field; does not contact ovary. The vagina bends broadly around distal end of cirrus sac before looping tortuously and entering genital atrium. The gravid uterus has 11 to 15 lateral branches on each side. (See figs. 3.10 and 3.11.)

Life Cycle

Gravid proglottids become detached and pass out with the feces of the host. Rupture of the proglottids releases embryonated eggs. When ingested by hares and rabbits, eggs hatch in the intestine. Oncospheres penetrate the intestinal wall, and make their way, via the blood and lymph channels, to the liver, where they grow rapidly, and in about a month the cysticerci break out of the liver into the coelom. After remaining free for a short time, they attach to the mesenteries and are enclosed in an adventitious cyst produced by the host. Dogs and other Canidae become infected when they swallow living cysticerci (fig. 3.12).

Humans serve as final hosts for the beef tapeworm, *Taeniarhynchus saginatus,* and the pork tapeworm, *Taenia solium.* In addition to occurring in different intermediate hosts, specific identification can be made on the scolices, the number of lobes of the ovary in mature proglottids, and the number of lateral branches on the uterus in gravid proglottids (fig. 3.13). In *Taeniarhynchus saginatus,* the scolex is unarmed, the ovary in the mature proglottid has no accessory, or third lobe, and the gravid uterus has 15 to 20 lateral branches on each side. In *T. solium,* the scolex is armed with 22 to 32 hooks of two sizes, the ovary in the mature proglottid has a small accessory third lobe, and the gravid uterus has seven to 13 lateral branches on each side. The eggs are indistinguishable. The cysticerci of *T. solium,* but not those of *T. saginatus,* may also occur in humans. When accidentally ingested, the eggs hatch and the oncospheres bore through the intestinal wall and enter the general circulation via the hepatic portal vein. They leave the blood vessels and develop into cysticerci in the muscles. Here they do little harm; but if they enter the nervous system and sense organs, serious damage may result.

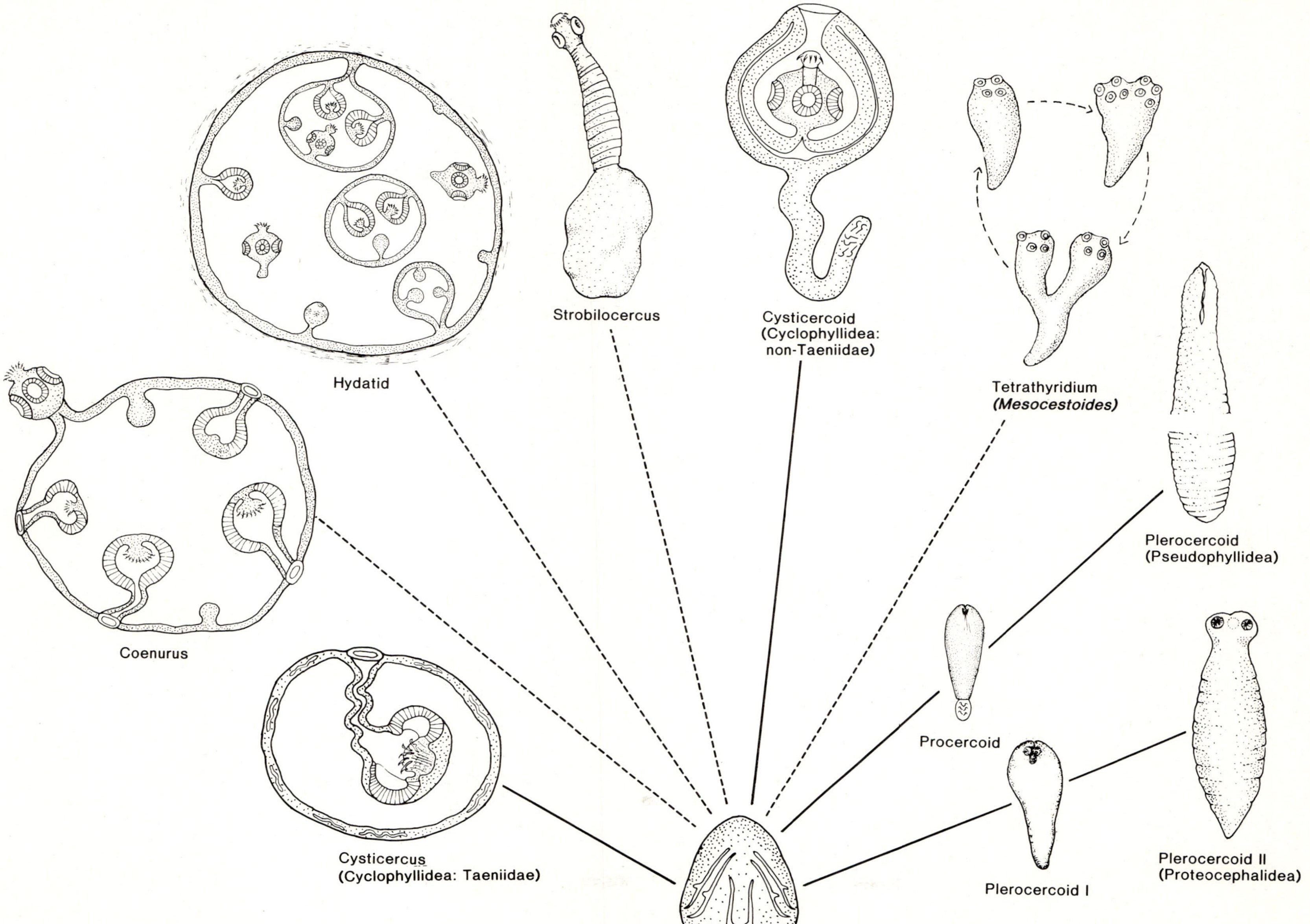

Fig. 3.9. Infective tapeworm metacestodes. The basic types are indicated by solid lines. The oncosphere is a larva, not a metacestode.

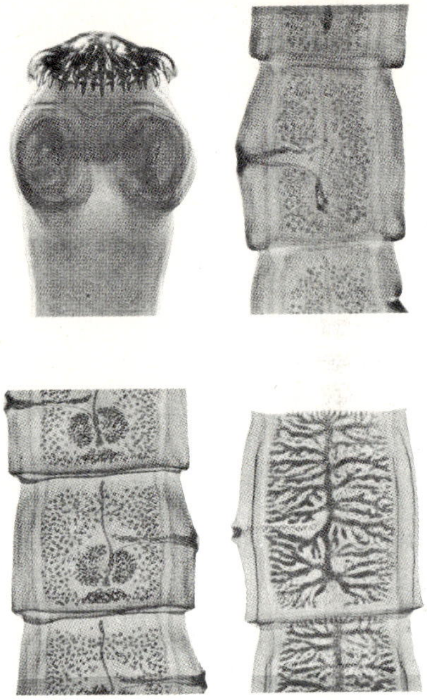

Fig. 3.10. *Taenia pisiformis.* (A) Scolex and neck region; (B) Immature region with male organs developed, female organs beginning to appear; (C) mature region showing both sets of sex organs; (D) gravid region with each proglottid more enlarged and with branched egg-filled uterus.

Taenia taeniaeformis

This species is of interest because of its metacestode, a **strobilocercus.** (See fig. 3.9.)

Description

Mature worms up to 60 cm long. The rostellum is short and armed with a double crown of 26 to 52 hooks of two sizes. The larger hooks are 294–429 μm (mean 380 μm) and the smaller ones 215–287 μm (mean 245 μm) long. The neck is very short. Vitelline gland is slightly elongate in transverse plane; does not contact ovary. The vagina is straight, without any loops or bends. The gravid uterus has five to 11 lateral branches on each side.

Life Cycle

Adults occur in the intestine of cats and related felids. Gravid proglottids passed in the feces contain infective eggs that when swallowed by the rodent intermediary, chiefly rats, mice, and muskrats, hatch in the small intestine. Freed oncospheres penetrate the intestinal wall, enter the hepatic portal system, and go to the liver,

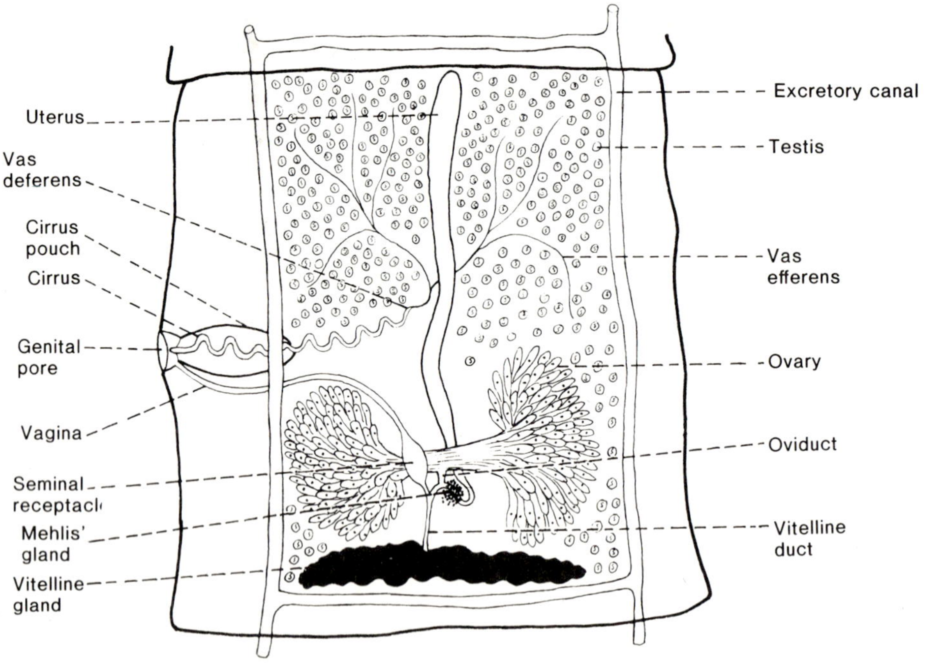

Fig. 3.11. *Taenia pisiformis,* mature proglottid.

where they develop into strobilocerci. In the liver, strobilocerci evaginate prematurely, resulting in a bladderworm with a single scolex followed by a series of immature proglottids and terminated by a small, vesicular bladder. When a liver with living strobilocerci is ingested by a cat or other suitable host, the bladder and some of the pseudostrobila are digested off, and new proglottids are formed from the neck.

Echinococcus granulosus

This is the smallest Taeniidae. (See fig. 3.14.) Adults consist of a scolex and neck followed by usually four but occasionally three or five proglottids. The small intestine of the final host, usually domestic and wild Canidae, may be "furred" with thousands of worms. The danger to humans from an infection is not with the adult

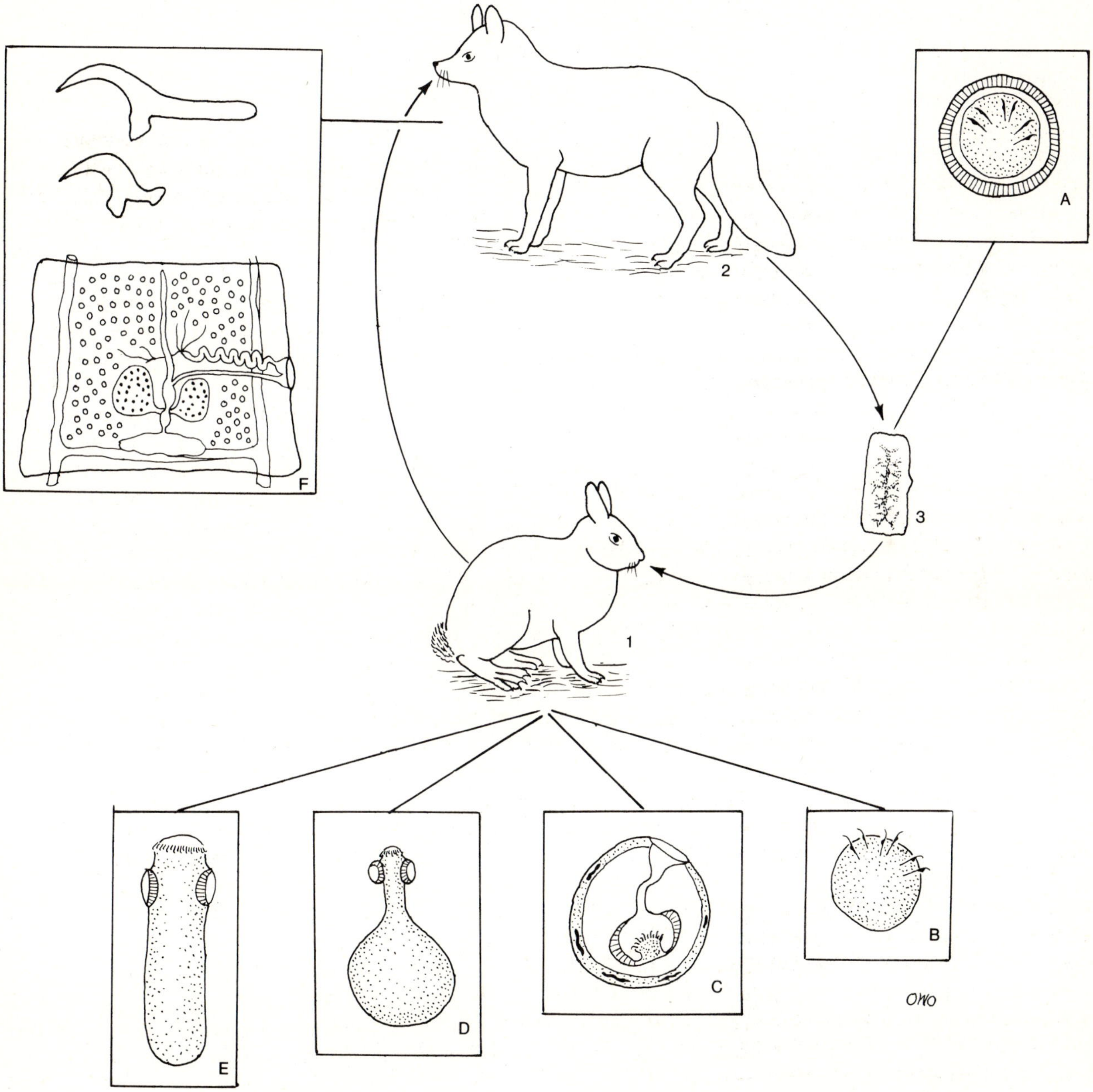

Fig. 3.12. Life cycle of *Taenia pisiformis,* typical of taenioid cycles except for hosts and kinds of metacestodes. (A) Embryonated egg; (B) oncosphere; (C) invaginated cysticercus; (D) evaginated cysticercus; (E) young strobila; (F) rostellar hooks and mature proglottid. (1) Rabbit intermediate host; (2) canine definitive host; (3) gravid proglottid voided with feces ruptures, scattering embryonated eggs.

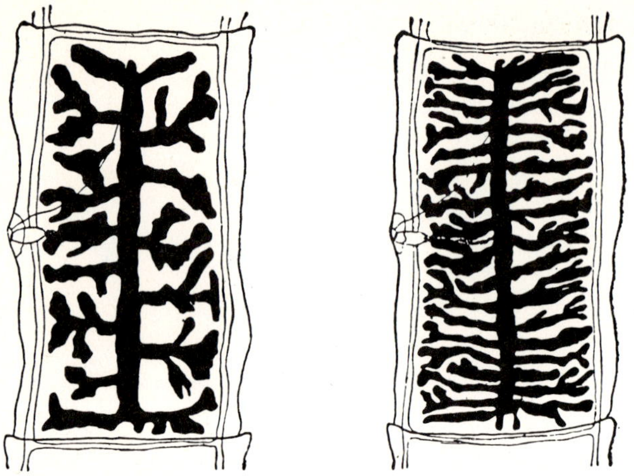

Fig. 3.13. Gravid proglottids of human taenias, showing lateral arms of central uterine stem. *Taenia solium*, left; *T. saginata*, right.

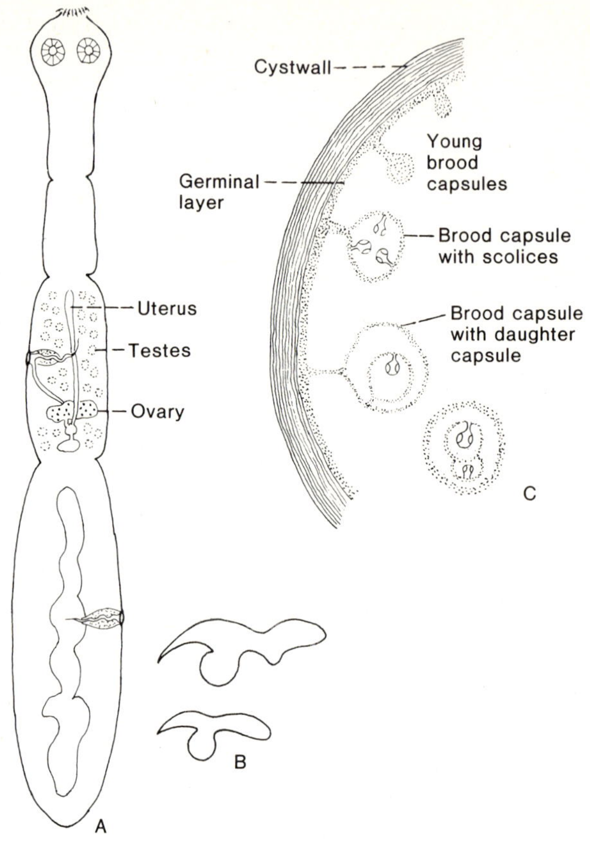

Fig. 3.14. *Echinococcus granulosus*. (A) Adult; (B) rostellar hooks; (C) section of hydatid cyst.

of this tapeworm, but with the larval **hydatid cyst,** which usually occurs in the liver. This is one of the most important human tapeworms throughout the world, and certainly the most dangerous. A few human cases of **unilocular hydatid disease** are diagnosed each year from the United States and Canada.

Description

Worms measure about 5 mm. There may be two immature proglottids, a mature and a gravid proglottid; or there may be only one immature proglottid and a gravid proglottid, and two proglottids in different stages of maturity. The rostellum is armed with a double crown of 30 to 40 hooks of two sizes. The larger hooks are 22 to 30 μm long and the smaller ones are 18 to 22 μm. There are about 40 to 60 testes to the proglottid.

Life Cycle

Adult worms occur in the intestine of dogs and related carnivores. The hydatid metacestode (see fig. 3.8) occurs in humans, sheep, deer, moose, and other large herbivorous mammals. Eggs passed by dogs hatch when ingested by a suitable intermediate host. As in the other Taeniidae, oncospheres enter the hepatic portal vein, whereupon they are carried in the bloodstream until becoming lodged in some capillary filter. The first filter is the liver, where the greatest number of hydatids occurs; the next filter is the lungs, where a smaller number is found, while still fewer reach more distant loci.

Fibrous tissue develops around the hydatid and grades off into normal tissue cells, which may already be undergoing pressure atrophy, due to the steady increase in the size of the cyst. The mother hydatid produces numerous scolices and single-layered brood capsules by budding of the germinal layer lining the inside. These are attached at first by a stalk, but many break loose and are free in the enclosed hydatid fluid. Brood capsules in turn produce scolices in a manner similar to that which takes place in the mother hydatid. Thus many thousands of scolices are formed (fig. 3.14).

Echinococcus multilocularis, which in the past has often been confused with *E. granulosus,* has a **multilocular** or **alveolar** type of hydatid cyst and utilizes a wider range of final hosts (dogs, foxes, and cats) than does the other species. While ungulates generally serve as intermediate hosts for *E. granulosus,* field mice and related rodents serve for *E. multilocularis.* The latter species, widely distributed in Europe and Eurasia, is known in North America from Canada and Alaska, and several north central states of the contiguous United States, where it is maintained sylvatically by foxes and wild mice (see Leiby et al., 1970).

Notes and Sketches

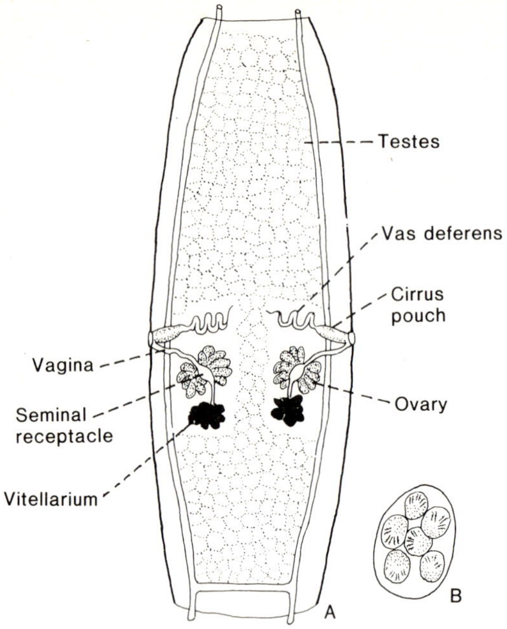

Fig. 3.15. *Dipylidium caninum.* (A) Mature proglottid; (B) egg packet containing eggs.

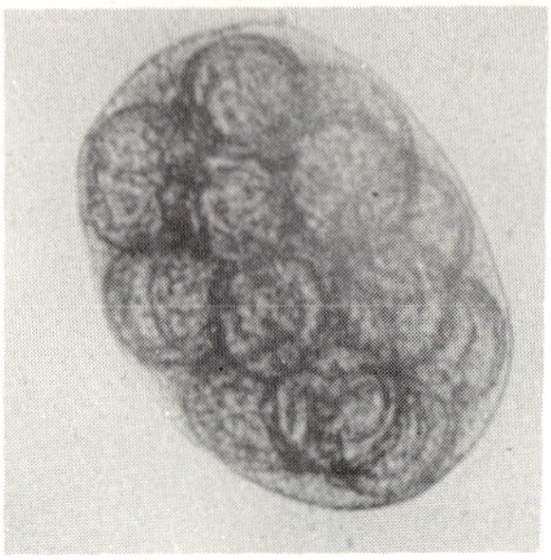

Fig. 3.16. *Dipylidium caninum,* eggs. In gravid proglottids eggs, measuring 35–60 μm, are bound together in egg-capsules, each containing 15–25 eggs.

Family Dilepididae

Dipylidium caninum

The best-known member of this family is the widely distributed *D. caninum,* the double-pored tapeworm of dogs, cats, and occasionally humans, especially children (fig. 3.15). Children probably become infected when their mouths are licked by a dog just after having "nipped" a flea. Infections in dogs can be recognized macroscopically by the occurrence on the feces of gravid proglottids, resembling cucumber seeds, which when freshly voided are active. Diagnosis is usually made on the evacuated proglottids, since fresh eggs are not ordinarily found in the feces of infected animals.

Description

Adults attain a length of 50 cm and a width of 3 mm. The scolex has four suckers and a retractable rostellum, armed with four to seven rows of rose thornlike hooks. There are two complete sets of reproductive organs, each opening in a common genital pore on the lateral margins of the proglottid. Numerous testes fill the space between the excretory canals. Bilobate ovaries and compact vitellaria, posterior to the ovaries, are in separate clusters. The uterus breaks up into egg-capsules, each containing 15 to 20 eggs (fig. 3.16), which remain intact after the proglottids disintegrate.

Life Cycle

Gravid proglottids voided with the feces, or leaving the host spontaneously, disseminate the egg-capsules. Intermediate hosts are the dog flea, *Ctenocephalides canis,* cat flea, *C. felis,* and the human flea, *Pulex irritans,* as well as the dog biting louse, *Trichodectes canis.* When the eggs are eaten by flea larvae, they hatch in the intestine, and oncospheres bore through into the hemocoel. Adult fleas have elongated mouthparts, adapted for piercing the skin and sucking blood, so they are incapable of ingesting eggs. Oncospheres develop little in larval fleas; considerable growth occurs in the pupal stage, and development of the **cysticercoid** is completed in adult fleas when they begin taking blood meals. In biting lice, oncospheres develop quickly to the infective cysticercoids. (See fig. 3.8.) The final host acquires the infection by swallowing infected adult fleas or biting lice. Prepatency is about a month.

Family Hymenolepididae

Vampirolepis nana

This species, commonly known as the dwarf tapeworm, is widely distributed and is common in rats and mice. It also occurs in humans, especially children. Authors are not in agreement on the identity of the species of worms found in humans and those found in rodents. Anatomically the rodent form, sometimes called *V. nana* var. *fraterna* or *V. fraterna,* is identical to the form from humans. Their life cycles are also identical. The rodent form can infect humans and vice versa, but development is best in the species of host from which the parent worms came. Similar differences occur in specimens obtained from wild rats and mice. The strain obtained from the former is equally infective for rats and

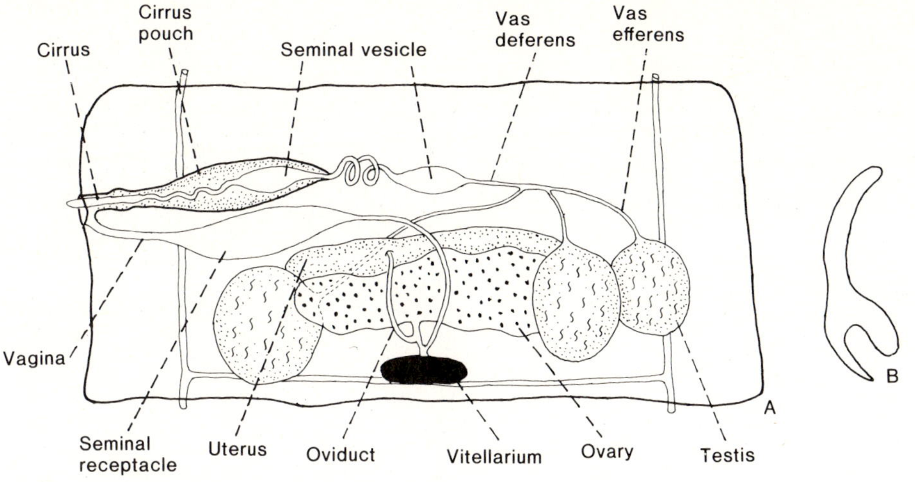

Fig. 3.17. *Vampirolepis nana.* (A) Mature proglottid; (B) rostellar hook.

mice, while the mouse strain is distinctly more infective for mice: it develops faster and to a larger size and produces a greater incidence and heavier worm burden in mice than in rats. It seems reasonable, therefore, to assume that the forms from mice, rats, and humans are the same species, *V. nana,* which has developed different physiological strains.

Prior to 1954 this species was placed in the genus *Hymenolepis.* But unlike that genus, *V. nana* has an armed rostellum, which justifies its placement in a different genus.

Description

Adults measure up to 10 cm in length and less than one mm in width. The globular scolex has a retractile rostellum with a crown of 20 to 30 hooks. The strobila consists of about 200 proglottids which are considerably broader than long. The three testes are arranged in a transverse line and separated by the ovary so that one testis is poral and two aporal. The uterus develops as a sac that practically fills the proglottids between the excretory canals. (See fig. 3.17.)

Life Cycle

The dwarf tapeworm, *V. nana,* is peculiar among cestodes in that no intermediate host is required (fig. 3.18), and eggs are infective as they are passed out with the feces. When eggs (fig. 3.19) are ingested by a final host, the oncospheres are freed in the intestine and burrow into the interior of the villi, where they develop into cysticercoids. In about four days postinfection, these escape into the lumen of the intestine, attach to the mucosa, and mature in about 15 days.

This species also may use an intermediate host. When the eggs are ingested by certain insects, such as flour beetles (species of *Tenebrio*) and flea larvae (*Ctenocephalides canis*), they hatch in the intestine and the oncospheres migrate to the hemocoel to develop into cysticercoids. When the infected insects are eaten by susceptible hosts, the cysticercoids are released, evaginate, attach to the intestinal wall, and develop to maturity.

Hymenolepis diminuta

This cosmopolitan parasite of rats, and occasionally other rodents, is infective to humans. It is not common in people, like *Vampirolepis nana,* because it requires an insect intermediate host. Many kinds of insects can serve, especially beetles. The arthropod becomes infected when it eats an egg (fig. 3.20); the rat when it eats the arthropod harboring a cysticercoid.

This tapeworm is much larger than *V. nana* and otherwise is easily differentiated from it by lacking an armed rostellum. The scolex does possess a small apical organ.

Because this parasite is easily maintained in laboratory rats and beetles, it is a handy organism for demonstrating a tapeworm life cycle. See page 278 for instructions. An entire book on this species was edited by Arai (1980).

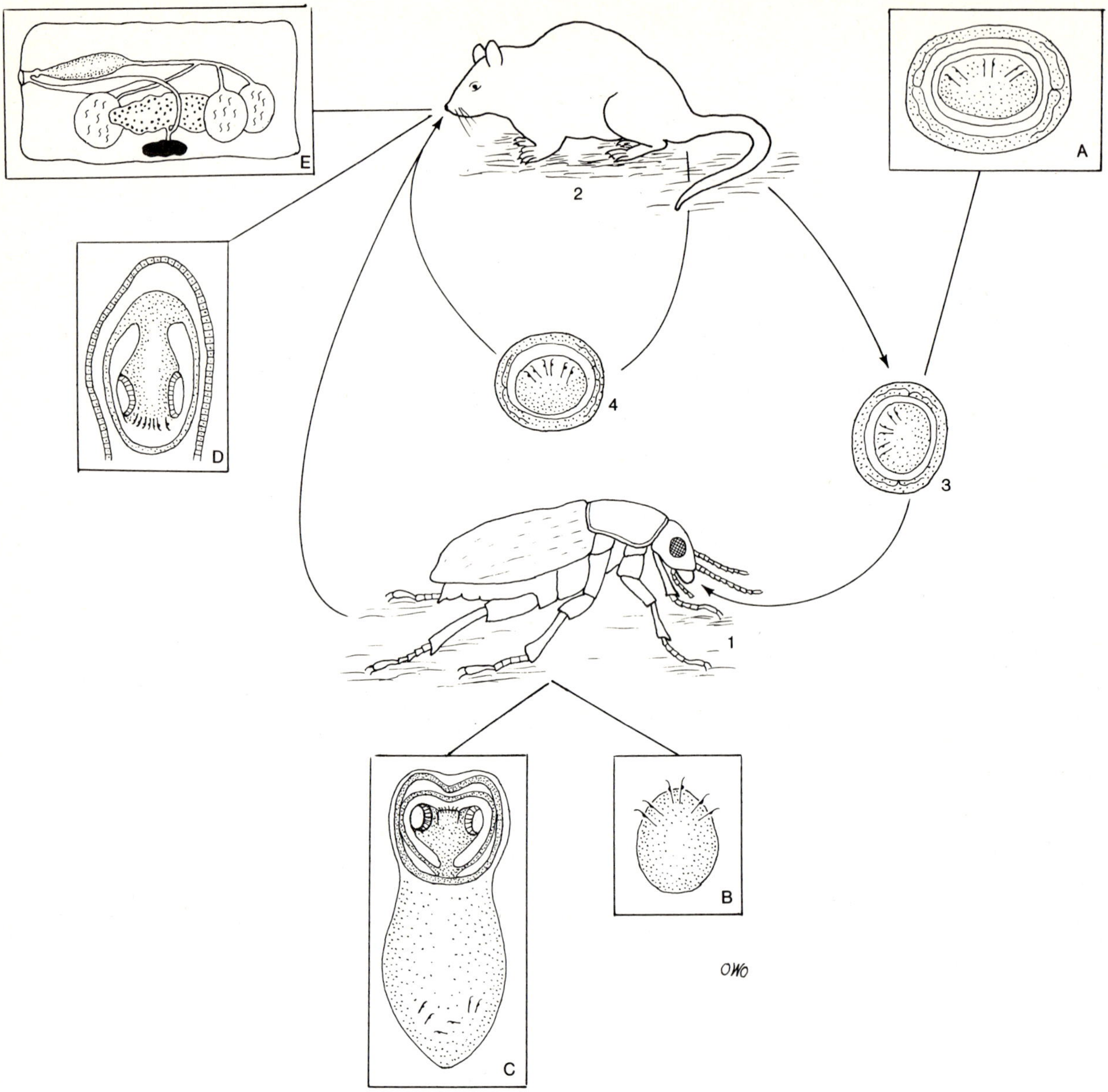

Fig. 3.18. Life cycle of *Vampirolepis nana,* a representative type of nontaenioid cyclophyllidean cestode in which the metacestode is a cysticercoid that develops commonly in arthropods. *V. nana* differs from other cestodes of this group in that the life cycle may be direct (2 and 4) or indirect (1, 2, 3). All other Hymenolepididae are indirect. (A) Embryonated egg with polar filaments; (B) oncosphere in beetle host; (C) tailed cysticercoid in beetle; (D) tailless cysticercoid from ingested egg (4) develops in intestinal villus of definitive host; (E) mature proglottid. (1) Arthropod intermediate host; (2) definitive host (rat, mouse); (3) egg when eaten infects beetle; (4) egg swallowed by rat hatches in gut, oncosphere (B) penetrates villus and develops into a tailless cysticercoid (D).

Family Anoplocephalidae

Moniezia expansa

One of the largest and best-known members of the family is the double-pored ruminant tapeworm, *M. expansa,* (fig. 3.21), which occurs commonly in sheep, goats, and cattle in most parts of the world. Lambs may become infected very early in life and pass gravid proglottids when they are six weeks old.

Description

Adults attain a length of 600 cm and a width of 1.5 cm. The scolex has four prominent suckers but lacks a rostellum and hooks. Proglottids are wider than long, and each contains two sets of reproductive organs with the genital pores situated on the lateral margins. The ovaries and the vitellaria, posterior to the ovaries, are in separate clusters. The numerous testes are scattered

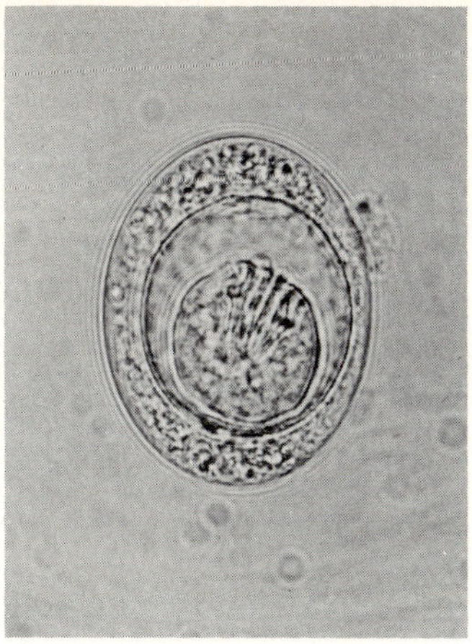

Fig. 3.19. *Vampirolepis nana,* egg. Spherical or subspherical; provided with two egg membranes closer together than are those of *H. diminuta,* internal one with two polar thickenings, from each arise four to eight threadlike filaments lying in the space between the shells; size 40–60 × 30–50 μm.

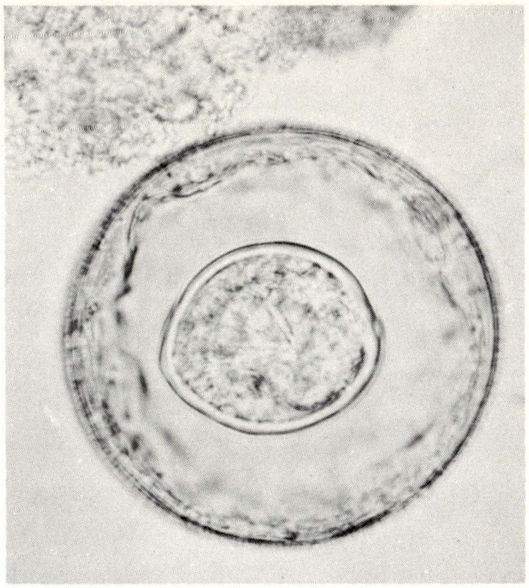

Fig. 3.20. *Hymenolepis diminuta,* egg. Ovoidal; provided with two widely separated egg membranes, internal one with a thickening at each pole but lacking polar filaments. Between the membranes is a colorless, elastic, gelatinous substance; size 72–85 × 60–80 μm.

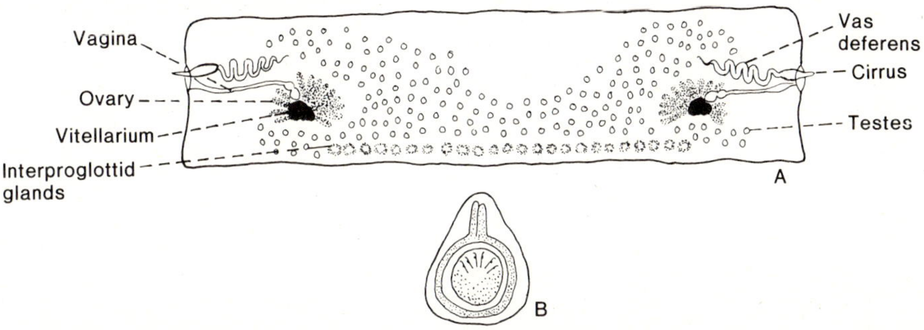

Fig. 3.21. *Moniezia expansa.* (A) Mature proglottid; (B) pyriform egg.

throughout the proglottid, or they may be concentrated toward the lateral fields. Situated along the posterior border of the proglottid is a row of **interproglottidal glands,** arranged around small pits. Eggs are variable in shape but often subtriangular, each with a bidigitate **pyriform apparatus** containing an oncosphere.

Life Cycle

Segments of gravid proglottids voided with the feces disseminate the embryonated eggs. When eggs are ingested by soil mites (in North America species of *Galumna, Oribatula, Scheloribates,* etc.), they hatch in the intestine. Oncospheres bore through the intestine into the hemocoel, where they develop into cysticercoids within a short time. When infected mites are swallowed along with the grass by grazing herbivorous animals, the cysticercoids are liberated in the intestine, where they attach and mature. After becoming gravid, the worms are lost after about three months in the host.

Family Davaineidae

Davainea proglottina

This small, almost microscopic tapeworm occurs in the duodenum of chickens, and occasionally other gallinaceous birds, throughout the world. It is seen best when the opened duodenum is floated in saline solution, with the gravid proglottids appearing as raised bumps on the mucosa.

Description

Adults measure up to 4 mm and usually consist of six or seven proglottids, exceptionally nine. The rostellum is armed with 86 to 94 small hooks, in two rows, and each of the suckers is armed with four or five rows of minute hooks along the margins. Genital pores alternate regularly and are situated far anterior on the lateral margin of the proglottids. Large spindle-shaped

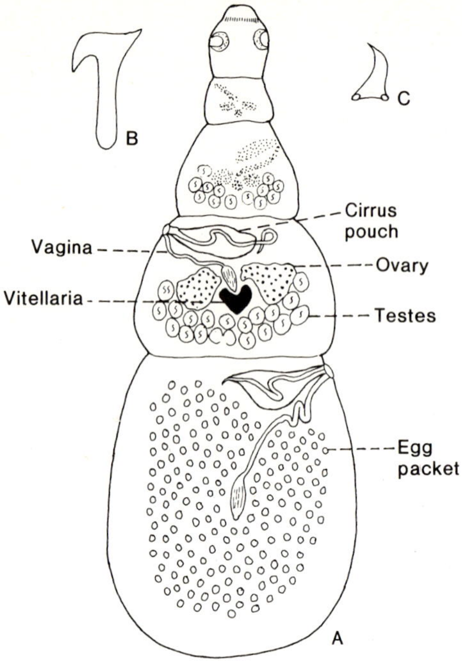

Fig. 3.22. *Davainea proglottina.* (A) Mature worm; (B) rostellar hook; (C) sucker hooklet.

cirrus pouch extends over two-thirds the width of proglottid. The cirrus is heavily armed with long spines. Uterus ephemeral, with eggs soon breaking out into the parenchyma. (See fig. 3.22.)

Life Cycle

When eggs, released from the gravid proglottids and voided in the host feces, are eaten by certain slugs and land snails, they hatch. Intermediate hosts are species of the slugs *Limax, Deroceras,* and *Arion* and of the land snails *Polygyra, Zonitoides,* and *Vallonia.* Oncospheres bore through the intestine to tissues, where they develop into infective cysticercoids within a month or less during the summer. Fowls become infected when they eat infected molluscs. Prepatency takes about 14 days.

Notes and Sketches

Key to Representative Families of Cestoidea*

Capital letters in parentheses refer to hosts: F–Fish, A–Amphibia, R–Reptiles, B–Birds, M–Mammals.

1 Large monozoic, linguiform cestodes with proboscis at anterior end but no scolex; testes scattered or in bands; ovary and genital pores at posterior end of body; uterus N shaped, opening near proboscis; vitellaria lateral, extend full length of body; eggs nonoperculate, embryos ten-hooked; in body cavity (Subclass CESTODARIA) (fig. 3.1) ....(F, rarely R) **Amphilinidae**
Small monozoic or polyzoic cestodes with scolex; embryos with six hooks (Subclass EUCESTODA) 2

2(1) Monozoic; scolex with grooves or depressions; adults with or without tail; testes either between lateral bands of vitelline follicles or surrounded by them and anterior to ovary and uterus; male and female genital pores and uterine pore midventral at anterior end of uterus (Order CARYOPHYLLIDEA) (fig. 3.2) ....(F) **Caryophyllaeidae**
Polyzoic, with segmentation usually distinct externally but occasionally visible only internally 3

3(2) Scolex with single well-developed apical sucker; body cylindrical; genital pores lateral (Order NIPPOTAENIIDEA) One family (fig. 3.23) ....(F) **Nippotaeniidae**
Scolex large with enormous glandular rostellum and four cup-shaped suckers, or small scolex without suckers but with a rostellum bearing ten hooks; external segmentation lacking; genital ducts without external openings (Order APORIDEA) One family (fig. 3.24) ....(B) **Nematoparataeniidae**
Scolex with four cup-shaped muscular suckers, sometimes with a fifth apical sucker or remnant of one; vitellaria in lateral bands; genital pore marginal (Order PROTEOCEPHALIDEA) (fig. 3.7) ....(F, A, R) **Proteocephalidae**
Scolex with four cuplike muscular suckers, with or without apical rostellum bearing one or more crowns of hooks; vitelline gland compact, usually behind ovary; genital pore lateral except in Mesocestoididae, where on midventral surface (Order CYCLOPHYLLIDEA) 4
Scolex typically with a dorsal and ventral bothrium (groove); permanent uterine pore on flat surface of proglottid; vitellaria scattered in proglottid (Order PSEUDOPHYLLIDEA) 16

4(3) Rostellum absent 5
Rostellum usually present** 9

5(4) Genital pores on midventral surface (fig. 3.25) ....(B, M) **Mesocestoididae**
Genital pores marginal, unilateral or alternating 6

6(5) Vitelline gland preovarian, bilobed; suckers commonly with outgrowth from margins; genital pores unilateral (fig. 3.26) ....(B, M) **Tetrabothriidae**
Vitelline gland postovarian 7

7(6) Segmentation apparent only posteriorly, body cylindrical; usually two testes; encapsulated eggs in several paruterine bodies per segment (fig. 3.27) ....(A, R) **Nematotaeniidae**
Segmentation distinct throughout; body usually flat 8

8(7) Gravid proglottids distinctly longer than wide; one set of reproductive organs per proglottid; ovary and vitelline gland branched, in anterior third of proglottid; testes mainly or entirely postovarian; gravid uterus with long median stem and irregularly shaped lateral lobes (fig. 3.28) .... (M, rodents) **Catenotaeniidae**
Gravid proglottids usually much wider than long; vitelline gland usually unbranched; gravid uterus a transverse tube, reticulate, or unstable; eggs with pyriform apparatus; single or double sets of reproductive organs (fig. 3.21) .... (R, B, M) **Anoplocephalidae**

9(4) Rostellum permanently extended, usually armed with two circles of hooks; ovary bilobed; vitellaria postovarian, gravid uterus a median stem with lateral branches; eggs with thick, striated shell (figs. 3.10, 3.11) .... (B, M) **Taeniidae**
Rostellum retractile within scolex; eggs without striated shell, gravid uterus not with median stem and lateral branches 10

10(9) Individual strobila with either male or female sex organs but not both, or partially separated, with genital organs of male in anterior parts and of female in posterior parts of same strobila (fig. 3.29) ....(B) **Dioecocestidae**
Monoecious forms 11

11(10) Testes large, one to four, usually three; genital pores unilateral; without or with single crown of rostellar hooks, uterus transverse (*Diplogynia* has double set of reproductive systems) (fig. 3.17) ....(B, M) **Hymenolepididae**
Testes small, usually more than four 12

*For more complete keys see Schmidt (1986).

**A few nonrostellate genera are included in families whose members are normally rostellate.

Fig. 3.23. *Nippotaenia chaenogobii.*

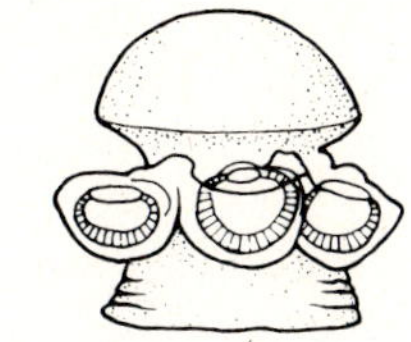

Fig. 3.24. *Nematoparataenia southwelli.*

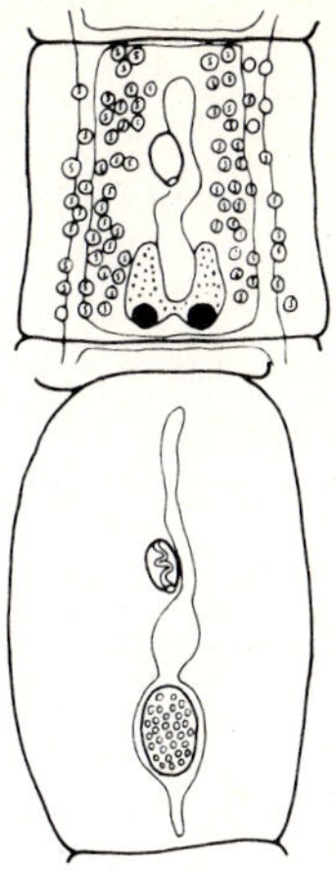

Fig. 3.25. *Mesocestoides lineatus.*

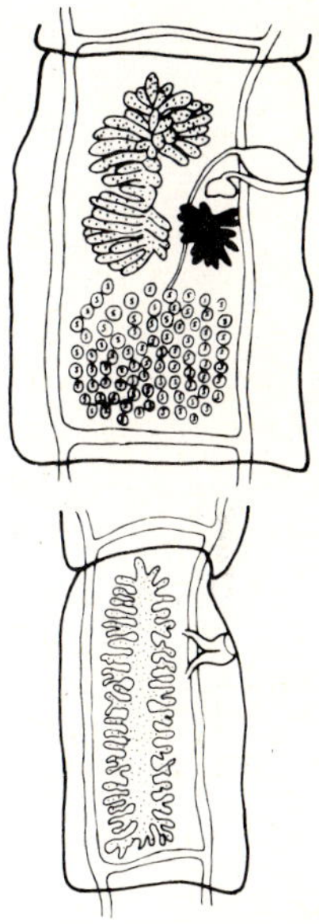

Fig. 3.28. *Catenotaenia peromysci.*

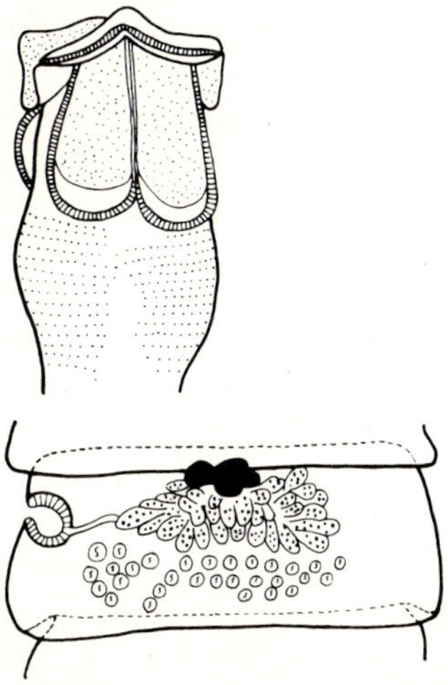

Fig. 3.26. *Tetrabothrius cylindricus.*

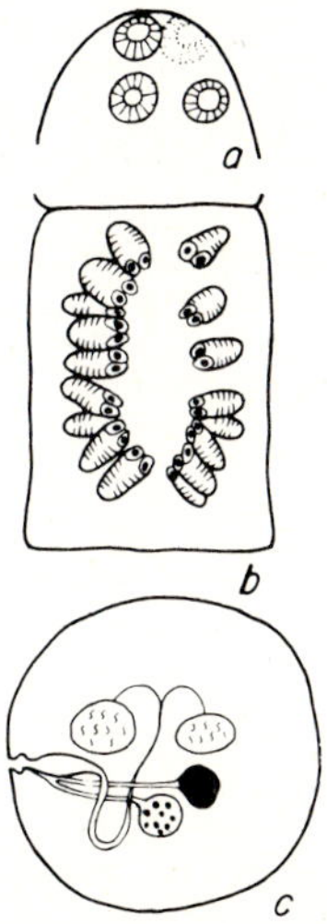

Fig. 3.27. (A) (B) *Baerietta baeri;* (C) *Nematotaenia dispar.*

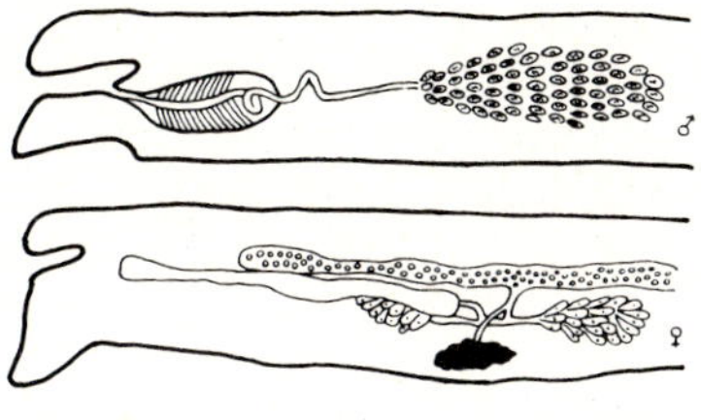

Fig. 3.29. *Dioecocestus fuhrmanni.*

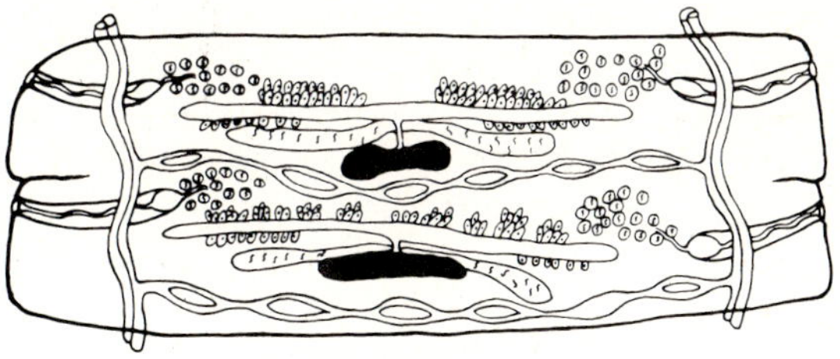

Fig. 3.30. *Diplophallus taglei.*

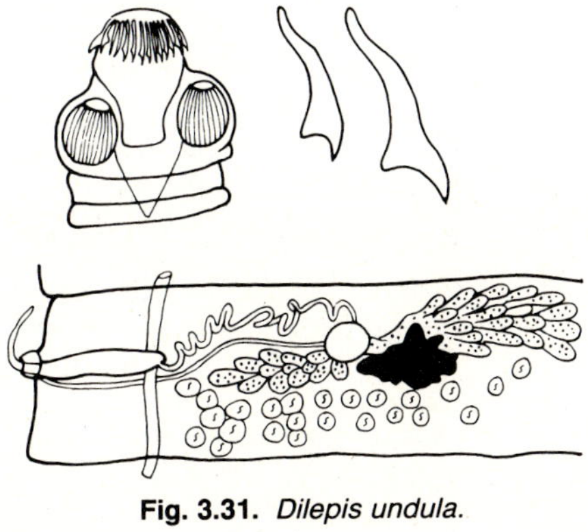

Fig. 3.31. *Dilepis undula.*

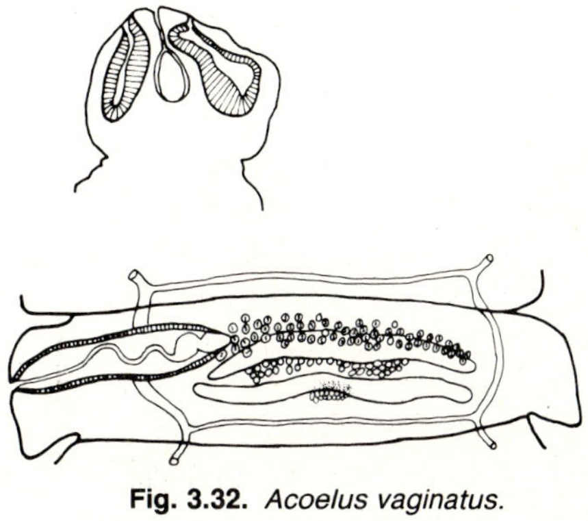

Fig. 3.32. *Acoelus vaginatus.*

Fig. 3.34. *Triaenophorus nodulosus.*

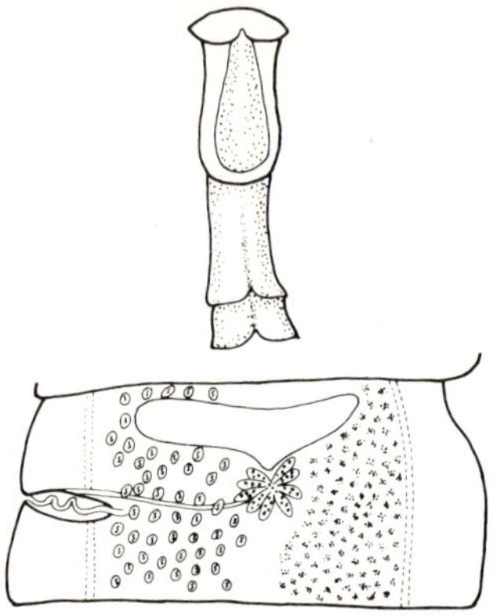

Fig. 3.35. *Abothrium crassum.*

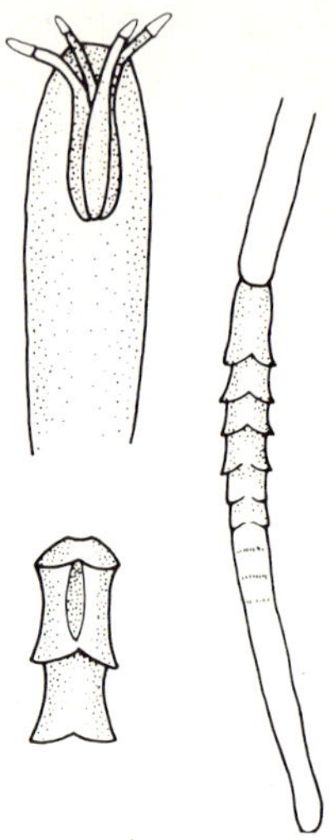

Fig. 3.36. *Haplobothrium globuliforme.*

Fig. 3.33. *Tatria duodecacantha.*

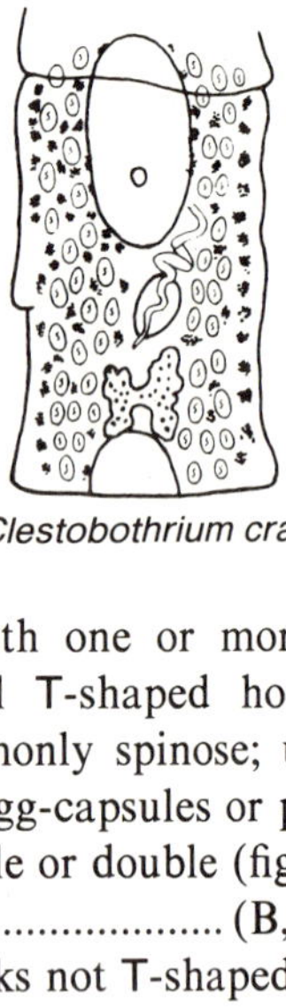

Fig. 3.37. *Clestobothrium crassiceps.*

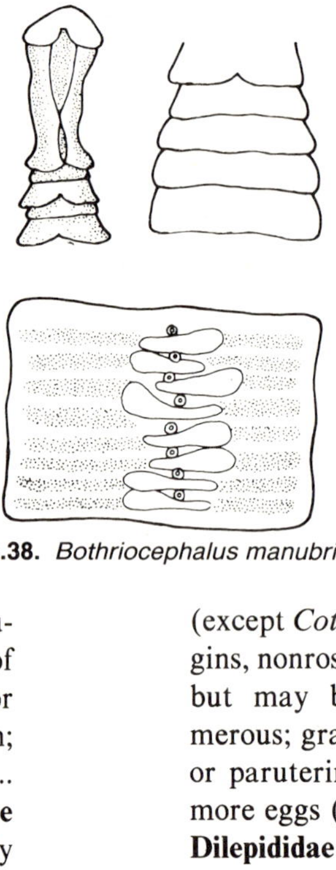

Fig. 3.38. *Bothriocephalus manubriformis.*

12(11) Rostellum with one or more circles of numerous small T-shaped hooks; margins of suckers commonly spinose; uterus saclike or replaced by egg-capsules or paruterine organ; genitalia single or double (fig. 3.2) .... (B, M) **Davaineidae**

Rostellar hooks not T-shaped; suckers usually not spinose 13

13(12) Male genital system with two separate cirrus pouches; testes few and in one or two groups; female genital system single, vaginal aperture lacking (fig. 3.30) ....(B) **Diploposthidae**

Male and female genital systems single (except *Diplopylidium* and *Dipylidium* in Dilepididae, where both male and female genital systems are double) 14

14(13) Vaginal apertures present and functional; rostellum usually present and armed with one to several circles of hooks of various forms, including rose-thorn shape; suckers aspinose (except *Cotylorhipis* with large hooks on margins, nonrostellate); genitalia commonly single but may be double; testes commonly numerous; gravid uterus a transverse, lobed sac, or paruterine organ, or capsules with one or more eggs (figs. 3.15, 3.31) ....(B, M) **Dilepididae**

Vaginal apertures absent 15

15(14) Vagina serving as seminal receptacle, no accessory pore developing later to replace vagina (fig. 3.32) ....(B) **Acoleidae**

Vagina absent, replaced in function later by an accessory canal opening to outside (fig. 3.33) .... (B) **Amabiliidae**

16(3) Genital pores marginal; uterine pore ventral (rarely dorsal) 17

Genital pores and uterine pore on same surface of proglottid or on opposite surfaces 18

17(16) Scolex subcuboidal or subpyramidal, bothria shallow, apex with or without four tridental hooks; uterine pore midventral, anterior to level of marginal genital pore; gravid uterus a tube with many coils; eggs operculate (fig. 3.34) ....(F) **Triaenophoridae**

Scolex subspherical, bothria well developed, apex without hooks; gravid uterus saclike; uterine pore median; eggs nonoperculate (fig. 3.35) ....(F) **Amphicotylidae**

18(16) Genital pores and uterine pore on same surface of proglottid 19

Genital pores and uterine pore open on opposite surfaces of proglottid 20

19(18) Small worms; scolex of primary segmented adult worm club shaped with four protrusible tentacles, replaced in secondary segmented adult by rectangular pseudoscolex that resembles first formed segment; cirrus spined; uterus with proximal coils and distal sac (fig. 3.36) ....(F) **Haplobothriidae**

Large worms, proglottids wider than long; scolex variable in shape, commonly compressed and with bothria weakly developed, without tentacles; uterus tubular throughout length, arranged in median loops rosettelike; genital system usually single (double in *Diplogonoporus* and *Cordiocephalus*); cirrus unspined; eggs operculate, unembryonated when laid (fig. 3.3) ....(F rarely, R, B, M) **Diphyllobothriidae**

20(18) Scolex with rounded apex, bothria deep and narrow or wide and shallow, eggs nonoperculate, embryonated when laid (fig. 3.37) .... (F) **Ptychobothriidae**

Scolex with a four-lobed, fleshy apex and long, oval, shallow bothria; eggs operculate, unembryonated when laid (fig. 3.38) ....(F) **Bothriocephalidae**

Review Questions

1. What are the three main body divisions of a typical tapeworm?
2. Consider the mode of segmentation of a tapeworm. How does it differ from that of an earthworm?
3. What is different between the first larva of a cestodarian and a eucestodan?
4. What is the obvious difference between the body of a caryophyllidean and a cyclophyllidean cestode? How do the vitellaria differ?
5. Name the three larval stages of a pseudophyllidean.
6. What is sparganosis? Which species is most likely to be involved in the United States? In the Orient?
7. What is the first intermediate host of *Proteocephalus?* Are there similarities in the biology and morphology of Proteocephalidea and Pseudophyllidea?
8. What are the two major morphological characteristics of the order Cyclophyllidea?
9. What is atypical of the scolex of *Taenia?* Is its rostellum retractable?
10. How are the egg membranes of Taeniidae different from other cestodes? The outer membrane of *Diphyllobothrium* eggs?
11. Consider the many variations of the cysticercus in Taeniidae. Are these different larval types also as different in their adult form?
12. *Echinococcus granulosus* and *E. multilocularis* both exist in the United States. What are the similarities and differences of their adult and larval stages?
13. *Dipylidium caninum* usually uses fleas as intermediate hosts. How can this be, considering that adult fleas are bloodsucking insects with piercing mouthparts?
14. What is a major morphological difference between *Hymenolepis* and *Vampirolepis?* Can you discuss a major difference between the life cycles of *H. diminuta* and *V. nana?*

References

Freeman, R. S. 1973. Ontogeny of Cestodes and Its Bearing on Their Phylogeny and Systematics. In *Advances in Parasitology,* ed. B. Dawes. Academic Press, New York, vol. 11, pp. 481–557.

Rees, G. 1967. Pathogenesis of adult cestodes. *Helminthol. Abstr.* 36:1–23.

Schmidt, G. D. 1970. *How to Know the Tapeworms.* Wm. C. Brown Company Publishers, Dubuque, Iowa, 266 pp.

Schmidt, G. D. 1986. *Handbook of Tapeworm Identification.* CRC Press, Inc., Boca Raton, Florida, 675 pp.

Slais, J. 1973. Functional Morphology of Cestode Larvae. In *Advances in Parasitology,* ed. B. Dawes. Academic Press, New York, vol. 11, pp. 395–480.

Smyth, J. D. 1963. The Biology of Cestode Life-Cycles. *Tech. Comm. No. 34, Commonwealth Bur. Helminthology,* St. Albans, Herts, England, 38 pp.

———. 1969. *The Physiology of Cestodes.* Oliver & Boyd, Edinburgh, 279 pp.

Smyth, J. D., and Heath, D. D. 1970. Pathogenesis of larval cestodes in mammals. *Helminthol. Abstr., Series A,* 39:1–23.

Stunkard, H. W. 1962. The organization, ontogeny, and orientation of the Cestoda. *Quart. Rev. Biol.* 37:23–34.

Voge, M. 1967 and 1973. The post-embryonic Development Stages of Cestodes. In *Advances in Parasitology,* ed. B. Dawes. Academic Press, New York, vol. 5, pp. 247–97; vol. 11, pp. 707–30.

Wardle, R. A., and McLeod, J. A. 1952. *The Zoology of Tapeworms.* The University of Minnesota Press, Minneapolis, 780 pp.

Wardle, R. A.; McLeod, J. A.; and Radinovsky, S. 1974. *Advances in the Zoology of Tapeworms.* The University of Minnesota Press, Minneapolis, 275 pp.

Yamaguti, S. 1959. Systema Helminthum. *The Cestodes of Vertebrates,* vol. 2, Interscience Publishers, Inc., New York, 860 pp.

Caryophyllidea

Calentine, R. L. 1964. The life cycle of *Archigetes iowensis* (Cestoda: Caryophyllaeidae). *J. Parasitol.* 50:454–58.

Mackiewicz, J. S. 1972. Caryophyllidea (Cestoidea): A review. *Exptl. Parasitol.* 31:417–512.

Mackiewicz, J. S. 1981. Caryophyllidea (Cestoidea): evolution and classification. *Adv. Parasitol.* 19:139–206.

Mackiewicz, J. S. 1982. Caryophyllidea: perspectives. *Parasitology* 84:397–417.

Pseudophyllidea

Corkum, K. C. 1966. Sparganosis in some vertebrates of Louisiana and observations on a human infection. *J. Parasitol.* 52:444–48.

Meyer, M. C. 1970. Cestoda zoonoses of aquatic animals. *J. Wildlife Dis.* 6:249–54.

———. 1972. The pattern of circulation of *Diphyllobothrium sebago* (Cestoda: Pseudophyllidea) in an enzootic area. *J. Wildlife Dis.* 8:215–20.

Mueller, J. F. 1974. The biology of *Spirometra. J. Parasitol.* 60:1–14.

Olsen, O. W., and Haas, W. R. 1976. A new record of *Spirometra mansoni,* a zoonotic tapeworm, from naturally infected cats and dogs in Hawaii. *Hawaii Med. J.* 35:261–63.

Vik, R. 1964. The genus *Diphyllobothrium.* An example of the interdependence of systematics and experimental biology. *Exptl. Parasitol.* 15:361–80.

von Bonsdorff, B. 1977. *Diphyllobothriasis in Man.* Academic Press, New York, 189 pp.

Wardle, R. A. 1935. Fish—tapeworm. *Bull. Biol. Bd. Canada,* no. 45, 25 pp.

Proteocephalidea

Brooks, D. R. 1978. Evolutionary history of the Cestode Order Proteocephalidea. *Systematic Zoology* 27:312–13.

Eure, H. 1976. Seasonal abundance of *Proteocephalus ambloplitis* (Cestoidea: Proteocephalidea) from largemouth bass living in a heated reservoir. *Parasitology* 73:205–12.

Fischer, H., and Freeman, R. S. 1969. Penetration of parenteral plerocercoids of *Proteocephalus ambloplitis* (Leidy) into the gut of smallmouth bass. *J. Parasitol.* 55:766–74.

———. 1973. The role of plerocercoids in the biology of *Proteocephalus ambloplitis* (Cestoda) maturing in smallmouth bass. *Canad. J. Zool.* 51:133–41.

Freze, V. I. 1965. Proteocephalata in Fish, Amphibians, and Reptiles. *Akad. Nauk SSSR,* Moscow, vol. 5, 597 pp. (Israel Program for Scientific Translations, 1969).

Hunter, G. W., III, and Hunter, W. S. 1929. Further experimental studies on the bass tapeworm, *Proteocephalus ambloplitis* (Leidy). In *Biological Survey of the Erie-Niagara System, Suppl. 18th Ann. Rept.,* 1928, New York State Conserv. Dept., Albany, pp. 198–207.

Cyclophyllidea

Abdou, A. H. 1958. Studies on the development of *Davainea proglottina* in the intermediate host. *J. Parasitol.* 44:484–88.

Abuladze, K. I. 1964. Taeniata of Animals and Man and Diseases Caused by Them. *Akad. Nauk SSSR,* Moscow, vol. 4, 549 pp. (Israel Program for Scientific Translations, 1970).

Arai, H. P. ed. 1980. *Biology of the Tapeworm* Hymenolepis diminuta. Academic Press, New York, 733 pp.

Enigk, K.; Sticinsky, E.; and Ergun, H. 1958. Die Zwischenwirte von *Davainea proglottina,* (Cestoidea). *Zeit. Parasitenk.* 18:230–36.

Freeman, R. S. 1952. The biology and life history of *Monoecocestus* Beddard, 1914 (Cestoda: Anoplocephalidae) from the porcupine. *J. Parasitol.* 38:111–29.

Gleason, N. N. 1962. Records of human infections with *Dipylidium caninum,* the double-pored tapeworm. *J. Parasitol.* 48:812.

Leiby, P. D.; Carney, W. P.; and Woods, C. E. 1970. Studies on sylvatic echinococcosis. *J. Parasitol.* 56:1141–50.

Leiby, P. D., and Kritsky, D. C. 1972. *Echinococcus multilocularis:* A possible domestic life cycle in central North America and its public health implications. *J. Parasitol.* 58:1213–15.

Pawlowski, Z., and Schultz, M. G. 1972. Taeniasis and Cysticercosis (*Taenia saginata*). In *Advances in Parasitology,* ed. B. Dawes. Academic Press, New York, vol. 10, pp. 269–343.

Rausch, R. L. 1967. On the ecology and distribution of *Echinococcus* spp. (Cestoda: Taeniidae), and characteristics of their development in the intermediate host. *Ann. Parasitol. Hum. Comp.* 42:19–63.

Smyth, J. D. 1964 and 1969. The Biology of the Hydatid Organisms. In *Advances in Parasitology,* ed. B. Dawes. Academic Press, New York, vol. 2, pp. 169–219; vol. 7, pp. 327–47.

Spasskii, A. A. 1951. Anoplocephalate Tapeworms of Domestic and Wild Animals. *Akad. Nauk SSSR,* Moscow, vol. 1, 783 pp. (Israel Program for Scientific Translations, 1961).

Stunkard, H. W. 1938. The development of *Moniezia expansa* in the intermediate host. *Parasitology* 30:491–501.

Venard, C. E. 1938. Morphology, bionomics, and taxonomy of the cestode *Dipylidium caninum. Ann. New York Acad. Sci.* 37:273–328.

Voge, M., and Heyneman, D. 1957. Development of *Hymenolepis nana* and *Hymenolepis diminuta* (Cestoda: Hymenolepididae) in the intermediate host *Tribolium confusum. Univ. California Publ. Zool.* 59:549–80.

Sources of Study Material for Cestoidea*

Diphyllobothrium: C,T,W
Diphyllobothrium erinacei: C
Diphyllobothrium latum, scolex: Tr,W

Dipylidium caninum: C,T,Rr,W

Echinococcus granulosus: C,T,W
Echinococcus granulosus, hydatid: C,W
Echinococcus multilocularis, alveolar cyst: C

Hymenolepis diminuta: C,T,W
Hymenolepis diminuta, cysticercoid: C,T,W

Moniezia expansa: C,T,W

Taenia sp.: Tr
Taenia pisiformis: C,T,W
Taenia pisiformis, cysticercus: C,T,W
Taenia seralis, section of coenurus: C
Taenia solium: C,T
Taenia solium, cysticercus: C,W
Taenia taeniaeformis: C

Taeniarhynchus saginatus: C,T
Taeniarhynchus saginatus, cysticercus in muscle: C

Vampirolepis nana: C,T,Tr,W

*The addresses of the sources are listed on p. 45.

Chapter 4 Phylum Acanthocephala

General Body Structure

This phylum consists of a small group of unsegmented, cylindrical, dioecious, endoparasitic worms lacking a digestive tract throughout their development. Anteriorly the body bears a retractable **proboscis,** armed with recurved hooks, that becomes embedded in the intestinal wall of the final host.

The body is divided into an anterior region, the **presoma,** and a posterior region, the **trunk,** which lies free in the intestinal lumen of the host. Externally the presoma consists of the **proboscis,** followed by the unspined **neck.** Attached inside the neck is the **proboscis receptacle,** into which the proboscis in most species is retractable. Associated structures of the presoma include the small **cephalic ganglion** in the receptacle, a pair of elongated **lemnisci** originating from the neck and hanging free in the body cavity, and a set of **retractor muscles.** The proboscis is everted by means of fluids under hydrostatic pressure by body contraction.

A pair of muscles, the **proboscis invertors,** extends back from the tip of the proboscis and inverts it upon contracting. Each of these muscles extends through the proboscis and the neck, and continues to the posterior extremity of the proboscis receptacle. Here they penetrate the receptacle, after which they are known as **receptacle retractors,** continue in the body cavity, and eventually attach to the trunk wall.

The remainder of the body comprises the trunk, containing the excretory (in one family) and reproductive systems, nerve fibers arising from the cephalic ganglion, and the body fluid. The **syncytial body wall** consists of an outer surface coat, a thick middle layer composed of three outer layers of fibrils and an inner layer containing a few nuclei, and an innermost double muscular layer. The **lacunar system** consists of dorsal and ventral, or lateral, canals that extend the length of the body and which are connected by a network or transverse system of small canals. The body cavity lacks a definite epithelial lining and is regarded as a **pseudocoelom.**

Extending from the base of the proboscis receptacle to the posterior end of the trunk is a hollow strand of single or double muscle, the genital **ligament,** which encloses the reproductive organs.

Reproductive Systems

The **male reproductive system** (fig. 4.1A) consists of a pair of **testes** and **genital ducts, cirrus,** and **accessory organs** (**cement gland** complex and copulatory apparatus). A single syncytial cement gland or several unicellular or syncytial glands, a **cement reservoir** in some species, and their ducts comprise the cement gland complex. The ducts of the cement glands, individually or collectively, enter the common genital duct. The copulatory apparatus consists of a **Saefftigen's pouch,** a muscular fluid reservoir, and an eversible **copulatory bursa,** with a **cirrus,** at the posterior end of the trunk. During copulation, the everted bursa fits over the posterior end of the female, and the cirrus enters the vagina. Semen is discharged into the uterus, after which the secretion from the cement gland plugs the vagina. Eversion of the bursa results from the hydrostatic pressure of fluid from the Saefftigen's pouch; inversion is by muscular action.

The female reproductive system (fig. 4.1B) consists of an **ovary, uterine bell, uterus,** and **vagina.** Early in development, the ovary breaks into **ovarian balls,** or floating ovarian follicles. The uterine bell is a bell-shaped structure with its large opening directed anteriorly and a small opening posteriorly, through which the mature eggs are passed into the uterus, whence they move into a short vagina and out through the gonopore. Immature eggs are returned to the body cavity or dorsal ligament (when present), through lateral pores near the base of the uterine bell. The mechanism by which the uterine bell "distinguishes" between mature and immature eggs, at least in *Polymorphus minutus,* is apparently based on their length (Whitfield, 1970).

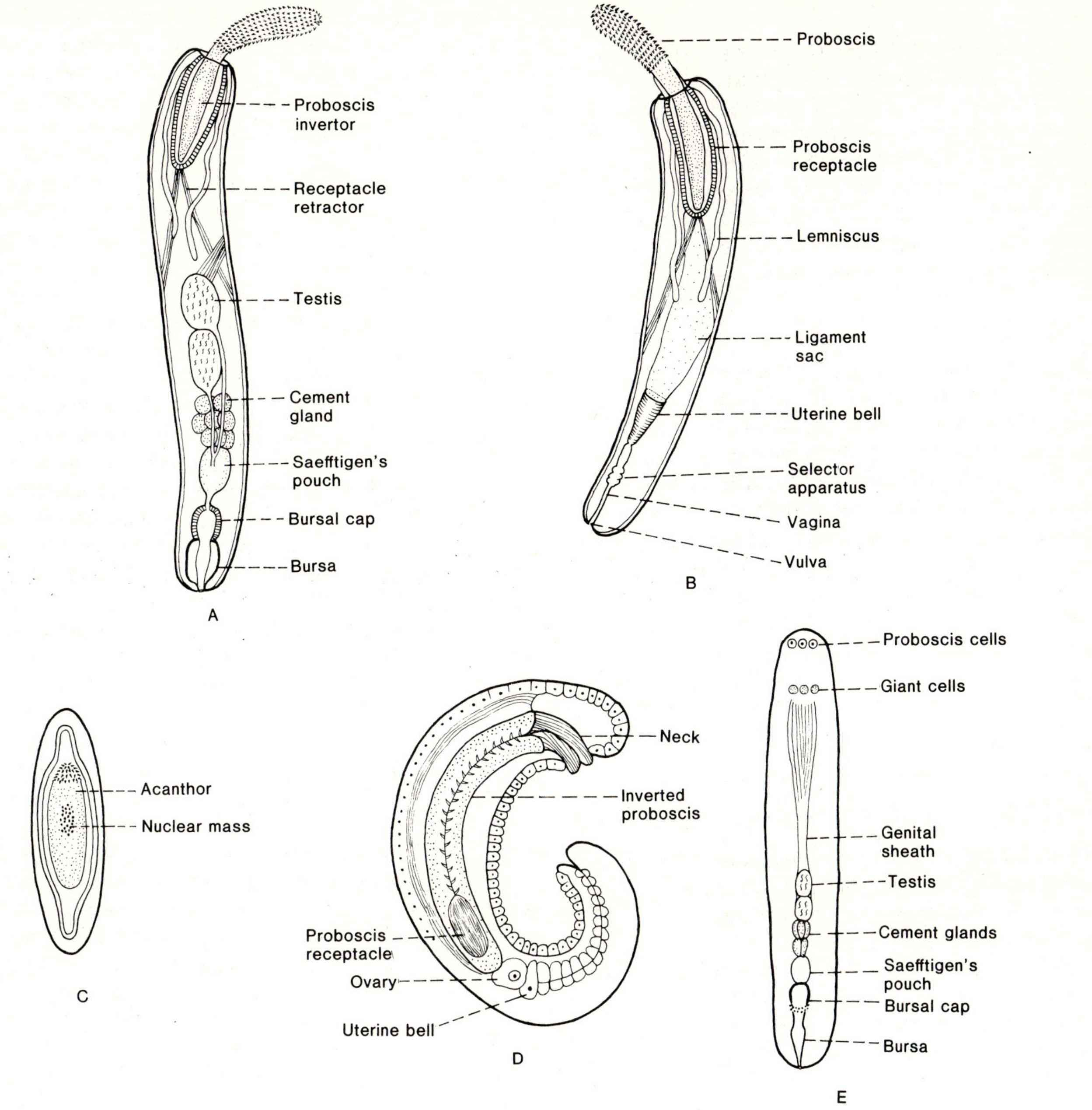

Fig. 4.1. Anatomy of an acanthocephalan, *Leptorhynchoides thecatus.* (A) Adult male; (B) adult female; (C) embryonated egg showing spiny acanthor with internal nuclear mass; (D) female cystacanth removed from cyst; and (E) male acanthella.

General Life Cycle

Eggs (fig. 4.1C) have three or four enveloping membranes and are embryonated when laid. They do not hatch until ingested by the proper crustacean or insect host. The fusiform larva, or **acanthor,** is covered with small spines and anteriorly, with hooks; inside are many small nuclei. After hatching, the acanthor penetrates the intermediate host's gut wall and reaches the body cavity. A progressive development ensues; the proboscis and its sac, the lemnisci, and the anlagen of the reproductive system appear. This preinfective stage is known as the **acanthella** (fig. 4.1D). After undergoing further development and encystment, it is known as **cystacanth** (fig. 4.1E). The sex of the now-infective larva is apparent, and, in fact, usually can be determined in the late acanthella; the proboscis is formed but inverted; and the larva is enclosed within a hyaline membrane.

Life cycles are basically similar in all species, following the sequence of larval stages. Adults are in vertebrates. Species occurring in terrestrial definitive hosts

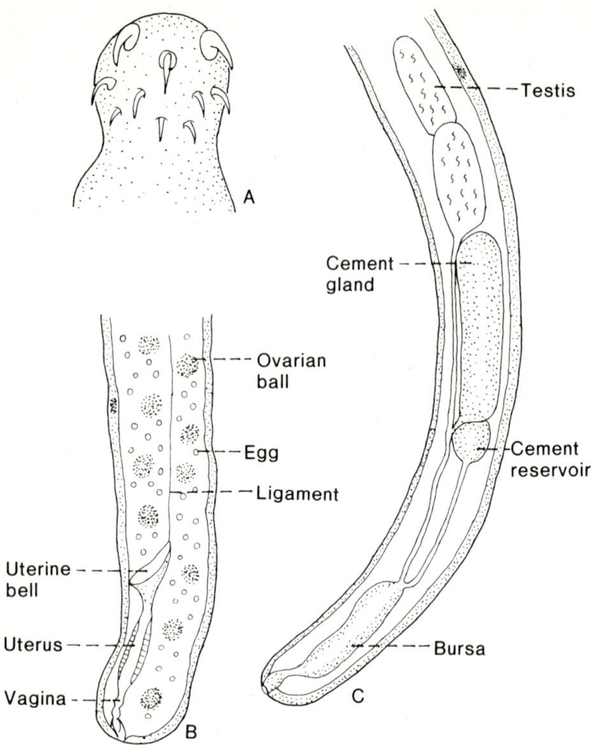

Fig. 4.2. *Neoechinorhynchus emydis.* (A) Proboscis; (B) caudal end of female (modified from Hopp, 1954.); (C) caudal end of male.

normally use terrestrial arthropod intermediate hosts; species in aquatic definitive hosts have crustacean intermediaries.

A paratenic host occurs in some species. Acanthocephala in fish-eating mammals (seals, sea lions) and birds (marine and freshwater) include a fish paratenic host in the life cycle. Larval stages develop in Crustacea. Cystacanths are transferred to plankton-feeding fish and finally to the piscivorous mammals and birds.

Class Eoacanthocephala

Order Neoechinorhynchidea

Body usually small; proboscis with few or many hooks; main lacunar canals median; nuclei of body wall few, usually large; proboscis receptacle single layered; ligament sacs persistent in females, in males cement glands syncytial, with a reservoir. Adults mostly in fish, rarely in turtles; cystacanths in Crustacea.

Family Neoechinorhynchidae

Neoechinorhynchus emydis

This species, one of the few thorny-headed worms occurring as common parasites of turtles in the United States, is of special interest because two hosts may be involved in the life cycle. *N. emydis* (fig. 4.2) or related species usually can be obtained from biological supply house turtles, which frequently are used in physiology teaching laboratories.

The following description is approximately the same for most species of *Neoechinorphynchus* in fishes.

Description

The proboscis, globular in shape, measures about 0.18 mm in length by 0.22 mm in width. It bears six longitudinal rows of three hooks each. There are six subcuticular nuclei, of which five are in the middorsal, and one in the midventral line near the anterior end. The longer of the lemnisci contains two nuclei, and the shorter contains a single nucleus.

Gravid females are 10.2 to 22.2 mm long by 0.8 to 1.3 mm wide. The genital pore is subterminal, opening on the ventral surface of the body. Adult males measure 6.3 to 14.5 mm in length and 0.5 to 1.0 mm in width. Reproductive organs occupy about the posterior half of the body. Testes are contiguous, in tandem, and immediately followed by the large, syncytial cement gland with eight large nuclei. Posterior to the single, syncytial cement gland is a large globular cement receptacle, with a thin-walled cement duct passing posteriorly to the bursa. Eggs ovoidal, 18–22 μm long.

Life Cycle

Adult females in the intestine of turtles (*Emys blandingi* and species of *Graptemys*) lay fully embryonated eggs that are voided with the feces. The ostracod, *Cypria maculata,* ingests the eggs, which hatch within eight to 12 hours. Upon release from the eggs, acanthors promptly penetrate the intestinal wall, arriving in the body cavity within 24 hours. After about 19 days, larvae have reached the definitive shape, the proboscis with its fully formed hooks is inverted in the proboscis receptacle, and the reproductive apparatus is developed to the extent that the sex can be readily determined. Completely developed unencysted larvae are present in ostracods 21 days postinfection. The proboscis is capable of full eversion and inversion, the reproductive structures are well developed, and the six giant subcuticular nuclei occupy the same position as in the adult worm.

Snails, *Campeloma rufum,* serve as paratenic hosts. They become infected by eating ostracods harboring the fully developed cystacanths. In the intestine of the snail, ostracods are digested and the larvae freed. Those which have developed to the infective stage penetrate the intestinal wall and burrow into the tissues, especially the foot, where cystacanths appear as opaque spots. The snail serves as the principal source of infection of the turtle, whose food is largely mollusks.

Other often available Neoechinorhynchidae and their hosts include *Neoechinorhynchus cylindratus* and *N. rutili,* freshwater fish; *N. chrysemydis,* turtles; *Octospinifer macilentus,* freshwater fish.

Notes and Sketches

Class Palaeacanthocephala

Order Echinorhynchidea

Body usually small; proboscis with few or many hooks, lacunar canals lateral; nuclei of body wall fragmented or ramified; proboscis receptacle double-layered; ligament sacs rupture in females; males normally with six or fewer cement glands with nuclear fragments. Adults mostly in fish and aquatic birds and mammals; cystacanths in Crustacea.

Family Rhadinorhynchidae

Leptorhynchoides thecatus

This species (fig. 4.3) parasitizes a variety of fish throughout the United States, but it occurs primarily in Centrarchidae: smallmouth bass (*Micropterus dolomieui*), largemouth bass (*M. salmoides*), and rock bass (*Ambloplites rupestris*). It differs from most of the rest of the family in lacking trunk spines, and in infecting freshwater rather than marine fishes.

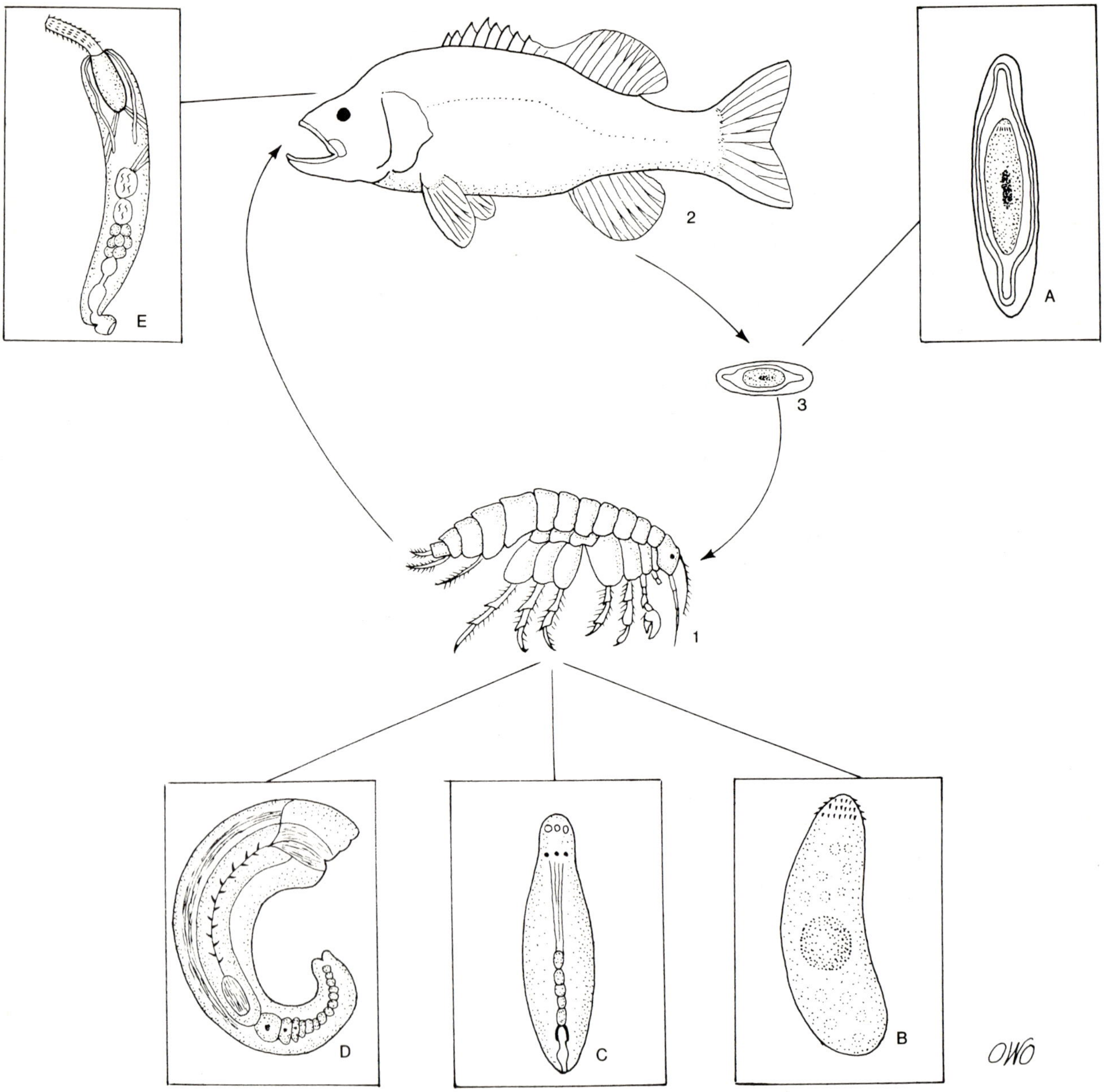

Fig. 4.3. Life cycle of *Leptorhynchoides thecatus*. (A) Egg containing acanthor; (B) acanthor in gut of amphipod; (C) acanthella in hemocoel of amphipod; (D) cystacanth in hemocoel of amphipod; (E) adult in enteral cavity of fish. Numbers 1 and 2 indicate intermediate and final hosts, respectively. (After DeGiusti [1949] from Meyer [1954].)

Description

When fully extended, the proboscis, armed with 12 longitudinal rows of 12 to 13 hooks each, is perpendicular to the long axis of the body; in extreme extension it may form an acute angle with the body. The hooks are surrounded throughout much of their length by an ensheathing cuticular collar. The shape and angle of the extended proboscis and the cuticular collar of the hooks are specifically diagnostic. Lemnisci are tubular to filiform, and considerably longer than the proboscis receptacle. The male has eight cement glands, closely compacted at the posterior border of the hind testis (fig. 4.3E). Eggs 80–100 × 24–30 μm.

Life Cycle

Eggs, fully developed when passed in the host feces, hatch within 45 minutes after being ingested by the amphipod *Hyalella azteca.* Acanthors soon make their way through the intestinal wall and come to rest beneath the serosal covering, where development proceeds to the acanthella stage. After about 14 days, during which there has been much increase in size and internal organization, acanthellas become detached and drop into the hemocoel. Cystacanths are infective about a month postinfection.

The cycle is completed when amphipods harboring cystacanths are eaten by appropriate fish hosts. Larvae freed in the stomach of the fish, enter the pyloric ceca, where they mature. Prepatency takes about eight weeks, but the males are sexually mature in about a month.

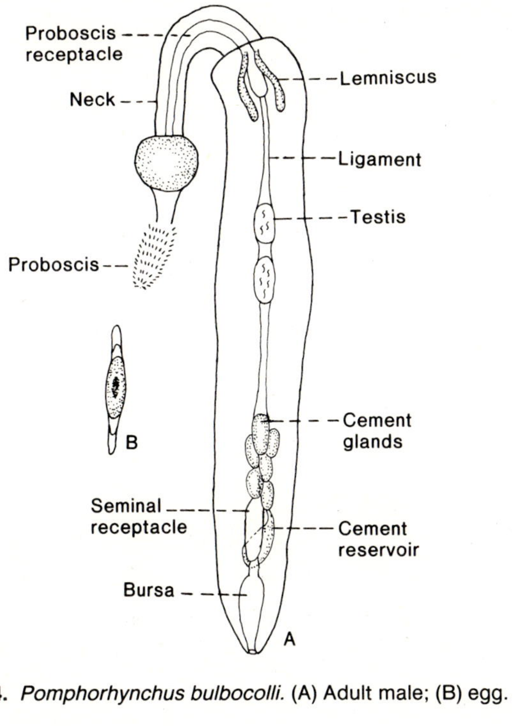

Fig. 4.4. *Pomphorhynchus bulbocolli.* (A) Adult male; (B) egg.

Family Pomphorhynchidae

Pomphorhynchus bulbocolli

This species (fig. 4.4), is a common parasite of the white sucker (*Catostomus commersoni*) and is occasionally found in closely related fish. The long cylindrical neck, between the trunk and the proboscis, distinguishes it from other species in fish. At the distal end of this thin, cylindrical neck there is a large, globular bulb, which is an accessory attachment organ, since the tissues of the host intestine grow around it. Macroscopically, the fibrous capsules surrounding the proboscis and bulb are clearly visible on the outer surface of the intestine.

Description

Proboscis long, almost cylindrical, armed with 12 longitudinal rows of 12 to 14 hooks each; posterior hooks more slender than anterior. Proboscis receptacle double-walled, inserted at base of proboscis and extending throughout the length of the neck. Lemnisci short, claviform. Testes oval, tandem, near middle of trunk. Cement glands six, rounded to oval. Eggs fusiform, 53–83 × 8–13 μm, with a prominent prolongation at each pole of middle shell.

Life Cycle

Eggs, fully developed when passed in the host feces, hatch when ingested by the amphipod *Hyalella azteca.* Recently released acanthors have 55 to 60 small accessory spines at the anterior end, and a central syncytial mass of nuclei in a surrounding syncytial area containing 18 giant nuclei. Within about five hours, the acanthors make their way through the intestinal wall of the host into the hemocoel.

After about 14 days of development, the central embryonic mass shows the primordia of the proboscis, proboscis sheath, anterior ganglion, gonads, and other reproductive structures. It is now an acanthella. Sixteen days postinfection, the sexes can be differentiated.

On the nineteenth day of development, the primordia of proboscis hooks can be seen within the still-inverted proboscis. The male genital system is made up of a pair of slightly overlapping testes, six cement glands, Saefftigen's pouch, cirrus, and a partially formed bursa. The female genital system consists of four to six embryonic ovarian masses within the ligament, uterine bell, and beginnings of the uterus and vagina. Basic structures of the adult worm are attained by the thirty-third day. After 35 to 37 days, the larva, now a cystacanth, is infective to the proper fish host.

Based upon sperm production, males reach sexual maturity in three weeks postinfection, whereas it takes ten weeks for females to produce viable eggs. At ten weeks, females reach a length of 14.5 mm, and males measure 10.6 mm.

Notes and Sketches

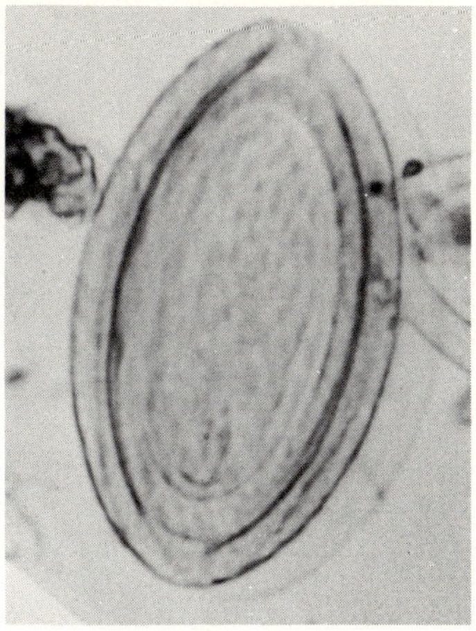

Fig. 4.5. *Macracanthorhynchus hirudinaceus,* egg. Provided with four embryonic membranes, the outer one rugose; fully embryonated when voided; acanthor has four pairs of aclid spines; ellipsoidal in shape; size 80–100 × 46–65 μm.

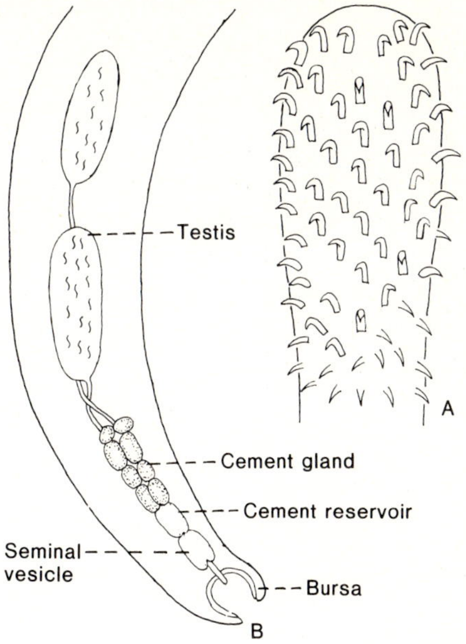

Fig. 4.6. *Moniliformis moniliformis.* (A) Proboscis; (B) caudal end of male.

Class Archiacanthocephala

Order Oligacanthorhynchida

Body medium to large; proboscis with few or many hooks; main lacunar canals median; nuclei in body wall few and large; proboscis receptacle complex or double-layered; ligament sacs persistent in females; males with eight uninucleate cement glands. Adults in terrestrial vertebrates; cystacanths in insects (grubs, cockroaches, etc.).

Family Oligacanthorhynchidae

Macracanthorhynchus hirudinaceus

Description

This species, the giant thorny-headed worm of swine, is the most widely known member of the phylum. This is because it has attained practically cosmopolitan distribution through human agencies, and the species has direct economic importance. The body is circular in cross section when preserved, tapering from a rather broad anterior region to a slender posterior end. Females may reach 48 cm, males 11 cm in length. The proboscis is globular, with 12 longitudinal rows of three hooks each. Lemnisci are flat, ribbonlike, and with a few giant nuclei anteriorly. Testes are elongate and located in the midbody or slightly anterior to it. The eight cement glands are elliptical, often arranged in pairs as an elonate series.

Life Cycle

Embryonated eggs (fig. 4.5) passed in the feces of swine are very resistant to adverse conditions, being able to survive for several years. When eggs are ingested by larvae of a number of species of Scarabaeidae beetles, including the genera *Cotinus* and *Phyllophaga,* hatching occurs within an hour. Released acanthors migrate through the intestinal wall into the body cavity. They usually remain attached to the outer surface of the gut for five to 20 days, after which they become free in the body cavity. In the meantime, the larval hooks have been lost and the primordia of the various organs of the adult have appeared, the most conspicuous of which is the proboscis with its hooks. At 24 C, cystacanths develop in experimentally infected grubs in 60 to 90 days, but the minimum developmental period in nature is nearer the upper limit. Infective cystacanths are essentially juvenile parasites, possessing the structures and organ systems of adult worms. After entering the final host, further development is merely an elaboration of the organ systems already formed.

Swine become infected by eating either the grubs or the adult beetles that harbor the infective cystacanths. Prepatency in swine is believed to be from two to three months, and patency may continue for ten months.

Family Moniliformidae

Moniliformis moniliformis

This species (*M. dubius* of some authors) is a common parasite of rats throughout the world, especially in regions where the cockroach intermediate hosts are present. (see fig. 4.6.)

Description

The body shows prominent sexual dimorphism, with females up to 32 cm, males 15 cm in length. Both sexes show conspicuous pseudosegmentation, except at the extremities. The proboscis is cylindrical and club-shaped, with 12 to 14 longitudinal rows of ten or 11 hooks, occasionally nine or 12 hooks, in each longitudinal series. Lemnisci are long and narrow, each with a few giant nuclei. Male organs are confined to the posterior nonannulated region of the body. Testes are very long, elliptical, in contact in young males, but may be widely separated in old males. Eight cement glands, pyriform in shape, are commonly arranged in an elongate series but rarely in pairs. Eggs 90–125 × 50–62 μm.

Life Cycle

When fully developed eggs in the host feces are ingested by cockroaches (*Periplaneta americana*), they hatch in the midintestine within 24 to 48 hours. Released acanthors, bearing six bladelike hooks, penetrate the intestinal wall and appear as minute iridescent specks on the outer surface in ten to 12 days, drop into the body cavity, and there continue their development. Infective cystacanths appear about seven to eight weeks after infection.

The cycle is completed when cockroaches harboring cystacanths are eaten by rats. Released cystacanths attach to the wall of the intestine and mature in five to six weeks.

Notes and Sketches

Key to Common Families of Acanthocephala

In this key and those like it which follow, numbers in parentheses indicate the alternative condition. For example, if the proboscis of worm in 1(2) has hooks instead of fine spines, proceed to 2(1).

Capital letters in parentheses refer to hosts: F-Fish, A-Amphibia, R-Reptiles, B-Birds, M-Mammals.

1(2) Proboscis with fine spines, not invaginable; without proboscis receptacle. Order APORORHYNCHIDEA (fig. 4.7) .... (B) **Apororhynchidae**

2(1) Proboscis with hooks, invaginable; with proboscis receptacle 3

3(6) Proboscis usually with few hooks; proboscis receptacle single layered; cement gland syncytial, single; mostly in fish, rarely in turtles .... NEOECHINORHYNCHIDEA 4

4(5) Trunk glabrous (fig.4.2) .... (F) **Neoechinorhynchidae**

5(4) Trunk spinose (fig. 4.8) ... (F) **Quadrigyridae**

6(3) Proboscis with few or many hooks; proboscis receptacle single or double layered; cement gland not syncytial, not single; not limited to fish 7

7(20) Proboscis usually with many hooks; proboscis receptacle double layered; cement gland divided into two or more lobes; egg fusiform, usually with polar prolongations of middle shell; embryo with hooks at one end only; mostly in fish, aquatic birds, and mammals .... Order ECHINORHYNCHIDEA 8

8(9) Trunk usually glabrous 10

9(8) Trunk usually spinose................................ 17

10(11) Cement glands four, occupying greater length of trunk; testes cylindrical (fig. 4.9) .... (F) **Fessisentidae**

11(10) Cement glands three to eight, not occupying greater length of trunk; testes not cylindrical 12

12(13) Neck very long, cylindrical or spirally twisted, with bulbous swelling (fig. 4.4) .... (F) **Pomphorhynchidae**

13(12) Neck short, if present, never cylindrical or spirally twisted, without bulbous swelling 14

14(15) Proboscis receptacle inserted anterior to base of proboscis and dividing the latter into two regions (fig. 4.10) ... (B, M) **Centrorhynchidae**

15(16) Cement glands usually compact or pyriform (fig. 4.11) .... (F) **Echinorhynchidae**

16(15) Cement glands usually tubular, long (fig. 4.12) .... (B, occ. R) **Plagiorhynchidae**

17(18) Trunk spines extensive or limited; proboscis hooks very numerous; cement glands two to eight, usually elongate (fig. 4.13) .... (F) **Rhadinorhynchidae**

18(19) Eggs with polar prolongation of middle shell (fig.4.14) .... (B, M) **Polymorphidae**

19(18) Eggs without polar prolongation of middle shell (fig. 4.15) .... (B) **Filicollidae**

20(7) Cement glands eight, uninuclear; egg oval to elliptical, without polar prolongations of middle shell; embryo with hooks at both ends and numerous minute spines elsewhere; in terrestrial vertebrates 21

21(22) Proboscis with few hooks, in short rows; protonephridial organ present .... (B, M) Order OLIGACANTHORHYNCHIDA **Oligacanthorhynchidae**

22(21) Proboscis with numerous hooks, mostly in longitudinal rows; protonephridial organ absent 23

23(24) Proboscis hooks divided into two groups of different size and shape (fig. 4.16) .... (B, M) Order GIGANTORHYNCHIDA **Gigantorhynchidae**

24(23) Proboscis hooks otherwise (fig. 4.6) ...... (M, occ. B) Order MONILIFORMIDA **Moniliformidae**

Review Questions

1. What are the three main body divisions of an acanthocephalan?
2. Name the two elongated organs that hang into the body cavity from the base of the neck.
3. Where is the brain of an acanthocephalan?
4. Describe the differences and similarities in the lacunar systems of the three classes of Acanthocephala.
5. Understand the structure and function of the following organs: (1) uterine bell, (2) copulatory bursa (3) Saefftigen's pouch, (4) cement glands, (5) ovarian balls, (6) giant nuclei.
6. Which orders of thorny-headed worms have the following structure of proboscis receptacle: (1) simple, single wall; (2) simple, double wall; (3) double, spiral wall; and (4) complex wall?
7. Name the three larval stages of Acanthocephala.
8. Which two classes of arthropods serve as intermediate hosts for this phylum?
9. How do vertebrates become infected with acanthocephalans?

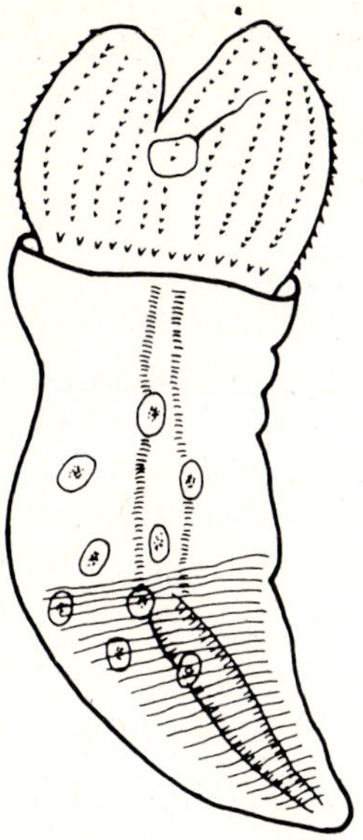

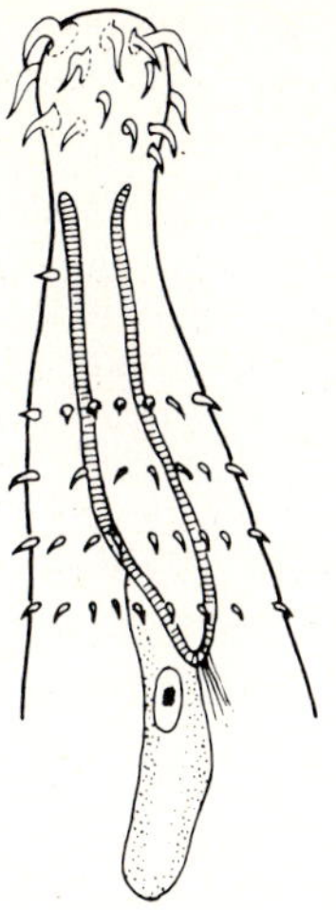

Fig. 4.8. *Quadrigyrus torquatus.*

Fig. 4.7. *Apororhynchus aculeatum;* female.

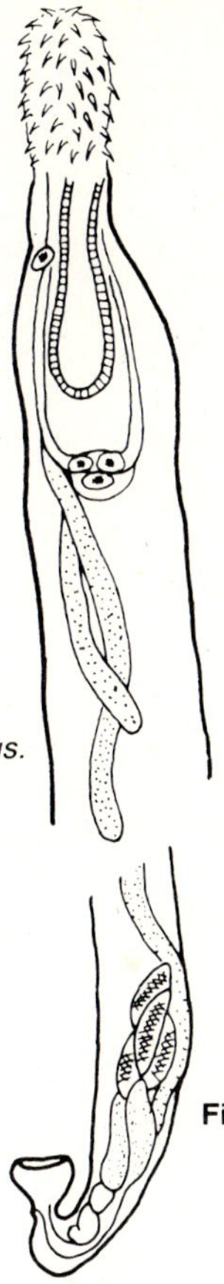

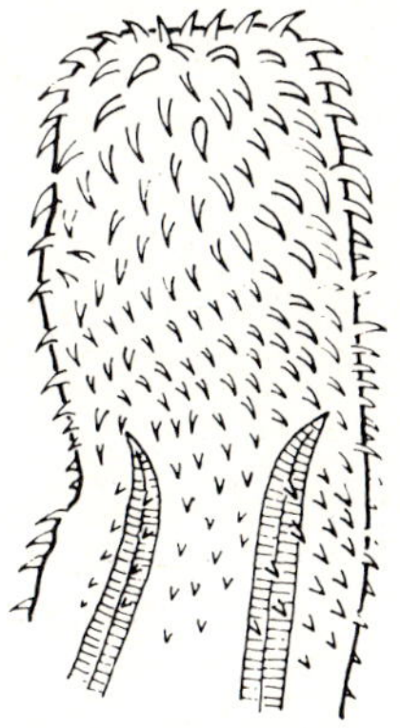

Fig. 4.10. *Centrorhynchus aluconis.*

Fig. 4.9. *Fessisentis fessus.*

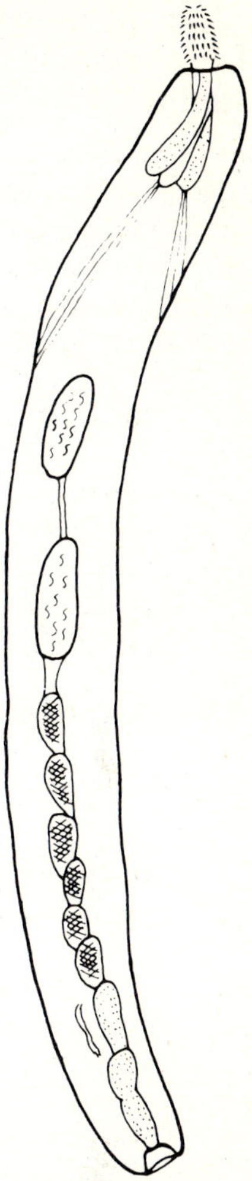

Fig. 4.11. *Echinorhynchus gadi.*

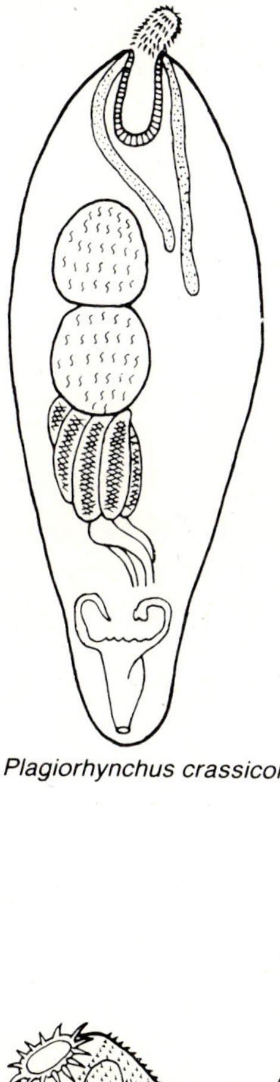

Fig. 4.12. *Plagiorhynchus crassicolle.*

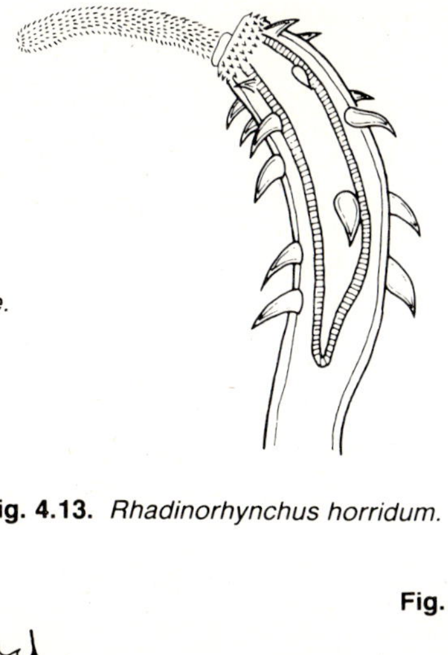

Fig. 4.13. *Rhadinorhynchus horridum.*

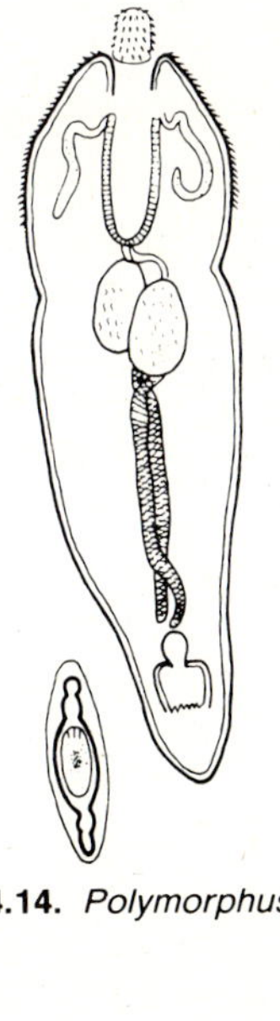

Fig. 4.14. *Polymorphus minutus.*

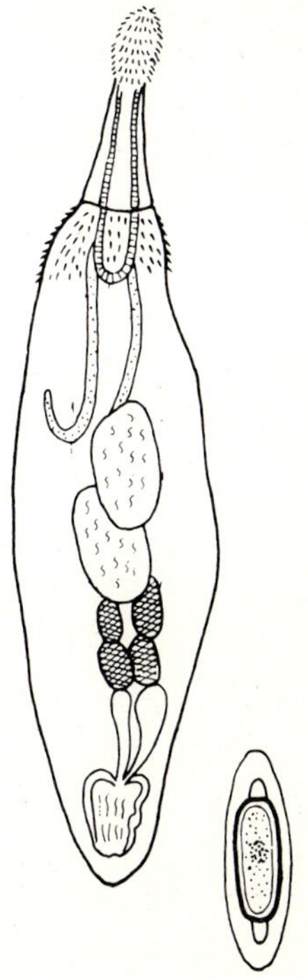

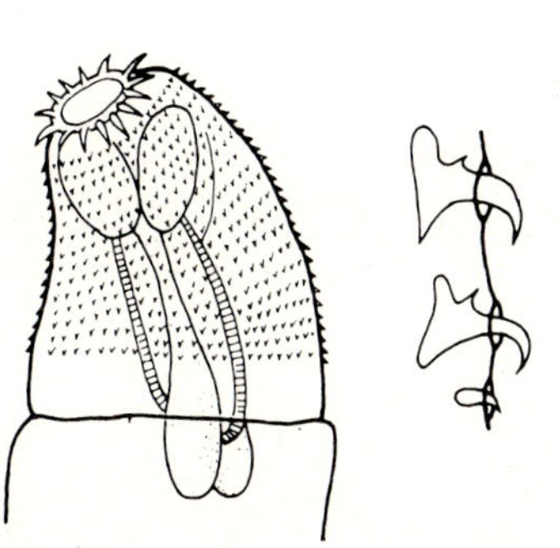

Fig. 4.16. *Gigantorhynchus echinodiscus.*

Fig. 4.15. *Filicollis anatis.*

References

General

Bullock, W. L. 1969. Morphological Features as Tools and Pitfalls in Acanthocephalan Systematics. In *Problems in Systematics of Parasites,* ed. G. D. Schmidt. University Park Press, Baltimore, pp. 9–43.

Crompton, D. W. T. 1970. *An Ecological Approach to Acanthocephalan Physiology.* Cambridge University Press, Cambridge, 125 pp.

Crompton, D. W. T., and Nickol, B. B., eds. 1985. *Biology of the Acanthocephala.* Cambridge University Press, Cambridge, 519 pp.

Moore, J. 1984. Parasites that change the behavior of their host. *Sci. Amer.* 250:108–115.

Nicholas, W. O. 1967 and 1973. The Biology of Acanthocephala. In *Advances in Parasitology,* ed. B. Dawes. Academic Press, New York, vol. 5, pp. 205–46; vol. 11, pp. 671–706.

Petrochenko, V. I. 1956 and 1958. Acanthocephala of Domestic and Wild Animals. *Akad, Nauk SSSR,* Moscow, vol. 1, 465 pp.; vol. 2, 478 pp.

Whitfield, P. J. 1970. The egg sorting function of the uterine bell of *Polymorphus minutus* (Acanthocephala). *Parasitology* 61: 111–26.

Yamaguti, S. 1963. *Systema Helminthum. Acanthocephala,* vol. 5. Interscience Publishers, Inc., New York, 423 pp.

Neoechinorhynchida

Fisher, F. M., Jr. 1960. On Acanthocephala of turtles, with the description of *Neoechinorhynchus emyditoides* n. sp. *J. Parasitol.* 46:257–66.

Hopp, W. B. 1954. Studies on the morphology and life cycle of *Neoechinorhynchus emydis* (Leidy), an acanthocephalan parasite of the map turtle, *Graptemys geographica* (Le Sueur). *J. Parasitol.* 40:284–99.

Echinorhynchida

DeGiusti, D. L. 1949. The life cycle of *Leptorhynchoides thecatus* (Linton), an acanthocephalan of fish. *J. Parasitol.* 35:437–60.

Jensen, T. 1953. The life cycle of the fish acanthocephalan, *Pomphorhynchus bulbocolli* (Linkins) Van Cleave, 1919, with some observations on larval development *in vitro. Diss. Abstr.* 12:607.

Schmidt, G. D., and Olsen, O. W., 1964. Life cycle and development of *Prosthorhynchus formosus* (Van Cleave 1918) Travassos, 1926, an acanthocephalan parasite of birds. *J. Parasitol.* 50:721–30.

Gigantorhynchida, Oligacanthorhynchida, Moniliformida

Kates, K. C. 1943. Development of the swine thorn-headed worm, *Macracanthorhynchus hirudinaceus,* in its intermediate host. *Amer. J. Vet. Res.* 4:173–81.

Kates, K. C. 1944. Some observations on experimental infections of pig with the thorn-headed worm, *Macracanthorhynchus hirudinaceus. Amer. J. Vet. Res.* 5:166–72.

King, D., and Robinson, E. S. 1967. Aspects of the development of *Moniliformis dubius. J. Parasitol.* 53:142–49.

Moore, D. V. 1946. Studies of the life history and development of *Moniliformis dubius* Meyer, 1933. *J. Parasitol.* 32:257–71.

Schaefer, P. W. 1970. *Periplaneta americana* (L) as intermediate host of *Moniliformis moniliformis* (Bremser) in Honolulu, Hawaii. *Proc. Helminthol. Soc. Washington* 37:204–7.

Schmidt. G. D. 1972. Revision of the class Archiacanthocephala Meyer, 1931 (Phylum Acanthocephala), with emphasis on Oligacanthorhynchidae Southwell et Macfie, 1925. *J. Parasitol.* 58:290–97.

Schmidt, G. D. 1973. Early embryology of the acanthocephalan *Mediorhynchus grandis* Van Cleave, 1916. *Tr. Am. Micro. Soc.* 92:512–16.

Wright, R. D. 1971. The egg envelopes of *Moniliformis dubius. J. Parasitol.* 57:122–31.

Sources of Study Materials for Acanthocephala*

Acanthocephala gen., sp., whole mount: C,W

Macracanthorhynchus hirudinaceus, eggs: C,T,W
Macracanthorhynchus hirudinaceus, preserved: C,W
Macracanthorhynchus hirudinaceus, sections in pig intestine: C,W

Live *Neoechinorhynchus* spp., almost always can be found in the intestines of turtles, available from biological supply houses. Other genera are common in waterfowl and fishes.

*The addresses of the sources are listed on p. 45.

Chapter 5 Phylum Nematoda

General Morphology

The body of nematodes is cylindrical except in the case of some females, in which it may be spindle shaped or twisted when gravid. Sexes are separate, with females generally larger than males. A **pseudocoel,** or false body cavity, is present (figs. 5.1A, and 5.2). The tail consists of that part of the body posterior to the anus.

Head

The anterior end of the body forms the head. The mouth is terminal and associated with it are various structures such as **lips, pseudolabia, odontia, cephalic papillae, amphids, collarettes,** and **cordons.**

The primitive number of lips is six, although this condition seldom is found on modern nematodes. On each of these triangular lips are two **sensory papillae,** one at the tip near the opening of the mouth and one at the base. Papillae at the tips of the lips form the inner circle, and those at the base constitute the outer circle. Four **cephalic papillae** are also found, one in each quadrant, near the lip bases.

Through fusion, in some species, the number of lips is reduced and a new arrangement established. When the two dorsal lips fuse and one lateral and one ventral on each side fuse, three lips occur (fig. 5.1G). This is the situation in the orders **Ascaridata** and **Oxyurata,** which often have a dorsal and two lateroventral lips. On the other hand, in other species all lips may be lost, and a forward shift of the buccal cavity lining produces two large lateral lips as occurs in some **Spirurata** (fig. 5.71). These new structures in the Spirurata are known as **pseudolabia.**

The lips may be lost entirely and not replaced by any other structures, as occurs in the **Camallanata** (fig. 5.60) and some Filariata (fig. 5.78). Other filarioids have some type of development such as a **circumoral elevation** (fig. 5.81) or toothlike projections known as **odontia,** or there may be lateral projections known as **helmets** (fig. 5.72).

While the number of labial papillae may be reduced by fusion or loss, the inner and outer circles are usually recognized. They are tactile in function.

The **amphids** consist of two pores, one on each side of the head near the base of the lips (fig. 5.81). They are chemoreceptors.

Tail

The posterior end of males varies in structure. In the **Strongylata** (fig. 5.1H), it is a bilateral, cuticular bursa supported by a system of paired muscular rays, and in the **Dioctophymata,** it is a fleshy, rayless, cup-shaped sucker (fig. 5.8C). **Ascaridata** have blunt, ventrally curved tails (fig. 5.18B),and **Oxyurata** have long, attenuated ones (fig. 5.24). The tails of **Spirurata** and **Filariata** are often coiled and somewhat tendrillike. Sensory papillae are present ventrally or laterally, or both, on the posterior end of the body (fig. 5.1I) except in the Strongylata. The tail of most females is very simple.

Body Wall

The body wall consists of three layers: the outermost **cuticle,** middle **hypodermis,** and innermost **somatic muscles** (fig. 5.20). The cuticle is a thin, transparent, elastic acellular layer covering the outside of the body. It extends inward at the mouth, anus, and vulva. The cuticle has several types of external markings. Fine transverse striations on the surface are present in most species. Strong annulations presenting the appearance of external segmentation occur in some species (fig. 5.66), and longitudinal ridges are present in others. Lateral expansions of the cuticle are called **alae** (fig. 5.30). They may be present at the anterior end of the body as **cervical alae,** the posterior end as **caudal alae,** or the full length of the body as **lateral alae.** Inflated areas of the cuticle in some species resemble an irregular arrangement of plaques or warts (fig. 5.63). A preanal circular structure having the appearance of a **sucker** is present in the males of some species (fig. 5.55). Expansions of the cuticle outward from the head are termed **helmets** or **hoods** (fig.5.72).

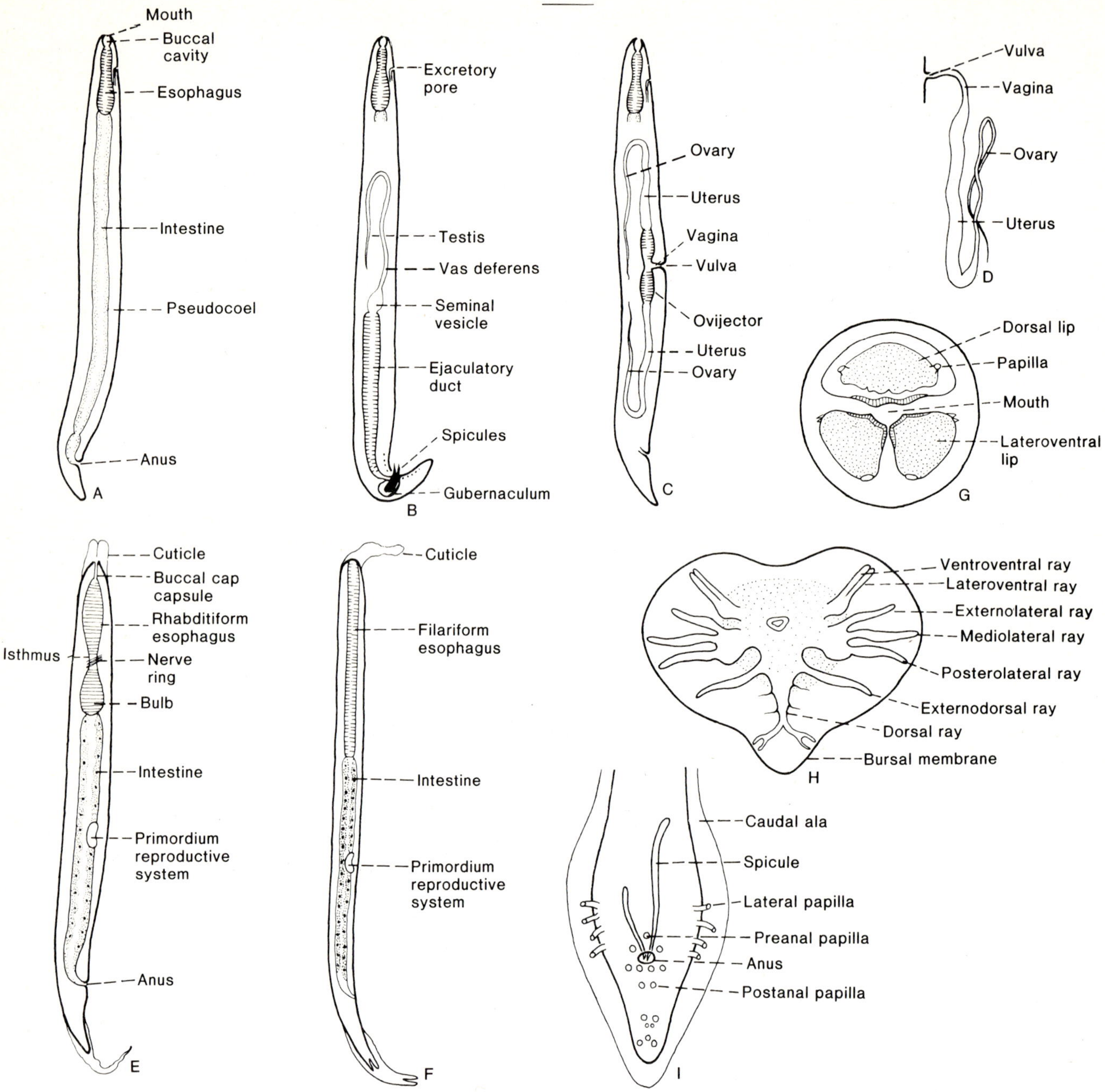

Fig. 5.1. Basic anatomy of nematodes. (A) Diagrammatic representation of digestive system; (B) diagrammatic representation of male reproductive system typical of *Ascaris* type; (C) diagrammatic amphidelphic reproductive system of a female nematode; (D) monodelphic reproductive system of a female nematode; (E) second-stage rhabditiform larval nematode, showing loosening cuticle of first-stage larva; (F) ensheathed third-stage infective filariform larva enclosed in cuticle of second-stage rhabditiform larva; (G) en face, or end, view of adult *Ascaris suum.* showing lips and papillae; (H) copulatory bursa of bursate male strongyle nematode (*Castorstrongylus castoris*) spread to show the basic arrangement of the rays; and (I) ventral view of a nonbursate male nematode (*Physaloptera* sp.), showing caudal alae and basic location of the various caudal papillae.

Several other cuticular structures are common. **Cordons** and **epaulets** are ribbon-shaped bands of cephalic cuticle that begin much as cephalic alae and then loop anteriorly on the ventral side where they may unite (fig. 5.74). Spines sometimes occur on the posterior margins of body annulations, forming spined **collarettes** (fig. 5.75), expanded **head balloons** (fig. 5.28), or simple rows on the body.

The outermost cellular layer of the nematode body is not a true epidermis since it is covered by cuticle. It is designated as the **hypodermis** and forms a thin syncytial layer lying between the cuticle and the somatic muscles. On the dorsal, ventral, and each lateral side of the body is an elongate thickening of the hypodermis that extends inward between the muscle cells into the body cavity. These are known as the **hypodermal chords** and may extend the full length of the body, or be present only in the anterior part (fig. 5.20). Dorsal and ventral thickenings contain longitudinal nerve trunks in some species, such as *Ascaris lumbricoides*.

The somatic muscle cells lie on the hypodermis and form the innermost layer of the body wall. They are separated into four major fields by the hypodermal chords.

The external openings include the mouth, anus, amphids, and excretory pore in both sexes (figs. 5.1A,B,C). Females possess a **vulva** in addition to the other openings (fig. 5.1C). It is located on the ventral side of the body anywhere from the anterior to just in front of the anus. In the **class Phasmidia,** two sensory organs, the **phasmids,** open through a minute pore in each side near the tip of the tail.

Digestive System

The alimentary canal is a more or less straight tube consisting of a **stoma (buccal capsule, vestibule), esophagus, intestine,** and **rectum.** In males, the rectum opens into a **cloaca.** The **anus** is ventral and usually subterminal (fig. 5.1A,E,F).

The stoma or **buccal cavity,** is the cavity into which the mouth opens (fig. 5.1A, E). It may be well developed with thick walls (fig. 5.40), rudimentary (fig. 5.8), or absent (fig. 5.24). When present, it may be at the extreme anterior end of the esophagus or partially surrounded by esophageal tissue.

The esophagus is an elongate muscular organ of variable shape. It may (1) be **rhabditiform,** in which there is an enlarged anterior portion connected by an isthmus, or a thin part, to a posterior bulb or enlargement (fig. 5.1E); (2) have a posterior **bulb** with a cuticularized valve inside (fig. 5.27); (3) be cylindrical and muscular (**filariform**) (fig. 5.1F); (4) be cylindrical and composed of a short anterior muscular and long posterior glandular part (fig. 5.68); (5) have a short anterior muscular part and a long posterior part consisting of a single row of cuboidal cells in which is embedded a capillary tube (figs. 5.5 and 5.7). The lumen of the esophagus is lined with cuticle and is triradiate, except in the capillary tube, with one ray directed ventrally and two dorsolaterally.

The intestine is a more or less straight, thin-walled tube composed of epithelial cells. The type of cells varies in the different groups of nematodes and is helpful in distinguishing them, especially in sections of tissue containing the worms.

The intestine may be composed of three parts: a short part next to the esophagus that is separated from the remainder by a constriction is the **ventriculus** (fig. 5.50), the intestine proper, and the prerectal part. In some species, an anteriorly or posteriorly directed **cecum** originates from the ventriculus. The **rectum** is the terminal part of the intestine and is lined with cuticle.

Male Reproductive System

The male reproductive system normally consists of three principal parts: the **testis** (there may be two in some species), the **seminal vesicle,** which is usually dilated and serves as a sperm storage organ, and the **vas deferens.** The terminal part of the vas deferens is generally a muscular structure that functions as an **ejaculatory duct** (fig. 5.1B).

Accessory structures are associated with the male reproductive system. They include one or two cuticularized **spicules** in a **spicular pouch** that opens into the dorsal part of the cloaca (fig. 5.1B). The spicules are protrusible and function in mating. Sometimes a **gubernaculum,** a thickening of the dorsal wall of the spicular pouch (fig. 5.1B), is present and serves as a guide in directing the spicules during protrusion. A somewhat complex structure formed by a thickening of the lining of the cloaca of lungworms is termed the **telamon** (fig. 5.48).

Female Reproductive System

The female reproductive system consist of **ovaries, oviduct, uterus,** and a cuticular lined **vagina** opening to the outside of the body through the **vulva,** which is usually located on the ventral side of the body. There may be one ovary, which is **monodelphic;** two ovaries, which is **amphidelphic;** or more than two, which is **polydelphic** (fig. 5.1C,D). In amphidelphic forms, the ovaries may be arranged opposite each other so that one lies in each end of the body, or they may be parallel and both lie in either the anterior or posterior parts of the body. The terminal part of the uterus may be muscular, forming an **ovijector** that forces the eggs into the vagina. The uterus of mature worms generally is filled with eggs or larvae.

Excretory System

The overt parts of the excretory system consist of the minute **excretory pore** opening ventral to the esophagus and its slender duct directed posteriorly (fig. 5.1B). The remaining parts of the system consist of ducts in the lateral chords. The arrangement of the ducts may vary considerably in pattern from the basic H-shaped type found in the oxyurids. In some forms, there is a gland that lies posterior to the cross bar of the H into which it empties. In others, parts of the basic H-shaped organ are missing.

Nervous System

The conspicuous **nerve ring** surrounding the esophagus about midway between the anterior and posterior ends represents the central nervous system (fig. 5.1E). From it, fibers extend (1) anteriorly to innervate the cephalic papillae of the lips, amphids, deirids (cervical papillae), and esophagus, and (2) posteriorly to the genital papillae, phasmids, and intestine. Peripheral nerves originate from the fibers and extend to the body muscles. Generally, only the nerve ring is seen.

General Biology

The life cycles of nematodes are of two basic types: direct or monoxenous (i.e., with only one host in the cycle) and indirect or heteroxenous (i.e., with two or more hosts in the cycle). The basic pattern of development is similar whether the life cycle is direct or indirect. Larvae (also known as juveniles) hatching from the eggs progress through a series of stages in their development. Beginning with the first stage, each one is separated by a molting of the cuticle. There are four larval stages, i.e., first, second, third, and fourth, followed by the adult. The third stage is infective to the final host. The pattern of growth of the larvae (L) and occurrences of the successive molts (M) may be expressed as follows:

$$\text{Egg} \rightarrow L_1 + M_1 \rightarrow L_2 + M_2 \rightarrow L_3 + M_3 \rightarrow L_4 + M_4 \rightarrow \text{Adult}$$

In many parasitic nematodes (orders **Rhabditata**), **Strongylata**) with a direct life cycle, the first, second, and third stages are free in the soil. The first two stages feed on organic material. The third stage, which retains the shed cuticle of the second stage as an enclosing sheath, is unable to ingest food (fig. 5.1E,F). In some species, the first (*Ascaris*) and second molts (some hookworms, strongyles, and trichostrongyles) take place inside the egg shell.

Third-stage larvae enter the final host when swallowed with food (trichostrongyles, strongyles), or by penetrating the skin (hookworms). The third and fourth molts are completed inside the final host.

Nematodes with an indirect life cycle have the same stages but reach the final host by means of a vector or an intermediate host, usually an arthropod, but other invertebrates are used. Vectors are transmitters of parasites. If the transmitter is essential in the life cycle of the parasite, it is a biological vector; if it is unesssential, it is a mechanical vector.

In oviparous forms with the indirect cycle, the eggs are embryonated when laid and in most cases hatch only when ingested by the intermediate host, in whose body the third or infective larval stage is reached through the necessary molts. In some of the lungworms, the eggs hatch on the ground or while still in the lungs or intestine of the final host and the larvae penetrate land snails, which serve as the intermediate host. Infection of the final host occurs when the intermediaries containing the infective larvae are swallowed. The third and fourth molts take place in the final host.

Eggs of parasitic nematodes vary in shape, size, thickness of shell, and external markings, as well as the stage of development when laid. Each ovum comprises an egg cell that is enclosed in three primary layers, which consist of: (1) the thin vitelline membrane surrounding the egg cell, (2) the middle thick egg shell, and (3) the outer protein layer, which may be stained yellow with bile. The stage of development when laid may be a single cell or with all succeeding stages up to and including ensheathed third-stage larvae. Some precocious eggs hatch in the intestine or lungs, in which cases larvae appear in the feces. Other hatch in the uterus of the female worms, and active larvae are born.

In ovoviviparous nematodes with an indirect cycle, the intermediaries, such as copepods, become infected by ingesting early larval stages free in the water, such as those of dracunculids, or mosquitoes and other bloodsucking arthropods feeding on animals containing **microfilariae** in the blood, which are the larval stages of filariae worms. Larvae develop to but not beyond the third, or infective stage, in the vector and intermediate hosts. Infection of the final hosts occurs when infected copepods are swallowed or when infective larvae are injected into or onto the body when mosquitoes and other hematophagous arthropods are feeding.

The principal types of basic life cycles of the parasitic species are represented in figure 5.9.

Notes and Sketches

Table 5.1. Abbreviated Classification of Nematodes Parasitic in Vertebrates.

Class	Order	Suborder	Superfamily	Family
PHASMIDIA	Rhabditata	Rhabditina	Rhabdiasoidea	Strongyloididae
			Rhabditoidea	Rhabdiasidae
	Strongylata	Strongylina	Ancylostomatoidea	Ancylostomatidae
				Uncinariidae
			Syngamoidea	Syngamidae
			Strongyloidea	Strongylidae
		Trichostrongylina	Trichostrongyloidea	Trichostrongylidae
			Heligmosomatoidea	Heligmosomatidae
		Metastrongylina	Metastrongyloidea	Metastrongylidae
			Protostrongyloidea	Protostrongylidae
			Pseudalioidea	Filaroididae
	Ascaridata	Ascaridina	Ascaridoidea	Ascaridae
				Toxocaridae
	Oxyurata	Oxyurina	Oxyuroidea	Oxyuridae
			Syphacioidea	Syphaciidae
		Heterakina	Heterakoidea	Heterakidae
				Ascaridiidae
	Spirurata	Spirurina	Spiruroidea	Spiruridae
				Tetrameridae
			Thelazioidea	Thelaziidae
				Spirocercidae
			Gnathostomatoidea	Gnathostomatidae
			Physalopteroidea	Physalopteridae
	Camallanata	Camallanina	Camallanoidea	Camallanidae
				Cucullanidae
			Dracunculoidea	Dracunculidae
				Philometridae
	Filariata	Filariina	Filarioidea	Filariidae
				Diplotriaenidae
			Onchocercoidea	Onchocercidae
				Dirofilariidae
				Dipetalonematidae
APHASMIDIA	Trichurata	Trichinellina	Trichinelloidea	Trichinellidae
				Trichuridae
				Capillariidae
				Trichosomoididae
	Dioctophymata	Dioctophymatina	Dioctophymatoidea	Diotophymatidae

The basic classification of nematodes parasitic in vertebrates is presented in table 5.1.

Class Aphasmidea

The class is characterized by having an esophagus consisting of a long chain of individual cuboidal cells or a cylindrical, muscular one, in which case the males possess a fleshy, cup-shaped, rayless, copulatory bursa. Phasmids are absent.

Order Trichurata

The esophagus consists of a short anterior muscular part followed by a long capillary tubule embedded in a long string of single cuboidal cells, the **stichosome.** It contains the single superfamily Trichuroidea with five families.

Family Trichinellidae

This family contains the species *Trichinella spiralis,* known as the **trichina** or porkworm.

Trichinella spiralis

This species was discovered as calcified cysts in the muscles of a human cadaver in 1835 and in swine in 1846. A fatal case of human trichinellosis in 1850 was traced to the ingestion of ham and pork sausage, thus establishing the relationship between swine and human infection. The common hosts are, in addition to humans and swine, rats, cats, dogs, and various wild rodents and carnivorous mammals, particularly foxes, bears, and bobcats. The incidence of trichinellosis in adult humans of the United States is 4.2%, according to Zimmermann et al. (1973), who examined more than 8,000 diaphragms from 48 states and the District of Columbia; adjustment for age distributions shows that approximately 2.2% of the population has detectable trichina infections.

Description

Adult worms are small, with the posterior end slightly thicker than the anterior part. The esophagus consists of an anterior muscular part, which terminates in a

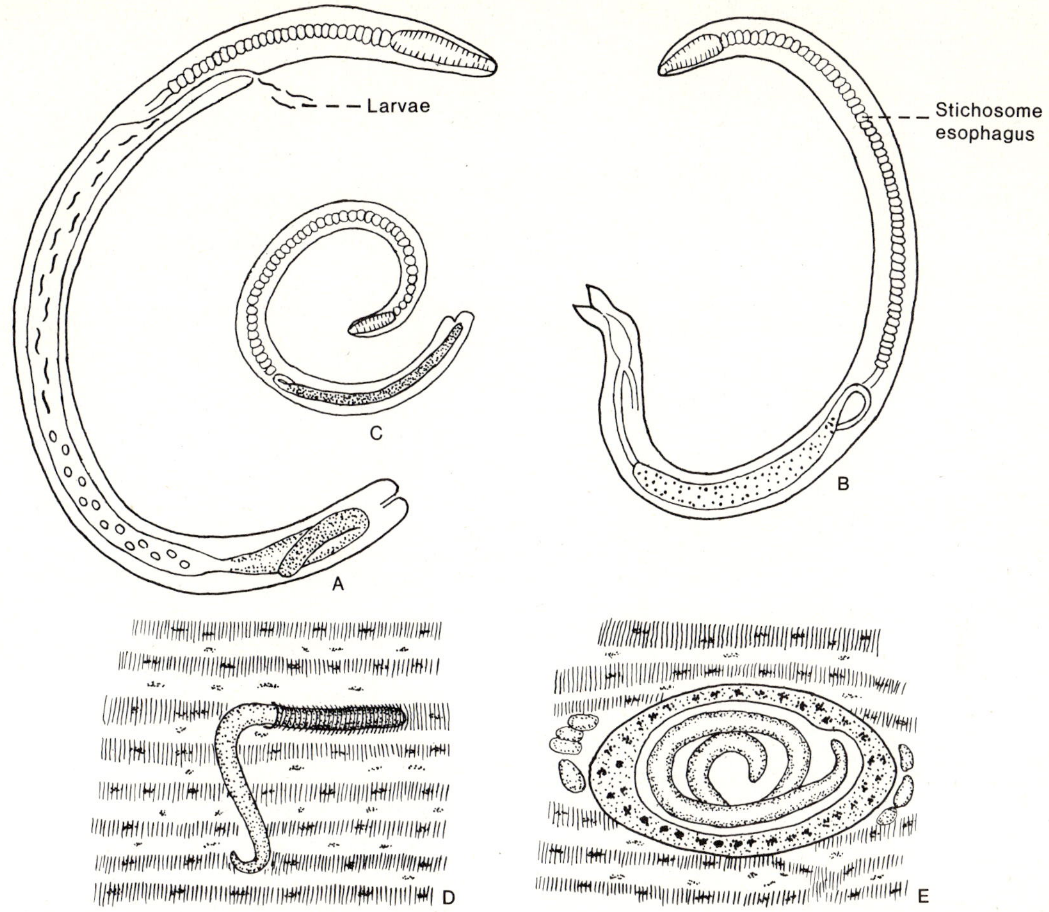

Fig. 5.2. *Trichinella spiralis.* (A) Adult female; (B) adult male; (C) newly born larva; (D) larva penetrating muscle fiber; (E) larva encapsulated in muscle, "nurse cell."

pseudobulb, followed by a constriction to a capillary tubule, that runs alongside a string of cuboidal cells, the stichosome, to enter the much larger intestine (fig. 5.2).

Males are up to 1.6 mm long and 50 μm in diameter. The posterior end of the body is blunt and bears a large, conical, **copulatory papilla** on each side of the terminal anus. The testis begins in the posterior fifth of the body as a large, elongate organ, continues forward to the posterior end of the esophagus, constricts, and bends backward, extending as a narrow tube to the cloaca. There is no spicule, although a **spicule sheath** (sac) can be extruded to perform as a spicule.

Females are up to 4 mm in length, with the greatest diameter exceeding that of the males. The posterior end is blunt and the anus terminal. The single ovary is in the posterior end of the body. It continues forward as a slender oviduct to about the junction of the middle and posterior thirds of the body, where it expands into a broad single uterus filled with eggs and hatched larvae. The vulva is in the esophageal, or stichosomal, region. These worms are ovoviviparous, with the larvae deposited in the host's intestinal mucosa.

Life Cycle

Infection of new hosts occurs when they (1) eat flesh containing live larvae, or (2) ingest feces containing larvae from a prior meal of trichinous meat, or adults that are passed spontaneously.

The transmission of *T. spiralis* is complex, involving either a **sylvatic** or an **urban cycle.** The sylvatic cycle includes wild carnivores, especially bears, wild boars, and seals as sources of human infection. The urban cycle involving swine, rats, and humans is common where swine are fed uncooked garbage, and it is the more common source of human infection. Figure 5.3 shows some of the sources of infection in both cycles and their interrelationships.

Once freed in the stomach, the larvae pass into the small intestine, where they reach sexual maturity within two or three days. Mating occurs, after which the females penetrate the intestinal lining. Fertilized eggs hatch within the female and leave her as larvae. Larviposition begins within four or five days after mating and from six to seven days postinfection.

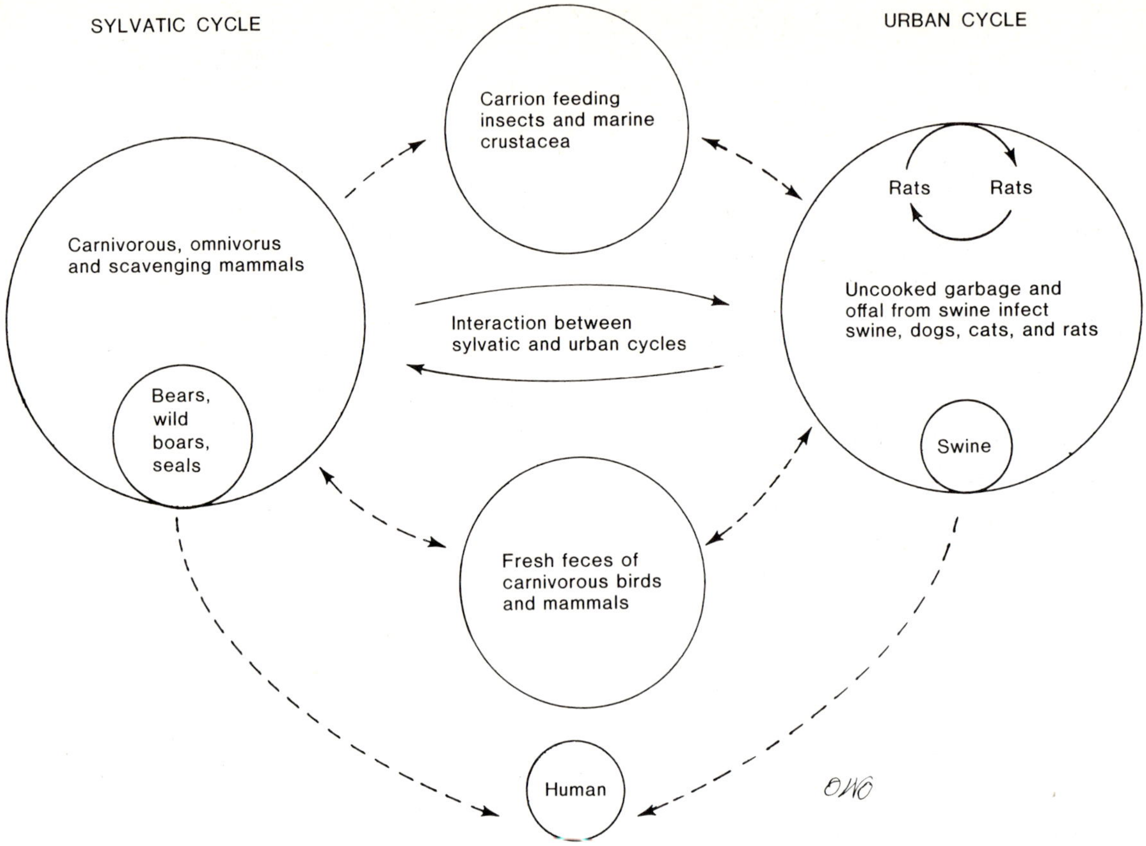

Fig. 5.3. The sylvatic and urban transmission cycles of *Trichinella spiralis,* with interaction between the two cycles. (Modified from Bessonov, 1972.)

Newborn larvae normally enter directly into either the lymph spaces or the hepatic portal system. Upon reaching the venous system, they are carried via the liver, heart, lungs, back to the heart, and through the arterial circulation to the various parts of the body, where they leave the capillaries and enter the skeletal muscles. The highest concentrations of larvae usually occur in the muscles of the diaphragm, tongue, larynx, masseters, intercostals, and orbits.

Once in the muscle fiber, the larvae increase in size and become coiled. Meanwhile, the muscle fibers that have been invaded by the larvae undergo degenerative changes, and each parasite becomes surrounded by a membranous cyst formed from muscular connective tissue. The cyst begins forming about the third week and is completed within a month (fig. 5.4).

Investigators are not in agreement on the number, time, and location of the molts, but it appears that *T. spiralis* has four molts and they occur in the intestine of the host. It is possible, as has been suggested, that the varying results have been influenced by: (1) strains of *T. spiralis* used, (2) species of experimental hosts, (3) intraspecific differences among experimental hosts, and (4) techniques employed.

Family Trichuridae

These are the **whipworms** and are medium to large in size. The esophageal part of the body is very slender and longer or shorter than the thick posterior part that contains the reproductive organs. The eggs always have a plug at each end. They are parasites of both the intestine and caecum of amphibians, reptiles, birds, and mammals.

Trichuris trichiura

This is the human whipworm that also occurs in swine and monkeys. Adults commonly live in the cecum but may occur in the appendix, colon, and rectum. The long, slender portion of the body is buried in the mucosa.

Description

The anterior end of the body is slender and about twice as long as the very much thicker posterior end. The esophagus consists of a short anterior, muscular portion and a long posterior portion made up of a capillary tubule running alongside a row of closely packed cells, reaching to the union of the two parts of the body. The mouth has no lips but is provided with a tiny **spear.** A

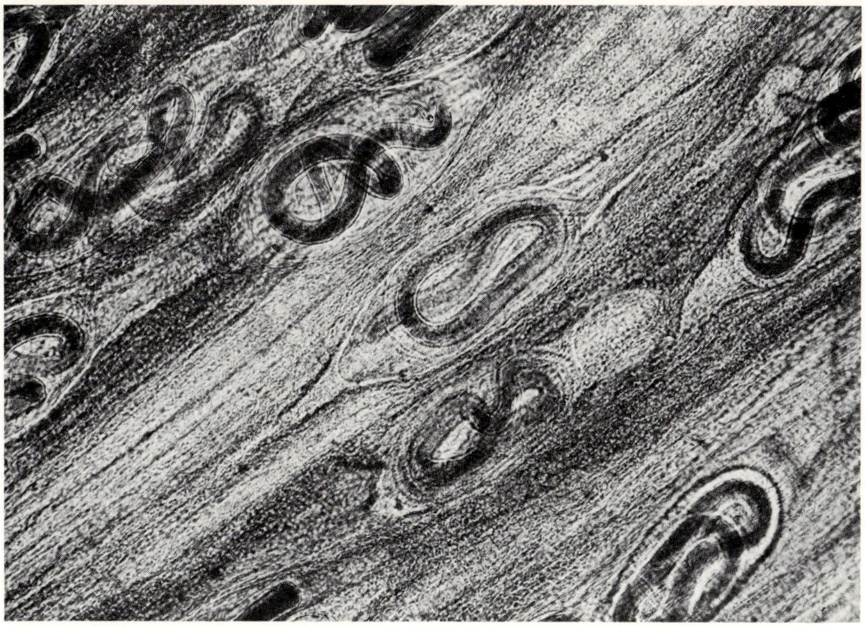

Fig. 5.4. *Trichinella spiralis,* encapsulated larvae in skeletal muscle.

bacillary band (a band of fine, low cells within the cuticle) extends the length of the ventral side of the esophageal region.

Males are up to 45 mm in length. The hind end is coiled ventrally as much as 360 degrees at times. The testis begins near the posterior end of the body and extends forward as a straight tube to the esophago-intestinal juncture where it bends abruptly back to form the thick, muscular ejaculatory duct. A second constriction is followed by a broad, slightly shorter cloacal tube that empties into an almost equally long spicule tube. A single, 2.5 mm long spicular protrudes through a retractable sheath with a bulbous end (fig. 5.5).

Females are up to 50 mm long and bluntly rounded posteriorly. The ovary is a slender, sinuous organ, beginning near the middle of the enlarged portion of the body and extending to the posterior extremity, where it sharply bends forward and quickly expands to form a narrow oviduct that opens into a broad uterus. The anterior end of the uterus narrows into a long vagina opening through the vulva, located near the posterior end of the stichosome.

Life Cycle

The life cycle is direct: eggs (fig. 5.6) reach the infective stage in about three weeks under favorable environmental conditions. They hatch in the alimentary canal of the definitive host. Larvae reach the caecum and mature in about a month. Eggs may remain viable in the soil for several years.

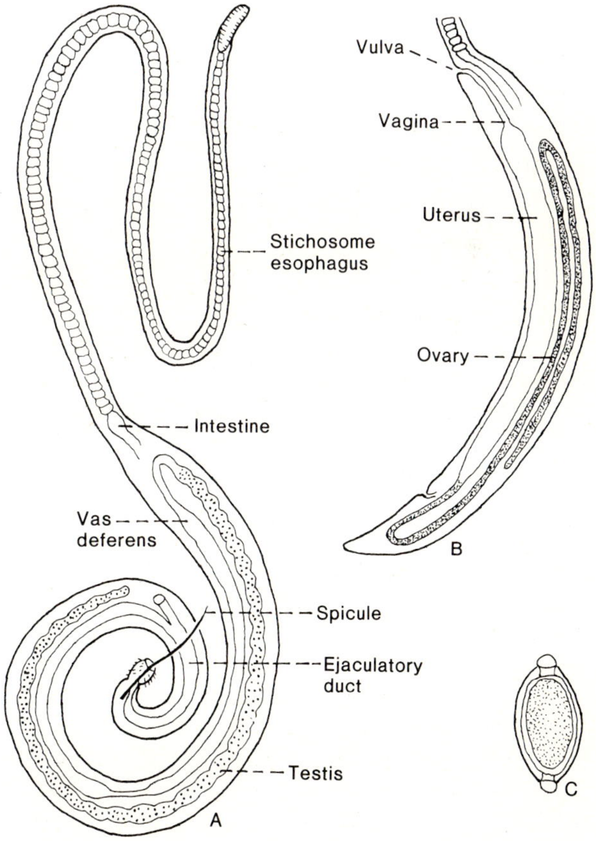

Fig. 5.5. *Trichuris trichiura.* (A) Adult male; (B) posterior portion of adult female; (C) unembryonated egg with clear terminal opercula.

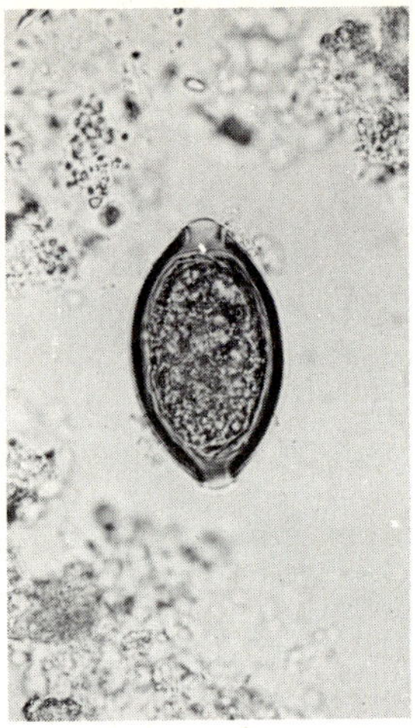

Fig. 5.6. *Trichuris trichiura.* Barrel shaped, with an outer and an inner shell, dark brown, with a distinct plug at each end; contents unsegmented; size 50–55 × 21–24 μm.

Other common species of whipworms include *T. ovis* of sheep, *T. opaca* of muskrats, *T. vulpis* of dogs and foxes, and *T. leporis* of rabbits. They are sufficiently similar to *T. trichiura* to replace it for anatomical studies.

Family Capillariidae

Members of this family are threadlike in having a slender esophageal region about equal to or slightly shorter than the thin posterior part of the body, which is only slightly thicker than the anterior part. It is a large family with numerous species, some of which occur in all the classes of vertebrates.

Capillaria columbae

This species occurs in the small intestine of domestic and wild pigeons, mourning doves, fowl, turkeys, and other birds in many parts of the world.

Description

The body is threadlike with the esophageal part a little shorter and only slightly thinner than the posterior portion. The esophagus consists of the short anterior muscular portion, the long, posterior capillary tube, and the long stichosome (fig. 5.7).

Males are up to 10.8 mm long by 0.06 mm in diameter. A narrow ventral and two well-developed lateral bacillary bands extend the length of the body. Each lateral one has a width equal to one-fourth to one-third the diameter of the body. The esophagus is up to 6 mm long. The posterior end of the body has a rounded, bursalike membrane supported by a pair of L-shaped processes, each of which has a terminal papilla. The spicule has an expanded base and a bluntly rounded tip; it is

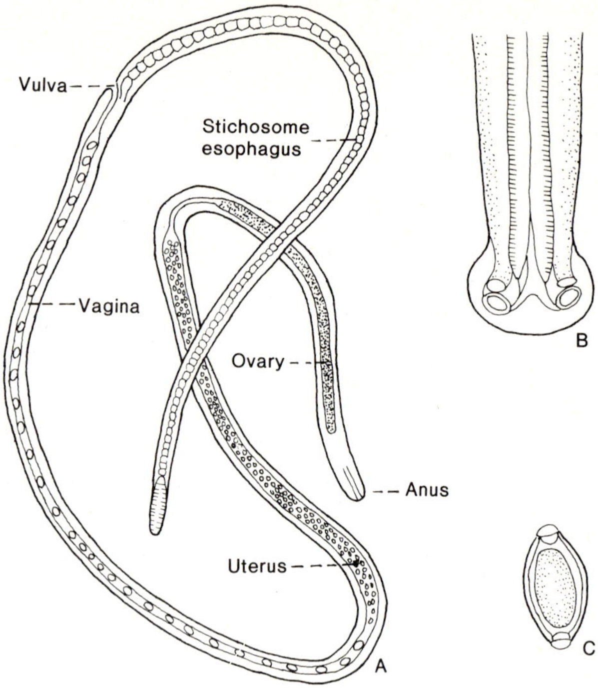

Fig. 5.7. *Capillaria columbae.* (A) Gravid female; (B) ventral view of caudal end of adult male; (C) unembryonated egg with clear operculum at each end.

up to 1.6 mm long by up to 12 μm in maximum width. The spicule-sheath is transversely wrinkled but devoid of spines.

Females attain lengths up to 19 mm and diameters up to 0.09 mm. The posterior end is bluntly rounded and dark colored. The vulva is slightly behind the esophago-intestinal junction. The vagina is directed posteriorly and is heavily muscled. The anterior lip of the vulva is slighly prominent, but there appears to be no protrusible membrane.

Eggs are unembryonated when laid, yellowish in color, vary somewhat from symmetrical lemon shaped to slightly asymmetrical, with a clear mucoid operculum at each end, and thick shells with a granular surface pattern. They measure 41–55 × 30–31 μm. The life cycle is direct and can be completed without the aid of an invertebrate intermediary.

Of the species with a monoxenous life cycle found in mammals, *Capillaria hepatica* is the most frequently encountered. It inhabits the liver of a variety of animals, primarily rodents, where its eggs accumulate without developing. Since the eggs require exposure to air to embryonate, preliminary passage through the intestine of a predator animal or death and decomposition of the original host is necessary to free them. Infection does not result from eating fresh, egg-burdened liver. Infection results when rodents swallow larvated eggs, which hatch in the small intestine, and

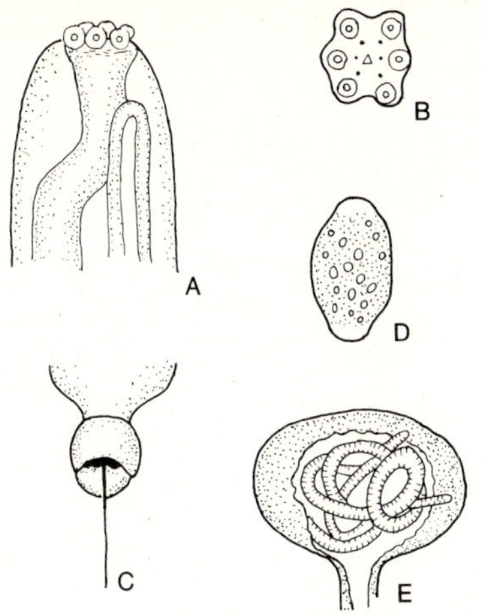

Fig. 5.8. *Dioctophyme renale.* (A) Anterior end of adult male, showing papillae and recurved part of testis; (B) en face view of head, showing arrangement of papillae around mouth; (C) caudal end of adult male, showing cuplike rayless copulatory bursa and single spicule; (D) unembryonated egg with pitted shell; (E) adult worm in kidney.

the larvae go to the caecum, enter the branches of the hepatic portal vein, and are carried to the liver, where they mature. They resemble the eggs of *T. trichiura* but have pitted shells.

Some commonly encountered Capillariidae with a heteroxenous life cycle, requiring earthworms as intermediaries, include *Capillaria caudinflata* in the lower intestine of chickens; *C. annulata* in the wall of the esophagus and crop of chickens; and *C. plica* of the urinary bladder of dogs, foxes, and wolves.

Order Dioctophymata

Members of the order are medium to large worms whose males have a fleshy, rayless, bell-shaped bursa and a single bristlelike spicule. The esophagus is muscular, cylindrical, and without a posterior bulb. Females have a single ovary.

Family Dioctophymatidae

Parasites of the kidney, occurring occasionally in the coelomic cavity of carnivorous mammals.

Dioctophyme renale

This, the **giant kidney worm,** normally is a parasite of mink but occurs in dogs, foxes, and other mammals, including humans. It occurs most commonly in the right kidney because the loop of the small intestine from which the larvae migrates lies close to or in contact with it. The kidney becomes a mere shell containing the coiled worm. Sometimes, they escape from the kidney and occur in the body cavity. (see fig. 5.8.)

Description

These are large cylindrical worms with the cuticle transversely striated. The hexagonal mouth is devoid of lips but is encircled by two series of six papillae each. The body is attenuated at both ends and bears a series of papillae along each lateral line.

Males are up to 45 cm long by up to 6 mm in diameter. The copulatory bursa is cup or bell shaped; its margin and inner surface are covered with minute papillae. The cloacal opening is near the center of the inside of the bursa. There is a single, slender spicule up to 6 mm long.

Females attain lengths of up to 100 cm and diameters up to 12 mm. The caudal end is rounded and the terminal anus crescent shaped. The vulva is anterior, being up to 7 cm from the anterior end of the body.

Eggs are barrel shaped with thick, deeply pitted shells and polar plugs, brownish yellow in color, and are in the two-celled stage when laid. They average $73 \times 46\ \mu m$.

Life Cycle

Eggs laid in the kidney are voided with the urine, develop to first-stage larvae at temperatures from 14 to 30 C; at room temperature it takes about 35 days. Embryonated eggs eaten by the aquatic oligochaete *Lumbriculus variegatus* hatch in the intestine, and the larvae migrate to the ventral blood vessel where development continues. The first molt occurs about 50 days after infection in oligochaetes at 20 C; the second molt occurs about 50 days later. Brown bullheads (*Ictalurus nebulosus*), and probably other fish, and frogs (*Rana* spp.) are believed to be important natural paratenic hosts in the circulation of *Dioctophyme.*

The mammalian host becomes infected by eating infected oligochaetes or paratenic hosts. In the final host, larvae penetrate the stomach wall and migrate to the liver. About 50 days postinfection, larvae leave the liver parenchyma and migrate directly to the kidney, where maturity occurs. Prepatency in mink is 154 days. The life cycle, which requires only one intermediary, takes about 9.5 months; 35 days for embryonation of the eggs, 100 days for larval development in the Oligochaeta, and 154 days for maturity in the final host.

Notes and Sketches

Class Phasmidia

The esophagus is club shaped, cylindrical, or with a posterior bulb, and muscular, or it may be muscular in the anterior and glandular in the posterior parts. A bilobed, copulatory bursa, when present, is supported by six pairs of lateral rays. Phasmids are present but difficult to see in adult worms.

Order Rhabditata

Females are the only known parasitic forms. The esophagus of the adult is muscular, cylindrical, and long or short.

Family Strongyloididae

The esophagus is long and slender in the parasitic females, there is no buccal capsule, and the vulva is in the posterior part of the body. Adult females are parasitic in the mucosa of the small intestine of reptiles, birds, and mammals.

There are two types of worms, the parasitic parthenogenic females in the small intestine, and the free-living males and females.

Strongyloides stercoralis

This human intestinal threadworm is cosmopolitan in distribution but occurs primarily in warm climates. (See fig. 5.37.)

Description

Parasitic females are small, slender, threadlike nematodes about 2 mm long, with both ends pointed. The vulva is near the junction of the middle and posterior thirds of the body. The anus is near the posterior end, leaving a short, pointed tail. The alimentary canal consists of a small buccal capsule, a long, slender filariform esophagus extending through the anterior third of the body, and a thin intestine. Cuticularized buccal spears appear in the anterior end of the esophagus. The short vagina divides into an anterior and a posterior uterus and ovary; each ovary begins near the middle of the body and extends anteriorly or posteriorly, and loops back upon itself to form the oviduct and uterus. Each branch of the uterus is filled with a short string of thin-shelled, embryonated eggs that hatch within the intestinal epithelium of the host. Parasitic males are unknown.

The free-living adults have a short rhabditiform esophagus. In the female, the vulva is near the middle of the body. The short vagina receives the anterior and posterior uteri, which contain a single string of thin-shelled, embryonated eggs. The smaller male has a long, rather thick testis that opens into an equally wide seminal vesicle that narrows into a terminal muscular ejaculatory duct. A pair of short spicules and a gubernaculum are present.

Life Cycle

Eggs of two kinds are produced by the parasitic females. The homogonic line has three chromosomes and produces rhabditiform first- and second-stage and filariform third-stage larvae with a notched tail. These filariform larvae are infective and enter the host through the skin and develop to parthenogenic females. The eggs that produce the heterogonic line are of two kinds. Those with one chromosome develop into free-living males and those with two chromosomes into free-living females. All stages of larvae hatching from the 1N and 2N eggs are rhabditiform, as are the adults. The eggs produced by the free-living females are 3N, and the larvae develop into infective larvae similar to those of the homogonic line.

There are two types of life cycles, each with different forms. They are the homogonic, or parasitic, in which a parthenogenic female produces larvae that enter the host or larvae that develop into heterogonic, or free-living males and females, whose larvae develop into the infective stage and enter the host. Upon contact with skin, the filariform larvae from either direct or indirect mode of development penetrate the epidermis, enter the small blood vessels, and are carried to the lungs. Here they break out of the capillaries into the alveoli and make their way via the trachea and throat to the intestine, where they molt twice and reach maturity in about two weeks.

Strongyloides ratti

This is a common intestinal parasite of colonies of white rats and populations of wild ones. (See fig. 5.9.)

Description

Parthenogenic females are up to 3.0 mm long. There are six lips. The finely pointed tail is about twice as long as the diameter of the body at the anus. The vulva is near the union of the third and last quarters of the body. There is a very shallow buccal capsule followed by a filariform esophagus one-fifth to one-third the total length of the body. There are two ovaries, one anterior with the reflexed part near the base of the esophagus and the other posterior and near the anus. The vagina is very short. Each uterus contains ten to 12 thin-shelled, embryonated eggs measuring 47–52 × 28–30 μm.

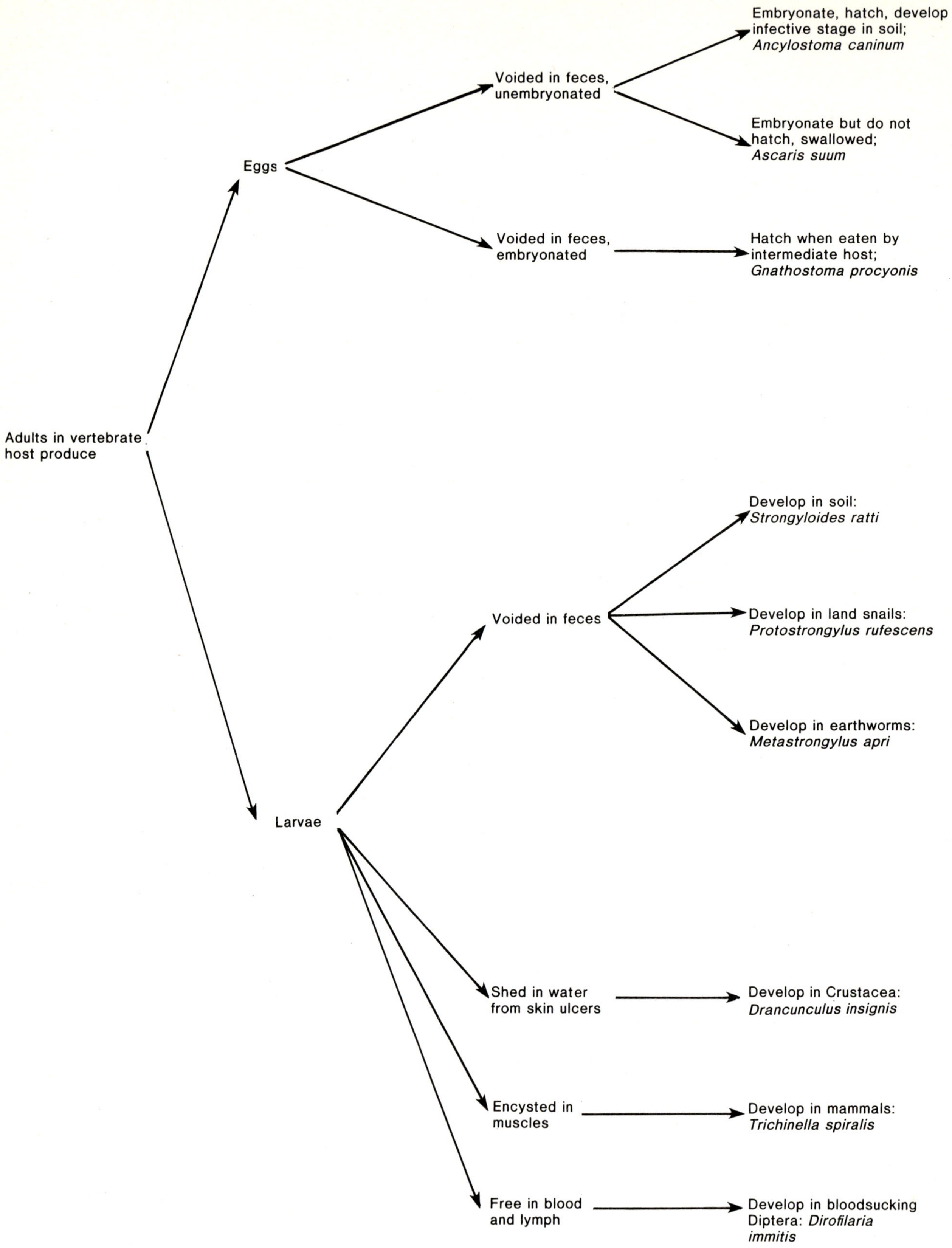

Fig. 5.9. Some basic types of life cycles of common parasitic nematodes. There are so many variations it is impossible to list all of them. Consult the classification presented in table 5.1.

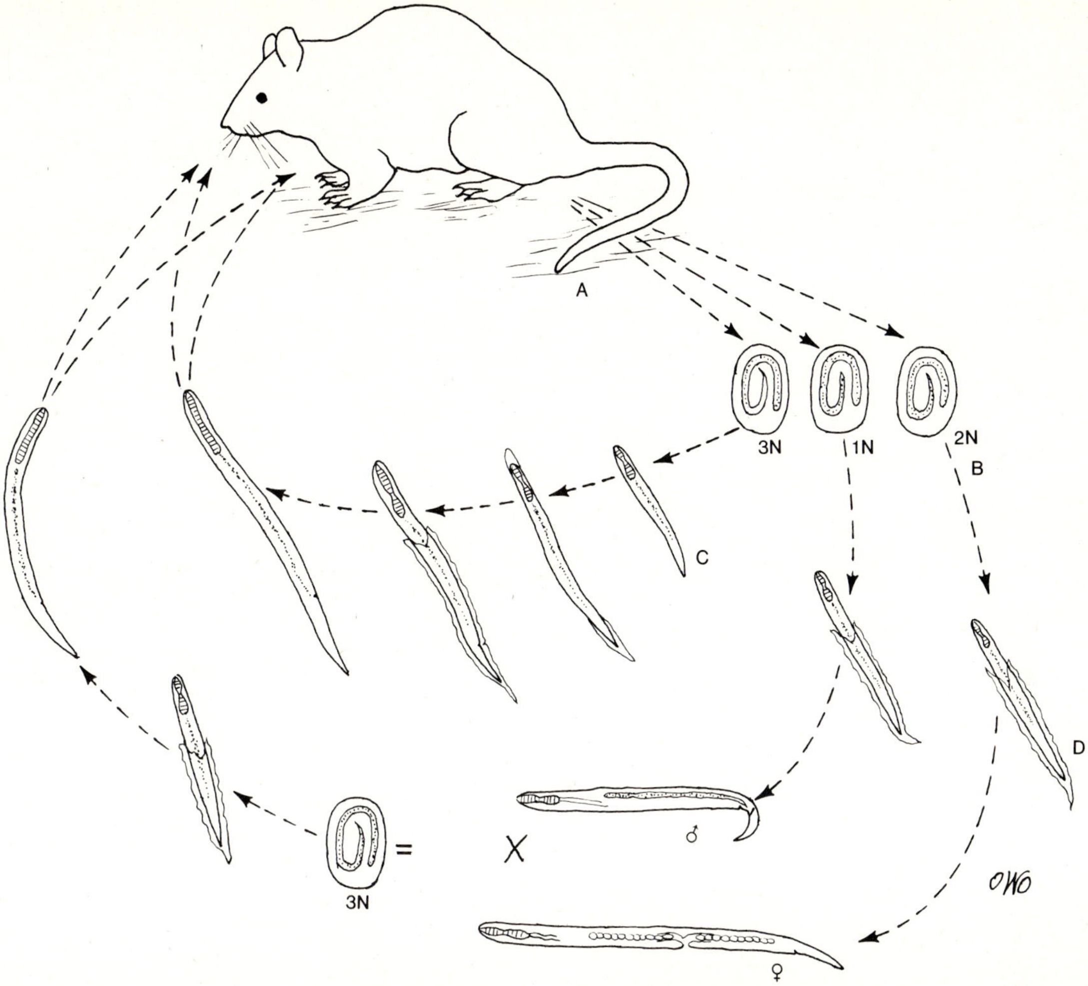

Fig. 5.10. Life cycle of *Strongyloides ratti.* (A) Definitive host harboring migrating larvae and adult parthenogenic females; (B) three types of eggs produced by parthenogenic females; (C) homogonic line from 3N egg showing two rhabditiform larvae and a filariform infective larva which may enter the rat by being ingested or by penetrating the skin; (D) heterogonic line where 1N eggs produce rhabditiform larvae and free-living males and 2N eggs free-living females. Mating males and females produce 3N eggs from which infective filariform larvae develop. This life cycle is basically similar for all species of *Strongyloides.* In some species such as *S. westeri* from horses and *S. ransomi* from swine, larvae migrating in the tissues enter the mammary glands and are transmitted to nursing young via the milk.

Life Cycle

Larvae of the homogonic line are 3N, rhabditiform in the first and second stages, and filariform in the third stage. (See fig. 5.10.)

Free-living stages are not well known, as studies have been directed mainly toward the parasitic phase. In general, they are short and plump worms. Adult males are around 0.8 mm long and have a ventrally curved, pointed tail. There is a pair of prominent, pointed spicules and a simple testis. Females have an anterior and posterior ovary.

Larvae of the heterogonic line developing from eggs produced by the parasitic females are all rhabditiform and develop into free-living adult males (1N) and females (2N). Their offspirng (3N) are rhabditiform in the first and second stages, and filariform and infective in the third stage.

In view of its ready availability, this is an ideal model for class use in demonstrating the anatomy and biology of this group of parasites. For life cycle details consult Spindler (1958), and Abadie (1963).

Family Rhabditidae

In this group, the buccal capsule is present and the esophagus is short. The vulva is near the middle of the body. Found in the lungs of amphibians and reptiles.

Rhabdias ranae

This species, the lungworm of frogs, is widespread in North America. (See fig. 5.36.)

Description

Adult parthenogenic females are up to 4.5 mm long. The circular mouth is surrounded by three groups of low papillae. There are two pairs of lateral postanal papillae. The vulva is up to 2.5 mm from the tip of the tail, and the anus is up to 0.3 mm. The buccal capsule is broadly crescent shaped, with the concave side posterior. A club-shaped esophagus up to 0.6 mm long is followed by a dark-colored intestine. A pair of long cervical glands is located at the esophago-intestinal junction, and three rectal glands are near the anus. Each ovary begins opposite the vulva, one extending anteriorly almost to the esophagus and the other posteriorly almost to the anus. Each bends back to form a short oviduct, seminal receptacle, and wider uterine portion that is filled with eggs. The vagina is extremely short. Embryonated eggs measure 40 × 75 μm, and the larvae are 0.3 mm long at the time of hatching.

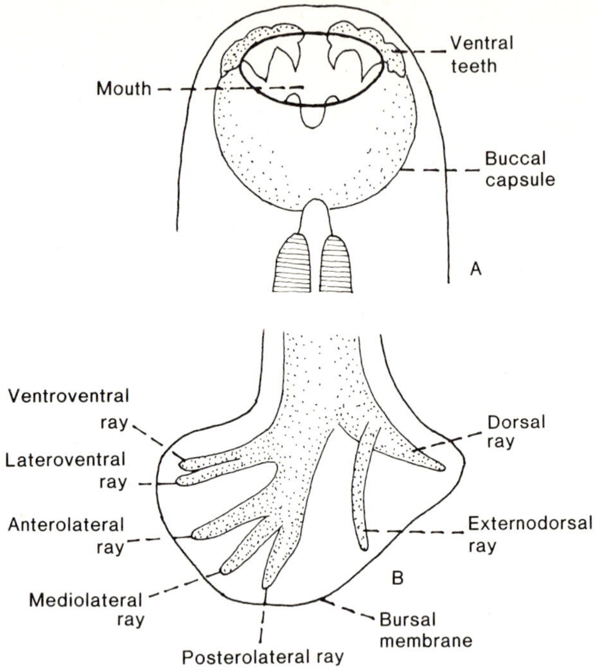

Fig. 5.11. *Ancylostoma duodenale.* (A) Anterior end of adult; (B) sinistral view of bursa, showing rays.

Order Strongylata

Males are characterized by a caudal membranous copulatory bursa supported by one dorsal ray, usually with a divided tip and six pairs of lateral rays. (See fig. 5.1H.) Some rays may be fused or lost. This order is known collectively as the bursate nematodes. Many species are of great medical and veterinary importance.

Family Ancylostomatidae

The **ventral cutting plates,** which surround the inner margin of the mouth, have teeth on their free margin. The head is bent sharply dorsad, giving a hooked appearance.

Ancylostoma duodenale

This is known as the Old World hookworm of humans and is of great public health importance. It occurs commonly in tropical and subtropical countries, bounded in general by the latitudes 36° N and 30° S.

Description

Refer to figure 5.11. Stocky worms with cuticle having fine transverse striations. A pair of cervical papillae, one on each side of the body, is located near the middle of the esophagus. The funnel-shaped buccal capsule is large and has a thick wall. There is a pair of ventral cutting plates, each with two large toothlike structures, and a small dorsal plate with a median indentation. A pair of internal teeth is located in the depth of the capsule. A short club-shaped esophagus opens through paired valves into the long tubular intestine. A pair of long cervical glands, each with a distinct nucleus, lies just posterior to the esophagus.

Males are up to 11 mm long. The **copulatory bursa** consists of a small **dorsal lobe** and two large **lateral** ones. Beginning with the single **dorsal ray,** whose tip is divided into several small digitations, the paired rays are as follows: **external dorsal rays** originating from each side of the base of the dorsal ray; **three lateral rays** arising from a single base are the **posterolateral, mediolateral,** and **externolateral** (sometimes called **anterolateral**), and two pairs of ventral rays, the **lateroventral** and the **ventroventral,** having a single basal stalk separate from that of the lateral rays. The posterio- and mediolateral rays originate from a single stalk. (See figs. 5.1H and 5.11B.)

The typical arrangement of the strongylid bursal rays can be represented roughly by placing both hands, palms down, on one's thigh with the thumbs together their full length and the fingers spread. The thumbs represent the dorsal ray, the index fingers the externodorsal rays; and the middle, ring, and little fingers the lateral complex of rays—postero-, medio-, and externolateral rays, respectively. With fingers lacking to represent the two ventral rays, they must be visualized.

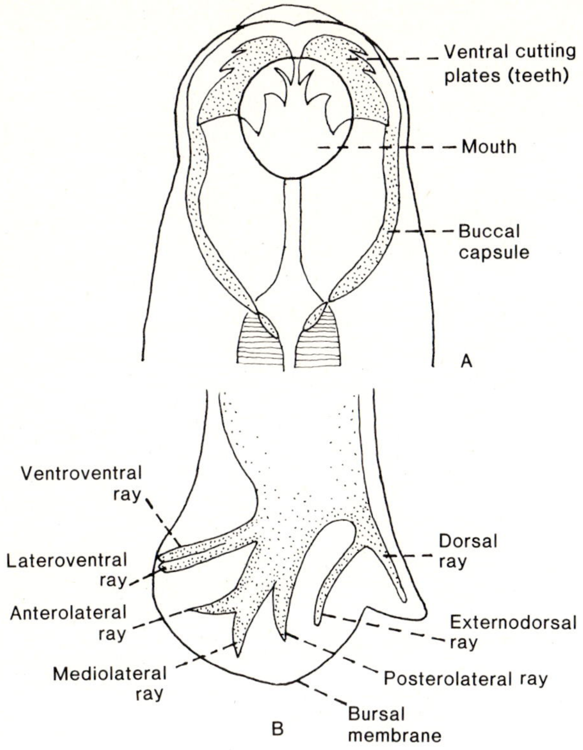

Fig. 5.12. *Ancylostoma caninum.* (A) Anterior end, showing mouth and teeth; (B) copulatory bursa, showing rays.

The reproductive system consists of the coiled, threadlike testis lying at the posterior ends of the cervical glands, followed by the posteriorly directed long, wide seminal vesicle, which empties into a long, muscular ejaculatory duct surrounded midway by cement glands. The ejaculatory duct narrows posteriorly and empties into the cloaca. A pair of long, cuticularized, brownish, needlelike spicules, associated with a gubernaculum in the dorsal wall of the cloaca, function in transferring seminal fluid from the male to the female.

Females are up to 13 mm long, and the tail ends in a small spine. The vulva is slightly behind the middle of the body. The two threadlike ovaries, one anterior to and the other posterior to the vulva, and their associated parts, the oviducts and uteri, are intricately wound about the intestine. Each uterus terminates in a slender muscular ovijector, and these join to form the short vagina. The ovijectors and terminal parts of the uteri are filled with elliptical, hyaline, thin-shelled eggs, typical of the Strongylina, in the early stages of cleavage. The eggs are similar to those of *Necator americanus* (see fig. 5.15.)

Life Cycle

Eggs, laid in the four-cell stage, pass in the feces onto the soil. Under favorable conditions of temperature and moisture, they hatch within 24 hours. After two molts, the larvae lose their typical rhabditiform esophagus and become strongyliform. Free-living larvae of hookworms in all stages are distinguishable from the larvae of *Strongyloides* by their longer and narrower buccal cavity, and in the strongyliform third stage by the unnotched, pointed tail. Third-stage infective larvae normally burrow through the skin, enter the lymphatics or veins, and are then carried by the bloodstream to the heart and then to the lungs. They break out of the capillaries into the alveoli, crawl up the bronchi and trachea to the pharynx, and finally are swallowed. In the small intestine, larvae undergo the third molt within three to five days. They grow rapidly, up to 5 mm, and the molt for the fourth and last time. Prepatency takes about six weeks.

Ancylostoma caninum

This is a common hookworm of dogs in temperate climates. It is larger than *A. duodenale* (males up to 12 mm and females 16 mm long). Each of the two dental plates has three teeth, of which the innermost is the smallest. Eggs are similar to those of *Necator americanus* (fig. 5.5). The life cycle is similar to that of *A. duodenale,* but transmammary infection is common. Larvae of *A. caninum,* together with those of *A. braziliense,* the adults of which occur in cats, cause cutaneous larval migrans or creeping eruption when they penetrate the skin of humans. (See fig. 5.12.)

Life Cycle

Eggs voided with feces develop and hatch in the soil, feces, or litter into first-stage **rhabditiform larvae** that molt to the second stage. By the end of a week in warm weather, a second molt occurs to produce third-stage **strongyliform infective larvae.**

Infection of dogs takes place when infective larvae burrow through the hair follicles or are swallowed (fig. 5.13). If entry is through the skin, larvae enter blood and lymph vessels and are carried by way of the heart to the lungs. In the latter, the larvae leave the capillaries, migrate up the trachea, and are swallowed. In the small intestine, they attach by means of the buccal capsule to the mucosa and develop to maturity in about two weeks. Some larvae return from the lungs by way of the pulmonary vein to the heart and enter the general arterial circulation. Those reaching the mammary

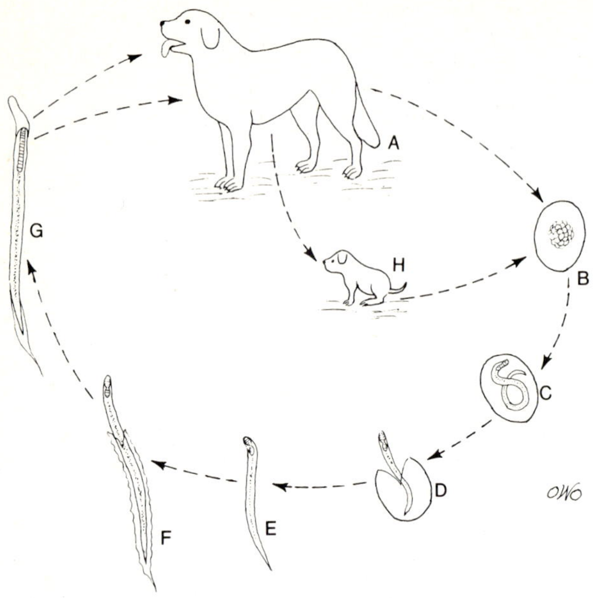

Fig. 5.13. Life cycle of *Ancylostoma caninum*, the dog hookworm, as a representative of the various species of hookworms. In this direct cycle, larvae infect the host by penetrating the skin or being swallowed. (A) Host; (B–G) development of larvae; (B) unembryonated egg; (C) unhatched first stage rhabditiform larva; (D) hatching; (E) free-living first-stage larva; (F) second-stage rhabditiform larva escaping from cuticle of first-stage larva; (G) infective third-stage strongyliform larva enters skin through hair follicles or is swallowed; (H) nursing pups may become infected by third-stage larvae in the milk. It is not known whether trans-mammary infection takes place with the species of hookworms that infect humans.

glands of pregnant bitches appear in the colostrum and are infective to nursing newborn pups. When infective larvae are swallowed, they develop directly in the small intestine without the extensive migration through the heart and lungs.

Family Uncinariidae

This family of hookworms is characterized by the large ventral **oral cutting plates** that are smooth along the free margin and semilunar in shape.

Necator americanus

This, the so-called **New World hookworm** of humans, is the predominant species in the United States and has a wide distribution in the tropical and subtropical regions of the world. (See fig. 5.14.)

In general, the anatomy is similar to *Ancylostoma duodenale* except that *N. americanus* is smaller and the free margin of the cutting plates is smooth and semilunar in shape. A pair of single **lancets** is at the bottom of the buccal capsule.

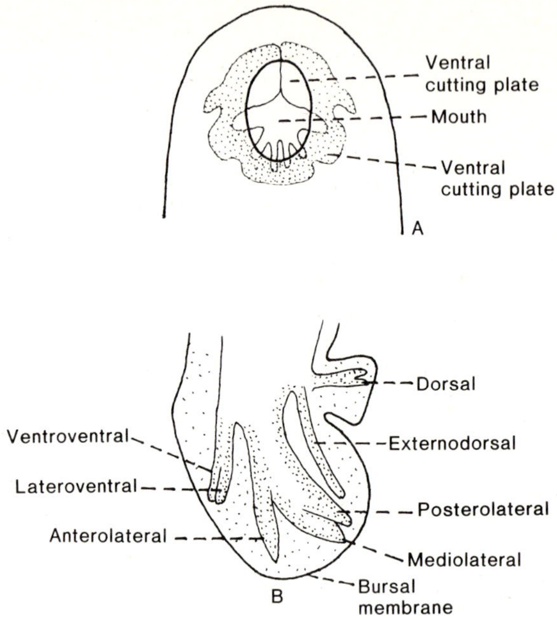

Fig. 5.14. *Necator americanus.* (A) Anterior end of adult, showing oval mouth and cutting plates; (B) left lobe of copulatory bursa, showing rays.

Males are up to 9 mm long. The bursa is much wider than long, with the dorsal lobe having a broad median indentation so that the dorsal ray is divided almost to its base into two widely divergent branches; the externodorsal rays are somewhat club shaped and near the laterals. The postero- and mediolateral rays originate from a common stalk, with the ventral rays arising from that same stalk. The spicules are long and slender. Their distal ends are fused and tipped with a barb.

Females are up to 11 mm long. The vulva is slightly anterior to the middle of the body. There is no spine at the tip of the tail. An egg is illustrated in figure 5.15.

Family Heligmosomatidae

This family has been chosen to represent the species of Trichostrongylina because of the ready availability of *Nematospiroides dubius* in laboratory mice, the ease with which material for study and experimentation can be obtained, and having a typical trichostrongyline life cycle. Other common Trichostrongylidae with similar life cycles include the ubiquitous **twisted stomach worm** (*Haemonchus contortus*), the **brown stomach worm** (*Ostertagia circumcinta*), and the **intestinal hairworm** (*Trichostrongylus colubriformis*) of sheep.

The Heligmosomatidae, considered by some authors as belonging to the Trichostrongylidae, have the following anatomical characters: The anterior end of the body has a cuticular expansion, and the copulatory

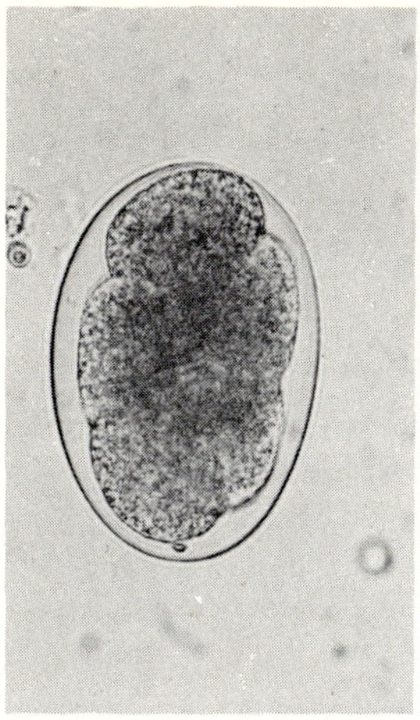

Fig. 5.15. *Necator americanus.* Ovoidal, with bluntly rounded ends, shell thin and hyaline; at discharge normally in early stages of segmentation (two- to eight-celled stage); size 65–75 × 36–40 μm. Ova of *Ancylostoma caninum,* indistinguishable from those of *N. americanus,* may be used as a substitute. Ova of *Ancylostoma duodenale,* otherwise identical, measure 56–60 × 35–40 μm.

bursa has large lateral lobes, which may be asymmetrical. The spicules are long and slender. Females have a single ovary and uterus, with the vulva near the anus. They are primarily parasites of rodents.

Nematospiroides dubius

This species (fig. 5.16) occurs commonly in the intestine of mice in colonies and in populations of wild mice.

Description

Live worms are red and coil in a spiral. The anterior end of the slender body is inflated. The cuticle has about 30 longitudinal striations. A buccal capsule is present but weakly developed.

Males are up to 10 mm long, have elongate prebursal papillae and a large asymmetrical bursa in which a dorsal lobe is lacking. Lateral and ventral rays originate from a large but short common trunk that divides into a large ventral and a small lateral branch. The large ventral rays are strongly divergent; the externo- and mediolateral rays are nearly parallel, but the posterolateral rays are divergent. Externodorsal rays originate near the base of a slender dorsal ray that is quadrifurcate at the tip. Spicules are slender, filariform, and up to 0.6 mm long. A gubernaculum is lacking.

Females are up to 21 mm long. There is a single reproductive organ with the vulva near the anus. The tail is fairly long, conical, and bears a small spine at the tip. Eggs 75–90 × 43–58μm.

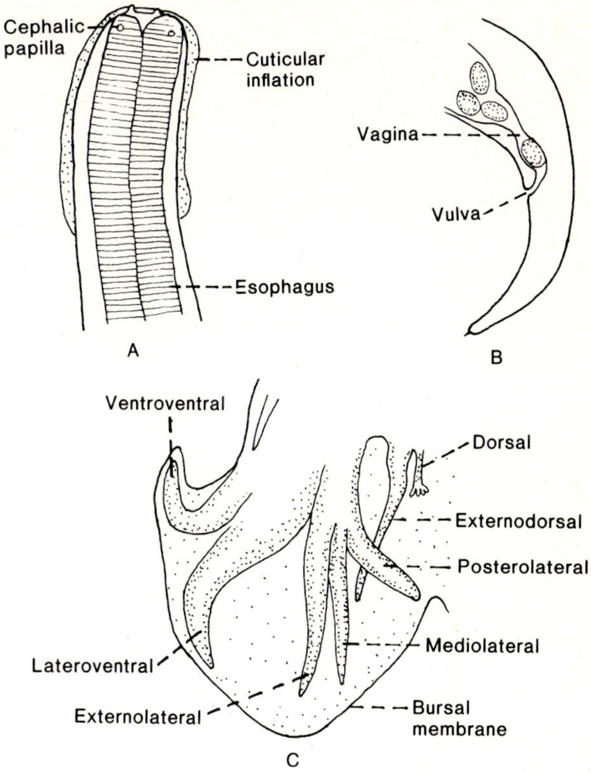

Fig. 5.16. *Nematospiroides dubius.* (A) Anterior end of adult; (B) posterior end of gravid female; (C) left lobe of bursa, showing rays.

Life Cycle

Eggs passed in the feces are in the eight to 16 cell stage, and hatching occurs about 24 hours after leaving the host. The free-living larval stages require about four days to become infective. Upon being ingested, the larvae lose their sheaths, and 24 to 48 hours later they have penetrated the intestinal mucosa and are lying near the longitudinal muscle layer of the gut wall. Here the third molt is completed in six to eight days, after which the fourth-stage larvae leave the mucosa and return to the intestinal lumen, where they undergo the last molt. The prepatent period is about nine days, and the cycle from egg to egg takes about 15 days. (See fig. 5.17.)

Families of Lungworms

Adults of this group occur in the respiratory system of mammals. They are long, filiform worms in which the buccal capsule is absent or rudimentary. The bursa is usually small and the rays atypical, being reduced in either size or number, often by fusion, or both. The vulva

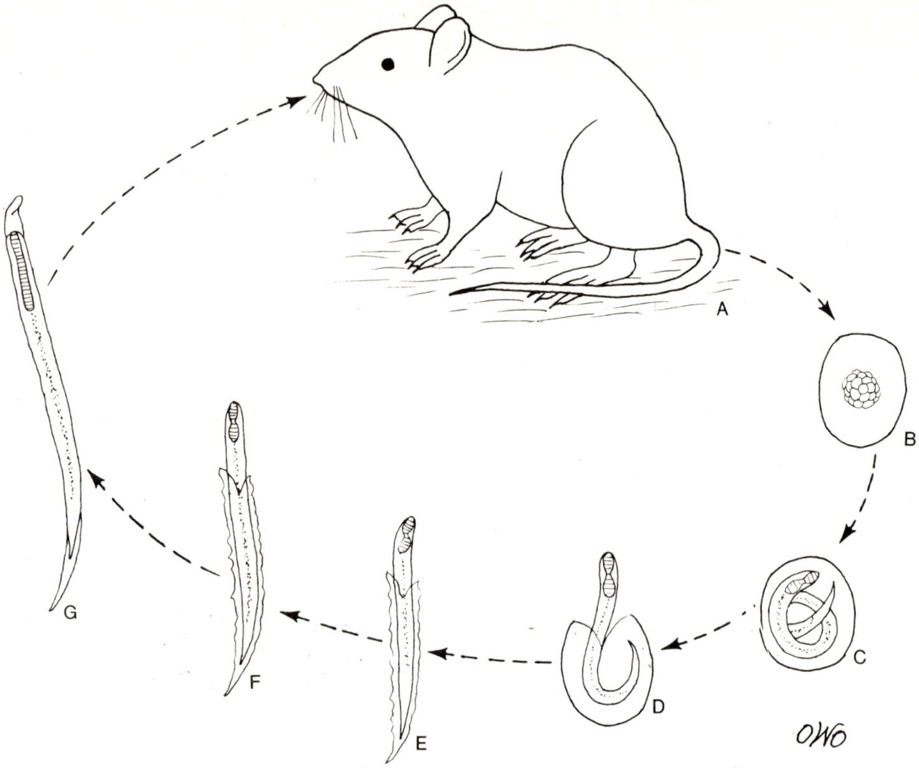

Fig. 5.17. Life cycle of *Nematospiroides dubius*, a typical trichostrongyle cycle. (A) Mouse host in whose intestine adult worms live; (B) unembryonated eggs voided in feces; (C) eggs embryonate in feces and litter, containing first-stage rhadbitiform larva; (D) hatching first-stage larva; (E) first-stage larva molting to form second-stage rhabditiform larva; (F) second-stage larva molting; (G) third-stage ensheathed filariform larva that infects mouse when swallowed. Third-stage infective trichostrongyle (*Haemonchus, Ostertagia, Trichostrongylus*, etc.) larvae of sheep and strongyle (*Strongylus*, etc.) larvae of horses climb moist grass blades and are swallowed with the forage by grazing animals.

is behind the middle of the body and the females are oviparous or ovoviviparous. The basic characteristics of a few common genera are given.

Dictyocaulus spp, (Dictyocaulidae) have a well-developed bursa in which the postero- and mediolateral rays and both ventral rays are fused except at the tips. The spicules are large, equal, and "sox shaped" (fig. 5.43). The vulva is near the middle of the body. They are parasites of equines and ruminants. The life cycle is direct.

Metastrongylus spp. (Metastrongylidae) have two lateral trilobed lips. The bursa has large lateral lobes. The ventral rays are distinctly separated from each other; the posterolateral rays are very large, long and separate; the mediolateral rays are short and thick, with the very small externolateral rays arising from the base; the dorsal ray is double and minute, as are the externodorsal rays (fig. 5.46). The spicules are long and delicate with transversely striated wings. The body of the female narrows abruptly behind the anus, and the vulva opens near it. They are parasites of swine. Earthworms serve as intermediate hosts.

Protostrongylus spp. (Protostrongylidae) males have a short bursa. The dorsal ray is very thick and bears six papillae on the ventral side; the externodorsal rays are spikelike; the postero- and mediolateral rays are fused except at the tip; the other rays are separate. The spicules are long with striated wings. A strong cuticularized arch, the telamon, strengthens the posterior end of the male body (fig. 5.48). The vulva is near the anus. They are parasites of sheep, goats, deer, big horns, hares, and rabbits. Land snails serve as intermediate hosts.

Notes and Sketches

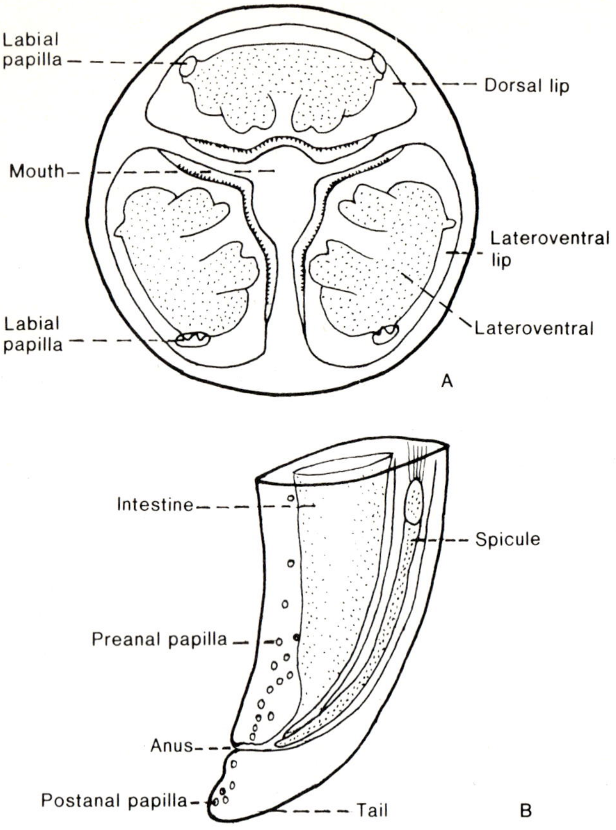

Fig. 5.18. *Ascaris suum.* (A) Adult, en face view; (B) caudal end of male, lateral view.

Order Ascaridata

Members of this order are generally stout worms, usually with three lips. The esophagus is muscular, with or without a caudal bulb, and may or may not have a ventriculus following the posterior end; there may or may not be a diverticulum at the anterior end. Spicules are small and equal or unequal. Parasitic in the alimentary canal of all classes of vertebrates.

Family Ascaridae

Esophagus is somewhat club shaped, lacking both a bulb and a ventriculus posteriorly. Male without precloacal sucker. Cervical alae usually absent (present in *Toxascaris*).

Ascaris suum

Ascaris suum (fig. 5.18) and *A. lumbricoides* occur in the small intestine of swine and humans, respectively. Anatomically, specimens recovered from the two hosts are virtually indistinguishable, but physiologically they differ in that the eggs from worms in swine do not readily infect humans, and vice versa.

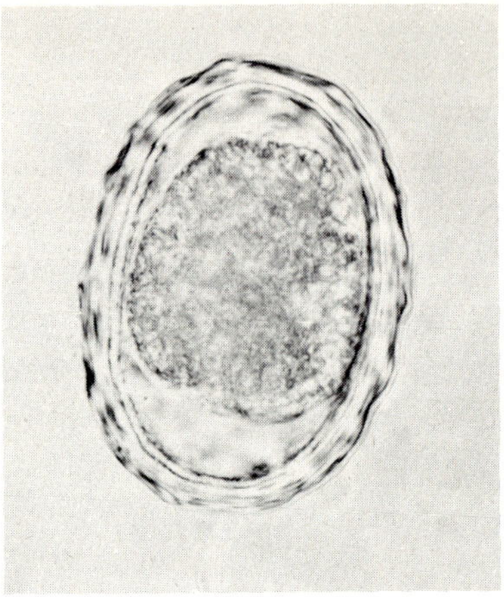

Fig. 5.19. *Ascaris lumbricoides.* The fertilized egg is broadly ovoidal, with a thick transparent and an outer, rugose, albuminous covering; size 45–70 × 35–50 μm. Unfertilized egg longer and narrower (more elliptical), and usually has a thinner shell and an irregular coating of albumen; size 88–94 × 38–44 μm.

Description

This is the largest of the swine intestinal nematodes: females up to 35 cm in legnth and about 5 mm in diameter; males up to 17 cm long by 3 mm in diameter. Males are distinguished by the ventrally curved tail, and two copulatory spicules, but no gubernaculum, whereas the females have a straight tail. The vulva is situated near the end of the first third of the body. The mouth is guarded by three lips: a dorsal and two lateroventral lips, each with two basal papillae. Each lip bears on its inner margin two rows of small denticles. The eggs (fig. 5.19), are identical with those of *A. lumbricoides.*

Directions for Study

Secure an entire preserved specimen from the preparation table. Females are larger than the males and have a straight, blunt tail, whereas that of the males is curved ventrally into a hooklike shape. Locate the anus near the posterior end on the ventral side of the body. Sometimes a pair of hyaline, needlelike spicules protrudes from the anus of the male. Examine the surface of the body to see the differentiation into the lateral, dorsal, and ventral surfaces. Note the fairly conspicuous light-colored stripes running along both sides of the body. These are the lateral lines. With a single-edged razor blade, cut the head off just behind the lips and at right angles to the longitudinal axis of the body, and mount in Berlese's fluid or glycerine for an en face view. Note the three roughly triangular lips arranged so there is a broad dorsal one and a pair of smaller lateroventral ones. Each lip bears two sensory papillae located near the outer margin of the base. The inner margin of each lip is finely denticulate. The small triangular buccal opening appears where the apexes of the lips come together.

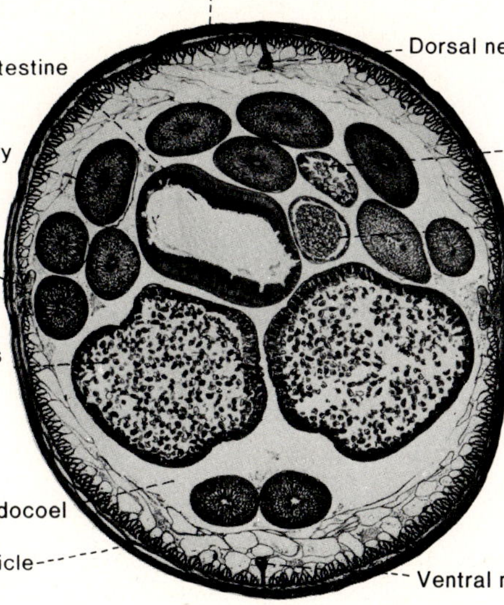

With fine pointed scissors open the specimen along the middle of the dorsal surface, being careful to avoid damage to the internal organs; then pin the worm out flat in a dissecting pan. The cavity which this slit exposes is the pseudocoel. The alimentary canal extends as a straight tube through the length of this cavity. The anterior end of the alimentary canal consists of a small, muscular esophagus. In the pseudocoel near the posterior end of the esophagus is a pair of large, yellowish, stellate cells, the coelomocytes. The mass of coiled tubules is the reproductive organs, either male or female. The genital opening of the female occurs midventrally near the union of the first and second quarters of the body, whereas in the male the reproductive organs share a common opening with the digestive system in the cloaca near the posterior end of the body.

The female reproductive organ is roughly Y-shaped, its branches becoming finer distally. The two free ends of the Y constitute the ovaries. The ovary, oviduct, and uterus form a continuous and very long tube representing the branches on either side of the Y. Externally, their boundaries are not discernible. The extent of each can be made out only through the study of sections. The stem of the Y, where the two branches meet, is a very short structure called the vagina. It leads to the vulva, which is located anteriorly, as stated above. Remove a portion of the uterus where it joins the vagina, tease apart with needles, and examine the eggs on a slide under the low power of a compound microscope. Note the thick shell. Focus carefully on the upper surface of an egg and observe the small irregular ridgelike elevations. Sometimes live and moving larvae are present in the eggs.

For the histology of the female reproductive system, study a cross section of an adult worm (fig. 5.20). Between the intestine and the body wall, four (ovary, oviduct, uterus, and intestine) types of structures appear. The ovary is the germinal zone and consists of a mass of protoplasm with an abundance of nuclei, or germinal vesicles, scattered through it. A short distance from the germinal zone, the protoplasm is formed around the vesicle and cells or ova are seen. This portion of the ovary is followed by a developmental zone in which the ova have become elongated and are arranged in wedgelike shapes around a central lumen or **rachis.** Further down, the rachis disappears as the ovary passes into the oviduct and the ova separate from each other, assuming a more or less oval form. In the uterus, the eggs are large, possess shells with surface ridges, and often show early stages of cleavage, or even larvae. The eggs are fertilized in the seminal receptacle, which is between the uterus and the oviduct. The intestine is somewhat flattened and lined with cells of uniform height.

The male reproductive organs are in the form of a long, single tube or thread. The free end of this thread is the testis. It joins an enlarged vas deferens, then a seminal vesicle, which extends to the posterior end of the body and unites with the alimentary canal to form a cloaca. Frequently, spicules, used in copulation, may protrude from the anus.

Cross sections show the basic structural pattern, characteristic of nematodes generally. The body wall consists of three parts: (1) the outer noncellular **cuticle** composed of three layers; (2) cellular **hypodermis** with a thickening on each lateral side and one each on the dorsal and ventral sides; these are the longitudinal **hypodermal chords;** the lateral chords contain the **excretory canals** and are more prominent than the dorsal and ventral ones; and (3) the innermost layer of bodywall muscle fibers; they are arranged in four principal groups between the chords. When the number of muscle fibers between the chords is great, as appears here, the arrangement is said to be **polymyarian,** and when there are few, it is **meromyarian.** If there are no chords present and the muscle fibers are small and closely packed, it is **holomyarian.** Recognition of these arrangements is useful in identifying cross sections of worms in tissues.

Life Cycle

The life cycle is shown in figure 5.21. Eggs passed in the feces are undeveloped and must undergo a period of incubation outside the host. Infection results from ingesting eggs containing second-stage larvae. Upon hatching in the small intestine of the host, the larvae burrow into the mucosa, enter the hepatic portal system, and are carried to the liver, then the heart, and eventually the lungs. Here they undergo the second and third molts, break out of the capillaries into the alveoli, and

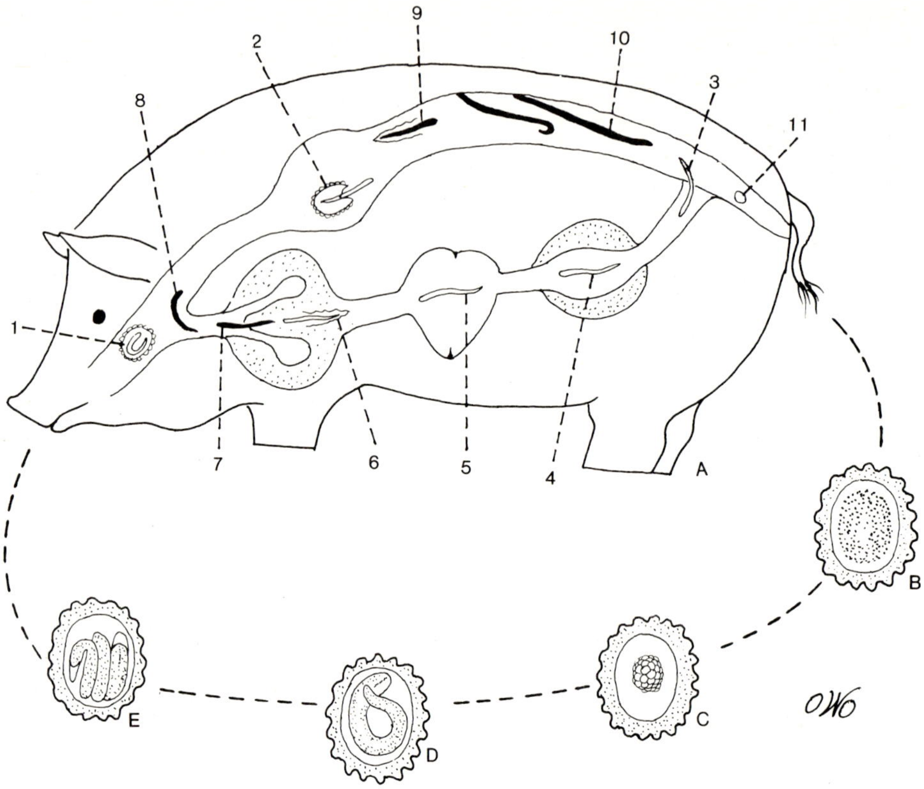

Fig. 5.21. Life cycle of *Ascaris suum* (similar to that of *A. lumbricoides*). (A) Swine with adults (10) in small intestine; (B) unembryonated eggs passed in feces; (C) eggs embryonate in soil and litter; (D) first-stage larvae; (E) second-stage larva inside cuticle of first stage does not hatch until swallowed by swine. (1) Egg containing second-stage larvae; (2) egg hatches in stomach; (3) larvae penetrates gut wall and enters hepatic portal system, passes through liver (4) and heart (5); (6) larva molts to third stage in lungs; (7) larva migrates up trachea, is swallowed (8) and goes to small intestine, molts final time (9) and develops into adult; (10) adult males and females; (11) eggs laid in gut and voided with feces. (Modified from Olsen, 1974.)

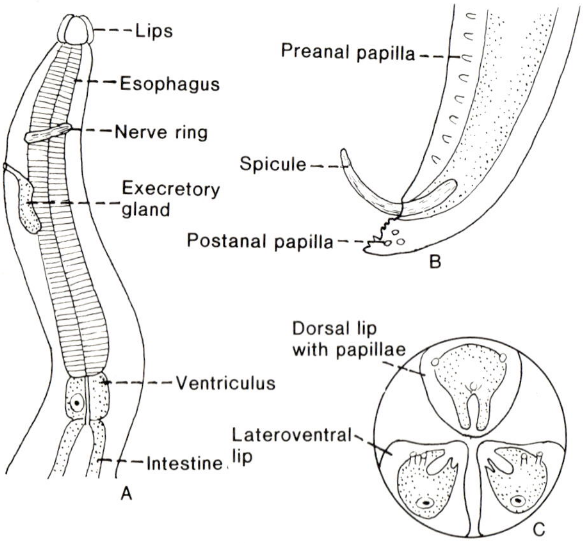

Fig. 5.22. Adult *Toxocara canis.* (A) Lateral view of anterior end; (B) lateral view of caudal end of male; (C) en face view.

make their way through the bronchi, trachea, pharynx, esophagus, and back to the small intestine, meanwhile having increased about ten times in length. Here they undergo their final molt and become mature, some eight weeks after ingestion of eggs.

Family Toxocaridae

Esophagus with posterior ventriculus. Cervical alae present: prominent in *Toxocara.*

Toxocara canis

This larage roundworm of carnivores (fig. 5.22) is cosmopolitan in distribution. Adults occur in the small intestine of dogs, foxes, and coyotes. They are often a serious problem in puppies, not only because the infection is so common but also because the infection is usually very heavy. Worms often enter the stomach and cause vomiting, with the vomitus containing many of them. On account of its general availability and resourcefulness in the various ways the host may acquire a patent infection, it is of special interest. Furthermore, larvae often infect human children who swallow the eggs.

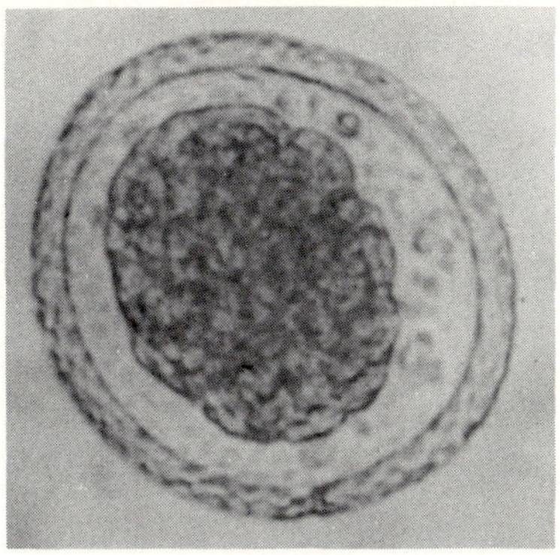

Fig. 5.23. *Toxocara canis.* Subglobose to ovoidal, densely granular internally and a rather thick pitted shell; size 80–90 × 70–80 μm.

Description

Adult females measure up to 18 cm long and males up to 10 cm. The posterior end is usually curved ventrally. There are broad cervical alae, which give the anterior end a spearhead shape, but caudal alae are lacking. The tail of the male is abruptly reduced in diameter and bears five pairs of lateral papillae. There are about 20 pairs of preanal lateral papillae. The vulva is situated in the first quarter of the body. The eggs are illustrated in fig. 5.23.

Life Cycle

The life cycle may be relatively simple, completed in one host, or it may be more complicated, involving two or more hosts. When the infective eggs are swallowed by the host, they hatch in the small intestine; a few of the second-stage larvae enter the lacteals (but most enter the hepatic portal system), and are carried to the lungs via the liver and heart. From this point on, the migratory behavior of the larvae varies depending upon the age and the species of the host.

In puppies three weeks and younger, the larve in the capillaries of the lungs enter the alveoli, migrate up the trachea, are swallowed, and go back to the small intestine, where they mature in four to five weeks.

In older dogs, five weeks and more, however, the larvae mainly return from the lungs to the heart and are dispersed via arterial circulation throughout the body, where they encapsulate in the various tissues.

Larvae embedded in the tissues of bitches become reactivated during pregnancy, and some reenter the bloodstream. Those that enter the placental circulation and make their way into the liver of the fetuses produce prenatal infection. In newborn puppies, third-stage larvae are present in the lungs; within another day, the larvae, having migrated up the trachea and been swallowed, are in the stomach; at three days after birth fourth-stage larvae are already in the intestine. In puppies so infected, patency is reached as early as four to five weeks after birth. Prenatal infection is apparently the usual mode of infection in dogs.

As a sequel to the prenatal mode of infection, Sprent (1961) has shown that postparturient infection of bitches may occur immediately after parturtion. Larvae migrate from the lungs of newborn pups into the intestine, but some of them fail to maintain a hold in the intestine and are discharged in the feces. Thus when the third-stage larvae, beyond the stage requiring migration through the lungs, are swallowed by the bitch, they develop to adults in the alimentary canal. The explanation for human infection with adult *T. canis* is that children ingest the advanced-stage larvae from the feces of a puppy or young dog.

Eggs swallowed by mice hatch in the small intestine. The larvae enter the hepatic portal system, are transported to the lungs via the heart, back to the heart, and out into the tissues. Many larvae are lodged in the brain but cause little disturbance of locomotion. This is in marked contrast to infection with larvae of *Baylisascaris columnaris,* in which locomotor disturbances are frequently observed in mice. When infected mice are eaten by dogs, the larvae are freed in the intestine and undergo the hepato-pulmono-tracheal migration to the small intestine, where they mature.

Thus there are four ways in which a dog may acquire a patent infection: (1) puppies ingesting eggs; (2) prenatally, resulting in high infection in young pups; (3) bitches swallowing third-stage larvae in the feces of their puppies; and (4) by ingesting tissues of animals harboring second-stage larvae. Since larvae have been observed in the milk of nursing bitches, it is possible that the transmammary route may serve as still another way in which patent infections are acquired.

When human beings swallow infective eggs, the larvae behave in much the same manner as in mice, causing a condition known medically and parasitologically as visceral larva migrans. The larvae occur primarily in the internal organs and musculature. For details of this complicated life cycle see Olsen (1974, plate 122), or Schmidt and Roberts (1985, fig. 26.8).

Some other common Toxocaridae with their hosts include: *Toxocara cati,* cats; *T. vitulorum* (*Neoascaris vitulorum* of some authors), cattle; and *Toxascaris leonina,* felids and canids.

Notes and Sketches

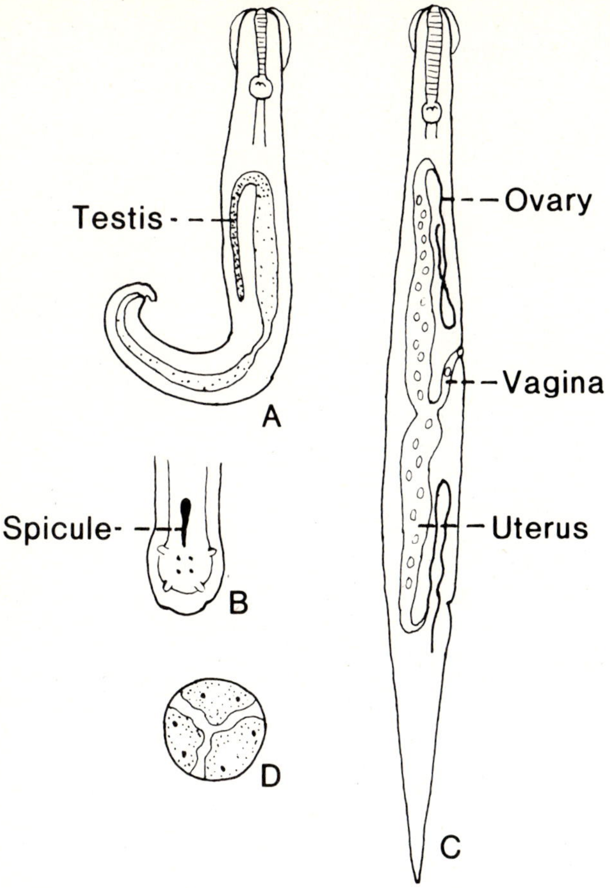

Fig. 5.24. *Enterobius vermicularis.* (A) Adult male; (B) ventral view of caudal end of male, showing spicule and papillae; (C) adult gravid female; (D) en face view, showing three lips, each with two papillae.

Order Oxyurata

Family Oxyuridae

Characterized by a nearly spherical, posterior bulblike enlargement of the esophagus; cervical spines are absent. A single spicule is present. Parasites of the large intestine of mammals.

Enterobius vermicularis

The human pinworm, *E. vermicularis* (fig. 5.24), is cosmopolitan in distribution but, unlike most helminthic infections, is relatively rare in the tropics. The pinworm is probably the most prevalent nematode of humans in the United States, occurring in about 50% of the children. Fortunately, however, it is more of a nuisance than a serious health problem.

Description

The mouth is surrounded by three lips, as in *Ascaris,* and there is no buccal capsule. The anterior end of the body has an inflated cuticle. The excretory pore, seen

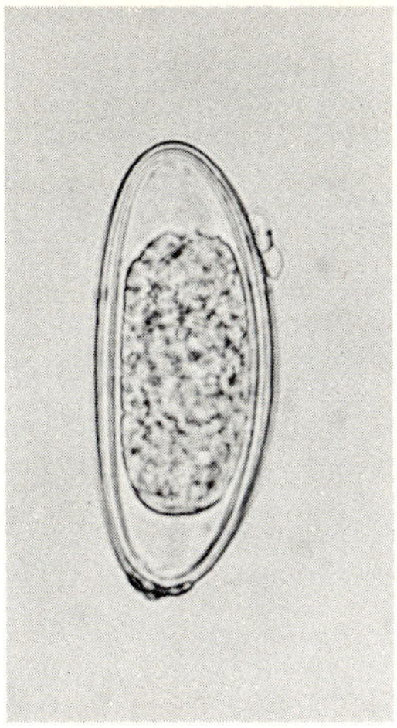

Fig. 5.25. *Enterobius vermicularis.* Egg in profile flattened on one side, rounded on the opposite side; a double membrane, containing a well-developed embryo; size 50–60 × 25–33 μm.

from the side, is near the middle of the esophagus. The basal end of the esophagus bears a distinct bulb with a valvular apparatus. The intestine is straight.

Males are up to 5 mm long. The tail is curved ventrally and the lateral caudal alae extend around the end. From the ventral view, the tail is broadly rounded and bears a pair of lateral adanal and postanal papillae and two pairs of ventral postanal ones. The single spicule with curved tip is 70 μm mm long. The threadlike testis begins somewhat posterior to the middle of the body, extends anteriorly, reflexes backward, forming the thin tubular vas deferens, as is followed by a thicker muscular ejaculatory duct.

Females measure up to 13 mm long. Gravid ones are somewhat spindle shaped and have a long, slender, pointed tail. The vulva is ventral, slightly anterior to the equator of the body, from which the narrow vagina extends backward to the middle of the body, where it divides into an anteriorly and posteriorly directed uterus. Each uterus extends to the end of the body, where it bends back on itself to form the threadlike ovary.

Life Cycle

At night gravid females migrate out through the anus, deposit their eggs (fig. 5.25) in the perianal region, and retreat to the rectum, but some worms are passed in the feces. Since eggs are seldom found in the feces, the standard method for their recovery involves pressing strips of adhesive tape to the perianal region (see p. 257). Upon ingestion of the infective eggs, larvae hatch and reach maturity in the caecal region. The life cycle is shown in figure 5.26.

Other common species of Oxyuridae include *Passalurus nonannulatus* of rabbits, *Oxyuris equi* of equines, and *Aspiculuris tetraptera* or *Syphacia muris* of laboratory rodents.

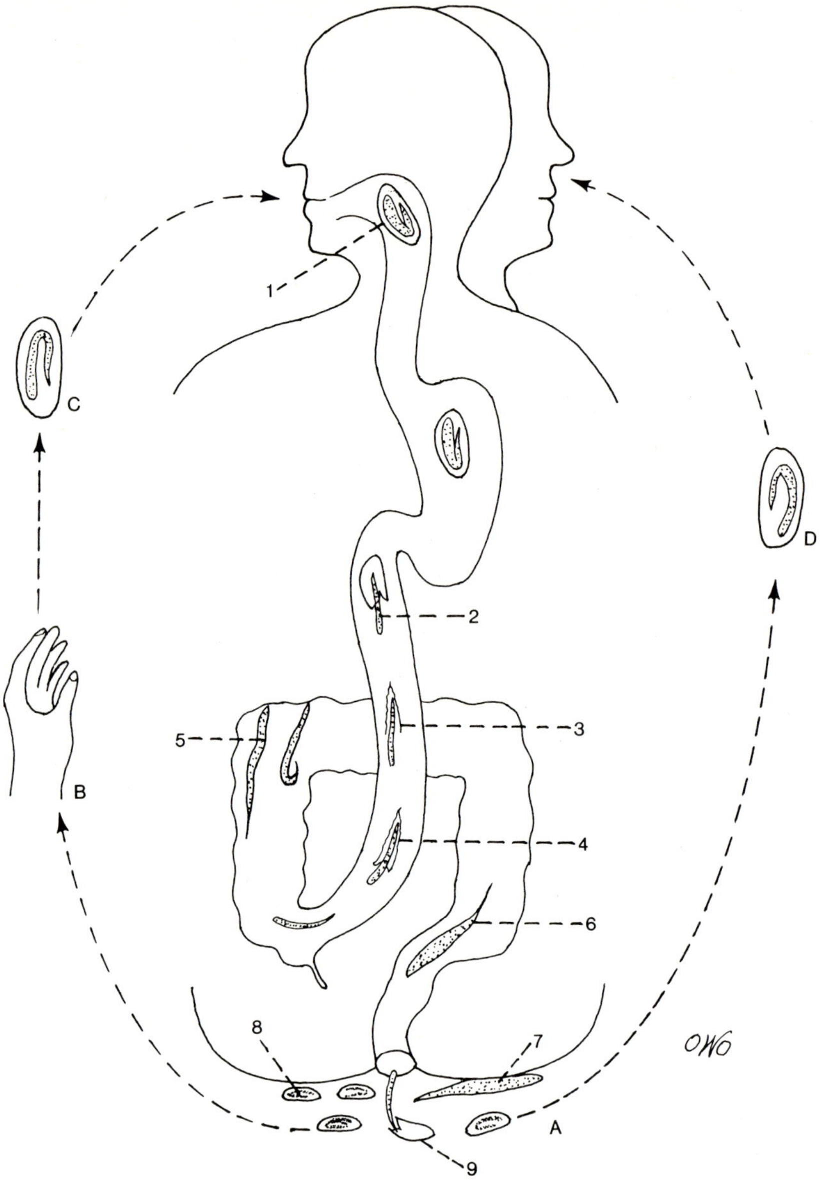

Fig. 5.26. Life cycle of human pinworm, *Enterobius vermicularis.* (A) Human host; (B) hand and fingers contaminated with eggs from scratching anal region to alleviate itching caused by ovipositing females (7); (C) embryonated eggs transported to mouth by contaminated fingers; (D) embryonated eggs borne by air currents in house or bed are inhaled. (1) Eggs enter mouth; (2) eggs hatch in small intestine; (3,4) larvae molt twice and enter large intestine to mature; (5) adult worms attach to mucosa of colon; (6,7) gravid females migrate through anus to perineal region to oviposit; (8) eggs embryonate; (9) some eggs hatch and larvae enter anus to infect by retrofection and develop to adults. (Modified from Olsen, 1974).

Family Syphaciidae

Characterized by males with both a spicule and a gubernaculum. Parasites of the caecum and large intestine of rodents.

Syphacia obvelata

This species (fig. 5.27), common in colonies of white mice, is an ideal model for laboratory study.

Description

Three lips surround the triradiate mouth. There are four cephalic papillae in the outer circle.

Males measure up to 1.6 mm long by 0.2 mm in diameter with the caudal end bent ventrally into a spiral. There are three cuticular projections, or **mamelons,** on the ventral surface of the body, with the middle one near the equator. The tail is 0.1 mm long. Ratio of the

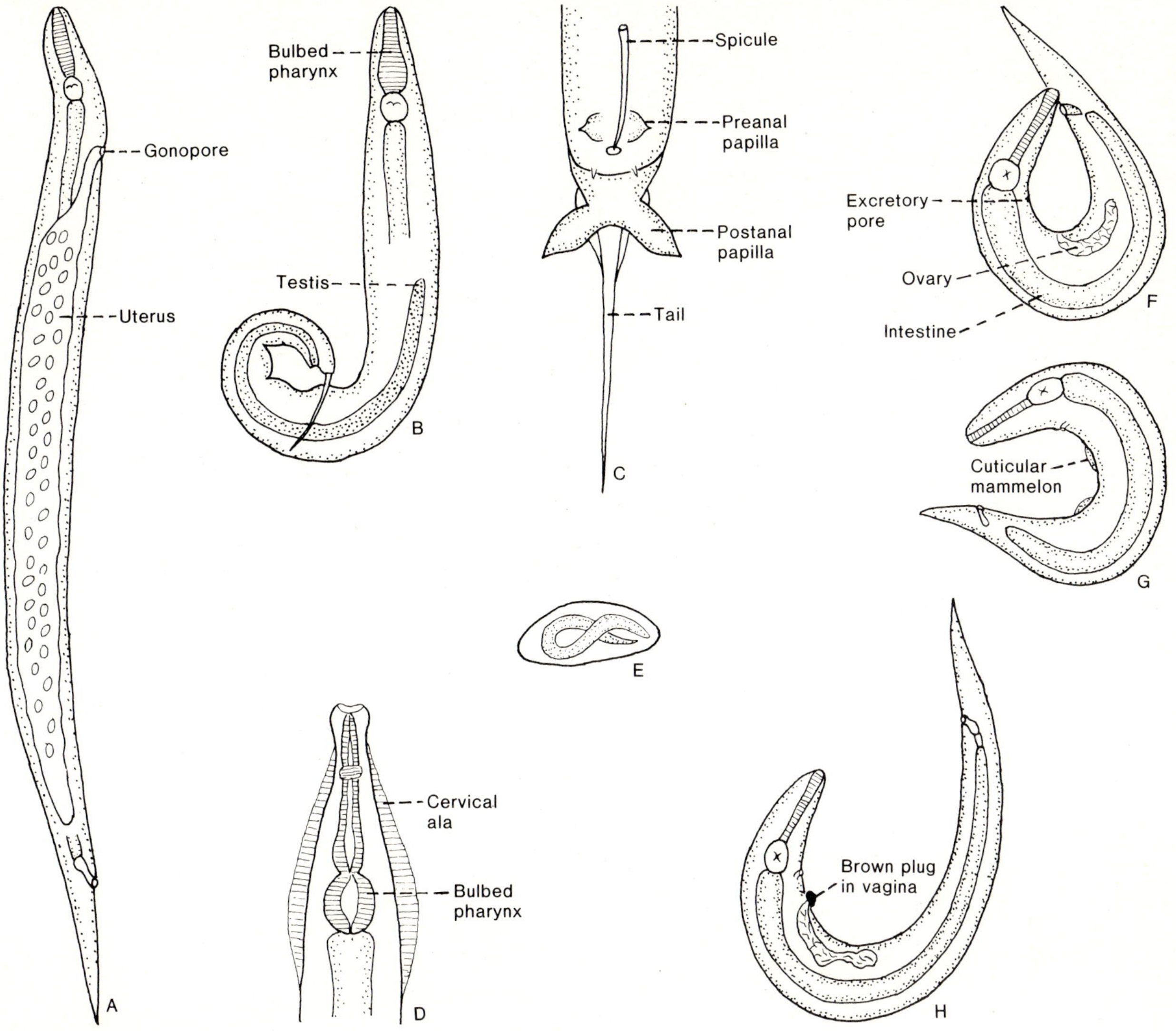

Fig. 5.27. *Syphacea obvelata.* (A) Gravid female; (B) adult male; (C) caudal end of adult male; (D) anterior end, ventral view, of adult; (E) embryonated egg; (F) 48-hour female larva; (G) 48-hour male larva; (H) 96-hour female larva.

distance from the anterior tip of the body to the excretory pore to the total body length is 1:4.4. The spicule is slightly curved and measures 90 μm mm long and 7 μm mm thick at the base; the ploughshare-shaped gubernaculum is located at the base of the spicule and is 40 μm mm long. There are two pairs of preanal and one pair of postanal papillae.

Females are up to 5.7 mm long by 0.3 mm in greatest diameter. The tail is 0.8 mm long. Ratio of the distance from the anterior end of the body to the excretory pore to the total body length is 1:9.9. The vulva is on a prominence about 0.9 mm from the anterior end of the body. Ovaries are in the middle third of the body; one is pointed anteriorly and the other posteriorly. The parallel oviducts extend posterior beyond the hind ovary and unite to form the long, single uterus that extends anteriorly to the vulva, located a short distance behind the esophagus. Eggs are flat on one side and average 134 × 36 μm. Eggs are unembryonated when laid.

Life Cycle

The life cycle is direct and of short duration. Infection is by mouth. Eggs may be picked up from the surroundings or licked from the perianal region. This life cycle is an ideal model for laboratory study of the Oxyurata nematodes, especially since it parallels that of *Enterobius vermicularis,* the human pinworm.

Notes and Sketches

Order Spirurata

Characterized by a cylindrical esophagus composed of a short anterior muscular and a long posterior glandular part. Spicules usually are unequal and dissimilar. The mouth is bordered by two large lateral **pseudolabia** or no lips at all. There may be caudal alae but no copulatory bursa. Females are oviparous. First- and second-stage larvae have a cylindrical esophagus. Parasites of the alimentary canal, tissues, and circulatory system. The life cycles are indirect, with arthropods always serving as intermediate hosts.

Family Tetrameridae

There is great sexual dimorphism in members of this family. Gravid females are globular, with the small pointed ends extending beyond the enlarged part of the body (*Tetrameres*) (fig. 5.68), or the enlarged, long axis of the body is coiled in a complex manner (*Microtetrameres*). The males are free in the lumen of the proventriculus, and the gravid females are in the proventricular glands of birds.

Microtetrameres centuri

Members of the genus are distinct from most Spiruroidea in the nature of the tightly coiled gravid females and the filariform males. Adults occur in the proventricular glands of meadowlarks (*Sturnella magna, S. neglecta*) and bronzed grackles (*Quiscalus versicolor*). Living females are bright red and tightly coiled so that the body presents a globose configuration.

Description

Males are up to 1.4 mm long, with the right spicule almost the length of the body and the left about one-tenth that of the other. There is no gubernaculum. There are two pairs of pre- and two pairs of postanal papillae. The tail is pointed.

Females are up to 1.4 mm long, the pointed ends extending beyond the body coils. Fixed specimens are pinkish and slightly longer than living ones. The buccal capsule is prominent and expanded midway between the anterior and posterior ends. An intestinal diverticulum is present at the esophago-intestinal junction. The vulva is near the anus. Embryonated eggs are thick shelled, flattened on one surface, with a thickening on at least one end, and measure about 35×51 μm in size.

Life Cycle

Eggs contain fully developed first-stage larvae when laid. They hatch when eaten by nymphs of certain species of *Melanoplus* grasshoppers. The larvae migrate into the hemocoel and undergo the first molt, beginning about the eighth day, which is followed closely by the second molt. By the end of the ninth day, the earliest third-stage larvae are present in the hemocoel and begin encystment in clear, thin-walled cysts. The worms are infective after 40 days at room temperature. When released from the grasshoppers in the intestine of the host, the larvae migrate forward to the crop, traveling beneath the keratinoid lining of the gizzard.

Other Tetrameridae whose life cycles are known to some extent include *Tetrameres americana* in poultry, pigeons, ducks, and geese, with grasshoppers and cockroaches as intermediate hosts; *T. crami* of ducks, with gammarid crustacea as intermediaries,; and *Microtetrameres helix* from crows, with grasshoppers and cockroaches as invertebrate hosts.

Family Gnathostomatidae

Characterized by a large cuticular head bulb covered with spines or strong, transverse striations and separated from the body by a constriction. The head bulb consists of four intercommunicating compartments called **ballonets.** The body may or may not be spined in the anterior portion. The species are parasitic in the stomach and intestine of fish, reptiles, and mammals.

Gnathostoma procyonis

This species (fig. 5.28) occurs in raccoons throughout the southeastern United States and may be as widespread as the host. Adults occur in the stomach, where they do great mechanical damage.

Description

They are large, stout worms with a head bulb bearing nine or ten transverse rows of 90 to 100 spines in each. The anterior half of the body is covered with large trident scales, which fuse toward the posterior end of the body to form encircling ridges or striations. The posterior part of the body is marked by coarse transverse wrinkles and noticeably less spinose, or aspinose. The esophagus is club shaped and about one-fifth the length of the body. There are four elongate **cervical sacs** attached inside the anterior of the body.

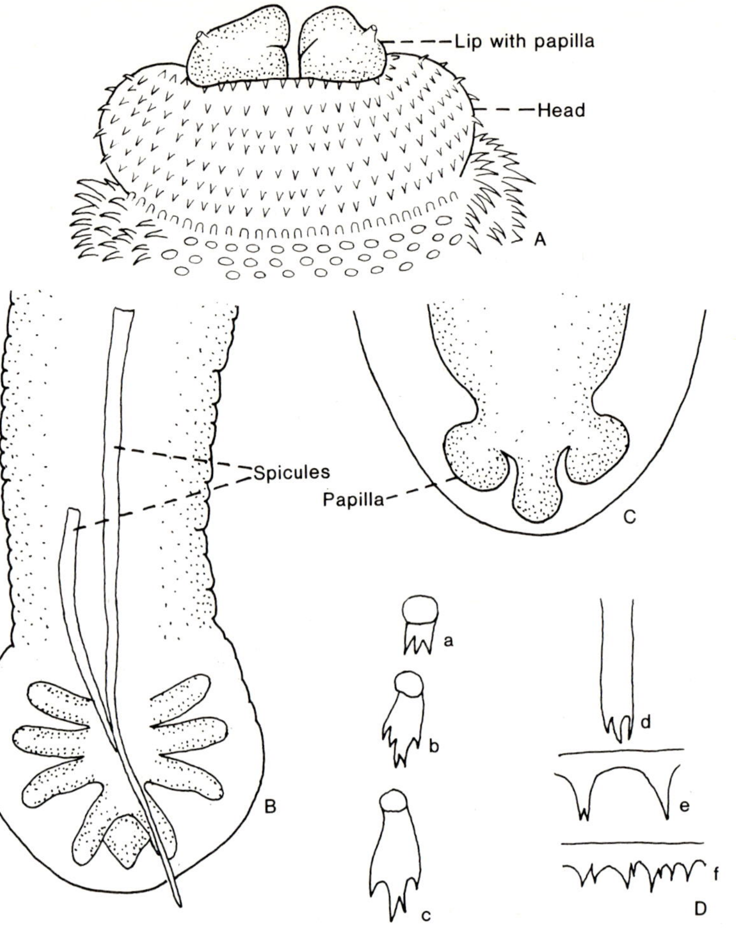

Fig. 5.28. *Gnathostoma procyonis.* (A) Anterior end of *G. spinigerum* (similar to *G. procyonis*); (B) caudal end of adult male; (C) caudal end of adult female; (D) body spines: (a) first row; (b) 1 mm back; (c) 5 mm back; (d) 10 mm back; (e) 12 mm back; (f) 12.5 mm back. (All except A modified from Chandler, 1942.)

Males are up to 19 mm long and 0.8 mm diameter where the cuticle is not inflated. The head bulb is up to 0.7 mm broad and 0.3 mm long. A bursalike expansion of the caudal end of the body is supported by four large, lateral, nipple-shaped papillae. Ventral papillae are absent. The right spicule is up to 3.0 mm long and the left 0.7 mm.

Females measure up to 26 mm in length and 2 mm in diameter. The head bulb is up to 0.8 mm broad and 0.4 mm long. The vulva is up to 8.5 mm from the tip of the tail. The ovijector extends forward, widens into a thin-walled tube filled with eggs that continues forward, then bends back and divides into two posteriorly directed uteri. Posterior end of body is bluntly rounded.

Eggs are brownish in color, with a thin, finely granulated shell, and bear a glassy, pluglike operculum at one end. They are unembryonated when passed in the feces and measure 64–76 × 37–41 μm.

Life Cycle

See figure 5.29. Eggs passed in the feces of raccoons soon embryonate, and the first-stage larva undergoes a partial molt, retaining the loosened cuticle. Hatching in the water begins in about 12 days, continuing for about two weeks. Sheathed larvae, with spikelike anterior ends, die within a few days unless eaten by the copepod first intermediate host. Nauplii of *Cyclops viridis, C. bicuspidatus,* and *Macrocyclops albidis* are more susceptible than adults. The first molt is completed immediately after larvae are eaten by copepods.

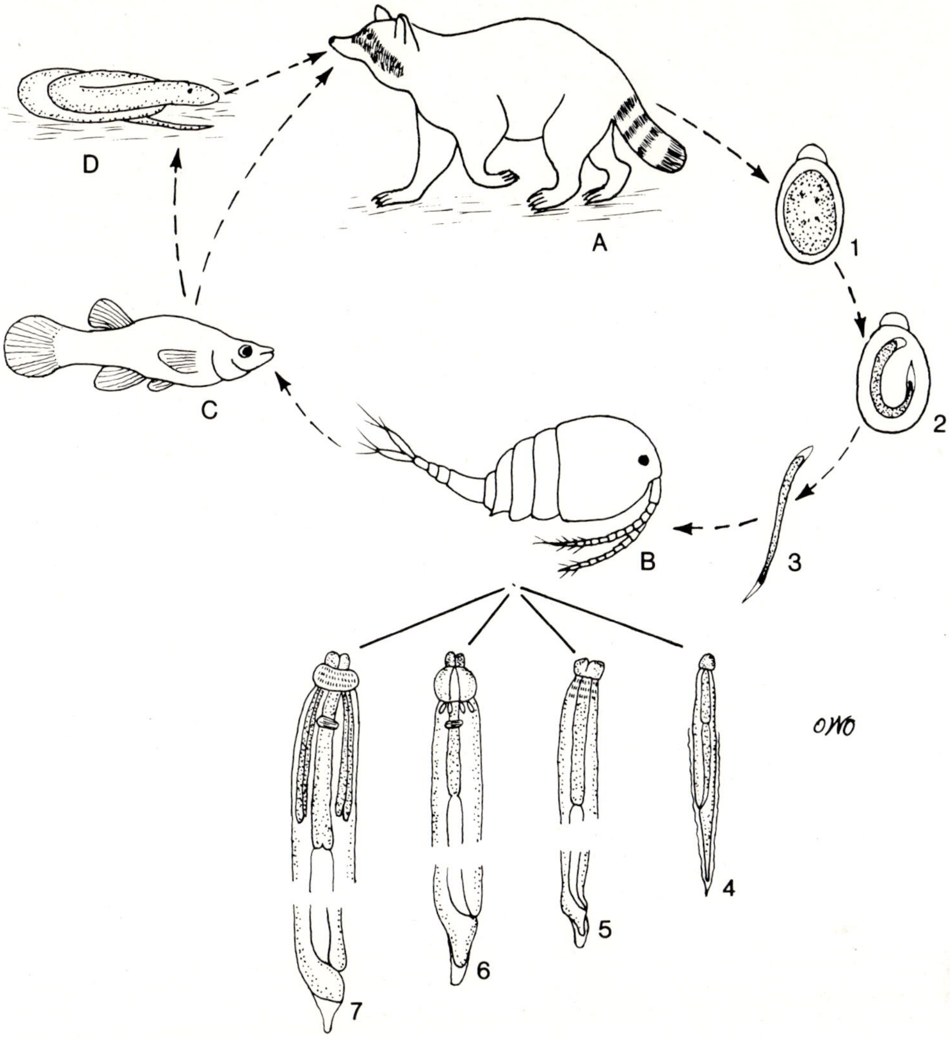

Fig. 5.29. Life cycle of *Gnathostoma procyonis,* an example with a first and second intermediate and a paratenic host. (A) Adult worms in stomach of raccoon; (B) *Cyclops* spp. first intermediate host; (C) fish second intermediate host; (D) snake paratenic host in which third-stage larvae occur and may be passed as such from snake to snake without further development. (1) Unembryonated eggs passed in feces; (2) larva develops and begins molt to second stage in egg; (3) egg hatches, freeing ensheathed larva into water; (4) larva completes first molt immediately when eaten by *Cyclops;* (5) developing second-stage larva (6) second molt to form third-stage larva; (7) fully developed third-stage larva infective to fish. When infected fish eaten by snakes, larvae accumulate in them without further development.

The second molt occurs after seven days development in copepods. At this time, third-stage larvae exhibit the gnathostome head bulb with its encircling rows of spines.

Third-stage larvae in copepods are infective to fish (guppies are used in experiments) but not to mammals (white mice). Larvae are free in tissues of fish where considerable growth in size takes place.

Fully developed third-stage larvae in fish are infective to raccoons. They develop directly to adult worms without prior migration to the liver, as occurs in some species. Adult worms thrust the head bulb into the stomach wall, leaving the body hanging free.

When fish infected with advanced third-stage larvae are eaten by water snakes (*Natrix rhombifera*) and others, the larvae migrate to the tissues, where they remain without further development. They may be passed from snake to snake.

These larvae are infective to raccoons, in which they develop to adult worms.

Thus the copepod and fish are true first and second intermediate hosts in which essential development takes place, and snakes, along with turtles, alligators, and some fish, are paratenic, or collector, hosts in which no development takes place. Being a part of the natural food of raccoons, their chances of survival are greatly enhanced.

Family Physalopteridae

The family is characterized by the large, simple, somewhat triangular lateral lips armed with one or more teeth on the free margin, large caudal alae, long-stalked papillae that appear to be supporting the alae, and sessile pre- and postanal papillae.

Physaloptera hispida

This species (fig. 5.30), which occurs in the pyloric region of the stomach in cotton rats (*Sigmondon hispidus*)* from the southeastern United States, is an excellent model of a Spirurata for anatomical studies of the adults and biological investigations of the larval stages in an arthropod intermediate host. The life cycle (fig. 5.31) requires a long time to complete.

Description

Adult females are pink, and the smaller males are white. There are two semicircular lips, each with a lateral amphid, one externolateral tooth and three associated internolateral teeth on each lip, together with one subdorsal and two subventral papillae. Deirids, or **cervical papillae,** and an excretory pore are at about the same level, slightly posterior to the nerve ring. There is no cuticular collar extending forward over the head.

Males are up to 42 mm long by 1.4 mm in diameter. The tail is flexed ventrally and bears prominent caudal alae supported by four pairs of long, evenly spaced, pedunculated papillae; there are three sessile, preanal papillae arranged like an inverted equilateral triangle, a transverse row of four papillae at the posterior margin of the anus, and three pairs of evenly spaced ones on the ventral side of the tail. The spicules are slightly dissimilar, unequal, and short; the right one is up to 0.6 mm long and the left up to 0.5 mm.

Females are up to 64 mm long and 2.0 mm in diameter. The uterus is amphidelphic, with the branches

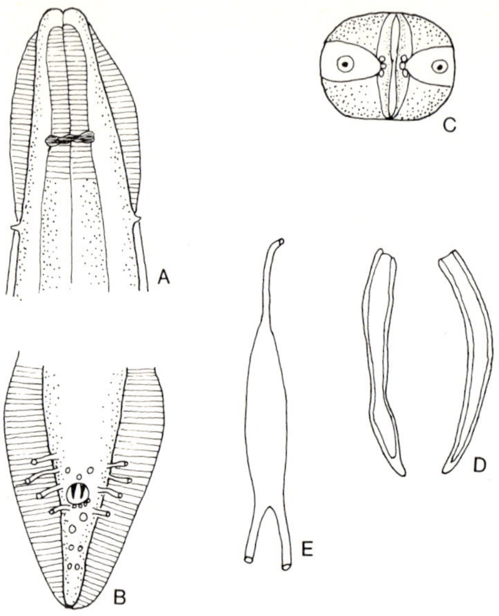

Fig. 5.30. *Physaloptera hispida.* (A) Anterior end showing cervical alae, cervical papillae, and muscular and glandular parts of esophagus; (B) ventral view of caudal end of male showing caudal alae, lateral, pre- and postanal papillae, and tips of spicules appearing in anal opening; (C) en face view showing vertical mouth surrounded by two lateral lips, two large amphids, and six small papillae; (D) spicules; (E) uterus with two branches of ovaries.

arising from the posterior end of the broad egg chamber, which follows the slender vagina. The vulva is near the junction of the first and second quarters of the body. The phasmids, midway between the anus and the tip of the tail, appear clearly as a pair of lateral pores. Eggs 24–30 × 40–52 μm and are embryonated when laid.

Life Cycle

Embryonated eggs containing first-stage larvae are voided with the feces. Hatching occurs only after ingestion by arthropod intermediaries, which include German cockroaches (*Blatella germanica*), earwigs (*Forficula auricularia*), and several species of ground beetles (*Harpalus*).

The first molt occurs 14–17 days after hatching, and the second-stage larvae promptly enter the epithelium of the hindgut, causing destruction of the epithelial cells. Within a week after entering the epithelium, the host has begun to produce connective tissue cysts that encapsulate one or several larvae. Growth of second-stage larvae is rapid inside the cysts where the second molt occurs 24–27 days after infection.

*The host rats of *P. hispida* as well as *Litomosoides carinii* (p. 170), may be trapped in the field or purchased from Rider Animal Co., Inc., Route 2, Box 270, Brooksville, Florida 33512.

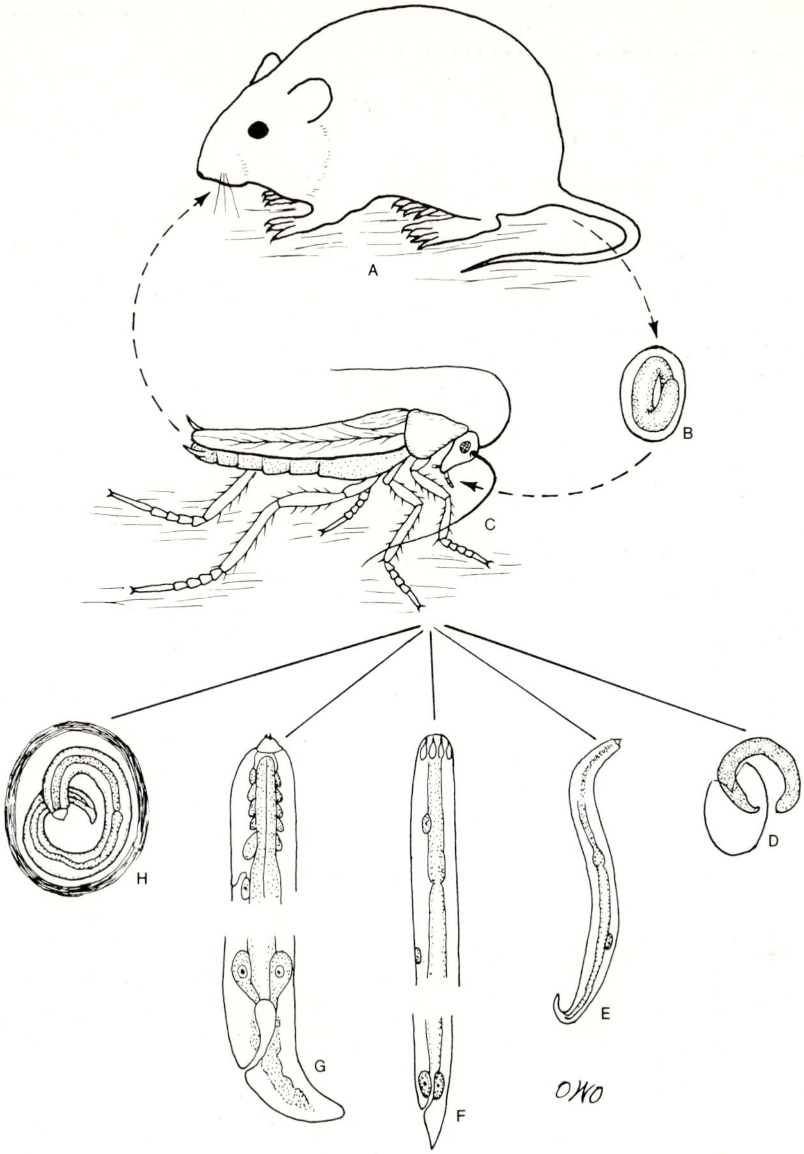

Fig. 5.31. Life cycle of *Physaloptera hispida.* (A) Cotton rat definitive host with adult worms in stomach ulcers; (B) embryonated eggs voided with feces; (C) cockroach (*Blatella germanica*) intermediate host; (D–H) developing stages in intermediate hosts; (D) egg hatches in intestine; (E) first-stage larva; (F) second-stage larvae; (G) third-stage larvae; (H) third-stage larva encysted in wall of midgut.

Third-stage larvae complete their growth 30–35 days after hatching, attaining a length of 1.2 mm. Cotton rats become infected readily by eating intermediate hosts containing fully formed third-stage larvae, which develop to sexual maturity and produce eggs within 72–90 days. Cotton rats may become infected from second-stage larvae, which later molt to the third stage and grow at a greatly reduced rate. Adult worms feeding in compact groups in the stomach wall produce ulcers.

Physaloptera hispida develops successfully in white rats.

Order Camallanata

Family Philometridae

The anterior extremity of the body lacks a cuticular shield, and the esophagus has no median constriction. Adults occur in the coelom, serous membranes, and connective tissues of fresh- and saltwater fish.

Philometroides nodulosa

Adult females occur in the skin of suckers (*Catostomus* spp.) and occasionally in other closely related fish. (See fig. 5.32.)

Description

The body is long, cylindrical, and with blunt extremities. The mouth, surrounded by three elevated lips, lies in a depression. The anterior end of the esophagus is slightly bulbous.

Males are up to 2.7 mm long and have a smooth cuticle. The mouth is surrounded by three small lips. The posterior end is blunt, rounded, and with two ventral and two dorsal swellings. Spicules are slender, with the right 130–161 μm long, and the left 137–164 μm. The gubernaculum is 48–59 μm long and bears a terminal barb.

Females are up to 45 mm long, and the body is covered by irregularly spaced cuticular bosses. The head bears three elevated fleshy lips and an inner ring of four and an outer ring of eight evenly spaced papillae. There is an anterior and a posterior ovary, and a single uterus packed with developing eggs and first-stage larvae. The vulva is atrophied. Intrauterine eggs are rounded to ovoid in shape and 44 $\times$ 43 μm in size.

Life Cycle

The life cycle (fig. 5.33) resembles that of other Camallanata in the requirement of a copepod intermediary and development in tissues of the final host. Females, gravid with fully developed first-stage larvae, lie in the subcutaneous tissues of the host, usually the cheeks. When the gravid females break out of the surrounding tissue, they burst and liberate many thousands of larvae. Motile larvae eaten by certain species of *Cyclops* undergo molts to the infective stage in about three weeks at room temperature. The cycle is completed when the infected copepods are eaten by suckers. The larval worms are liberated in the host's intestine and migrate anteriorly to their eventual location. Fertilization occurs during migration, after which the males die.

Philometroides huronensis also occurs in white suckers. Adult females are mainly in the pectoral fins and males in the peritoneum around the swim bladder. Gravid females are long, up to 73.6 mm. First-stage larvae liberated from the prolapsed and ruptured uterus have a buttonlike swelling at the tip of the threadlike tail, a feature not present in *P. nodulosa*.

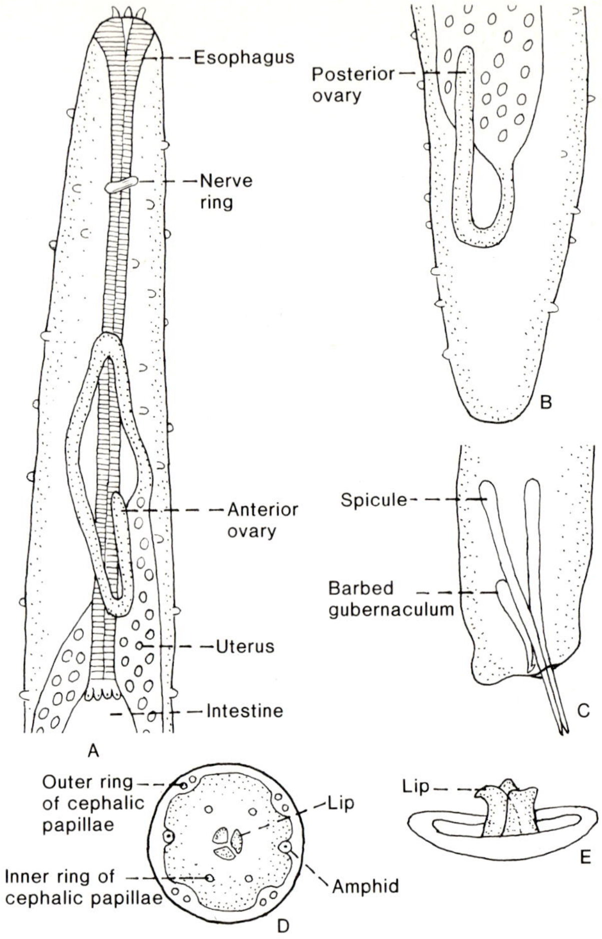

Fig. 5.32. *Philometroides nodulosa.* (A) Anterior end of adult female; (B) caudal end of same; (C) caudal end of adult male; (D) en face view of adult male; (E) lateral aspect of lips. (Modified from Dailey, 1966.)

The right spicule of the male is 136 μm, the left 128 μm, and the gubernaculum 72 μm long, the last with a terminal barb.

The life cycle is basically similar to that of *P. nodulosa.*

Order Filariata

Families Onchocercidae and Dirofilariidae

These two families, representatives of the large group of filarial worms, are slender, delicate parasites of the body cavities, air sacs, tissues, and circulatory system of amphibians, reptiles, birds, and mammals. Many

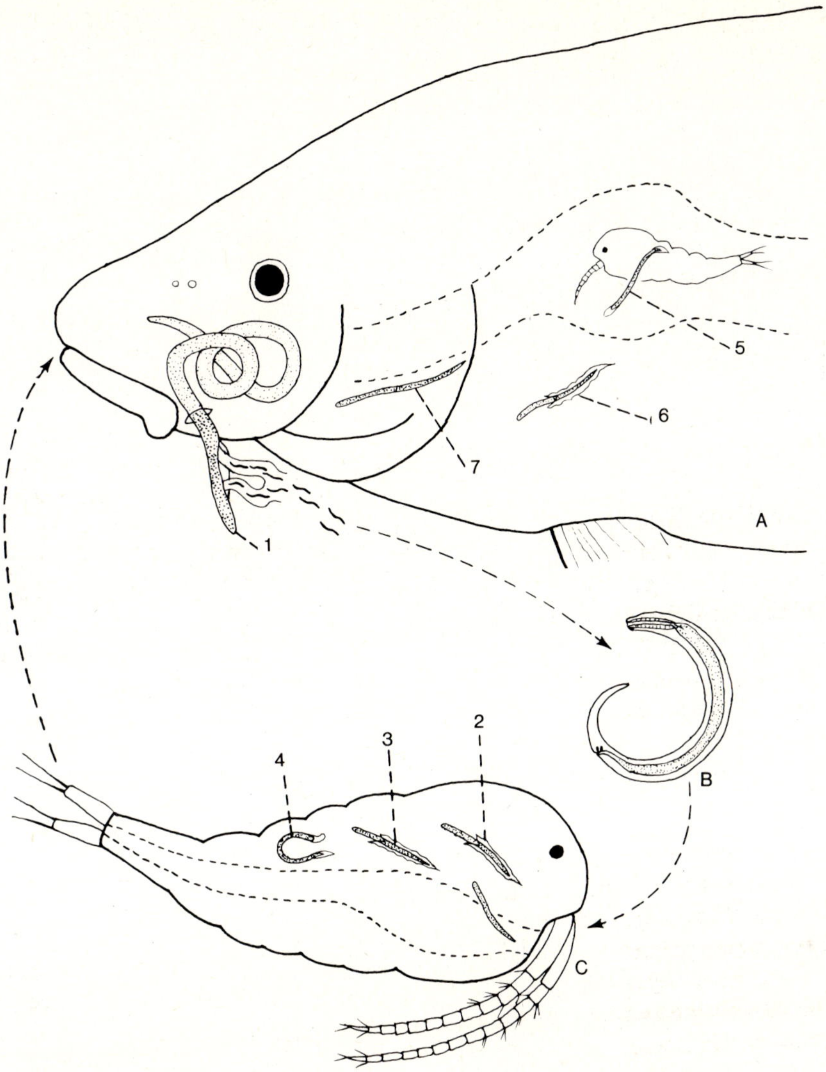

Fig. 5.33. Life cycle of *Philometroides nodulosa,* an example of a dracunculid cycle. (A) Adult females in cheeks of white sucker; (B) first-stage larva in water; (C) copepod intermediate host ingests larvae. (1) Adult females extend anterior end of body through ulcers, the uterus prolapses, and larvae are liberated; (2) larvae ingested by copepod enter hemocoel from intestine and molt; (3) second-stage larva molting; (4) third-stage ensheated larva in hemocoel; (5) larvae escape from digested copepod; (6) larvae molt whole migrating in fish tissues; (7) grows to adult males and females.

species of this group are of great medical and veterinary importance because of the serious and fatal diseases produced in their vertebrate hosts, including humans and domestic animals.

Females are ovoviviparous, giving birth to larvae known as **microfilariae,** which in the Onchocercidae are **sheathed** and in the Dirofilariidae **unsheathed.** Microfilariae may appear periodically with peak numbers in the blood at given times during the day or night, thus being diurnal or nocturnal, or they may be present continuously. Such conditions are said to be of **periodic** or **continuous** occurrence in the blood.

Family Onchocercidae

This family is characterized by a rudimentary stoma without lips or spines. There is no ring around the mouth nor one connecting the buccal capsule and esophagus. The latter usually is not divided into a short anterior muscular and a long posterior glandular part. Caudal alae generally are absent, but if present, they are small and not supported by papillae. Microfilariae are periodic. Parasitic in the connective tissue, heart, arteries, or body cavities.

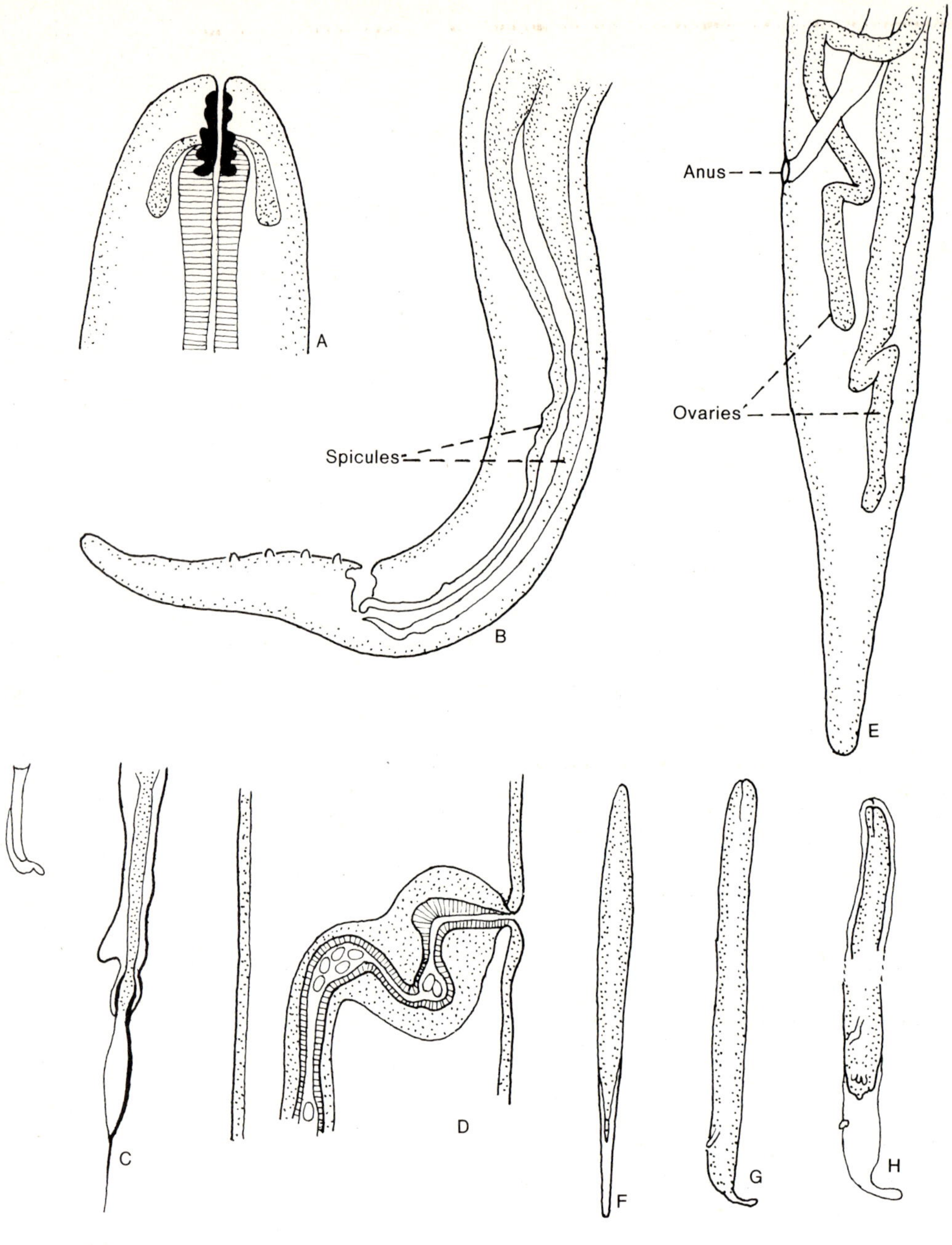

Fig. 5.34. *Litomosoides carinii.* (A) Anterior end of adult female; (B) posterior end of adult male; (C) short right and long left spicules; (D) vulva and vagina; (E) posterior end of adult female; (F–I) larvae; (F) larva in embryonic sheath; (G) first-stage larva; (H) ensheathed second-stage larva; (I) ensheathed third-stage larva (G–I in mite). (Modified from Cross and Scott, 1947.)

Litomosoides carinii

This species (fig. 5.34) is prevalent in cotton rats (*Sigmodon hispidus*) throughout their range in the southern half of the United States, especially in the southeastern states. It is the most useful model for studying filarial worms in the laboratory. Cotton rats are easily obtained, reproduce well in captivity, have a high incidence of natural infection, especially in the enzootic area, and are easily infected experimentally. The mite intermediate host is prevalent and easily reared in great numbers on white rats, which, if necessary, may be used in lieu of cotton rats as final hosts. Having a relatively short life cycle, all the stages of a filarial worm with a bloodsucking arthropod intermediate host can be demonstrated in the laboratory. (See p. 284.)

Description

Adults are slender, cylindrical worms tapering to a bluntly rounded head and a slender conical tail. The buccal capsule is up to 23 μm long with cylindrical, heavily sclerotized walls encircled by thickened outer rings to which the esophageal muscles attach. The mouth is without lips or surrounding papillae. The esophagus is muscular throughout, up to 660 μm long, and is straight except for a bulbous anterior enlargement that surrounds the buccal capsule.

Males measure up to 28 mm long by 140 μm in greatest diameter. The tail, 180 μm long, is curled ventrally and tapers gradually in the posterior half of its length to a rounded tip. There are no preanal papillae, but four pairs of small, nearly equally spaced latero-ventral papillae are located in the anterior half of the tail. Spicules are unequal and dissimilar. The shorter, right one is sclerotized and trough shaped for up to 80 μm, about three-fourths its length, at which point there is an elbowlike, delicate, membranous projection up to 30 μm long; the longer left spicule has a total length of up to 295 μm; the anterior two-fifths is trough shaped and sclerotized, followed by a portion partly sclerotized and partly membranous that ends in a delicate filament. The testis is located in the region of the posterior part of the esophagus and continues directly caudad as a straight vas deferens and seminal vesicle; the latter divide into two small tubes that empty into the cloaca.

Females are up to 100 mm long and have a tapering tail. The vulva, located a distance about twice the length of the esophagus from the anterior end of the body, is surrounded by a large muscular bulb, followed by the posteriorly directed, thick-walled muscular vagina (the single, expanded part of the anterior end of the uterus formed by the union of the two uteri), and two long, thin-walled, intricately looping uteri that extend to the posterior end of the body, and then loop forward to the anterior end, each terminating in a posteriorly directed threadlike ovary. Uteri are filled with eggs, and toward the proximal ends, with developing and sheathed microfilariae.

Anatomical landmarks of the microfilariae have not been described and located in percentages of the total body length. Stain blood smears from infected cotton rats with Giemsa or other suitable blood stain (see p. 256). Locate the landmarks, and designate in percentages the positions of: (1) first body cells in the anterior part of the body, (2) nerve ring, (3) excretory pore, (4) excretory cell, (5) R_{14} cells, (6) anus, and (7) last body cell in tail.

Life Cycle

Adult worms lie free on the serosal surfaces of the pleural cavities of the rat host, where the sheathed larvae are discharged. From the pleural cavities, the microfilariae enter the circulation and appear continuously in the peripheral blood vessels. When infected blood is ingested by the tropical rat mite (*Ornithonyssus bacoti*), microfilariae burrow through the intestinal wall into the hemocoel. By about the eighth day, larvae become typically sausage shaped and molt for the first time. The second molt occurs a day or two later, and by the fourteenth to fifteenth days the slender third-stage larvae are infective. Upon injection into rats, larvae migrate to the pleural cavity. Here they undergo the fourth and fifth molts and mature. Prepatency is about 75 days.

Family Dirofilariidae

Parasites of the heart and connective tissues of mammals, they are long, slender worms with a simple mouth devoid of lips. Caudal alae of the male are narrow and supported by large papillae. The vulva is located far anterior and microfilariae are without a sheath.

Dirofilaria immitis

This is the heartworm of dogs that occurs in the right side of the heart and in the pulmonary artery, often in large intertwined clusters. It is a dangerous parasite because up to 100 of these large worms may occur in a single animal, seriously interfering with the flow of blood. Infections occur in dogs over a wide area in the United States, but they are especially prevalent along the Atlantic seaboard and in the southern half of the United States. In addition, foxes, wolves, coyotes, and cats are infected. Larvae and adults have been reported in humans.

Description

Dirofilaria immitis (fig. 5.35) are very long, slender worms with a small circular mouth and six weakly developed cephalic papillae. The esophagus is short and narrow, and the anterior muscular and posterior glandular parts are indistinctly separated.

Males are up to 200 mm long by nearly 1 mm in diameter, with a tapering, spirally coiled tail. Caudal alae are narrow and inconspicuous. Generally there are five pairs of large preanal papillae that support the alae and six pairs of smaller postanal papillae. The left spicule is up to 375 μm long and has a sharp elbowlike bend slightly posterior to the middle; the terminal portion ends in a narrow point; the right spicule is up to 229 μm long and basically arcuate in shape. A gubernaculum is lacking.

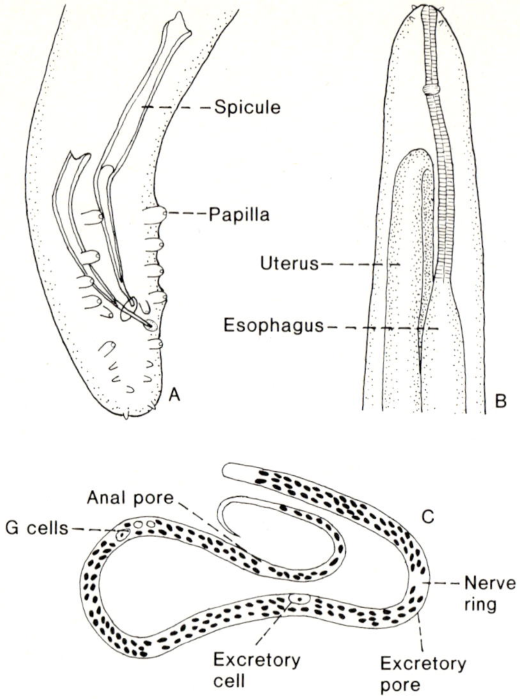

Fig. 5.35. *Dirofilaria immitis.* (A) Caudal end of adult male; (B) anterior end of adult female; (C) microfilaria. Sausage and third-stage larvae generally similar to those of *Wuchereria bancrofti.* (Adapted from Mönnig, 1947.)

Females are up to 310 mm long by up to 1.3 mm in diameter. The tail is blunt, and the vulva is located just behind the posterior end of the esophagus. The uteri and ovaries fill the body of gravid individuals. The vagina and terminal parts of the uteri are packed with microfilariae.

Life Cycle

Unsheathed microfilariae are liberated in the blood of the heart and pulmonary artery and are carried throughout the body. They may, however, be subperiodic, showing a minor nocturnal peak. When infected blood is ingested by the yellow fever mosquito *Aedes aegypti,* several house mosquitoes (e.g., *Culex pipiens, C. tarsalis* and *C. territans*), and the plasmodia-transmitting species *Anopheles maculipennis* and *A. quadrimaculatus,* the microfilariae migrate from the intestine to the Malpighian tubules, penetrate the cells at the tips, and develop to the sausage stage. After about two weeks, the larvae leave the Malpighian tubules and migrate anteriorly in the mosquitoes, going through the thoracic muscles and into the proboscis. In the meantime, they have developed to the third-stage infective larvae. Transmission of larvae occurs while the vectors are feeding on the dog host. The young worms undergo development in various extravascular tissues before entering the blood vessels, migrating to the heart about ten days postinfection. The prepatent period is eight to nine months.

A Dipetalonematidae, *Dipetalonema reconditum,* occurs in the connective tissues of dogs in this country, and its microfilariae appear concurrently with those of *Dirofilaria immitis* in the peripheral blood. In general, they are smaller, have a small cephalic spine, and the landmarks have different locations, in terms of percentage, in the body. Vectors are fleas (*Ctenocephalides canis, C. felis*) and lice (*Heterodoxus spiniger*). Consult Newton and Wright (1956) for the anatomical landmarks used in separating the two species of microfilariae.

Notes and Sketches

Key to Common Families of Parasitic Nematodes

Capital letters in parentheses refer to hosts: F-fish, A-Amphibia, R-Reptiles, B-Birds, M-Mammals.

1 Esophagus club shaped, cylindrical, or with posterior bulb and muscular, or with anterior muscular and posterior glandular portions; bilobed copulatory bursa when present always membranous and supported by six pairs of lateral rays; phasmids present; excretory canal lined with cuticle; males with paired genital papillae. Class **Phasmidia** 2

Esophagus with long posterior portion consisting of a chain of cuboidal cells or cylindrical and muscular, in which case males possess a fleshy, cup-shaped rayless copulatory bursa; phasmids absent; excretory duct not lined with cuticle; males usually without paired caudal papillae. Class **Aphasmidia** 60

2(1) Females only known from parasitic forms; esophagus of adult muscular, cylindrical, and short or long. Order **RHABDITATA,** Suborder **Rhabditina** 3

Both sexes parasitic and known 4

3(2) Esophagus short; buccal capsule present; vulva near middle of body; in lungs (fig. 5.36) (A) Superfamily **Rhabditoidea** .... **Rhabditidae**

Esophagus long; buccal capsule absent; vulva in posterior part of body; in intestine (fig. 5.37) (R, B, M) Superfamily **Rhabdiasoidea** .... **Strongyloididae**

4(2) Males with cuticular bilobed copulatory bursa supported by paired rays; esophagus more or less club shaped. Order **STRONGYLATA** . 5

Copulatory bursa absent; esophagus muscular or part muscular and part glandular 17

5(4) Buccal capsule large, with thick wall; body thick. Suborder **Strongylina** 6

Buccal capsule rudimentary or absent; filiform body 11

6(5) Mouth with ventral toothed or smooth cutting plates; head bent dorsally. Superfamliy **Ancylostomatoidea** 7

Mouth without cutting plates; head straight... 8

7(6) Cutting plates with smooth margin; in intestine (fig. 5.14) .... (M) **Uncinariidae**

Cutting plates with teeth on margin; in intestine (figs. 5.11 and 5.12) .... (M) **Anclostomatidae**

8(6) Buccal capsule with thickened rim, small teeth in base, mouth wide; small males often permanently attached to large females; in trachea (fig. 5.38) (B, M) Superfamily **Syngamoidea** .... **Syngamidae**

Buccal capsule without thickened rim; in alimentary canal or kidneys. Superfamliy **Strongyloidea** 9

9(8) Buccal capsule short, ring shaped; transverse groove on ventral side of body anterior to excretory pore; in caecum, colon (fig. 5.39) .... (M) **Oesophagostomidae**

No ventral transverse groove; buccal capsule globular or cup shaped 10

10(9) Buccal capsule globular, usually with longitudinal dorsal groove and well-developed leaf crowns; copulatory bursa long; in intestine (fig. 5.40) .... (M) **Strongylidae**

Buccal capsule cup shaped; copulatory bursa short; dorsal groove lacking; in kidneys (fig. 5.41) .... (M) **Stephanuridae**

11(5) Bursa and rays well developed, rays slender and not enlarged or fused; anterior end of the body usually inflated; body with longitudinal ridges; eggs thin shelled, unembryonated in uterus; in alimentary canal and lungs. Suborder **Trichostrongylina** 12

Bursa less developed or absent, rays fused or abnormally large; body without anterior inflation or longitudinal ridges; uterus contains eggs with larvae or first-stage larvae enclosed in a thin membrane; in respiratory tract. Suborder **Metastrongylina** 14

12(11) Genitalia of female single; spicules long, slender; vulva near anus; in intestine (fig. 5.16) (M) Superfamily **Heligmosomatoidea** .... **Heligmosomatidae**

Genitalia of female double; spicules short, stout; vulva not near anus. Superfamily **Trichostrongyloidea** 13

13(12) Buccal capsule small; spicules simple, sox shaped; bursa short; in lungs (fig. 5.43)......... .... (M) **Dictyocaulidae**

Buccal capsule rudimentary or absent; spicules complex, with crests, or long and filiform; bursa with lateral lobes well developed; in alimentary canal (figs. 5.42, 5.44) .... (A, R, B, M) **Trichostrongylidae**

14(11) Bursa reduced to two elongate lateral lobes projecting ventrally at almost right angles to body; spicules short, membranous; vulva near anus at extremity of tubular process; first-stage larva in thin membrane in uterus; in bronchi, blood vessels, heart, frontal sinuses (fig. 5.45) (M) Superfamily **Pseudalioidea** .... **Filaroididae**

Bursa not reduced to two elongate lateral lobes, symmetrical or absent 15

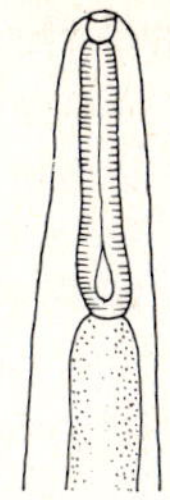

Fig. 5.36. *Rhabdias ranae*

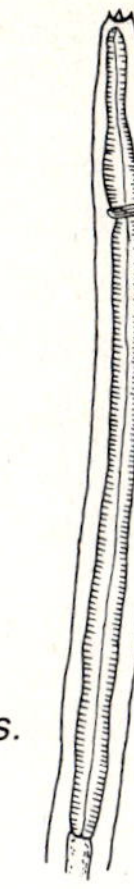

Fig. 5.37. *Strongyloides stercoralis.*

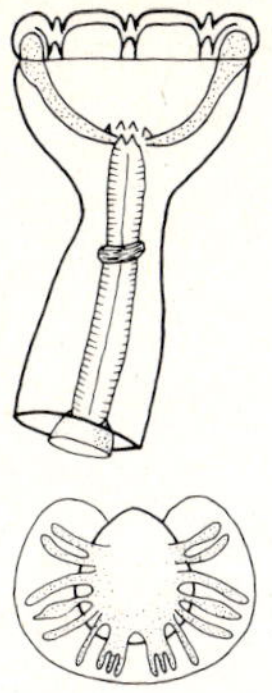

Fig. 5.38. *Syngamus trachea.*

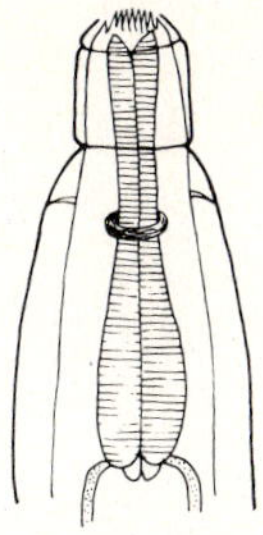

Fig. 5.39. *Oesophagostomum columbianum.*

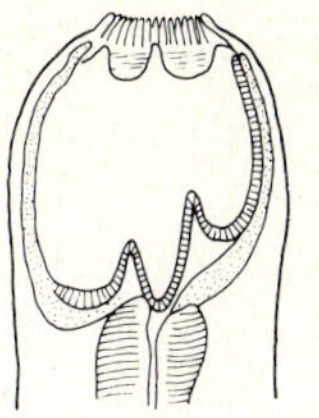

Fig. 5.40. *Strongylus equinus.*

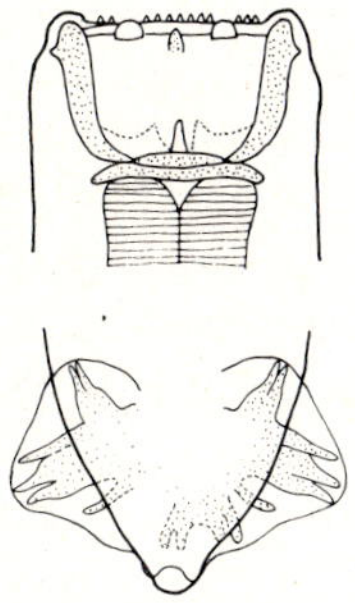

Fig. 5.41. *Stephanurus dentatus.*

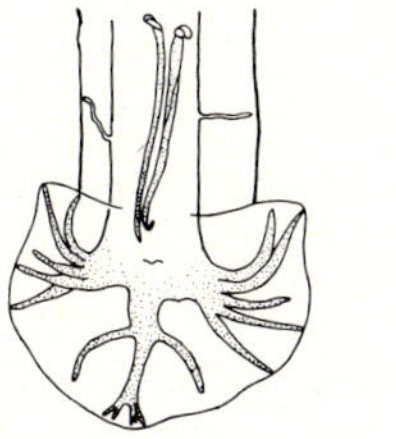

Fig. 5.42. *Viannaia hamata.*

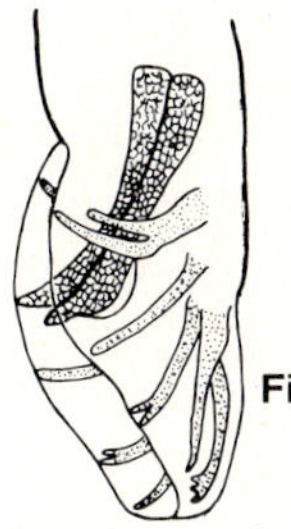

Fig. 5.43. *Dictyocaulus filaria.*

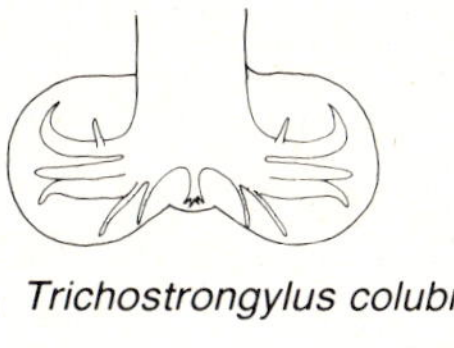

Fig. 5.44. *Trichostrongylus colubriformis.*

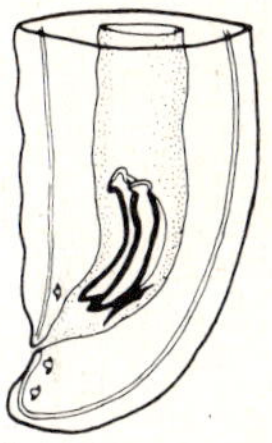

Fig. 5.45. *Filarioides osleri.*

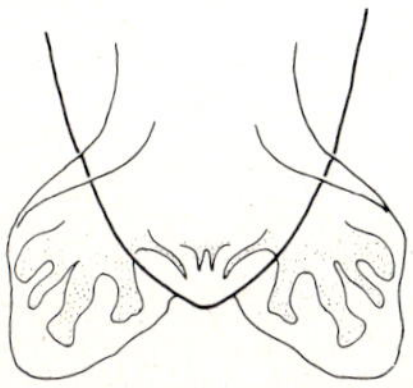

Fig. 5.46. *Metastrongylus elongatus.*

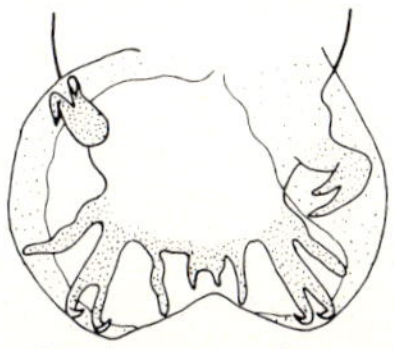

Fig. 5.47. *Angiostrongylus gubernaculum.*

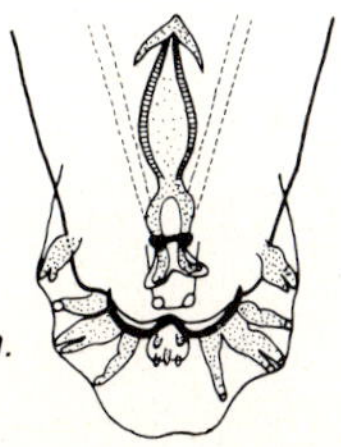

Fig. 5.48. *Protostrongylus rufescens.*

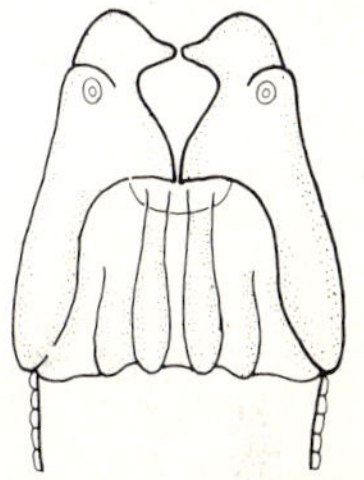

Fig. 5.49. *Heterocheilus tunicatus.*

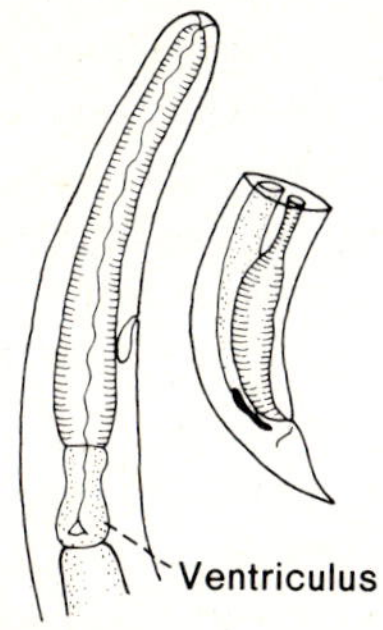

Fig. 5.50. *Oxyascaris oxyascaris.*

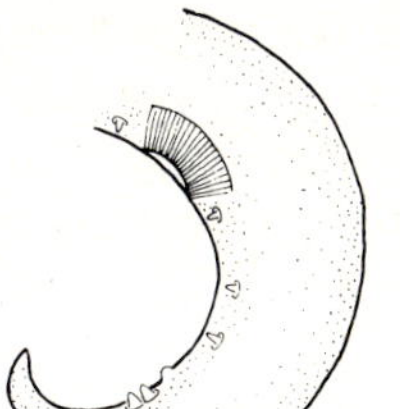

Fig. 5.51. *Quimperia lanceolata.*

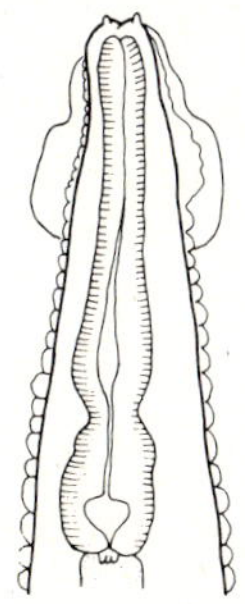

Fig. 5.52. *Heteroxynema wernecki.*

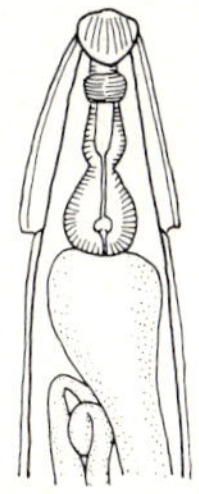

Fig. 5.53. *Aspiculuris tetraptera.*

Fig. 5.54. *Aspidodera scoleciformis.*

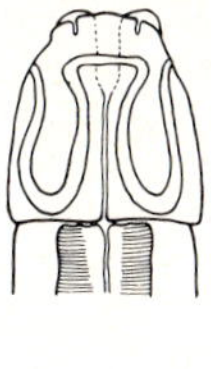

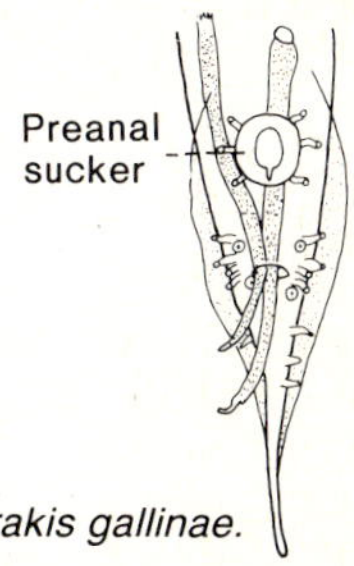

Fig. 5.55. *Heterakis gallinae.*

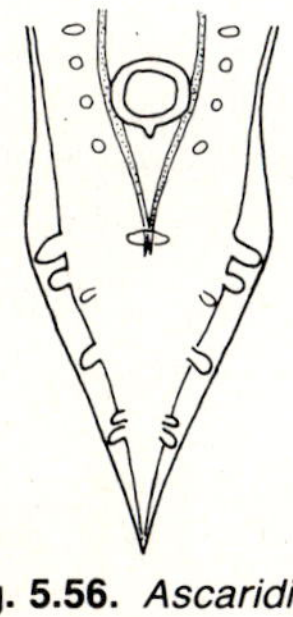

Fig. 5.56. *Ascaridia sp.*

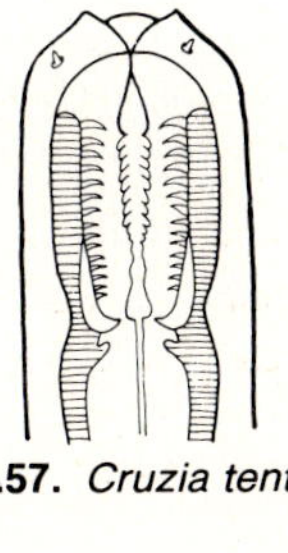

Fig. 5.57. *Cruzia tentaculata.*

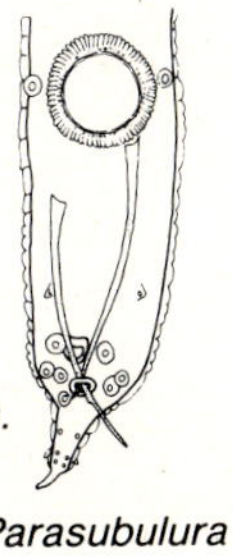

Fig. 5.58. *Parasubulura gerardi.*

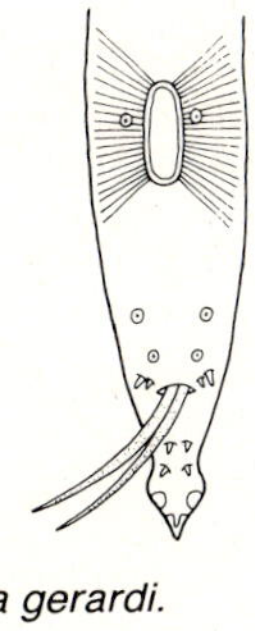

Fig. 5.59. *Subulura acutissima.*

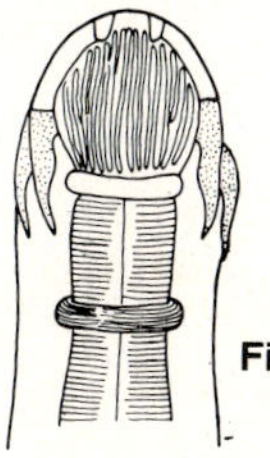

Fig. 5.60. *Camallanus lacustris.*

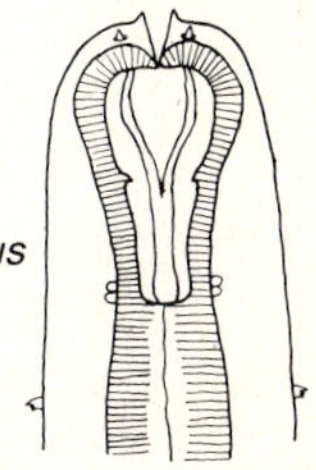

Fig. 5.61. *Cucullanus schubarti.*

15(14) Bursa well developed with large lateral lobes; dorsal ray double, externolateral ray lobed at tip; spicules long; buccal capsule small; in respiratory system (fig. 5.46) (M) Superfamily **Metastrongyloidea** .... **Metastrongylidae**
Bursa small, dorsal ray enlarged and single with one to several ventral papillae or with tip divided into two or more branches; spicules short; buccal capsule absent or rudimentary. Superfamily **Protostrongyloidea** 16

16(15) Spicules long, slender, without alae: buccal capsule absent; gubernaculum present or absent; in circulatory or respiratory systems (fig. 5.47) .... (M) **Angiostrongylidae**
Spicules somewhat stout, long, tubular, and with striated alae in distal portion; buccal capsule absent or rudimentary; gubernaculum present (rarely absent); in respiratory and circulatory systems (fig. 5.48) ....(M) **Protostrongylidae**

17(4) Esophagus muscular throughout, cylindrical or with a basal bulb; spicules equal in length when present. Order ASCARIDATA 18
Esophagus cylindrical, usually with short anterior, muscular and long posterior glandular parts; spicules usually different in shape and almost always unequal in size; esophageal glands uni- or multinucleate. Order **SPIRURATA** 32

18(17) Esophagus more or less cylindrical. Suborder **Ascaridina** 19
Esophagus with distinct basal bulb. Suborder **Oxyurina** 23

19(18) Esophagus more or less cylindrical (sometimes with small posterior bulb), with unusual modifications such as a ventriculus with or without a caecum; a ventriculus and intestinal caecum, or only an intestinal caecum but no ventriculus; three lips, usually with interlabia; in alimentary canal (fig. 5.49) (B,M).................... .... Superfamily **Heterocheiloidea. Heterocheilidae**
Esophagus almost club shaped, without ventriculus (except *Toxocara, Neoascaris*) or intestinal caecum; three lips without interlabia; large intestional worms. Superfamily **Ascaridoidea** 20

20(19) Esophagus with ventriculus or glandular bulb 21
Esophagus without ventriculus 22

21(20) Lips greatly reduced; intestine with voluminous pyriform, cuticularized rectum; in intestine (fig. 5.50) .... (A,R) **Oxyascarididae**
Lips well developed; intestine without enlarged, cuticularized rectum; in intestine (fig. 5.22) .... (M) **Toxocaridae**

22(20) Lips well developed, esophagus never divided into different parts; vulva in anterior half of body; preanal sucker absent; in intestine (fig. 5.18) .... (R, B, M) **Ascaridae**
Lips small or absent; esophagus may be divided into narrow anterior and broader posterior parts; vulva in posterior half of body; preanal sucker usually present and without cuticularized rim; in intestine (fig. 5.51) .... (F) **Quimperiidae**

23(18) No preanal sucker. Suborder **Oxyurina** 24
Preanal sucker present. Suborder **Heterakina** 27

24(23) With a gubernaculum and a single spicule; in intestine (fig. 5.27) (M) Superfamily **Syphacioidea** .... **Syphaciidae**
With or without single spicule but without a gubernaculum, with or without cuticular ornamentations. Superfamily **Oxyuroidea** ... 25

25(24) Single spicule; no cervical spine; in large intestine (fig. 5.24) .... (M) **Oxyuridae**
Without spicule (if present, short) 26

26(25) Cuticular ornamentation in precloacal area; in intestine (fig. 5.52) .... (M) **Heteroxynematidae**
Preanal ornamentation lacking; in intestine (fig. 5.53) .... (M) **Aspicularidae**

27(23) Preanal sucker circular, well formed, with cuticularized rim. Superfamily **Heterakoidea** 28
Preanal sucker (pseudosucker) formed by preanal muscles, without a cuticularized rim. Superfamily **Subuluroidea** 30

28(27) Cervical cordons present; precloacal muscles strongly developed; in intestine (fig. 5.54) .... (M) **Aspidoderidae**
Cervical cordons absent 29

29(28) Esophagus with well defined posterior bulb, with short narrow anterior portion; male and female tails long; lateral caudal alae in males; small worms; in intestine (fig. 5.55) .... (A, R, B, M) **Heterakidae**
Esophagus without bulb or short, narrow anterior portion; male and female tails conical; without caudal alae in males; medium-size worms; in intestine (fig. 5.56) .... (B) **Ascaridiidae**

30(27) Intestine with anterior diverticulum; in caecum, colon (fig. 5.57) .... (A, R, M) **Cruziidae**
Intestine simple, without diverticulum 31

31(30) Preanal sucker circular, with cuticularized crown of about 70 leaflike elements and equal number of buttonlike basal elements; buccal capsule cylindrical, thin cuticularized walls; in intestine (fig. 5.58) ..... (M) **Parasubuluridae**

Preanal sucker without cuticularized elements; precloacal muscles well developed, forming elongated pseudosucker; vestibule usually with basal teeth; in intestine (fig. 5.59) .... (B, M) **Subuluridae**

32(17) Esophageal glands uninucleate; internal and external circle of cephalic papillae present; mouth well developed or rudimentary; larvae with large pocketlike phasmids; intermediate hosts copepods. **Order Camallanata** 33

Esophageal glands multinucleate; stoma rudimentary or well developed, with or without pseudolabia; oviparous or ovoviviparous; intermediate hosts usually insects or copepods 36

33(32) Internal circle of papillae reduced in size, external circle partially fused; stoma well developed; oviparous or ovoviviparous. Superfamily **Camallanoidea** 34

Internal circle of papillae well developed; external circle of eight separate papillae; stoma rudimentary; ovoviviparous. Superfamily **Dracunculoidea** 35

34(33) Buccal capsule globular or consisting of two large lateral cuticularized, shell-like parts; esophagus with anterior muscular and posterior glandular portions; vagina directed posteriorly; ovoviviparous; in alimentary canal (fig. 5.60) .... (F, A, R) **Camallanidae**

Esophagus muscular throughout with posterior club-shaped swelling and anterior dilation into a false buccal cavity; preanal sucker in males; vagina directed anteriorly; ovoviviparous; in intestine (fig. 5.61) .... (F) **Cucullanidae**

35(33) Anterior extremity with cuticular thickening or shield; esophagus long and cylindrical with constriction at level of nerve ring; vulva immediately behind head; tail of female conical, slightly coiled; spicules equal; in tissues (fig. 5.62) .... (R, B, M) **Dracunculidae**

Anterior extremity without cuticular shield; esophagus short, cylindrical, and without median constriction; tail of female rounded, straight; spicules equal; in tissues (fig. 5.32) .... (F) **Philometridae**

36(32) Stoma well developed with or without pseudolabia, or rudimentary and with pseudolabia; vulva usually near or posterior to middle of body; generally oviparous; intermediate hosts usually nonbloodsucking insects, occasionally copepods; in alimentary canal, orbital and nasal cavities, respiratory system. Order **Spirurata** 37

Stoma rudimentary without pseudolabia; vulva in anterior part of body; oviparous or ovoviviparous; in air sacs, peritoneal cavity, tissues, blood vessels; intermediate hosts bloodsucking insects and mites. Order **Filariata** 51

37(36) Pseudolabia absent (except *Physocephalus*) Superfamily **Thelazioidea** 38

Two well-developed pseudolabia present .. 42

38(37) With cuticular plaques over anterior part of body; in wall of esophagus and stomach, slender worms (fig. 5.63) .... (B, M) **Gongylonematidae**

Without cuticular plaques 39

39(38) With cuticular hooklike spines arranged in rows or circles over whole or anterior portion of body; in intestine (fig. 5.64)..................................... .... (R, B, M) **Rictulariidae**

Without cuticular spines 40

40(39) Buccal capsule long, cylindrical, composed of cuticular ridges in form of rings; in stomach (fig. 5.65) .... (M) **Ascaropsidae**

Buccal capsule smooth 41

41(40) Caudal alae absent; in orbital and nasal sinuses, lungs (fig. 5.66) (B, M) **Thelaziidae**

Caudal alae present; mouth surrounded by six masses of parenchyma; caudal papillae not numerous; in esophagus, stomach, aorta, lungs (fig. 5.67) .... (M) **Spirocercidae**

42(37) Pseudolabia lobed 43

Pseudolabia not lobed 49

43(42) Cephalic papillae posterior to pseudolabia; interlabia present or absent. Superfamily **Spiruroidea** 44

Eight partially fused double cephalic papillae on pseudolabia. Superfamily **Gnathostomatoidea** 47

44(43) Female fusiform or coiled to give fusiform shape; vulva near anus; males much smaller than females; females imbedded in crypts of proventriculus, males free in lumen (fig. 5.68) .... (B) **Tetrameridae**

Sexes more or less equal in size, females not fusiform 45

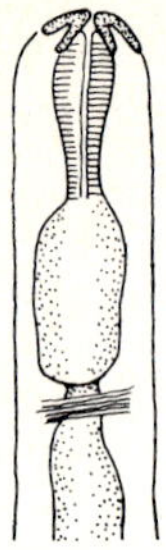
Fig. 5.62. *Dracunculus medinensis.*

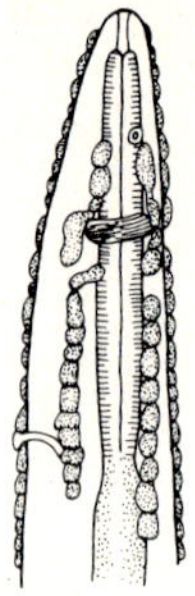
Fig. 5.63. *Gongylonema pulchrum.*

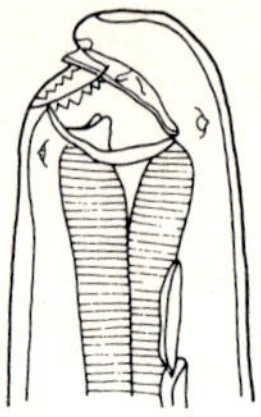
Fig. 5.64. *Rictularia affinis.*

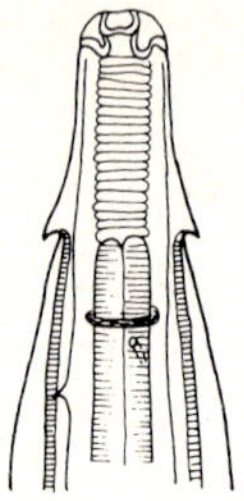
Fig. 5.65. *Physocephalus sexulatus.*

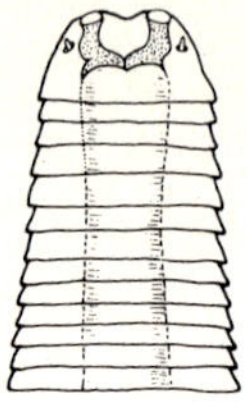
Fig. 5.66. *Thelazia rhodesi.*

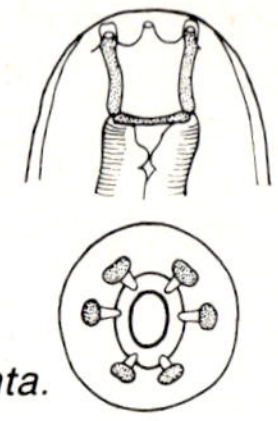
Fig. 5.67. *Spirocerca sanguinolenta.*

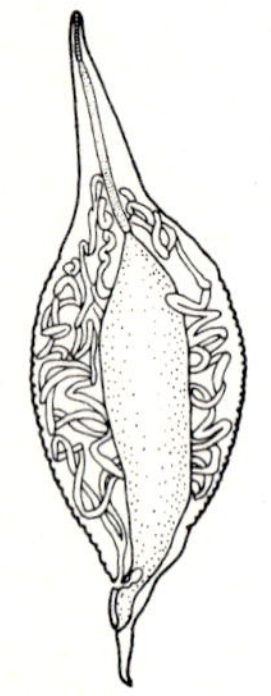
Fig. 5.68. *Tetrameres fissipina.*

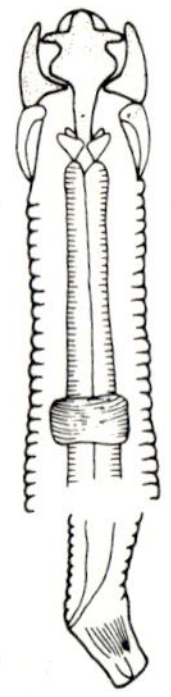
Fig. 5.69. *Hedruris androphora.*

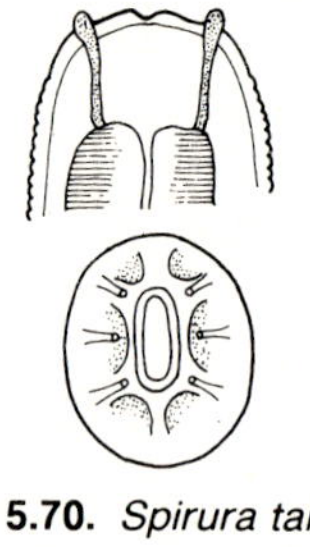
Fig. 5.70. *Spirura talpae.*

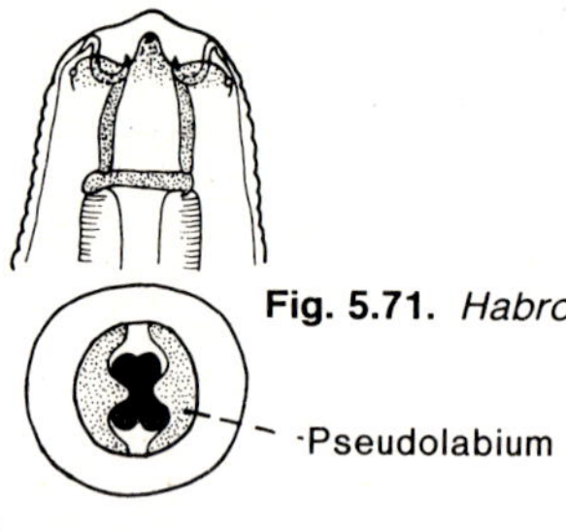

Fig. 5.71. *Habronema muscae.*

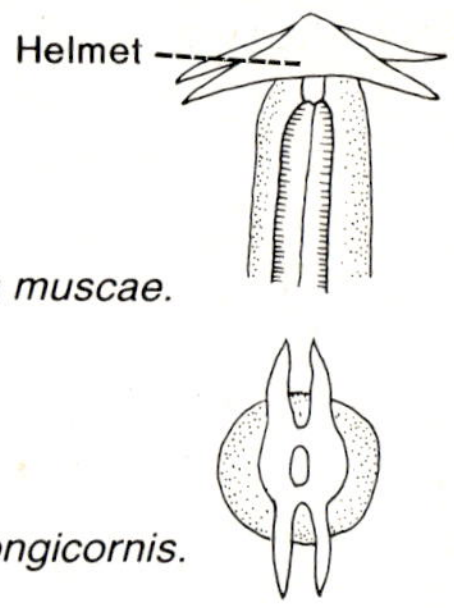

Fig. 5.72. *Schistorophus longicornis.*

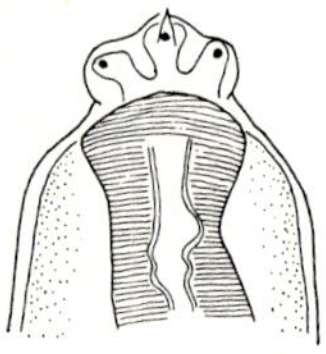
Fig. 5.73. *Spiroxys contortus.*

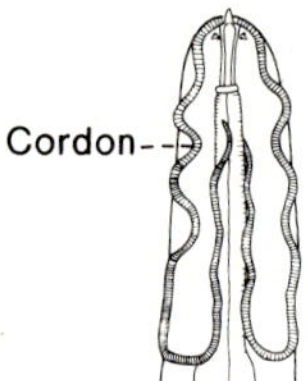

Fig. 5.74. *Acuaria spiralis.*

Fig. 5.75. *Seuratia shipleyi.*

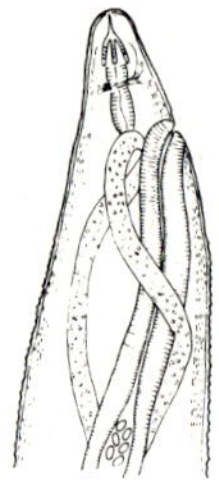
Fig. 5.76. *Diplotriaena ozouxi.*

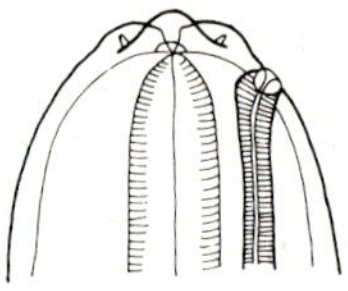
Fig. 5.77. *Filaria martis.*

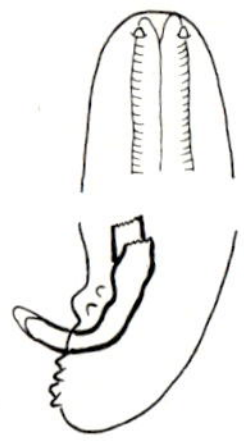
Fig. 5.78. *Aprocta semenovi.*

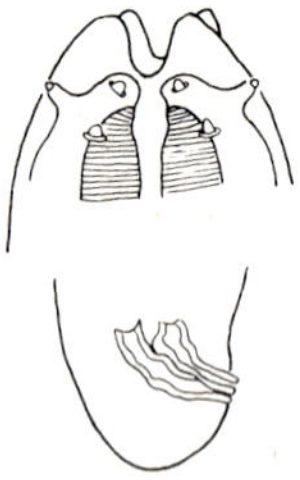
Fig. 5.79. *Tetracheilonema tringae.*

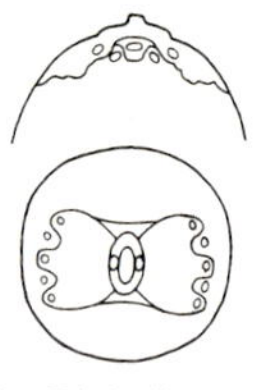
Fig. 5.80. *Dicheilonema rheae.*

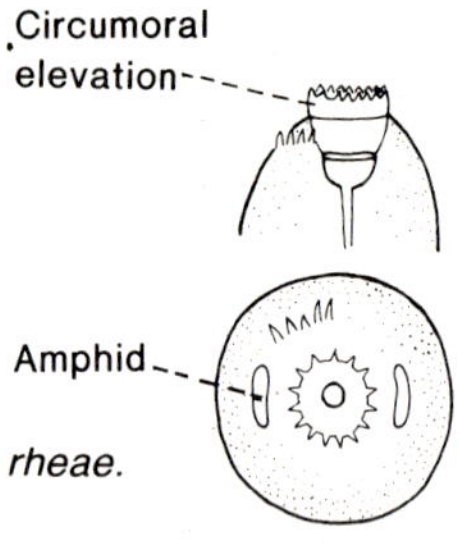

Fig. 5.81. *Stephanofilaria stilesi.*

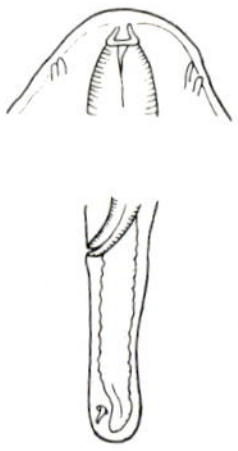
Fig. 5.82. *Oswaldofilaria bacillaris.*

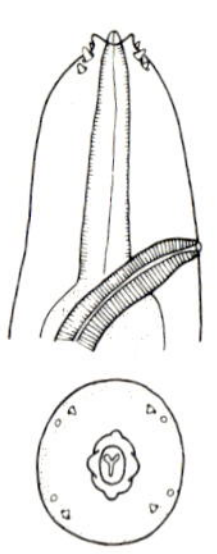
Fig. 5.83. *Setaria equina.*

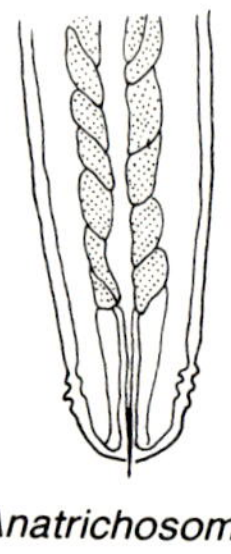
Fig. 5.84. *Anatrichosoma cynamologi.*

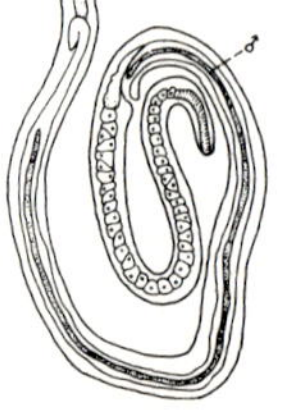
Fig. 5.85. *Trichosomoides crassicauda.*

Fig. 5.86. *Hystrichis tricolor.*

45(44) With four specialized lips, dorsal and ventral ones in form of isosceles triangles; male always rolled about female; festooned cuticularized ring at anterior extremity of apparently undivided esophagus; spicules equal; vulva near anus; tail of female invaginated to form groove with projecting hook; in intestine (fig. 5.69) .... (A, R) **Hedruridae**
Without four lips, cuticular esophageal ring, or invaginated tail; spicules unequal; vulva near middle of body 46

46(45) Interlabia and cuticular flange absent; tail of male conical; in intestine (fig. 5.70) .... (R, B, M) **Spiruridae**
Interlabia usually present; cuticular flange present on one or both sides; tail of male coiled spirally; in stomach (fig. 5.71) .... (B, M) **Habronematidae**

47(43) Head with prominent backward pointing appendages of feathered or unfeathered appearance; in alimentary canal (fig. 5.72) .... (R, B) **Schistorophidae**
Head without appendages 48

48(47) Head inflated into prominent spined or striated bulb; vulva in posterior half of body; in alimentary canal (fig. 5.28) .... (F, R, B, M) **Gnathostomatidae**
Head without inflation; vulva near middle of body; in stomach (fig. 5.73) .... (R) **Spiroxyidae**

49(42) Anterior end without cordons; cuticle usually reflected forward over lips to form a cephalic collar; pseudolabia large, simple, triangular, and with one or more apical teeth; caudal alae wide, joined ventrally anterior to cloaca, supported by long papillae; in alimentary canal (fig. 5.30) (P, A, R, B, M) Superfamily **Physalopteroidea Physalopteridae** ....
Anterior end of body with cordons (raised cuticular ridges), hood, or spined collarette extending caudad; buccal capsule long, cylindrical; caudal alae narrow, lateral. Superfamily **Acuarioidea** 50

50(49) Simple cordons present; in gizzard and intestine (fig. 5.74) .... (B, M) **Acuariidae**
Spined cephalic collar present; in intestine (fig. 5.75) .... (B) **Seuratidae**

51(36) Females oviparous: eggs thick shelled, embryonated when laid. Superfamily **Filarioidea** 52
Females ovoviviparous; microfilariae in thin membranous sacs or free of membrane. Superfamily **Onchocercoidea** 56

52(51) With cuticularized trident on each side of anterior end of esophagus; in air sacs and body cavity (fig. 5.76) .... (B) **Diplotriaenidae**
Without tridents 53

53(52) Mouth simple without any surrounding structures 54
Mouth with cuticularized peribuccal ring with lateral expansions or surrounded by four cone-shaped elevations 55

54(53) Spicules unequal and dissimilar, vulva near mouth; caudal alae often present in male; in tissues (fig. 5.77) .... (M) **Filariidae**
Spicules equal and similar, vulva distinctly removed from region of mouth; caudal alae present or absent; in orbital and nasal cavities, subcutaneous tissues (fig. 5.78) .... (B) **Aproctidae**

55(53) Mouth suurounded by four conical cuticular elevations; eight papillae in external circle and four small ones in internal circle; in air sacs and tissues (fig. 5.79) .... (B) **Tetracheilonematidae**
Mouth surrounded by raised cuticularized peribuccal ring with a pair of lateral extensions; in air sacs and subcutaneous tissues (fig. 5.80) .... (R, B) **Dicheilonematidae**

56(51) Mouth surrounded by a ring of many minute cuticularized spines or with only six large papillae in external circle; amphids large; male tail with numerous cloacal papillae; in subcutaneous tissues (fig. 5.81) .... (M) **Stephanofilariidae**
Mouth not surrounded by ring of spines or six large papillae 57

57(56) Buccal capsule rudimentary or absent; with ring attaching it to anterior extremity of esophagus; in body cavity, tissues, heart (fig. 5.82) .... (M) **Dipetalonematidae**
No ring connecting buccal capsule and esophagus 58

58(57) Cuticularized peribuccal ring with medial or lateral protrusions, or mouth surrounded by lateral plates; tail of female regularly long and ornamented with tubercles or one pair of lateral projections; in body cavity (fig. 5.83) .... (M) **Setariidae**
No peribuccal ring 59

59(58) Caudal alae well developed, supported by large cloacal papillae; male tail short (less than twice the diameter of body at level of anus); esophagus usually divided; in tissues, blood vessels, heart, body cavity (fig. 5.35) .... (M) **Dirofilariidae**

Caudal alae absent, small, or well developed, but when well formed not supported by elongate papillae; esophagus usually not divided; tail long or short; in connective tissues, heart, arteries, body cavity .... (B, M) **Onchocercidae**

60(1) Small threadlike worms with esophagus composed of short, anterior muscular part and a long posterior chain of cuboidal cells with a capillary tubule running through them; caudal end of males without a bursa. Order **TRICHURATA,** Superfamily **Trichuroidea** 61

Large worms; males with fleshy, cup-shaped copulatory bursa without supporting rays; esophagus cylindrical, muscular. Order **DIOCTOPHYMATA** 65

61(60) Spicule and spicular sheath present, spicule only present, or, if spicule absent, spicular sheath present 62

Spicule and spicular sheath absent 64

62(61) Spicule present, spicular sheath absent; small worms; vulva posterior to base of esophagus; ovoviviparous; in connective tissue (fig. 5.84) .... (M) **Anatrichosomatidae**

Spicule and spicular sheath present, or, if spicule absent, sheath present; large to medium-size worms 63

63(62) Esophageal part of body distinctly thinner and much longer than thick posterior part; spicule and sheath present; in alimentary canal (fig. 5.5) .... (M) **Trichuridae**

Esophageal part of body only slightly smaller in diameter and about equal in length to posterior part; spicule and sheath may be present, or spicule may be absent, in which case sheath is present; in alimentary canal (B, M), in liver and lungs (M) .... **Capillariidae**

64(61) Males minute, in vagina or uterus of female; eggs with thick shell, bioperculate, fully embryonated when laid; in urinary tract (fig. 5.85) .... (M) **Trichosomoididae**

Male free, small but not minute, with two conical processes on caudal end; females embedded in intestinal mucosa, ovoviviparous; adults in intestine, larvae in muscles (fig. 5.2) ..... (M) .... **Trichinellidae**

65(60) Six head papillae in one circle; vulva anterior; in kidney and body cavity (M) Superfamily **Dioctophymatoidea** .... **Dioctophymatidae**

Twelve or more head papillae in two circles; vulva near anus; in forestomach (fig. 5.86) (B) Superfamily **Eustrongyloidea** .... **Eustrongylididae**

Review Questions

1. Describe the primitive nematode head. Include the locations of the labial papillae, cephalic papillae, and amphids.
2. Diagram a cross section of a nematode, labeling the intestine, cuticle, hypodermis, lateral line canals, longitudinal nerve cords, muscle layer, and reproductive organs.
3. Define the following terms: alae, cordon, hood, epaulet, phasmid, cloaca, ventriculus, gubernaculum, ovijector, amphidelphic, microfilaria.
4. Compare the types of esophagi in all orders of parastic nematodes. How do they show a progressive evolutionary picture? How do the esophagi of larvae show a relationship between the Rhabditata and the Strongylata?
5. What is the basic structure of the nematode nervous system?
6. Describe the molting pattern of nematodes. Which stage is usually infective to the definitive host?
7. What is unique to the Trichurata in the esophagus? The excretory system? The tail? The egg?
8. Describe the life cycle of *Strongyloides.*
9. Compare the basic life cycles of Strongyloidea, Trichostrongyloidea, and Metastrongyloidea. Which has an intermediate host? Which usually has an enlarged, complex buccal capsule?
10. Describe the heart-lung migration of *Ascaris, Ancylostoma,* and *Toxocara.*
11. Using examples, compare and contrast visceral larva migrans and dermal larva migrans.
12. What is the single characteristic by which one may identify a nematode from a human as a pinworm?
13. What are the four ways in which a dog can become infected with *Toxocara canis?*
14. Compare the esophagi of the Spirurata, Camallanata, and Filariata. Describe the evolutionary progression in their life cycles, including arthropod hosts and final locations in their definitive hosts.

References

Anderson, R. C.; Chabaud, A. G.; and Willmott, S., eds. 1974–1983. CIH Keys to the Nematode Parasites of Vertebrates. Commonwealth Agricultural Bur., Farnham Royal, Bucks, England. No. 1. General Introduction. Glossary of Terms by S. Willmott, Keys to Subclasses, Orders, and Superfamilies by A. G. Chabaud, 17 pp.; No. 2. Keys to Genera of the Ascaridoidea by G. Hartwich, 15 pp.; No. 3. (3 parts) Keys to Genera of the Order Spirurida by A. G. Chabaud. 116 pp. No. 4. Keys to the genera of Oxyuroidea by A. J. Petter and J. C. Quentin,

30 pp. No. 5. Keys to the genera of the superfamily Metastrongyloidea by R. C. Anderson, 40 pp. No. 6. Keys to the genera of the superfamilies Cosmocercoidea, Seuratoidea, Heterakoidea, and Subuluroidea by A. G. Chabaud, 71 pp. No. 7. Keys to the genera of the superfamily Strongyloidea by J. R. Lichtenfels, 41 pp. No. 8. Keys to the genera of the superfamilies Ancylostomatoidea and Diaphanocephaloidea by J. R. Lichtenfels, 26 pp. No. 9. Keys to the genera of the superfamilies Rhabditoidea, Dioctophymatoidea, Trichinelloidea and Muspiceoidea by F. C. Anderson and O. Bain, 26 pp. No. 10. Keys to genera of the superfamily Trichostrongyloidea by M. C. Durette-Desset, 69 pp.

Anya, A. O. 1976. Physiological Aspects of Reproduction in Nematodes. In *Advances in Parasitology,* ed. B. Dawes, Academic Press, New York, vol. 14, pp. 267–351.

Bird, A. F. 1971. *The Structure of Nematodes.* Academic Press, New York, 318 pp.

Croll, N. A. 1976. *The Organization of Nematodes.* Academic Press Inc., London, 439 pp.

Lee, D. L. 1968. *The Physiology of Nematodes.* Oliver & Boyd, Edinburgh, 154 pp.

Levine, N. D. 1980. *Nematode Parasites of Domestic Animals and of Man.* 2nd ed. Burgess Publishing Company, Minneapolis.

Olsen, O. W. 1974. *Animal Parasites: Their Biology and Ecology.* 3rd ed. University Park Press, Baltimore, 525 pp.

Schmidt, G. D., and Roberts, L. S. 1985. *Foundations of Parasitology.* 3rd ed. The C. V. Mosby Co., St. Louis, 775 pp.

Skrjabin, K. I. et al. 1949–1952. Key to Parasitic Nematodes, vols. 1–3. *Akademii Nauk SSSR,* Moscow.

Skrjabin, K.I. et al. 1953–1986. Essentials of Nematodology, vols. 1–27. *Akademii Nauk SSSR,* Moscow.

Yamaguti, S. 1961. Systema Helminthum. *The Nematodes of Vertebrates.* vol. 3, pts. 1 and 2. Interscience Publishers, Inc., New York 1261 pp.

Rhabditata

Abadie, S. H. 1963. The life cycle of *Strongyloides ratti. J. Parasitol.* 49:241–48.

Spindler, L. A. 1958. The occurrence of the intestinal threadworms, *Strongyloides ratti,* in the tissues of rats, following experimental percutaneous infection. *Proc. Helminth. Soc. Washington* 25:106–11.

Strongylata

Ehrenford, F. A. 1954. The life cycle of *Nematospiroides dubius* Baylis (Nematoda: Heligmosomidae). *J. Parasitol.* 40:480–81.

Fahmy, M. A. M. 1956. An investigation on the life cycle of *Nematospiroides dubius* (Nematoda: Heligmosomidae) with special reference to the free-living stages. *Zeit. Parasitenk.* 17:394–99.

Miller, T. A. 1971. Vaccination against the Canine Hookworm Diseases. In *Advances in Parasitology,* ed. B. Dawes. Academic Press, New York, vol. 9, pp. 153–83.

Nichols. R. L. 1956. The etiology of visceral larva migrans. II. Comparative larval morphology of *Ascaris lumbricoides, Necator americanus, Strongyloides stercoralis,* and *Ancylostoma caninum. J. Parasitol.* 42:363–99.

Olsen, O. W., and Lyons, E. T. 1965. Life Cycle of *Uncinaria lucasi* Stiles, 1901 (Nematoda: Ancylostomatidae) of fur seals, *Callorhinus ursinus* Linn., on the Pribilof Islands. Alaska. *J. Parasitol,* 51:689–700.

Stone, W. M., and Girardeau, M. 1968. Transmammary passage of *Ancylostoma caninum* larvae in dogs. *J. Parasitol.* 54:426–29.

Ascaridata

Beaver, P. C. 1949. Methods of pinworm diagnosis. *Amer. J. Trop. Med.* 29:577–87.

———. 1969. The nature of visceral larva migrans. *J. Parasitol.* 55:3–12.

Chan, K. F., and Kopilof, S. 1958. The distribution of *Syphacia obvelata* in the intestine of mice during their migratory period. *J. Parasitol.* 44:245–46.

Douvres, F. W.; Tromba, F. G.; and Malakatis, G. M. 1969. Morphogenesis and migration of *Ascaris suum* larvae developing to fourth stage in swine. *J. Parasitol.* 55:689–712.

Greve, J. H. 1971. Age resistance to *Toxocara canis* in ascarid-free dogs. *Amer. J. Vet. Res.* 32:1185–92.

Hussey, Kathleen. 1957. *Syphacia muris vs S. obvelata* in laboratory rats and mice. *J. Parasitol.* 43:555–59.

Spirurata, Camallanata, and Filariata

Bailey, W. S. 1972. *Spirocerca lupi:* a continuing inquiry. *J. Parasitol.* 58:3–22.

Crichton, V. F. J., and Beverley-Burton, M. 1975. Migration, growth, and morphogenesis of *Dracunculus insignis* (Nematoda: Dracunculidea). *Canad. J. Zool.* 53:105–13.

Dailey, M. D. 1967. Biology and morphology of *Philometroides nodulosa* (Thomas, 1929) n. comb. (Philometridae; Nematoda) in the western white sucker (*Catostomus commersoni*). *Diss. Abstr.* 28:1266–67.

Duke, O. L. 1971. The Ecology of Onchocerciasis in Man and Animals. In *Ecology and Physiology of Parasites.* ed. A. M. Fallis. University of Toronto Press, Toronto, pp. 213–22.

Ivashkin, V. M.; Sobolev, A. A.; and Khromova, L. A. 1971. Camallanata of Animals and Man and Diseases Caused by Them. *Akad. Nauk SSSR.* Moscow, vol. 22, 381 pp. (Israel Program for Scientific Translations, 1977.)

Kessel, J. F. 1965. Filarial infections of man. *Amer. Zool.* 5: 79–84.

Mueller, R. 1971. *Dracunculus* and Dracunculiasis. In *Advances in Parasitology,* ed. B. Dawes. Academic Press, New York, vol. 9, pp. 73–151.

Nelson, G. S. 1962. *Dipetalonema reconditum* (Grassi, 1889) from the dog with a note on its development in the flea, *Ctenocephalides felis* and the louse, *Heterodoxus spiniger. J. Helminthol.* 36:297–308.

———. 1964. Factors Influencing the Development and Behavior of Filarial Nematodes in their Arthropodan Hosts. In *Host-Parasite Relationships in Invertebrate Hosts,* ed. Angela Taylor. Blackwell Scientific Publications, Oxford, pp. 75–119.

———. 1966. The pathology of filarial infections. *Helminthol. Abstr.* 35:311–36.

Sawyer, T. K.; Rubin, T. K.; and Jackson, R. F. 1965. The cephalic hook in microfilariae of *Dipetalonema reconditum* in the differentiation of canine microfilariae. *Proc. Helminthol. Soc. Washington* 32:15–20.

Schell, S. C. 1952. Studies on the life cycle of *Physaloptera hispida* Schell (Nematoda: Spiruroidea) a parasite of the cotton rat (*Sigmodon hispidus littoralis* Chapman). *J. Parasitol,* 38:462–72.

Skrjabin, K. I.; Sobolev, A. A.; and Ivashkin, V. M. 1967. Spirurata of Animals and Man and the Diseases Caused by Them. *Akad. Nauk SSSR.* Moscow, vol. 16, pt. 4 (Thelazioidea), 610 pp. (Israel Program for Scientific Translations, 1971.)

Sonin, M. D. 1966 and 1968. Filariata of Animals and Man and Diseases Caused by Them. *Akad. Nauk SSSR.* Moscow, vol. 17, pt. 1 (Aproctoidea), 365 pp.; vol. 21, pt. 2 (Diplotriaenoidea), 411 pp. (Israel Program for Scientific Translations, 1974, 1975.)

Taylor, A. E. R. 1960. The development of *Dirofilaria immitis* in the mosquito *Aedes aegypti. J. Helminthol,* 34:27–38.

Uhazy, I. S. 1977. Development of *Philometroides huronensis* (Nematoda: Dracunculoidea) in the intermediate and definitive hosts. *Canad. J. Zool.* 55:265–73.

———. 1977. Biology of *Philometroides huronensis* (Nematoda: Dracunculoidea) in the white sucker (*Catostomus commersoni*). *Canad. J. Zool.* 55:1430—41.

Williams, R. W. 1946. The laboratory rearing of the tropical rat mite, *Liponyssus bacoti* (Hirst). *J. Parasitol* 32:252–56.

———. 1948. Studies on the life cycle of *Litomosoides carinii,* filariid parasite of the cotton rat, *Sigmodon hispidus litoralis. J. Parasitol.* 34:24–43.

Trichurata

Bessonov, A. S. 1972. *Epizootology (epidemiology) and prophylaxis of trichinelliasis.* pt. 1. Vil'nyus, USSR: Izdatel'stov "Mintis". 304 pp. [Russian].

Gould, S. E., ed. 1970. *Trichinosis in Man and Animals.* Charles C. Thomas, Springfield, Ill., 540 pp.

Robinson, H. A., and Olsen, O. W. 1960. The role of rats and mice in the transmission of the porkworm, *Trichinella spiralis* (Owens, 1835) Railliet, 1895. *J. Parasitol.* 46:589–97.

Skrjabin, K. I.; Shikhobalova, N. P.; and Orlov, I. V. 1957. Trichocephalidae and Capillariidae of Animals and Man and the Diseases Caused by Them. *Akad. Nauk SSSR.* Moscow, vol. 6, 599 pp. (Israel Program for Scientific Translations, 1970.)

Wright, K. A. 1961. Observations on the life cycle of *Capillaria hepatica* (Bancroft, 1893) with a description of the adult. *Canad. J. Zool.* 39:167–82.

Zimmermann, W. J., and Hubbard, E. D. 1969. Trichiniasis in wildlife of Iowa. *Amer. J. Epidemiol.* 90:84–92.

Zimmerman, W. J.; Steele, J. H.; and Kagan, I. G. 1973. Trichiniasis in the U.S. Population, 1966–1970. Prevalence and epidemiologic factors. *Health Service Repts.* 88: 606–23.

Zimmerman, W. J., and Zinter, D. E. 1971. The prevalence of trichiniasis in swine in the United States, 1966–1970. *Health Serv. Ment. Health Adm. Health Repts.* 86: 937–45.

Dioctophymata

Hallberg, C. W. 1953. *Dioctophyma renale* (Goeze, 1782), a study of the migration routes to the kidneys of mammals and resultant pathology. *Trans. Amer. Micr. Soc.* 72:351–63.

Mace, T. F., and Anderson, R. C. 1975. Development of the giant kidney worm, *Dioctophyma renale* (Goeze, 1782) (Nematoda: Dioctophymatoidea). *Canad. J. Zool.* 53: 1552–68.

Sources of Study Material for Nematoda*

Slides

Order Trichurata

Capillaria hepatica, eggs in liver: T,W
Trichinella spiralis, adult: C,T,Tr,W
Trichinella spiralis, larvae: C,T,Tr,W
Trichuris trichiura, eggs: C,T,Tr
Trichuris, in section of appendix: T
Trichuris, adults: C,T,W

Order Rhabditata

Strongyloides stercoralis, adult: C,T,W
Strongyloides stercoralis, in section of intestine: W
Strongyloides, larvae: C,W

Order Strongylata

Ancylostoma braziliense: C,T
Ancylostoma caninum, adult: C,T,Tr,W
Ancylostoma caninum, eggs: C
Ancylostoma caninum, larvae: C,W
Ancylostoma duodenale, adult: T,Tr,W
Ancylostoma duodenale, eggs: C,T
Ancylostoma duodenale, larvae: C
Angiostrongylus cantonensis, adult: C
Necator americanus, adult: T
Necator americanus, eggs: C, Tr
Necator americanus, larvae: C,W

*The addresses of the sources are listed on p. 45.

Order Ascaridata

Ascaris lumbricoides, cross sections: C,T,Tr,W
Ascaris lumbricoides, eggs: C,T,W
Ascaris lumbricoides, larvae in lung: C,T,W
Ascaris megalocephala, cross section: C,Tr
Toxocara, eggs: C

Order Oxyurata

Enterobius vermicularis, adult: C,T,Tr,W
Enterobius vermicularis, eggs: C,T,Tr,W
Enterobius vermicularis, in section of appendix: C,T
Heterakis papillosa, adult: C
Passalurus ambiguans, adult: C

Order Filariata

Dipetalonema perstans, microfilariae in blood: C
Dirofilaria immitis, microfilariae in blood: C,T,Tr,W
Dirofilaria immitis, in section of lung: T
Loa loa, microfilariae in blood: W
Onchocerca volvulus, microfilariae in tissue smear: C,T,W
Onchocerca volvulus, sections of adults in nodule: C,T,W
Wuchereria bancrofti, microfilariae in blood: T

Live Material

Trichinella spiralis, in mouse tissue: W

Chapter 6 Phylum Annelida

Class Hirudinea

Hirudinea, the **leeches,** belong to the phylum **Annelida.** They are distinguished from other annelids by having: (1) a sucker or suckerlike depression at the anterior end of the body and a well-developed sucker at the posterior end, (2) a constant number of body segments, (3) no setae or parapodia (except in *Acanthobdella*), and (4) a body cavity largely filled with muscles and connective tissue.

General Morphology

Leeches vary in length from a few mm to more than 25 cm and are elongate, egg shaped or leaflike in form. Since leeches have no caudal growth, the number of segments being fixed, increased in length results from the subdivision and elongation of the **annuli** into which the segments are divided. The number of annuli per segment in the midbody region is typical for any given species, three and five being commonest; but the number of annuli comprising the terminal segments is reduced. **Eyes** commonly vary from one to five pairs, placed anteriorly. The **clitellar region,** which secretes the egg cocoon, extends from segment X through XII and contains the male and female genital openings.

The **digestive tract** is a tube from mouth to anus and is divided into **buccal chamber, pharynx, esophagus, stomach** (or **crop**), **intestine,** and **rectum.** Structure of the anterior portion constitutes one of the most important taxonomic characters for dividing the class into two orders: aproboscidal (order Arhynchobdellae) and proboscidal (order Rhynchobdellae) leeches. In the first group the **buccal sinus** is a restricted chamber housing the jaws (when present), and in the latter a space around the pharynx, which is thereby freed as a protrusible **proboscis.** The esophagus connects the pharynx with the stomach, which in predaceous species is tubular and simply chambered, without lateral caeca or with only a posterior pair. In bloodsucking leeches, the stomach becomes a spacious crop with lateral, paired caeca for the storage of food between feedings, and the saliva contains the anticoagulant **hirudinin,** which keeps the food fluid until digestion. A sphincter separates the storage stomach from the intestine, the principal digestive and absorptive region. The intestine, usually a straight, somewhat chambered tube, may in some species bear four paired caeca laterally. A simple rectum leads to the dorsally situated anus.

Food relationships are varied. Leeches feed on oligochaete worms, crustaceans, insect larvae, molluscs, and all classes of vertebrates. Most subsist upon body fluids, involving all gradations from true predation to temporary and permanent parasitism.

Reproduction

In all leeches, the internal sexual organs are paired, but the terminal parts and genital openings are single. In the male reproductive system, the **testes,** enclosed in coelomic sacs, are multiple, usually in segmental pairs. Laterally from each testis, a small **sperm duct** enters a larger duct, which becomes enlarged and often much convoluted anteriorly to form a **seminal vesicle** or **epididymis.** Each of the two epididymides leads to a muscular, often bulbous **ejaculatory duct,** which opens into the median **atrium,** variously provided with glands. The female system consists of a pair of elongated **ovisacs,** which contain the true ovaries and unite anteriorly to form a common **oviduct,** which opens externally. In those forms in which fertilization occurs through copulation, this common passage serves as a **vagina.**

Reproduction in leeches is exclusively sexual. Fertilization occurs in two forms, depending upon whether a male copulatory organ is present. In the Rhynchobdellae and the Erpobdellidae, sperm transfer is accomplished by **spermatophores;** these are implanted by one leech onto the body of another. Sperms, forced out of the spermatophores, penetrate the tissues and enter the ovaries, resulting in fertilization. In the Hirudinidae and the Haemadipsidae, direct copulation occurs, the sperm being transmitted by the cirrus of one animal into the vagina of another. Eggs are usually deposited in **cocoons** soon after fertilization. Cocoons may be borne by the parent leech, attached to a host animal, to a foreign object, or deposited in damp earth. Development is direct.

Leeches As Parasites

Throughout the Mediterranean area and the Middle and Far East regions, certain leeches are injurious to animals and humans. Small, fasted leeches, taken in with drinking water, reach the laryngopharynx region, or may enter the nasopharynx, and become attached. By sucking blood, they increase greatly in size, cause congestion and inflammatory swelling, and may occlude the passages involved. In some countries this condition is known as **halzoun.** Other portals of body entry may involve the anus, external auditory canal, and urogenital sinus of both sexes. Attacks by land leeches (Haemadipsidae) are relatively common among pedestrian travelers through southern Asia and adjacent Sunda Islands.

The Piscicolidae occur as temporary or permanent parasites, chiefly on marine and freshwater fish, upon whose body fluids they feed. Not only may the host be killed or left in an emaciated condition as a result of heavy infestation, but the value of commercial and game fish may be materially impaired as a result of the inflammation developing where the leeches have attached. These breaks in the mucosal covering, resulting from attachment, may serve as a portal of entry for microbial organisms, which cause secondary infection. The glossiphoniid *Theromyzon* commonly enters the nares of waterfowl, often resulting in asphyxiation.

Parasitologically, leeches are of dual concern: some are parasites and others are involved in the life cycle of other parasites. Leeches are vectors of protozoans such as *Trypanosoma, Cryptobia, Haemogregarina,* and probably others; of fish, amphibians, and reptiles.

They also serve as hosts, both intermediate and final, for some digenetic trematodes.

Order Rhynchobdellae

Marine and freshwater leeches with colorless blood, with an extensible proboscis, without jaws. The mouth is a small median opening situated within the anterior sucker, or rarely upon its front edge. Copulation by spermatophores.

Family Piscicolidae

Piscicola milneri

See figure 6.1. The anterior sucker, slightly narrower than the body at its widest point, has two pairs of distinct eyes arranged in the form of a trapezoid. Segments in the midregion of the yellowish colored body are characterized by large, brownish stellate flecks, arranged in five longitudinal rows, which extend dorsally, frequently meeting with those from the opposite side in the ganglionic region of the segment. Middorsally these flecks form a slightly acute angle and spread out laterally, where they are frequently continuous with those of the adjacent segments. Although these flecks are distributed over the entire length of the body, the triangular arrangement, so characteristic of the midregion, is lost anteriorly and posteriorly. The posterior sucker, equal to body width at its widest point, has a crown of ten or 12 punctiform spots situated near the margin. Ratio of body length to width about 12:1. There are 11 pairs of lateral pulsating vesicles. Complete segments are divided into 14 annuli. The male gonopore is larger than that of the female and is located anterior to the latter.

There are six pairs of metamerically arranged testes. The vasa deferentia continue anteriorly and dorsally slightly beyond the level of the male gonopore and then curve ventrally before uniting to form the muscular atrium. Ejaculatory ducts are located at the level of the gonopores. Paired ovisacs, which extend posteriorly to the second pair of testes, unite anteriorly to form a common oviduct, which opens two annuli posterior to the more conspicuous male gonopore. Gonopores are within segments XI and XII.

The stomach, the largest division of the alimentary canal, expands into segmental chambers which alternate between consecutive pairs of testes. Slightly posterior to the last pair of testes, in XIX, the alimentary canal divides into intestine (dorsally) and postcaecum. The intestine, characterized by four pairs of lateral caeca, is followed by a rectum that empties posteriorly and dorsally through the anus. The postcaecum, situated ventral to the intestine, has four fenestrae. Except for the occasional properly engorged specimen, serial sections are necessary for an understanding of the alimentary canal. Parasitic on various freshwater fish.

The life cycle of *Piscicola milneri* has not been studied, but with the possible exception of the time of reproduction it can be expected to be similar to that of *P. salmositica,* which parasitizes Salmonidae throughout the Pacific Northwest. Becker and Katz (1965) found that *P. salmositica* breeds during the fall and winter, thus adapting its reproductive activity to the presence of salmon in fresh water rather than to the usual rise in water temperature in spring.

Leeches appear in early fall attached to their hosts as they ascend into the hatchery and natural spawning areas. From late September through January, during a period of feeding, growth, and reproduction, the leech populations attain great abundance. By late winter, subsequent to the death of their hosts, the leeches have become virtually nonexistent. Only the small, immature leeches are believed to survive the summer and return attached to spawning salmon the following fall.

Other often available Piscicolidae with their hosts include *Piscicola punctata, Myzobdella luqubris,* and *Cystobranchus verrilli,* all from various freshwater fishes.

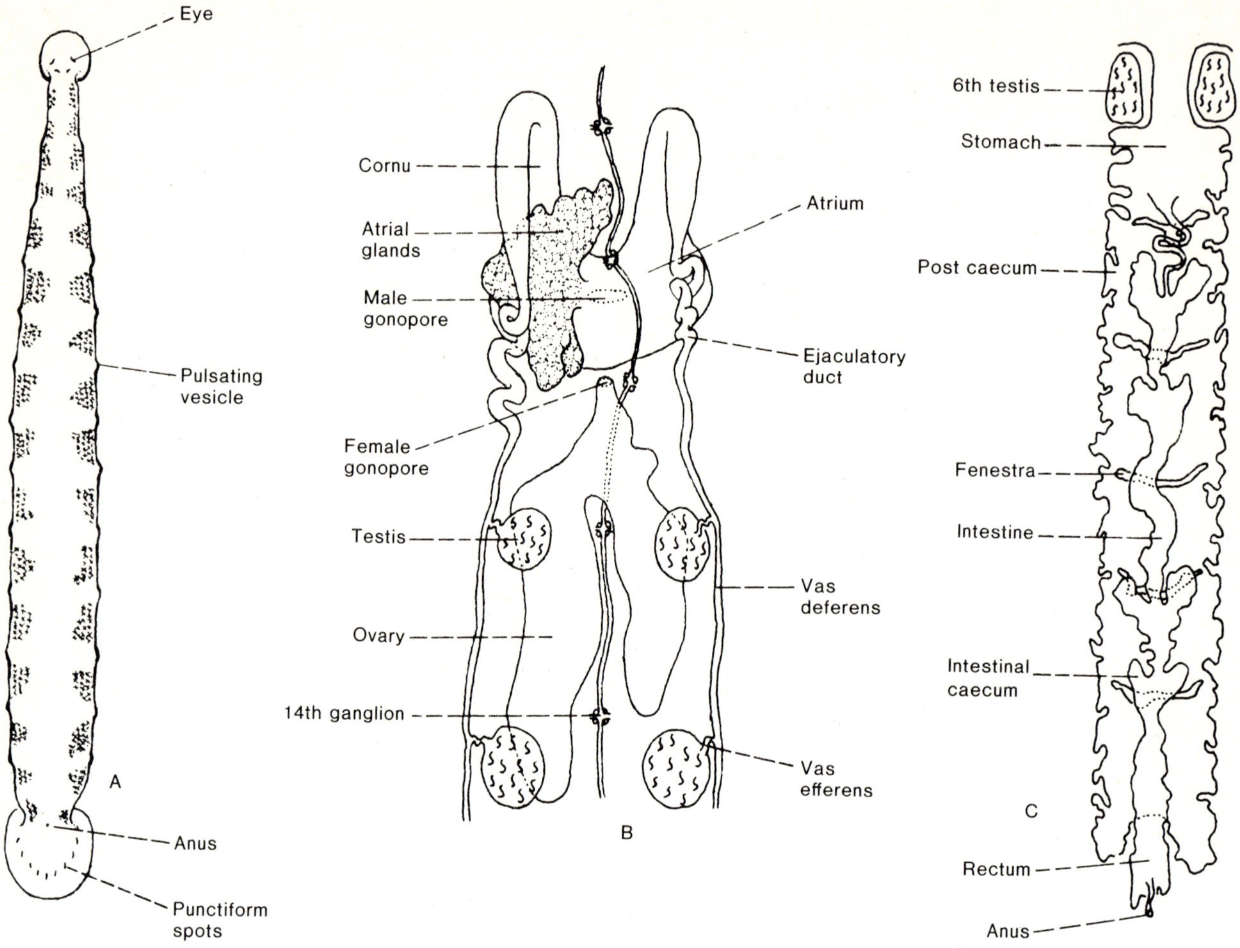

Fig. 6.1. *Piscicola milneria.* (A) External; (B) reproductive system; (C) posterior portion of alimentary tract. All dorsal views.

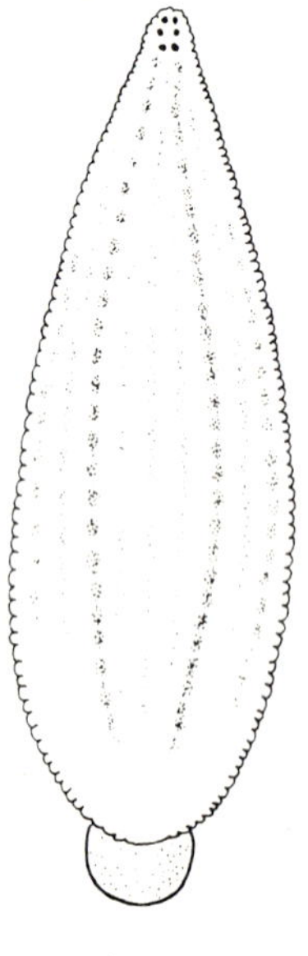

Fig. 6.2. *Glossiphonia camplanata.*

Key to Families of North American Hirudinea

1(4) Mouth a small pore on anterior sucker through which the proboscis protrudes; blood colorless .... Order RHYNCHOBDELLAE 2

2(3) Body frequently divided into trachelosome and urosome, usually long and narrow, little flattened; complete segments usually more than three-annulate (fig. 6.1A). .... **Piscicolidae**

3(2) Body not divided into trachelosome and urosome, usually flattened and much wider than head; complete segments usually three-annulate (fig. 6.2) .... **Glossiphoniidae**

4(1) Mouth large, opening from behind into entire sucker cavity; pharynx fixed, no proboscis; blood red .... Order ARHYNCHOBDELLAE 5

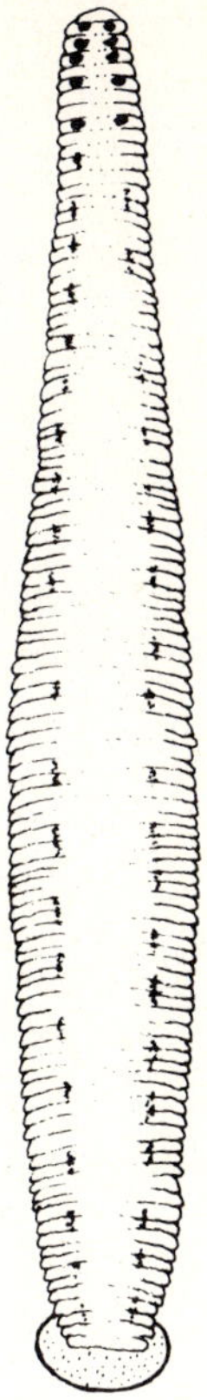

Fig. 6.3. *Hirudo medicinalis.*

Fig. 6.4. *Erpobdella punctata.*

5(6) Pharynx a suction bulb, not reaching to clitellum; eyes five pairs, arranged in an arch; segments five-annulate; muscular jaws usually present; testisacs large, metameric, usually ten (fig. 6.3) .... **Hirudinidae**

6(5) Pharynx a crushing tube extending to clitellum; eyes three or four pairs in separate labial and buccal groups; segments five-annulate but often further divided; three muscular pharyngeal ridges but no true jaws; testisacs small and numerous, in grape-bunch arrangement (fig. 6.4) .... **Erpobdellidae**

Notes and Sketches

Review Questions

1. How do leeches differ from other annelids?
2. What is the difference between segments and annuli?
3. What is the major difference between the Arhynchobdellae and the Rhynchobdellae?
4. Are all leeches parasitic? Name the anticoagulant in the saliva of some leeches.
5. Discuss reproduction in leeches.
6. Are leeches vectors of other parasites? Intermediate hosts?

References

Barrow, J. H. 1953. The biology of *Trypanosoma diemyctyli* (Tobey). I. *Trypanosoma diemyctyli* in the leech, *Batrachobdella picta* (Verrill). *Trans. Amer. Micr. Soc.* 72:197–216.

Bartonek, J. C., and Trauger, D. L. 1975 Leech (Hirudinea) infestations among waterfowl near Yellowknife, Northwest Territories. *Canad. Field-Natur.* 89:234–43.

Becker, C. D., and Katz, M. J. 1965. Distribution, ecology, and biology of the salmonid leech, *Piscicola salmositica* (Rhynchobdellae: Piscicolidae). *J. Fish. Res. Bd. Canada* 22:1175–95.

Corkum, K. C., and Beckerdite, D. W. 1975. Observations on the life history of *Alloglossidium macrobdellensis* (Trematoda: Macroderoididae) from *Macrobdella ditetra* (Hirudinea: Hirudinidae). *Amer. Midl. Natur.* 93:484–91.

Khan, R. A. 1976. The life cycle of *Trypanosoma murmanensis* Nikitin. *Canad. J. Zool.* 54:1840–49.

Khan, R. A., and Meyer, M. C. 1976. Taxonomy and biology of some Newfoundland marine leeches (Rhynchobdellae: Piscicolidae). *J. Fish. Res. Bd. Canada* 33:1699–714.

Mann, K. H. 1962. *Leeches (Hirudinea), Their Structure, Physiology, Ecology, and Embryology.* Pergamon Press, New York, 201 pp.

Moore, J. P. 1959. Hirudinea. In *Ward and Whipple's Freshwater Biology,* ed. W. T. Edmondson. 2nd ed. John Wiley and Sons, New York, pp. 542–57.

Stuart, J. 1982. Hirudinoidea. In *Synopsis and Classification of Living Organisms,* ed. Sybil P. Parker. McGraw-Hill, New York, vol. 2, pp. 43–50.

Vojtek, J.; Opravilova, V.; and Vojtkova, L. 1967. The importance of leeches in the life cycle of the order Strigeidida (Trematoda). *Folia Parasitol. (Praha)* 14:107–19.

Sources of Study Materials for Hirudinea*

Slides

Leech, whole mount: C, T, Tr, W
Leech, cross section: C, T, Tr, W

Preserved

Leech, various sizes in alcohol: C, W
Leech, latex injected; in alcohol: C

Living

Leech, various sizes and species: C, W
Hirudo medicinalis: C, W

*The addresses of the sources are listed on p. 45

SECTION 3 Phylum Arthropoda

The phylum Arthropoda is the largest and most varied in the animal kingdom (table 7.1). Arthropods are bilaterally symmetrical dioecious animals, have a chitinous exoskeleton, and a body divided into separate rings or segments, each of which may bear jointed appendages. They possess a complete digestive tract, except in a few barnacles (Cirripedia), with mouthparts adapted for biting, chewing, or sucking.

Insects and other arthropods are of great parasitological importance with respect to humans and animals. They may be parasitic, living more or less permanently on or in the body of the host, and they may serve as intermediate hosts in the life cycle of parasites and vectors of pathogens.

Species of parasitological importance belong to the classes: **Crustacea** (copepods, amphipods, crabs, crayfishes), **Arachnida** (ticks and mites), **Pentastomida** (tongue worms), and **Insecta** (fleas, lice, flies, biting midges, mosquitoes, and true bugs).

Table 7.1. Abbreviated Classification of Phylum Arthropoda

Subphylum	Class	Order	Suborder	Family
Crustacea	Copepoda	Siphonostomatoidea		Caligidae
				Lernaeopodidae
		Cyclopoida		Lernaeidae
		Poecilostomatoidea		Ergasilidae
	Branchiura	Argulidea		Argulidae
	Malacostraca	Amphipoda		Gammariidae
				Caprellidae
		Isopoda		Asellidae
				Oniscidae
				Gnathiidae
				Bopyridae
		Decapoda		Pinnotheridae
				Parthenopidae
Chelicerata	Arachnida	Acari	Metastigmata	Ixodidae
				Argasidae
			Astigmata	Sarcoptidae
				Psoroptidae
				Knemidokopidae
				Pyroglyphidae
			Prostigmata	Demodicidae
				Trombiculidae
				Psorergatidae
			Mesostigmata	Laelaptidae
				Dermanyssidae
				Rhinonyssidae
Uniramia	Insecta	Siphonaptera		Ceratophyllidae
				Pulicidae
				Tungidae
		Anoplura		Pediculidae
				Phthiriidae
				Haematopodidae
				Linognathidae
		Mallophaga	Ambycera	Menopodidae
				Gyropidae
			Ischnocera	Trichodectidae
				Philopteridae
		Diptera	Nematocera	Psychodidae
				Ceratopogonidae
				Simuliidae
				Culicidae
			Brachycera	Tabanidae
			Cyclorrhapha	Muscidae
				Hippoboscidae
				Glossinidae
				Calliphoridae
				Gasterophilidae
				Hypodermatidae
				Oestridae
				Cuterebridae
		Hemiptera		Reduviidae
				Cimicidae

After Schmidt and Roberts (1985).

Chapter 7 Subphylum Crustacea

The Crustacea comprise a group of mandibulate, usually aquatic, Arthropoda, typically possessing two pairs of antennae. Parasitologically, certain taxa are of dual concern: some as parasites and others as intermediate hosts in the life cycle of certain helminths. While the **Ostracoda, Amphipoda,** and **Isopoda** each contain a few parasitic species, they, along with the **Decapoda,** are certainly more important as intermediate hosts. The **Branchiura,** all of which are parasitic, infest principally marine and freshwater fish; the barnacles (**Cirripedia**) contain numerous species parasitizing various marine animals. But of the taxa of Crustacea, the **Copepoda,** both as intermediate hosts and as parasites, are of greatest importance.

Class Copepoda

The free-living copepods abound in both salt and freshwater and may be collected with the aid of a plankton net in open waters throughout the year. The body is divided into two regions, a wider anterior division known as the **cephalothorax** and a rather narrow posterior one, the **abdomen.** The abdomen ends in a pair of **caudal rami,** which are provided with terminal, laterally situated setae. The cephalothorax can be further divided into two regions, the **head** with two pairs of antennae and three pairs of mouthparts, and a **thorax** with a pair of maxillipeds, four pairs of biramous swimming appendages, and a fifth uniramous, much reduced pair.

In the parasitic species the external body shape varies considerably, often showing little similarity to that of their free-living progenitors. The body metamerism is usually less prominent and may be totally obscured. Thoracic appendages may be reduced in number or absent. Mouthparts are usually specialized for piercing and sucking, and the second antennae are usually specialized as prehensile organs. Thus, the cephalic region is often greatly modified in appearance.

A characteristic feature of both free-living and parasitic species is the presence of a pair of egg sacs in females. Most species are dioecious, with males usually smaller than females.

Family Lernaeidae

Lernaea cyprinacea

Mature females (fig. 7.1A) attach to the skin of the host by a highly modified head buried in the epidermis, with the posterior part of the body hanging free. Because of the shape of the head and the elongate vermiform shape of the protruding body, species of *Lernaea* are commonly referred to as **"anchor worms."**

The body consists of a cephalothorax, free thorax, and abdomen. The head and thorax bear the diminutive appendages of these parts. There is a median tripartite eye. Anteriorly the cephalothorax expands into large **cephalic horns** (fig. 7.1B), which are soft and leathery in texture. The cylindrical neck is soft and enlarges gradually to form the trunk, which bears on the ventral side the vulvae. The abdomen is short, rounded, and terminates bluntly. A pair of elongated to conical egg sacs is attached ventrally at the genital pore near the posterior end of the abdomen.

The life cycle and developmental stages of the parasitic copepods are basically similar to those of the free-living species. As one stage molts and metamorphoses into the next stage, the larvae increase in size, number of body segments, and appendages.

Eggs hatch into **nauplii,** which possess three pairs of appendages and a pair of spines posteriorly. Larval stages are motile, temporary parasites that feed on the host's body fluids. Nauplii molt and become **metanauplii,** which are characterized by an increase in the number of spines and body segmentation.

Males do not advance beyond the fifth **copepodid stage** and do not become permanent parasites; females do so only after copulation, which occurs during the fifth copepodid stage. After copulation the males disappear; the females penetrate the host tissue and attain a permanently fixed position. They then increase in length, develop cephalic horns, and become adults.

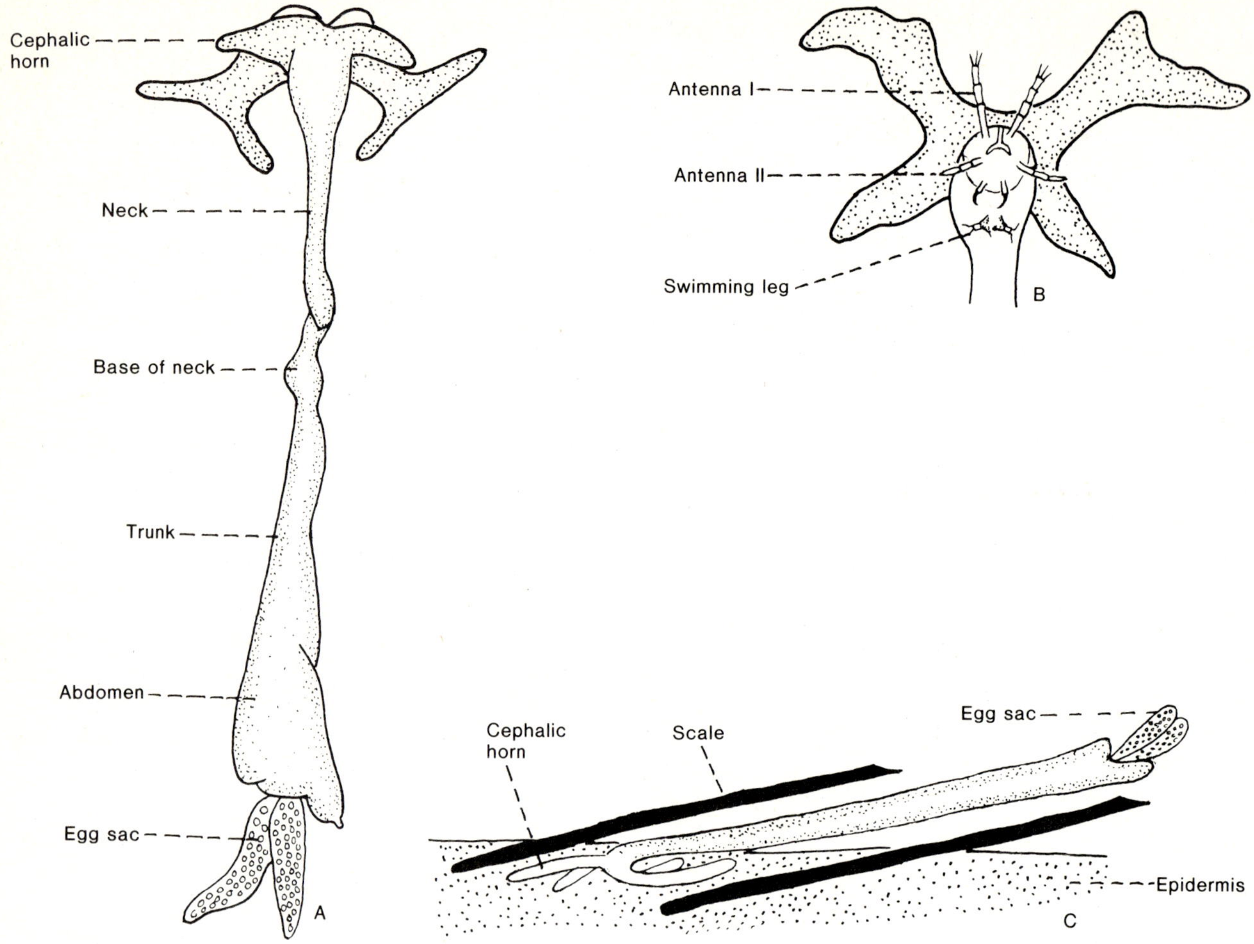

Fig. 7.1. *Lernaea cyprinacea.* (A) Adult female; (B) cephalic horns with anterior end of cephalothorax and degenerated appendages; (C) adult embedded in fish epidermis between two scales.

After reaching maturity, they form egg sacs filled with eggs, completing the life cycle. Adult females are usually embedded at the base of the fins (fig. 7.1C), particularly the dorsal fin, but they attach in the nares and on other parts of the body. Under normal conditions, when the infestation is usually light, *L. cyprinacea* and other parasitic copepods cause relatively little damage, but when abundant, as is likely to be the case under crowded hatchery conditions, they may become a serious menace to their host. Not only may the fish be killed or left in an emaciated condition as a result of heavy infestation, but the value of the commercial and game fish may be materially impaired as a result of the hemorrhagic lesions and tumorlike tissue at the site of attachment. The breaks in the outer mucosal covering of the fish's skin may serve as a portal of entry for microbial organisms, often resulting in secondary infection.

Epizootics and mortalities of significant economic importance among fish due to massive infestations have been reported by Tidd (1934). There are also reports of heavy infestations in lakes and streams (Haley and Winn, 1959).

Class Branchiura

In this taxon of marine and freshwater crustaceans, the body is depressed dorsoventrally; the carapace extends laterally and posteriorly so that all or a portion of the segmented thorax and abdomen are free. There is a pair of ventral compound eyes and a dorsal median simple eye on the carapace; ventrally, each lateral lobe of the carapace contains two respiratory areas, whose arrangement and shape are useful in specific identifications. Ventrally, the first antennae are armed with claws and are prehensile; the second antennae are uniramous. In a longitudinal groove extending forward mesiad between the antennae is a preoral **sting,** which is retractile into a sheath. Posterior to and somewhat larger in diameter than the preoral sting is the proboscis, with the mouth. The first maxillae are modified

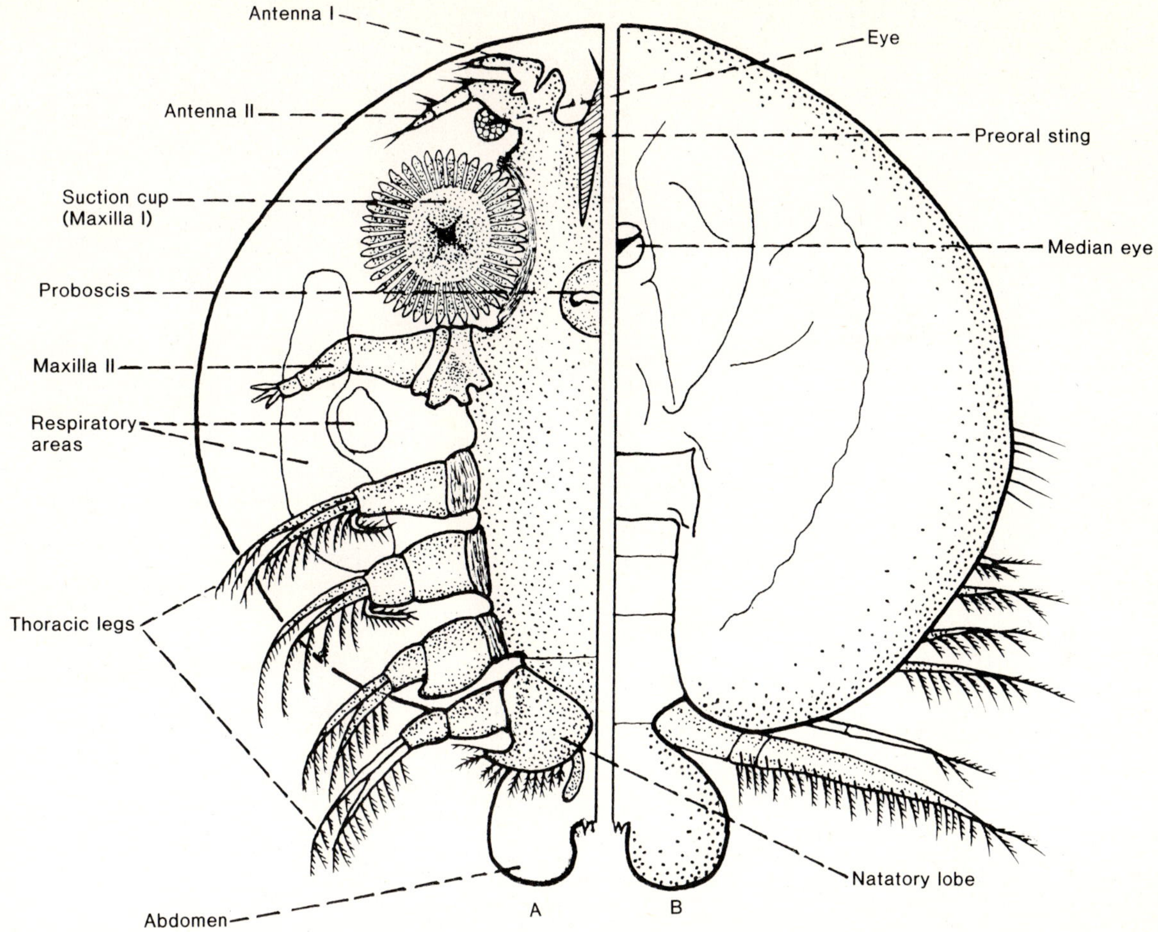

Fig. 7.2. *Argulus catostomi*, ventral (A) and dorsal (B) view of female.

as a pair of large **sucking cups,** and the second pair consists of simple, five-jointed leglike appendages terminating in a stout hook. There are four pairs of three-jointed, setose walking legs. The abdomen is a single, bilobed segment. Elongate testes or rounded seminal receptacle, depending on the sex, are situated ventrally on the abdomen.

Branchiurans are periodic parasites on fish, occasionally on tadpoles, with little host specificity. Specimens may be found anywhere on the body surface of the host and in the mouth and gill chamber. Attachment to the host is by the hooked antennae and sucking cups, or discs. As in the parasitic copepods above, in the case of heavy infestations the fish's vitality is weakened from the loss of blood, and the mechanical injury resulting to the host tissue may be equally, or even more, serious.

Both males and females can swim freely, and they leave their hosts regularly at the breeding season. Depending on the species, females deposit 30 to 600 eggs singly in rows, on plants and other fixed objects. Eggs are ovoidal in shape, about 450 μm long, and covered by a gelatinous capsule that secures each one firmly to the substrate.

The nauplius, metanauplius, and in some species the early copepodid stage occur in the egg, emerging in a late copepodid stage. Upon finding a host, the copepodid larva attaches and feeds for five to six weeks before becoming sexually mature. Adults may live free from the host for up to 15 days.

Depending on the species and temperature, completion of the life cycle requires from 40 to 100 days. The lifespan of females may be as long as 18 months, but males live for less than a year.

The subclass is represented in North America only by *Argulus,* the fish louse. Some common species with their usual fish hosts include: *Argulus alosae: Alosa, Salvelinus; A. appendiculosus: Stizostedion, Ictalurus; A. catostomi: Catostomus,* Cyprinidae; *A. flavescens: Amia, Micropterus.* (See fig. 7.2.)

Review Questions

1. What are the main divisions and subdivisions of a crustacean's body?
2. Describe the life cycle and development of an anchor worm.
3. In what ways does *Argulus* differ from *Lernaea*?
4. What is a major effect of infestation of parasitic crustaceans on fish?

References

Askew, R. R. 1971. *Parasitic Insects.* American Elsevier Publishing Co., New York, 316 pp.

Beesley, W. N. 1973. Control of Arthropods of Medical and Veterinary Importance. In *Advances in Parasitology,* ed. B. Dawes. Academic Press, New York, vol. 11, pp. 115–92.

Bowman, T. C., and Abele, L. G. 1982. Classification of the recent Crustacea. In *The Biology of Crustacea,* ed. L. G. Abele. Systematics, the fossil record, and biogeography. Academic Press, New York, vol. 1, pp. 1–27.

Horsfall, W. R. 1962. *Medical Entomology: Arthropods and Human Diseases.* Ronald Press Company, New York, 467 pp.

James, M. T., and Harwood, R. F. 1969. *Herms's Medical Entomology.* 6th ed. Macmillan Company, New York, 484 pp.

Leclercq, M. 1969. *Entomological Parasitology.* Pergamon Press, New York, 158 pp.

Orkin, M.; Maibach, H. I.; Parish, L. C.; and Schwartzman, R. M. 1977. *Scabies and Pediculosis.* J. B. Lippincott Company, Philadelphia, 203 pp.

Philip, C. B., and Burgdorfer, W. 1961. Arthropod Vectors as Reservoirs of Microbial Disease Agents. In *Annual Review of Entomology,* E. A. Steinhaus and R. F. Smith, eds. Annual Reviews, Inc., Palo Alto, Cal., vol. 6, pp. 391–412.

Smith, K. G. V., ed. 1973. *Insects and Other Arthropods of Medical Importance.* British Museum (Natural History), London, 561 pp.

Yamaguti, S. 1963. *Parasitic Copepoda and Branchiura of Fishes.* Interscience Publishers, New York, 1104 pp.

Copepoda

Grabda, J. 1963. Life cycle and morphogenesis of *Lernaea cyprinacea* L. *Acta Parasitol Polonica* 11:169–99.

Haley, A. J., and Winn, H. E. 1959. Observations on a lernaean parasite of freshwater fish. *Trans. Amer. Fish. Soc.* 88:128–29.

Tidd, W. M. 1934. Recent infestations of goldfish and carp by the "anchor parasite," *Lernaea carassii. Trans. Amer. Fish. Soc.* 64:176–80.

Branchiura

Cressey, R. F. 1972. The genus *Argulus* (Crustacea: Branchiura) of the United States. *Biota of Freshwater Ecosystems,* Washington, D.C., no. 2, 14 pp.

Meehean, O. L. 1940. A review of the parasitic Crustacea of the genus *Argulus* in the collections of the United States National Museum. *Proc. U.S. Nat. Mus.* 88:459–522.

Wilson, C. B. 1944. Parasitic copepods in the United States National Museum. *Proc. U.S. Nat. Mus.* 94:529–82.

Sources of Study Materials for Crustacea*

Slides

Argulus sp.: Tr, W
Cyclops sp.: C. T. Tr, W
Gammarus sp.: C, T, Tr
Ergasilus labricis: T

Preserved

Argulus: C
Cyclops: C
Gammarus: C, W
Oniscus (sowbug): C, W

Living

Amphipods (*Gammarus,* etc.): C, W
Copepods (*Cyclops,* etc.): C, W
Isopods (*Armadillidium,* etc.): C, W

*The addresses of the sources are listed on p. 45.

Notes and Sketches

Chapter 8 Order Acari

The Acari includes the **ticks** and **mites,** which differ from insects in normally having three pairs of legs as larvae and usually four pairs of legs in the nymphal and adult stages, and in lacking antennae. The head, thorax, and abdomen are fused into one body region, and the mouthparts are more or less distinctly set off from the rest of the body on a false head, the **capitulum** or **gnathosoma.**

Normally the life cycle includes four stages: the **egg, larva, nymph,** and **adult.** The eggs are usually deposited under the surface of the soil or in crevices or, in some parasites, under the skin of the host; some species are ovoviviparous. One or more nymphal stages occur between the larva and adult, and the nymph resembles the adult except in its smaller size and the absence of the genital opening.

The Acari is important from the medical and veterinary point of view because some of its members are carriers of diseases and infections affecting humans and domestic animals, while others may cause dermatitis and allergic reactions. Still other species are important to the parasitologist because they occur as internal parasites in the lungs and air sacs of snakes, birds, and mammals.

Ticks and mites, although belonging to the same order, can be separated by the following characteristics. Ticks are usually 3 mm or more in length; have leathery body covering, with few short hairs; a toothed **hypostome** in some stage of life cycle (usually in adult); and a **Haller's organ** (sensory pore on tarsus of first leg). Mites, on the other hand, especially those of medical importance, are rarely more than 1 mm in length and when unfed have a white or pale yellow color. Their body texture is often membranous or heavily sclerotized, frequently with many long hairs; hypostome without teeth; and lacking a Haller's organ.

Ticks: Suborder Metastigmata

With a pair of **spiracles** posterior or lateral to the third or fourth coxae, located in a **stigmal** or **spiracular plate,** without an elongated tube; Haller's organ present on the tarsus of the first leg; hypostome modified as a piercing organ.

The role of ticks in the human economy merits special attention, for not only are they annoying pests, but in temperate and tropical regions they surpass all other arthropods in the number and variety of disease-producing agents which they transmit to humans and domestic animals. As vectors of agents of human disease, they run mosquitoes a close second. Ticks are known to transmit five groups of microbial organisms: (1) spirochaetal, such as **relapsing fever;** (2) rickettsial, such as **Rocky Mountain spotted fever;** (3) bacterial, such as **tularemia;** (4) viral, such as **Colorado tick fever;** and (5) protozoan, such as **Texas fever.** They also produce **tick paralysis,** which is probably caused by a neurotoxic substance in the tick's saliva. *Dermacentor andersoni* has been referred to as a "veritable Pandora's box of disease-producing agents." It is a vector of the agents of Rocky Mountain spotted fever, tularemia, Colorado tick fever, and **Q fever,** and produces tick paralysis in both humans and animals. It is also a vector of **anaplasmosis** in cattle.

Synopsis of Families of Metastigmata

Scutum present; in males it extends over the entire dorsal surface, in females only a portion of the anterior surface; capitulum terminal and visible from the dorsal surface .... **Ixodidae**

Scutum absent; capitulum ventral and usually concealed beneath the anterior margin .... **Argasidae**

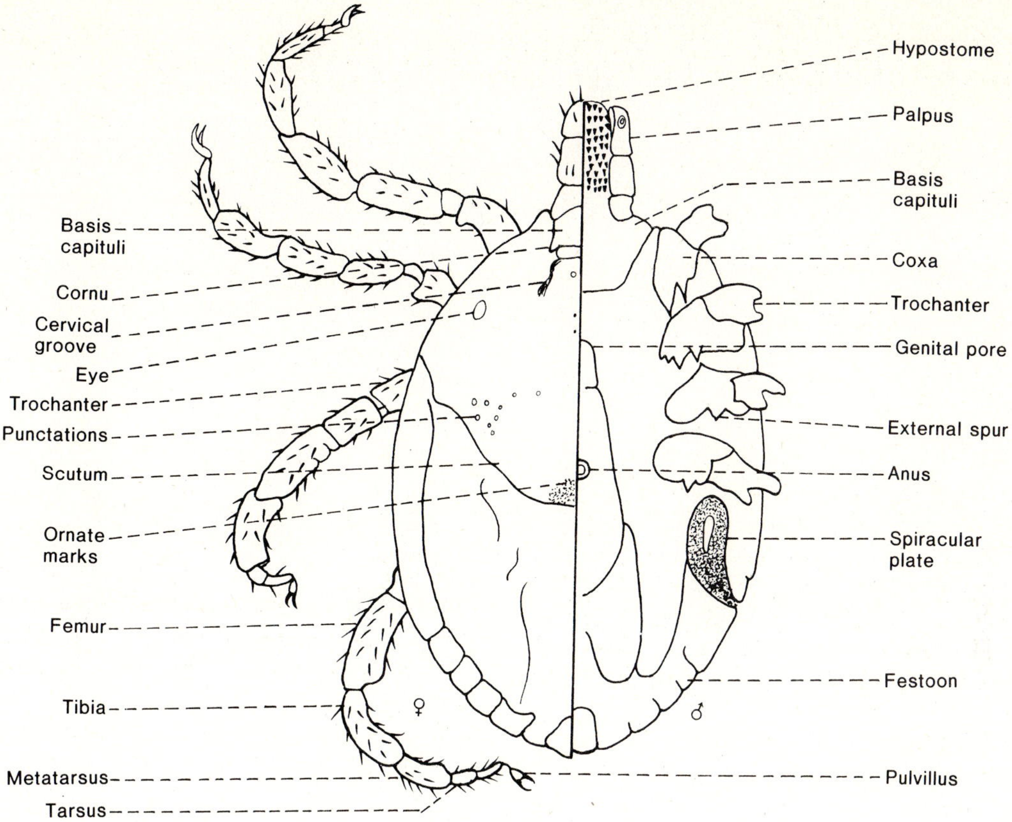

Fig. 8.1. Dorsal view of female and ventral view of male ixodid ticks.

In the United States, Ixodidae (**hard ticks**) (fig. 8.1) are much more abundant than Argasidae (**soft ticks**), cause greater annoyance, and are far more important as vectors of disease-producing agents to humans and animals.

Family Ixodidae

Key to Genera of Ixodidae

1(2) Second segment of palp twice as long as wide .... *Amblyomma*
2(1) Second segment of palp as long as wide 3
3(4) Anal groove in front of anus .... *Ixodes*
4(3) Anal groove behind anus 5
5(6) Second segment of palp laterally produced .... *Haemaphysalis*
6(5) Second segment of palp not laterally produced 7
7(8) Basis capituli laterally produced 9
8(7) Basis capituli not laterally produced 11
9(10) Festoons present .... *Rhipicephalus*
10(9) Festoons absent .... *Boophilus*
11(12) Eleven festoons present .... *Dermacentor*
12(11) Seven festoons present .... *Anocentor*

Dermacentor variabilis

This species, a vector of the agents of Rocky Mountain spotted fever and tularemia, is widely distributed east of the Rocky Mountains and also occurs on the Pacific Coast. Dogs are the preferred host of the adult, although it feeds readily on other large mammals, including humans. Adults are commonly found in the spring clinging to low vegetation, waiting to attach to a passing host. Males remain on the host for an indefinite time, alternately feeding and mating. Females feed, mate, engorge, and drop off to lay thousands of eggs.

The morphology and biology of this species will serve as models for the other hard ticks, which are basically similar.

Upon hatching, larvae remain close to the soil and mouse burrows. Larvae attach to and feed on mice and other small mammals, after which they drop off and seek a concealed niche for molting. Nymphs are similar in appearance to the larvae but have four instead of three pairs of legs and a pair of spiracular plates. They attach to small mammals, feed, drop to the ground, and molt. The entire life cycle, typical for three-host ticks, requires from four months to a year.

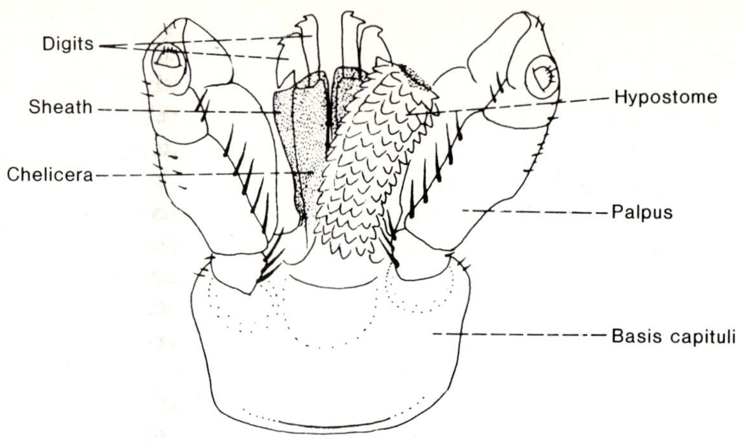

Fig. 8.2. Capitulum of *Dermacentor variabilis*, with hypostome slightly displaced to left to show chelicerae. Ventral view.

The adult is oval in contour and somewhat flattened dorsoventrally (fig. 8.1). Anteriorly there is a deeply concave region of the scutum or dorsal shield into which is fitted the capitulum (fig. 8.2), comprising the following six mouthparts (two of which are paired): a median **hypostome,** which has six rows of large, backwardly directed teeth on its ventral surface; a pair of **chelicerae** lying close together immediately above the hypostome, each chelicera consisting of a cylindrical shaft, which projects for a considerable distance into the body of the tick, and armed distally with a pair of serrated digits that can be moved by muscular action, and each is protected by a chitinous sheath into which it can be withdrawn when not in use; and a pair of four-segmented **palpi,** inserted anterolaterally on the **basis capituli.** The basis capituli is the dense basal part articulating with the body.

Directly back of the head lies the **scutum** or **dorsal shield.** This is small in the female but covers or almost covers the entire body in the male. **Festoons** are the uniform rectangular areas into which the posterior margin of the body is divided. On the venter note the following: the **genital pore,** situated just posterior to the basis capituli; **genital grooves,** starting at either side of the gonopore and running posteriorly; the **anus,** situated in the median line, posterior to the coxae; **postanal groove,** a curved transverse furrow immediately posterior to the anus in some genera. The **spiracular plates** or **spiracles,** paired openings into the respiratory organs, are situated posterolaterally, posterior to coxae IV. These plates vary greatly in shape and configuration in the different species. In *D. variabilis,* the plate has a dorsal prolongation and the **goblets** (beadlike structures under the plate) are very numerous and small; in *D. albipictus,* it lacks a dorsal prolongation and the goblets are few and large.

The six-segmented **legs** consist of the **coxa,** the immovable segment to which the rest are attached, **trochanter, femur, tibia, metatarsus,** and **tarsus** bearing a stalk with claws.

Other widely distributed hard ticks with their common hosts include *Dermacentor albipictus,* deer, elk, horse, moose; *Haemaphysalis leporispalustris,* rabbits; and *Rhipicephalus sanguineus,* dogs.

Family Argasidae

This is the family of "soft ticks," species of which parasitize birds and mammals, sometimes transmitting serious spirochaete and virus diseases. They can be distinguished from the hard ticks by the following features:

1. There are no festoons or scutum.
2. The capitulum in nymphs and adults is subterminal and cannot be seen in dorsal view. It lies within a groove called the **camerostome,** which extends over the capitulum to form the **hood.**
3. The pedipalps are freely articulated and leglike.

Soft ticks have habits like bedbugs, hiding during the day and emerging at night to feed on sleeping hosts. Usually they are not found on an active host. Most species are found in dry habitats, lurking in bird nests and roosts, loose soil, and crevices. Females lay their eggs in their hiding places. The emerging larva and several subsequent nymphal stages have the feeding habits of their parents. Most argasids can go months or even years without feeding.

Argas persicus

Species of *Argas* are parasites of birds and bats, although most will bite humans. The **fowl tick,** *Argas persicus,* and closely related species are important parasites of domestic fowl and other birds. Their population can increase in a henhouse to the extent that their nocturnal raids can debilitate or even kill chickens.

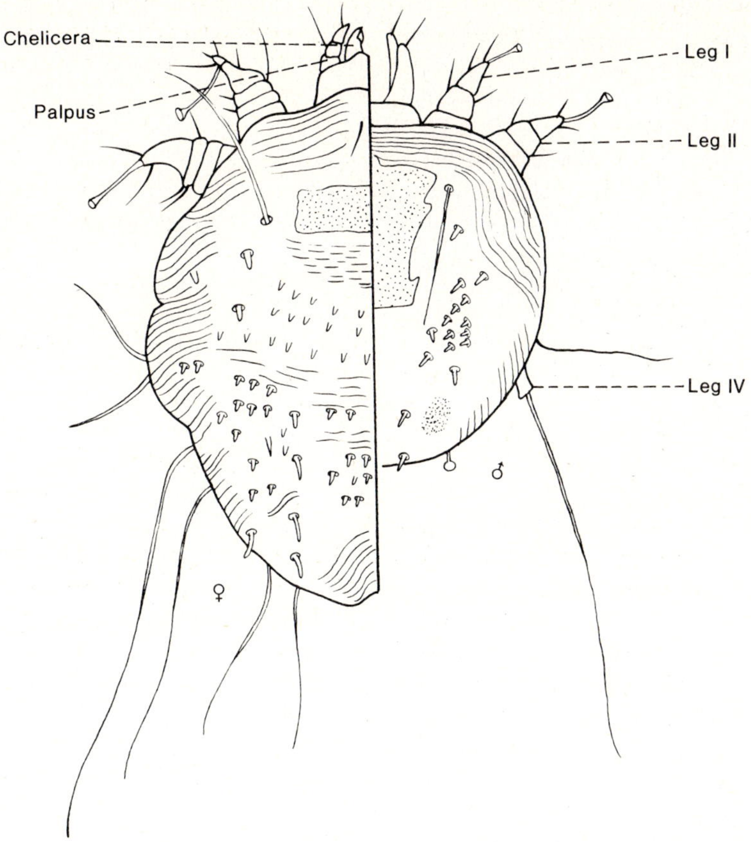

Fig. 8.3. *Sarcoptes scabiei;* dorsal views of female and male.

Examine a prepared slide of *A. persicus,* comparing its morphology with that of a hard tick. A slide of the **spinose ear tick,** *Otobius megnini,* may also be available for study. Although a large tick, it is a nymph, for the adults have vestigeal mouthparts and do not feed.

Mites: Suborder Astigmata

Without spiracles, but with a system of tracheae opening through stigma and porose areas on various parts of the body; chelicerae usually scissorslike, for chewing; palpi simple; anal suckers often present in the male.

These are minute, rounded, short-legged, flattened mites, causing **itch, mange** or **scabies** in humans and various animals. They spend their entire life on the host. Transmission is usually by direct contact, either with infested animals or with objects with which the hosts have been in contact.

Burrowing and feeding of the **itch mite,** *Sarcoptes scabiei* (fig. 8.3), of humans cause intense itching, inflammation, and swelling. Fertilized females cause the most trouble due to their tunneling just beneath the surface of the skin, laying eggs as they progress. The burrows ooze serum that hardens into scabs. Different varieties attack livestock, dogs, and rabbits.

Notoedres cati causes very severe and occasionally fatal mange in cats. The mites burrow into the skin of the face and ears and sometimes spread to the legs and around the genitals. Burrowing of the mites causes a constant intense itching, and scratching and rubbing done by the host will often produce open sores which frequently become infected. Scabs are formed as with *S. scabiei.*

The **ear mange mite,** *Otodectes cynotis,* infests dogs and cats. The mites do not burrow into the skin but occur deep in the ear canal near the eardrum, where they feed upon tissue fluids. Intense irritation results, and the canal becomes filled with inflammatory products and modified ear wax, as well as mites. The host animal scratches and rubs its ears and shakes its head or holds its head to one side.

Psoroptes mites cause mange or scab in livestock and rabbits. All the *Psoroptes* mites of cattle, sheep, goats, horses, and rabbits are considered as varieties of *P. equi.* They live on the surface of the skin on any part of the body that is thickly covered with hair. They feed upon serum or lymph, continually migrating to the periphery of the enlarging lesions. Scabs are formed from serum, dirt, and blood.

A condition known as **scaly-leg** in chickens and turkeys is caused by *Knemidokoptes mutans*. They apparently get onto the feet of the birds from the ground, since the lesion usually develops from the toes upward. They pierce the skin beneath the scales, causing the scales to separate from the skin and the feet and legs to swell and become deformed. Closely related to the scaly-leg mite is the **depluming mite** of chickens and other birds (*K. laevis* var. *gallinae*). These mites, which are embedded in the tissues or scales at the base of the quills, cause a loss of feathers over the back and sides.

Suborder Prostigmata

A pair of spiracles on or near the capitulum, but occasionally the stigmata may be absent; palpi usually free; chelicerae usually modified for piercing. These include, among others, *Demodex folliculorum* (fig. 8.4), the **hair follicle mite,** and species of *Leptotrombidium,* the **chiggers** or **redbugs,** and other Trombiculidae.

The hair follicle mite, *D. folliculorum*, an almost wormlike form, is found in the pores of man, especially around the nose and eyelids. The short and stumpy legs are three-jointed. Setae are lacking on the body. The palpi are closely appressed to the tiny rostrum. It has a typical **trombidiform life cycle,** the **egg, larva, protonymph, deutonymph,** and **adult.** The entire life cycle takes about 14 days and is spent on the host. The deutonymph is the distributive stage (Spickett, 1961), and transmission is by direct contact. Numerous forms from various animals have been described as different species, but they are very similar and their specificity is questionable.

Although the attacks of a number of mites are responsible for dermatitis and allergic reactions in humans, the most important ones are the chiggers of the genus *Leptotrombidium.* In the United States, these are common in the southern states, the Mississippi valley, and certain areas of California. They are most abundant in woody and grassy areas, swamps, and particularly where wild rodents and birds occur, such as berry patches. Only the larval stage feeds on vertebrates, the nymphal and adult stages being free-living. Chiggers attach especially in areas where the clothing fits tightly, such as the tops of shoes and socks, the waist region where belts and underwear are fastened, and under the armpits.

Larvae inject saliva into the host tissue, forming a **stylostome,** or feeding tube, through which the larva sucks up the semidigested tissue debris. After engorgement, the larva drops off and molts.

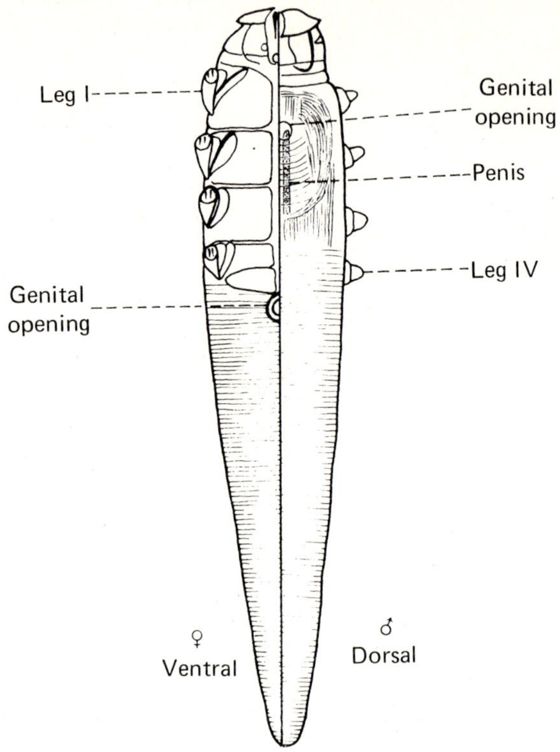

Fig. 8.4. *Demodex folliculorum,* ventral view of female and dorsal view of male.

Chiggers are not known to transmit any agents of diseases in the United States, but their attacks produce violent itching, which when scratched may result in secondary infections. In Southeast Asia and the Southwest Pacific, larval chiggers transmit the rickettsia causing **scrub typhus,** also known as **tsutsugamushi fever.** This disease is often fatal, and it took a heavy toll among troops during World War II. Since the larvae feed on humans or other vertebrates only once in their life, the rickettsiae are transmitted transovarially from infected adults through the egg to the "chigger" larval stage, which feeds on humans and rodents. The postlarval stages are predatory, feeding on small arthropods and their eggs in the soil.

Suborder Mesostigmata

Body well chitinized, with dorsal and ventral plates; capitulum small, anterior; one pair of spiracles lateral to legs, usually associated with an elongate tube or **peritreme,** or if absent, highly specialized parasites of the respiratory tract of vertebrates. Representatives with their hosts include: *Dermanyssus gallinae* (fig. 8.5),

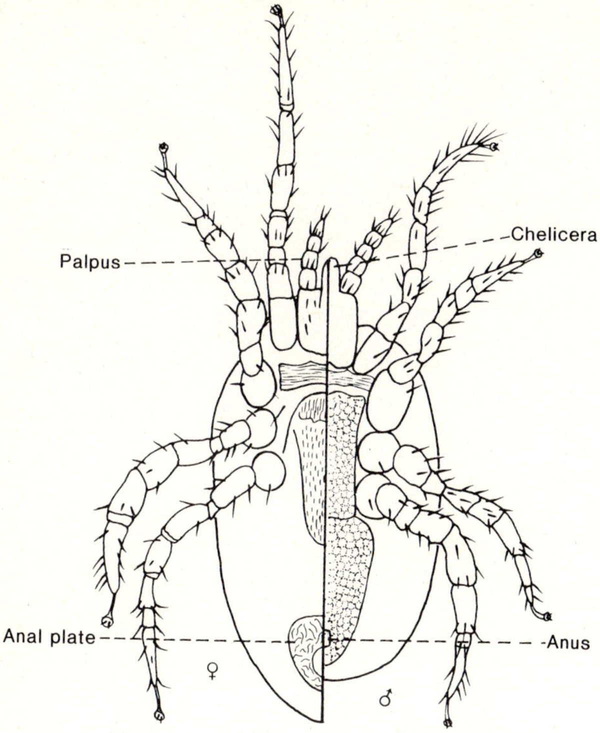

Fig. 8.5. *Dermanyssus gallinae;* ventral views of female and male.

chicken mites; *Ornithonyssus bacoti,* tropical rat mite, the vector of *Litomosoides carinii,* a filarial nematode parasitic in rodents (p. 170); *O. bursa,* tropical fowl mite; *O. sylviarum,* northern fowl mite; and *Echinolaelaps echidninus,* spiny rat mite, which serves as the final host for *Hepatozoon muris,* a haemogregarine of the rat. Mites become infected by feeding on infected rats, but cannot transmit the infection in this manner, as rats acquire the infection only by ingesting infected mites.

Suborder Cryptostigmata

In addition to the above parasitic mites, the **oribatid,** or **beetle mites** are of parasitological concern because they serve as intermediate hosts for Anoplocephalidae cestodes. Eggs of the tapeworms are ingested by the mites, in which they develop to the infective cysticercoid stage in several months. Animals feeding on the vegetation on which infected mites are crawling can easily acquire the infection, and the tapeworms develop to the adult stage within a few weeks.

Notes and Sketches

Review Questions

1. What are the stages in the life cycle of a tick?
2. What are the morphological differences between mites and ticks?
3. What are some diseases transmitted by ticks?
4. Name the morphological differences between hard and soft ticks.
5. What are the parts of the capitulum?
6. Name the segments of a tick's leg.
7. Compare the feeding habits of Argasidae with Ixodidae.
8. Name the orders of mites, with basic characteristics and an example of each.

References

Acari

Arthur, D. R. 1962. *Ticks and Disease.* Row, Peterson and Company, Evanston, 445 pp.

———. 1965 and 1970. Feeding in Ectoparasitic Acari with Special Reference to Ticks. In *Advances in Parasitology,* ed. B. Dawes. Academic Press, New York, vol. 3, pp. 249–98; vol. 8, pp. 275–92.

Baker, E. W., and Wharton, G. W. 1952. *An Introduction to Acarology.* The Macmillan Company, New York, 465 pp.

Baker, E. W.; Evans, T. M.; Gould, D. J.; Hull, W. B.; and Keegan, H. L. 1956. A Manual of Parasitic Mites of Medical or Economic Importance. *Nat. Pest Control Assoc. Tech. Publ.,* New York, 170 pp.

McDaniel, B. 1979. *How to Know the Ticks and Mites.* Wm. C. Brown Company Publishers, Dubuque, Iowa.

Schmidt, G. D., and Roberts, L. S. 1985. *Foundations of Parasitology,* 3rd ed., Mosby College Publishing, St. Louis, pp. 715–741.

Spickett, S. G. 1961. Studies on *Demodex folliculorum* Simon (1842). *Parasitology.* 51:181–92.

U.S. Department of Agriculture. 1965. Manual on Livestock Ticks. *Agr. Res. Service* 91–49, 142 pp.

Sources of Study Material for Acari*

Slides

Amblyomma americanum: C, T, Tr, W
Amblyomma cajanense: T
Amblyomma maculatum: T
Argus persicus: C, T, Tr, W
Boophilus annulatus: T, Tr, W
Cnemidocoptes gallinae: T
Dermatophagoides sp.: W
Demodex folliculorum: T, W
Demodex, in skin: T, Tr, W
Dermacentor andersoni: C, T, W
Dermacentor andersoni, eggs: C
Dermacentor andersoni, larvae: C, W
Dermacentor andersoni, nymph: W
Dermacentor variabilis: C, T, Tr, W
Dermanyssus gallinae: C, T
Haemaphysalis leporispalustris: T
Hyalomma savignyi: T
Liponyssus bacoti: T
Ornithonyssus bursa: T, W
Otobius megnini: T
Psoroptes cuniculi: T
Rhipicephalus sanguineus: C, T, Tr
Rhipicephalus sanguineus, larva: T
Sarcoptes, in skin: T, Tr, W
Sarcoptes scabei: T, W

Preserved

Argas sp.: C
Demodex canis, follicle mites in skin (red mange): C
Dermacentor andersoni: C, W
Dermacentor variabilis: C, W
"Tick": C

*The addresses of the sources are listed on p. 45.

Chapter 9 Class Insecta

Members of this class are air-breathing arthropods with distinct head, thorax, and abdomen, except in some larvae and some modified adults. They have a single pair of antennae, three pairs of thoracic legs, and usually one or two pairs of wings in the adult stage. As chiefly terrestrial animals, insects have become adapted to the greatest variety of conditions and have become so successful that they rank highest among metazoan animals both in species and in individuals. Both as predators and as parasites upon plants and animals, and in their relations to the spread of etiologic agents of disease, they occupy a position of economic importance not excelled by any other animal group.

Order Siphonaptera

Fleas are wingless, laterally compressed insects. They are armed with backwardly directed spines, long stout spiny legs, and short clubbed antennae that lie in grooves along the side of the head. Mouthparts are elongated, adapted for piercing the skin, and sucking blood.

Their life cycle includes a complete metamorphosis, with **egg, larva, pupa,** and **adult** stages. The female lays her eggs in the host's nest or bedding, or in the host's fur or feathers, whence they drop into the nest or onto the ground. The full quota of eggs is not laid all at once but singly or in batches over a considerable period of time, punctuated by blood meals which are necessary for their development. The length of the entire life cycle may be as short as a month for some individuals in hot, moist climates, but may be much longer, even in other individuals of the same brood. The adults infest mammals and birds and feed only on blood, although they may spend considerable time in the burrow or nest of their host instead of on its body. Of the known species, about 95% occur on mammals and only about 5% on birds.

Most fleas have preferred hosts, but many of them feed upon a variety of animals and bite people readily in the absence of their normal host species. Thus, humans are always liable to flea infestations from dogs or cats, or even from rats living in the house or adjacent structures. In early summer, fleas on their hosts lay large numbers of eggs, which drop off into cracks of the flooring or into debris. Here the larvae hatch and feed, mature, and transform into pupae. Then about mid-summer, often about the time the occupants of the infested dwelling return from their summer vacation, large numbers of the second generation of adults appear. In the summer, fleas of cats and dogs, particularly if stray pets are around, will breed out of doors in vacant lots, under buildings, and similar situations.

Fleas are of importance as pests and in connection with the transmission of disease-producing agents. By their insidious attacks on humans and domestic animals, they cause irritation, blood loss, and discomfort. In addition to their role as pests, fleas are of greater importance because of their connection with the transmission, both among their reservoir hosts and to humans, of the etiologic agents of two human diseases of outstanding importance—plague and murine, or endemic, typhus.

Plague, caused by the bacterium *Yersinia pestis* (formerly *Pasturella pestis*), has been a scourge of humans since early times. Of the two chief clinical types of plague, **bubonic** and **pneumonic,** the former is the commoner and much less severe. It is characterized by swellings ("bubos") of the lymph glands, especially those of the groin and armpits. When untreated, the fatality rate may exceed 50%. The pneumonic form is a highly contagious and usually fatal pneumonia-like disease, spread directly among people by sputum or droplets coughed up by the sick. It occurs secondarily to the bubonic type when the plague bacteria become localized in the lungs.

Bubonic plague ordinarily results from the bite of an infected flea, but the disease may be contracted by direct contact or the bite of an infected rodent. Fleas become infected when feeding on the blood of a diseased host. The bacteria multiply so rapidly in the proventriculus and stomach of the flea that an obstruction is formed. As these "blocked" fleas attempt to feed on a human or a rodent, the blood, which cannot pass beyond the obstruction, becomes mixed with the plague bacilli. When the contaminated blood is regurgitated into the bite-puncture or other skin abrasion, infection results. All rodent fleas are not equally susceptible to stomach blockage, and hence different species are not equally dangerous as plague vectors. The oriental rat flea, *Xenopsylla cheopis,* is the most important vector of urban plague because of its great tendency to become blocked, its ability to feed on both rodents and humans, and its great abundance in urban centers. It occurs in most of the coastal ports of the United States and has been found in a number of inland states.

There are two epidemiological types of plague: **urban plague,** the classical form of the disease and the type usually contracted in cities where people have close contact with domestic rats and fleas; and **sylvatic plague,** which is more often contracted by people in rural areas who have contact with wild rodents or their fleas.

Murine typhus is primarily a disease of murine rodents, which in North America include only rats and mice, and is caused by *Rickettsia typhi.* It is transmitted among rodents by their fleas, lice, and possibly mites, and occasionally to humans from rats and mice, presumably by their fleas. The most likely mode of transmission is by contamination of the bite-wound or abraded skin with the infected feces or crushed fleas, especially *X. cheopis.* According to Jellison (1959), transmission by direct bite remains a possibility, and infection by inhalation and ingestion is probable. In the southern United States, murine typhus is common among Norway and roof rats without evidence of ill effect. Fleas are not harmed by the rickettsia and presumably remain infected for life but do not transmit the organism to their progeny. While 5,400 cases of murine typhus were officially reported in the United States in 1944, the disease has since all but disappeared in this country as a result of the vigorous control programs of domestic rodents and their fleas.

In addition to the two above-mentioned familiar pathogens that are transmitted by fleas, they transmit a rat hemoflagellate, *Trypanosoma lewisi* (see p. 6), and serve as intermediate hosts for certain tapeworms that occasionally occur in humans, especially children. One of these is *Dipylidium caninum* (see p. 108) of dogs and cats, the cysticercoids of which occur in *Ctenocephalides canis, C. felis,* and *Pulex irritans.* The worm eggs are ingested by larvae, but the cysticercoids finish their development in the adult fleas. Two other tapeworms, *Hymenolepis diminuta* and *V. nana* of rats and mice, which occasionally are found in people, may use certain fleas as intermediate hosts. A nematode, *Dipetalonema reconditum,* found in the subcutaneous tissues of dogs, is transmitted by species of *Ctenocephalides.*

Family Pulicidae

Ctenocephalides felis

This species, the **cat flea,** and the closely allied **dog flea,** *C. canis,* are now almost cosmopolitan, having followed humans and their domestic pet animals over most of the world. The cat flea is usually more abundant and generally more widely distributed than the dog flea. In both species, the eyes are large and deeply pigmented; both **genal** and **pronotal combs** are heavily pigmented and sharply pointed. Typical specimens of the two species of *Ctenocephalides* can be readily separated by the shape of the head and the character of the genal combs. In *C. felis* (figs. 9.1 and 9.2D), the head is quite long and pointed, and the first and second genal spines are of approximately equal length; in *C. canis* the head is relatively short and rounded, and the first genal spine is shorter than the second.

Xenopsylla cheopis

This species, the **oriental rat flea** (fig. 9.2C), has been introduced throughout much of the world with the Norway and roof rats. It is established throughout most of North America. When rats are killed, these fleas leave their hosts and readily bite people. Medically, it is undoubtedly the most important flea in the world, for a large percentage of the human cases of bubonic and pneumonic plague that have occurred during the many historical pandemics and epidemics of this disease may be attributed to the presence of this flea. Also, it is the chief vector of the agent of murine typhus.

Eyes well developed; genal and pronotal combs lacking; **ocular bristle** situated anterior to eye (cf. *Pulex*). Females can be easily distinguished by the large deeply pigmented **spermatheca.**

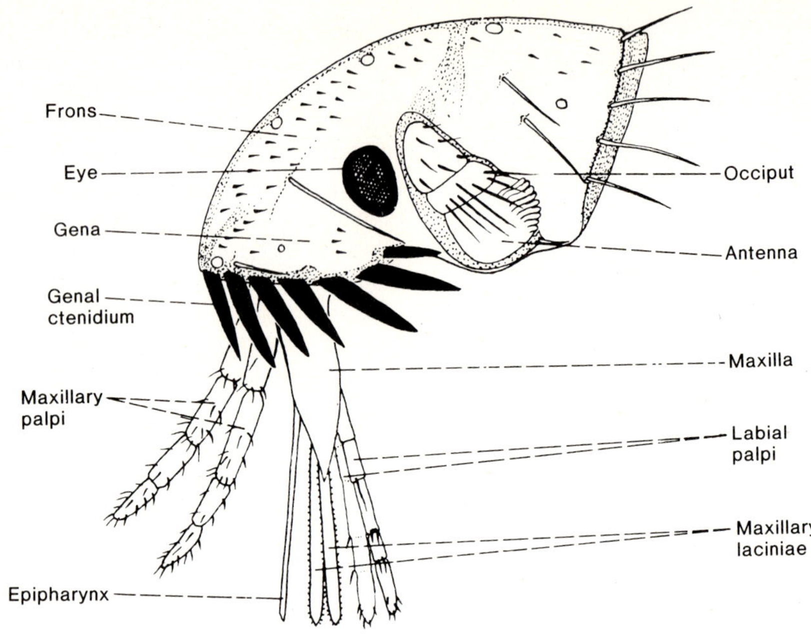

Fig. 9.1. Head of *Ctenocephalides felis,* male, lateral view.

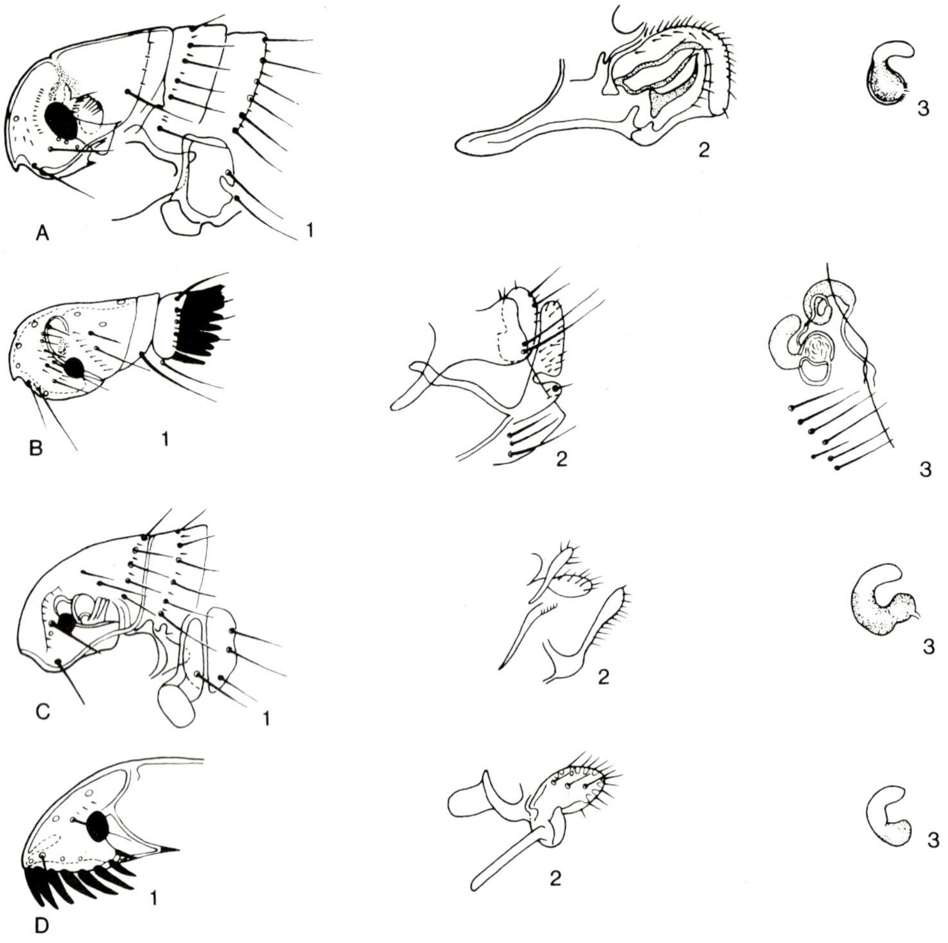

Fig. 9.2. Taxonomic characteristics of four common and important species of fleas. (A) *Pulex irritans;* (B) *Nosopsyllus fasciatus,* showing pronotal comb; (C) *Xenopsylla cheopis;* and (D) *Ctenocephalides felis,* showing genal comb. (1) Head; (2) clasper of male; and (3) spermatheca of female.

Pulex irritans

This species, the **human flea,** though found on a wide variety of other hosts in nature, especially dogs, is widespread throughout the warmer regions of the world. Of all the species of fleas, *P. irritans* (fig. 9.2A) is probably the most nearly cosmopolitan in distribution. In addition to being a serious pest of humans and serving as an intermediate host of *Dipylidium caninum,* the human flea has been found infected with plague bacteria in nature.

Eyes large and deeply pigmented; genal comb absent, or represented by one inconspicuous tooth; pronotal comb absent; ocular bristle situated ventral to eye.

Family Ceratophyllidae

Nosopsyllus fasciatus

This species, the **northern rat flea,** is found commonly on domestic rats and house mice throughout North America. It does not bite humans readily and is found more commonly in temperate regions, where plague is less likely to be present. It may be of importance in the transmission of the plague bacterium among rats. While relatively unimportant in the transmission of plague to people, it is an efficient vector of murine typhus. Eyes well developed; genal comb lacking; pronotal comb with long, dark brown spines. (See fig. 9.2B.)

Key to Common Families of Siphonaptera

1(2) The three thoracic segments combined, shorter than the first abdominal tergite (fig. 9.3) .... **Tungidae**

2(1) The three thoracic segments not combined, longer than the first abdominal tergite 3

3(4) Abdominal tergites II through VII, each with only one row of setae (fig. 7.2A) ... **Pulicidae**

4(3) Abdominal tergites each with more than one row of setae 5

5(6) Lacking genal comb; eyes usually well developed (fig. 9.2B) .... **Ceratophyllidae**
Eyes present and pigmented (fig. 9.4) .... **Amphipsyllinae**
Eyes usually present (vestigial in *Dactylopsylla* and *Foxella*) (fig. 9.2B) .... **Ceratophyllinae**
Eyes absent (fig. 9.5) .... (*Dolichopsyllus,* only genus) **Dolichopsyllinae**

6(5) Usually with genal comb; eyes not well developed 7

7(10) Usually with genal comb; with two spines (on each side) 8

8(9) Genal comb with two spines, situated at the tip; eyes vestigial (on bats) (fig. 9.6) .... **Ischnopsyllidae**

9(8) Genal comb with two or four spines, distinct, not overlapping, situated behind the eyes; eyes much reduced (fig. 9.7) .... **Leptopsyllidae**

10(7) Usually with genal comb, with more than two spines (if only two, overlapping and not distinct, and situated near the eyes); eyes absent or vestigial, not large and heavily pigmented (fig. 9.8) .... **Hystrichopsyllidae***

*Included are a few genera lacking genal comb, e.g., *Atyphloceras, Callistopsyllus, Catallagia, Conorhinopsylla, Delotelis, Megarthroglossus, Saphiopsylla, Trichopsylloides.*

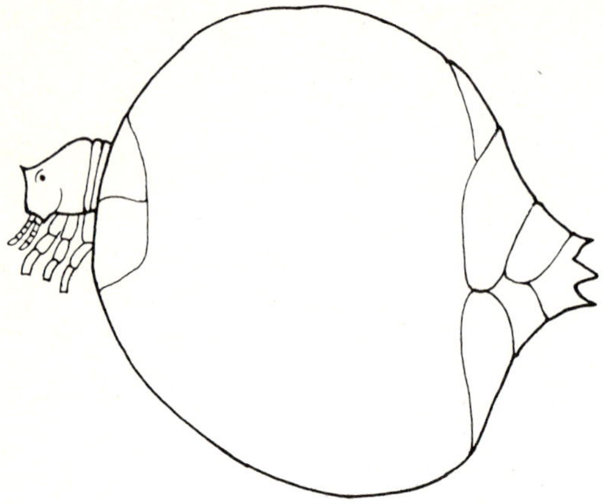

Fig. 9.3. *Tunga penetrans, engorged female.*

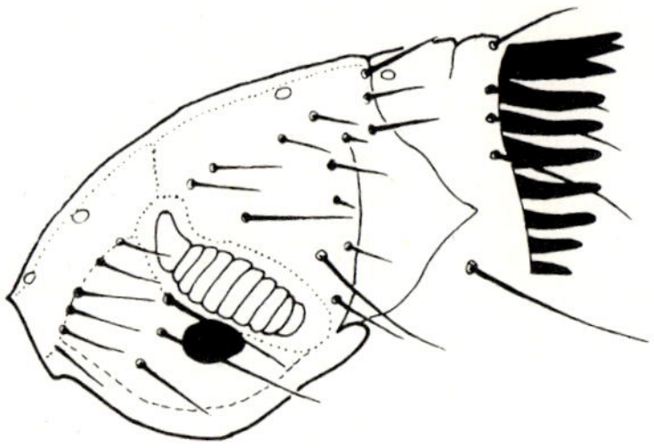

Fig. 9.4. Head of *Amphipsylla sibirica.*

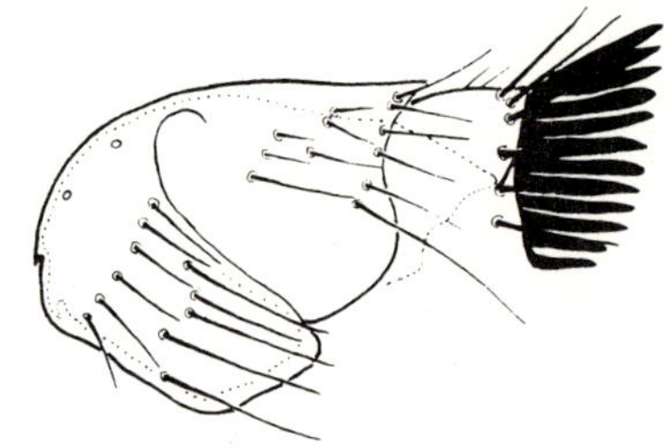

Fig. 9.5. Head of *Dolichopsyllus stylosus.*

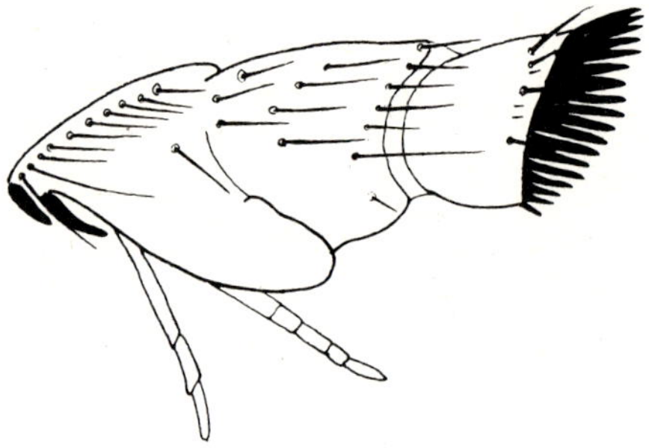

Fig. 9.6. Head of *Eptescopsylla vancouverensis.*

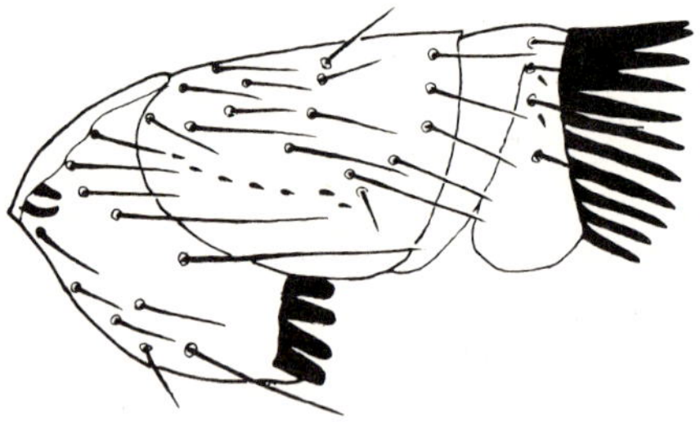

Fig. 9.7. Head of *Leptopsylla segnis.*

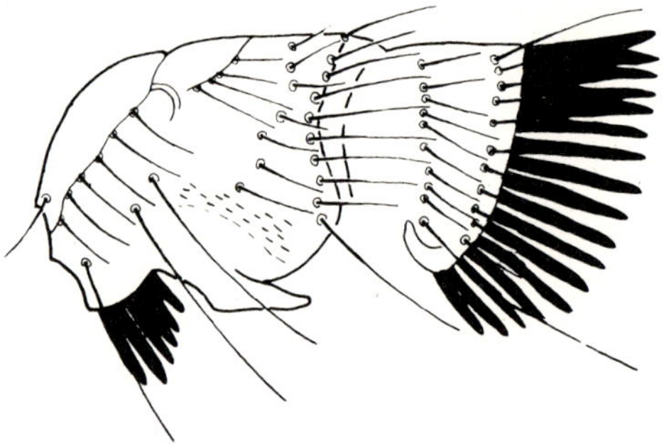

Fig. 9.8. Head of *Hystrichopsylla dippiei.*

Notes and Sketches

Review Questions

1. How many antennae do insects have?
2. Do any arthropods besides insects possess wings?
3. Do fleas parasitize mostly mammals or birds?
4. Name two important diseases of humans transmitted by fleas.
5. What are genal and pronotal combs? Are they present in the human flea?
6. What two species of fleas are most important in transmission of plague to humans?

References

Siphonaptera

Ewing, H. E., and Fox I. 1943. The fleas of North America. *U.S. Dept. Agr., Misc. Publ. No. 500,* 142 pp.

Holland, G. P. 1949. The Siphonaptera of Canada. *Canad. Dept. Agr., Pub.* 817, *Tech. Bull.* 70, 306 pp.

Jellison, W. L. 1959. Fleas and Disease. In *Annual Review of Entomology,* E. A. Steinhaus and R. F. Smith, eds. Annual Reviews, Inc., Palo Alto, Cal., vol. 4, pp. 389–414.

Jellison, W. L.; Locker, B.; and Bacon, R. 1953. A synopsis of North American fleas, north of Mexico, and notice of a supplementary index. *J. Parasitol.* 39:610–18.

Lehane, B. 1969. *The Compleat Flea.* The Viking Press, New York.

Pratt, H. D., and H. E. Stark. 1973. Fleas of public health significance and their control. *U.S. Department of Health, Education, and Welfare Pub. No. (CDC) 75–8267.* U.S. Government Printing Office, Washington, D.C.

Rothschild, M. 1965. Fleas. *Sci. Am.* 213(6):44–53.

Rothschild, M. 1975. Recent advances in our knowledge of the order Siphonaptera. *Ann. Rev. Entomol.* 20:241–59.

Rothschild, M., Schlein, Y., Parker, K., Neville, C., and S. Sternberg. 1973. The flying leap of the flea. *Sci. Am.* 229(5):92–100.

Sources of Study Material for Siphonaptera*

Slides

Ctenocephalides canis: C, T
Ctenocephalides felis: W
Ctenocephalides sp.: C, Tr, W
Ctenocephalides, egg: C
Ctenocephalides, larva: C, W
Echidnophaga gallinacea: C, T
Pulex irritans: C, T, Tr, W
Xenopsylla cheopis: T, Tr, W

*The addresses of the sources are listed on p. 45.

Chapter 10 Lice (Anoplura and Mallophaga)

Lice are wingless, dorsoventrally flattened insects, ectoparasitic on birds and mammals. They are parasitic throughout their entire life cycle, even their eggs being firmly attached to the hairs or feathers of their hosts. The immature stages have the same habit of life as the adults and resemble them closely, so that there is little metamorphosis. There are three nymphal instars before reaching the adult stage. They have only short three- to five-segmented antennae, sometimes hidden in a recess on the head; reduced or no compound eyes; and no ocelli. Thoracic segments are sometimes indistinct, legs are short but stout; and the abdomen has from five to eight distinct segments.

Authorities are not in agreement as to whether the sucking lice of mammals and the chewing lice of birds and mammals should be placed together in the order Phthiraptera, or the former in the suborder Anoplura and the latter in the suborder Mallophaga, or the suborders be given ordinal status. Since the two taxa are anatomically distinct and have different habits and host preferences, they are here considered as separate orders.

Order Anoplura

Anoplura have highly specialized mouthparts adapted for piercing the skin of their hosts and sucking blood, which when not in use are retracted into a pouch. The head is narrower than the thorax, the three segments of which are fused together, without any clear division of segments. The number of spiracles has been reduced to not more than one pair on the thorax and six pairs on the abdomen. The single tarsal joint bears only one claw, and in most species the tip of the tibia is widened and drawn out into a "thumblike process," opposable to the claw. All are ectoparasitic on mammals and very specific in their host requirements.

Humans are infested with two species of sucking lice: the **crab louse,** *Phthirus pubis,* and the **human louse,** *Pediculus humanus.* The latter is commonly regarded as comprising two anatomically similar biological races or possibly species, but occurring on different body locations of the host. Those normally clinging on the inner garments of a person, except when feeding on the body, are sometimes referred to as *P. humanus humanus,* and the others, inhabiting the head, are known as *P. h. capitis.* But Ferris (1951) prefers to use the name *P. humanus* to cover the population as a whole, with the addition of the vernacular names "body louse" and "head louse" for those forms when the occasion demands. Except for the body louse, which attaches its eggs to the fibers of the underwear, lice attach their eggs to the hairs of the host. Sucking lice have been intimately associated with humans from time immemorial. They are most common in times of stress, such as war, famine, and disaster, when people are forced to live under crowded conditions and facilities for cleanliness are inadequate. Anoplura are of importance in two ways: their attacks on humans and domestic animals result in irritation and loss of blood, and they are involved in the transmission of microbial organisms causing human disease.

They inject an irritating saliva into the host during feeding, resulting in considerable itching. A person's first exposure to lice results in little or no discomfort, but sensitivity occurs after about seven to ten days, when the average individual develops an intense itching from the feeding of the lice. Heavy infestations may lead to scratching and secondary infections.

The body louse is the vector of **epidemic typhus,** which is highly fatal during epidemics, and **relapsing fever** of Europe, Africa, and Asia. The first is caused by *Rickettsia prowazeki* and the second by the spirochete, *Borrelia recurrentis.* Lice become infected with the rickettsial agent by ingesting blood from a diseased person; the parasites multiply in the midgut epithelium of the louse and are voided with the feces. People usually acquire the infection by the contamination of the bite wound or abraded skin with the infected feces or crushed lice. Lice frequently leave the typhus patient when high fever develops and attack other human hosts, causing a rapid spread of the disease. In the case of the

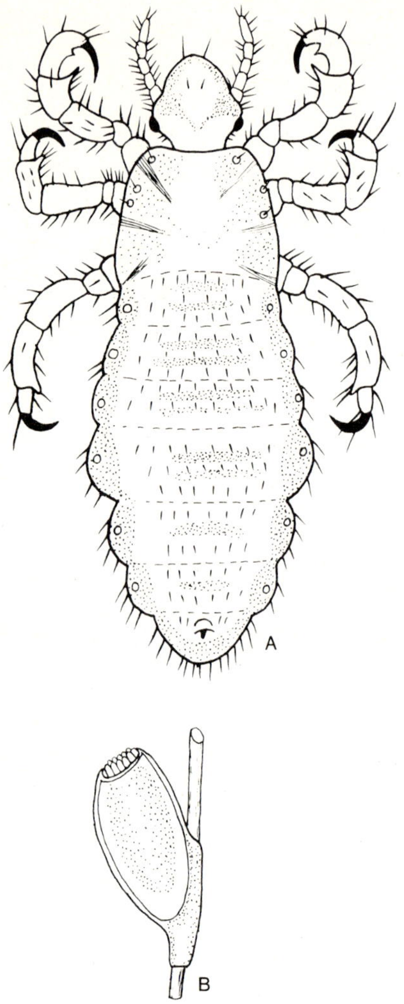

Fig. 10.1. Body louse, *Pediculus humanus.* (A) Adult male; (B) egg attached to body hair.

spirochete causing relapsing fever, which is likewise acquired by the louse with the ingested blood of a positive patient, it multiplies throughout the body of the louse. Neither the feces nor the bites of the lice are infective, transmission occurring only by crushing lice and inoculating the bitten or abraded skin with the infective body fluid.

Family Pediculidae

Pediculus humanus

The human louse (fig. 10.1) bears a pair of prominent eyes, a pair of five-jointed antennae, and mouthparts which are withdrawn and not visible. The three thoracic segments, each with a pair of legs terminating in a hooklike claw, are fused together. The abdomen is elliptical and consists of seven segments. The margins are festooned and chitinized to form deeply pigmented plates, the **tergal** and **paratergal plates,** on which six pairs of spiracles are situated. In the male, the abdomen is rounded posteriorly and the genital organ, the

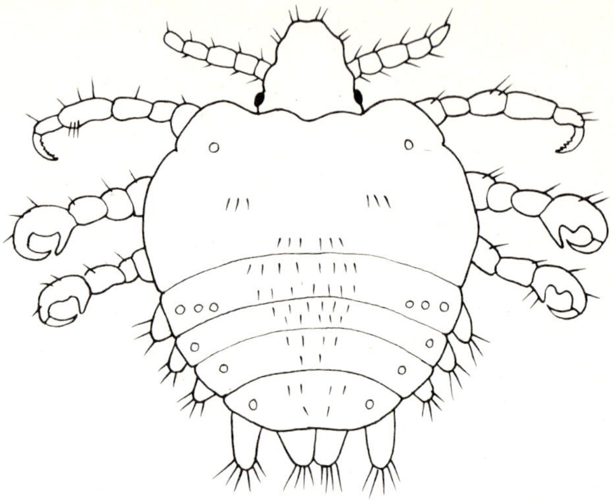

Fig. 10.2. Adult crab louse, *Phthirus pubis,* female.

aedeagus, is easily visible and usually extruded; in the female the terminal portion of the abdomen is deeply cleft. The life cycle may be completed in about 18 days.

Phthirus pubis

The **crab louse** (fig. 10.2) is characterized by its relatively short head, which fits into a broad depression in the thorax. The latter is broad and flat and fused with the abdomen. The first pair of legs is slender and terminated by a straight claw, the second and third pairs are thicker and provided with powerful claws for clinging to hairs. The first three abdominal segments are fused into one, bearing three pairs of spiracles, followed by three tandem arranged segments with as many pairs of spiracles. The last four abdominal segments bear wartlike processes laterally, known as tubercles, the last of which is the longest. The life cycle takes about 15 days.

Key to Common Families of Anoplura

1(2) Abdomen with paratergal plates (a row of dorsal plates on each side of abdomen) 3

2(1) Abdomen without paratergal plates 7

3(4) Paratergal plates with their apical parts projecting free from body (on rodents, a few on insectivores and primates) (fig. 10.3) .... **Hoplopleuridae**

4(3) Paratergal plates without free apices 5

5(6) Eyes present, pigmented; head not retracted deeply into thorax (on primates, including humans) (fig. 10.1) .... **Pediculidae**

6(5) Eyes absent, but with prominent ocular lobes posterior to the antennae (*Pecaroecus* from peccaries, an aberrant genus, lacks ocular lobes); head deeply retracted into the thorax (on Artiodactyla and Equidae) (fig. 10.4) .... **Haematopinidae**

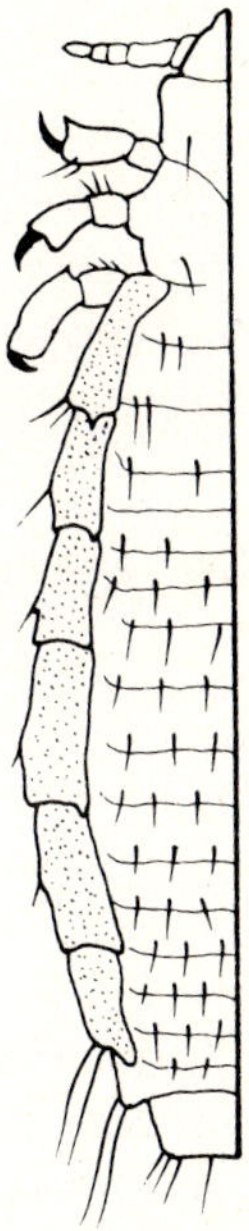

Fig. 10.3. *Hoplopleura hesperomydis,* female.

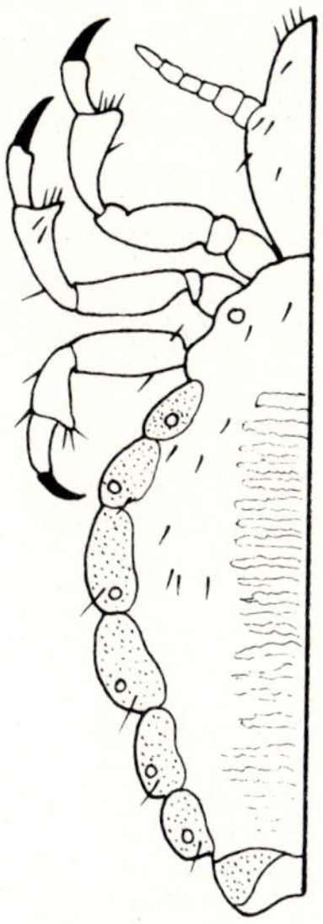

Fig. 10.4. *Haematopinus suis,* female.

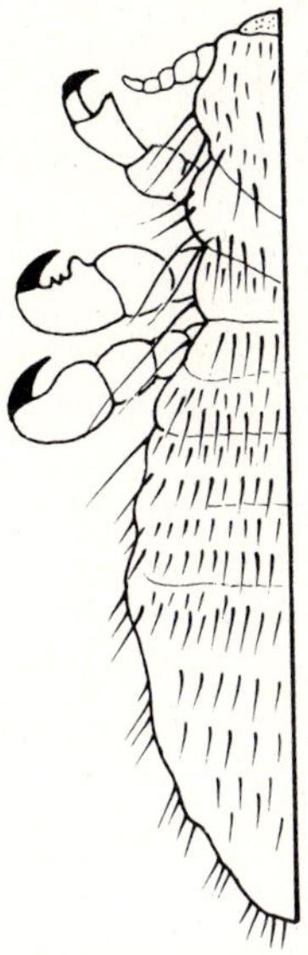

Fig. 10.5. *Antarctophthirus callorhini,* female.

Fig. 10.6. *Linognathus setosus,* female.

7(8) Body thick and stout, with heavy spines and, in some cases, scales; legs, at least two pairs, with stout undivided tibiotarsus (on marine mammals) (fig. 10.5) .... **Echinophthiriidae**

8(7) Body with setae but not scales, first pair of legs smaller than the others (on Artiodactyla and Canidae) (fig. 10.6) .... **Linognathidae**

Order Mallophaga

These **chewing** or **biting lice,** like their sucking counterpart, are wingless, dorsoventrally flattened insects that are parasitic throughout their entire life cycle, even the eggs being firmly attached to the feathers or hairs of their hosts. However, they differ from the Anoplura in several respects: the mandibles are heavily chitinized; the head is very broad, usually as wide or wider than the thorax; the thorax is divided into at least two distinct segments; the last of the two tarsal joints bears one or two claws, and only rarely is the tip of the tibia so formed that it is opposable to the claw; and most are ectoparasites of birds, but several genera infest mammals only.

The "biting" ascribed to the Mallophaga refers to their manner of feeding and not to any wound they may inflict on the host, since in most species the food consists of feathers and hairs, though some are found to have an admixture of blood in their diet. Whatever their feeding habits may be, however, they are most annoying pests because of the irritation produced by their constant crawling and nibbling. Nearly all birds are infested with these lice, and domestic poultry and pigeons suffer particularly from them, owing to the crowded conditions under which these birds live. While less widely distributed on mammals, domestic animals are attacked by various species of *Trichodectes.* Dogs, cats, horses, cattle, sheep, and goats may suffer considerable loss of condition if badly infested with chewing lice.

Suborder Amblycera
Family Menoponidae

Menopon gallinae

The common **shaft louse** of chickens (fig. 10.7) is a small pale species, about 1.5 mm long. The head is broadly triangular, strongly enlarged on the temples and evenly enlarged behind. Head without ventral sclerotized processes situated lateral and posterior to the mandibles. Prothorax smaller than the head, meso- and metathorax united. *Menopon* is a large genus and well represented on galliform birds.

Other species on chickens include: *Menacanthus stramincus* (the **body louse**); *Lipeurus caponis* (the **wing louse**); *Goniocote gallinae* (the **brown chicken louse**); and *Cuclutogaster heterographus* (the **head louse**).

Suborder Ischnocera
Family Trichodectidae

Trichodectes canis

The **dog biting louse** (fig. 10.8), which occasionally also infests other canines, is most troublesome on puppies. It is characterized by a broad head, greater in width than length, and a broadly rounded preantennary region. The antennae are sexually dimorphic; in the male they are shorter than the head. There are six pairs of abdominal spiracles.

This species may serve as the intermediate host of the double-pored tapeworm, *Dipylidium caninum,* of dogs, cats, and occasionally humans, especially children (see p. 108). Another species of dog louse is *Heterodoxus longitorsus. Felicola subrostrata* is found on cats.

Key to Common Families of Mallophaga

1(4) Maxillary palps present; antennae arising from ventral portion of head, usually concealed in grooves **Suborder AMBLYCERA** 2

2(3) Tarsi with one claw or none (on guinea pigs) (fig. 10.9) .... **Gyropidae**

3(2) Tarsi with two claws (on birds) (fig. 10.7) .... **Menoponidae**

4(1) Maxillary palps absent; antennae arising from or near lateral margin of head, not concealed in grooves **Suborder ISCHNOCERA** 5

5(6) Antennae five-segmented, tarsi with two claws (on birds) (fig. 10.10) .... **Philopteridae**

6(5) Antennae three-segmented, tarsi with one claw (on mammals) (fig. 10.8) ...... **Trichodectidae**

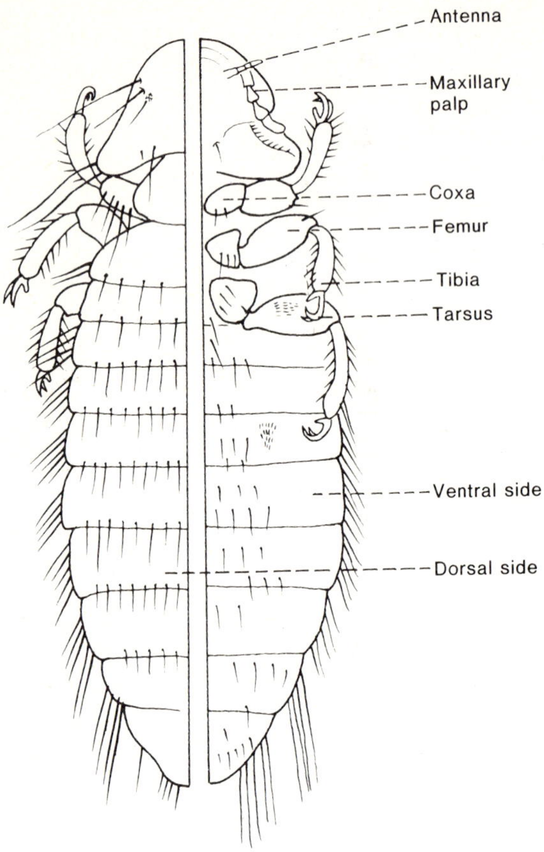

Fig. 10.7. *Menopon gallinae,* dorsal and ventral side of female.

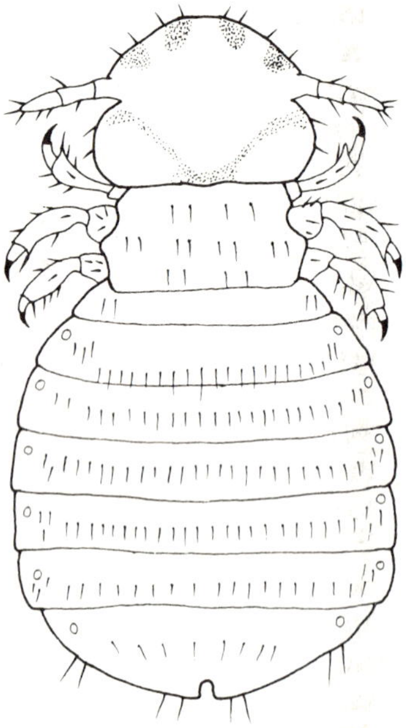

Fig. 10.8. *Trichodectes canis,* dorsal view of female.

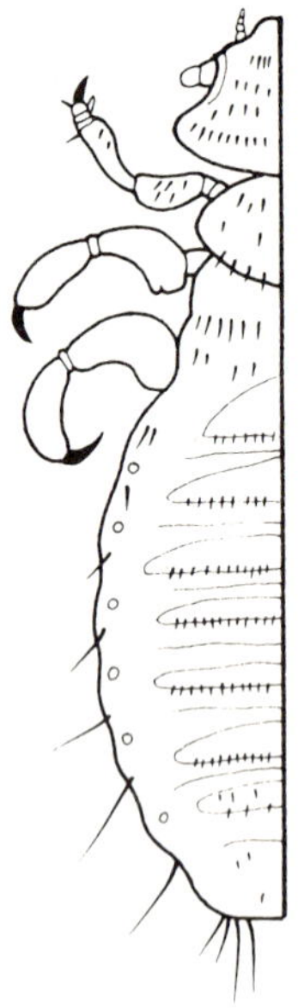

Fig. 10.9. *Gyropus ovalis,* female.

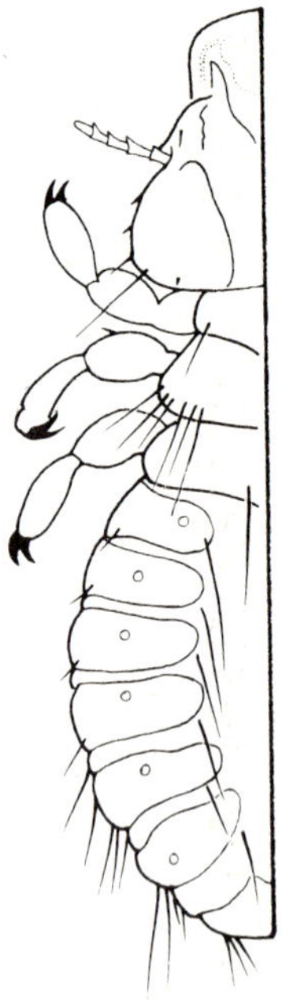

Fig. 10.10. *Philopterus dentatus,* female.

Notes and Sketches

Review Questions

1. Name the basic morphological differences between sucking and biting lice.
2. Which order is most common on birds? Which on mammals?
3. Name the three common lice on humans.
4. What diseases of humans are transmitted by lice? Are they acquired by the bite of the insect?
5. How many tarsal claws per leg do lice have?

References

Lice (Anoplura and Mallophaga)

Clay, T. 1970. The Amblycera (Phthiraptera: Insecta). *Bull. Br. Mus. Nat. Hist. (Entomol.)* 25:73–98.

Ferris, G. F. 1951. The sucking lice. *Pacific Coast Ent. Soc. Mem.* 1:1–320.

Kim, K. C., and Ludwig, H. W. 1978. The family classification of the Anoplura. *System. Entomol.* 3:249–84.

Pratt, H. D., and Littig, K. S. 1973. Lice of public health importance and their control. *U.S. Department of Health, Education, and Welfare Pub. No. (CDC) 77–8265.* U.S. Government Printing Office, Washington, D.C.

Weyer, F. 1960. Biological relationships between lice (Anoplura) and microbial agents. In *Annual Review of Entomology,* eds. E. A. Steinhaus and R. F. Smith. Annual Reviews, Inc., Palo Alto, Cal., vol. 5, pp. 405–20.

Sources of Study Material for Lice*

Slides

Anoplura

Haematopinus asini: T
Haematopinus eurysternus: C, T, W
Haematopinus eurysternus, egg: W
Haematopinus suis: C, T, Tr
Haematopinus suis, egg: C
Pediculus humanus capitus: C, T, W
Pediculus humanus capitus, egg: C, T
Pediculus humanus capitus, nymph: C
Pediculus humanus humanus: C, T, Tr, W
Pediculus humanus humanus, egg: C, W
Pediculus humanus humanus, nymph: C
Phthirus pubis: T, Tr, W
Phthirus pubis, egg: C, W
Phthirus pubis, nymph: W
Polyplax spinulosa: C

Mallophaga

Bovicola bovis: T, W
Columbicola columbae: T, W
Linognathus piliferus: T
Lipeurus variabilis: C
Lipeurus caponis: C
Menopon gallinae: C, Tr, W
Menopon stramineus: T
Trichodectes canis: C, T, W

*The addresses of the sources are listed on p. 45.

Chapter 11 Order Diptera

Members of this order possess one pair of functional wings, and rudiments of a second pair in the form of short, knobbed organs known as **halteres.** Mouthparts are modified for piercing and sucking, or for lapping. Metamorphosis is complete. Larvae are legless and often wormlike, and frequently occur in tissues or cavities of humans and animals, causing **myiasis.** There are three suborders, based on antennal characters of the adults.

Of all the taxa of arthropods that are involved as vectors of disease and infectious organisms the Diptera, or true flies, are of paramount importance. The order includes a great variety of bloodsucking forms responsible for the transmission of organisms that cause such important diseases as malaria, yellow fever, dengue, trypanosomiasis and leishmaniasis, encephalitis, and filariasis, as well as a number of nonbloodsucking forms of sanitary interest, such as the ordinary houseflies and blowflies.

Suborder Nematocera

Antennae many segmented, each segment similar and of approximately equal diameter throughout its length, often quite long, longer than the length of the head and thorax combined. There is no **arista.** The palpi are four- or five-segmented.

Family Psychodidae

Members of this family, commonly known as **sandflies** (fig. 11.1), are small, mothlike flies found in nearly all warm and tropical climates. Their bodies and wings are hairy and the antennae are long, consisting of 12 to 16 segments.

Eggs of *Phlebotomus,* laid in batches of about 50, are deposited in dark, moist crevices of rocks or concrete walls, damp cracks in shaded soil, caves, or other such places where there is sufficient moisture. Eggs hatch in about a week, and the larvae transform into

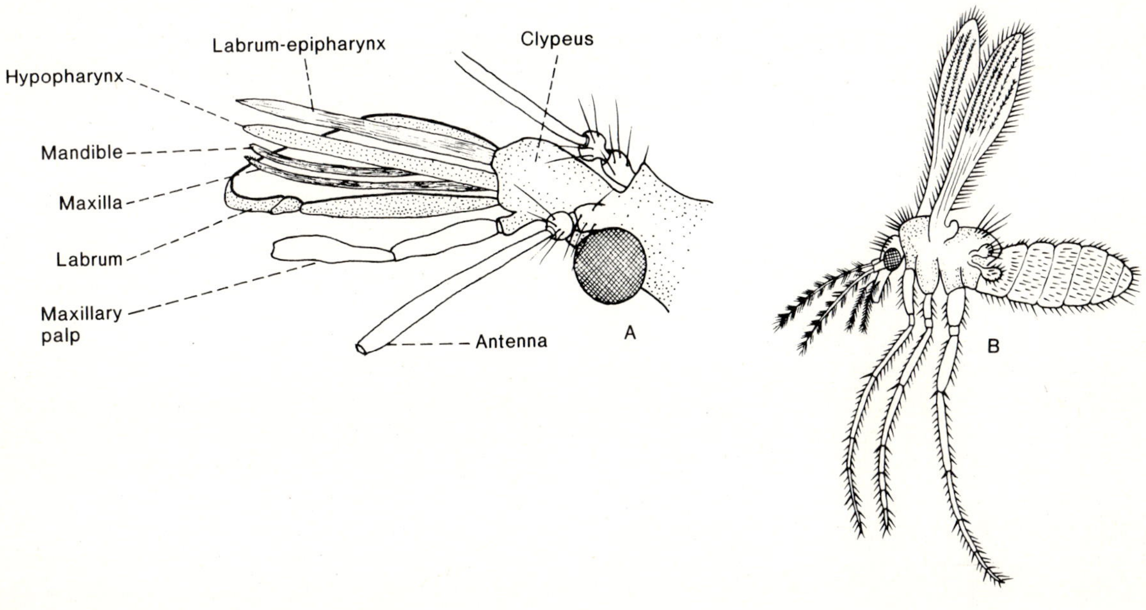

Fig. 11.1. *Phlebotomus,* a sandfly. (A) Mouthparts of female; (B) adult female.

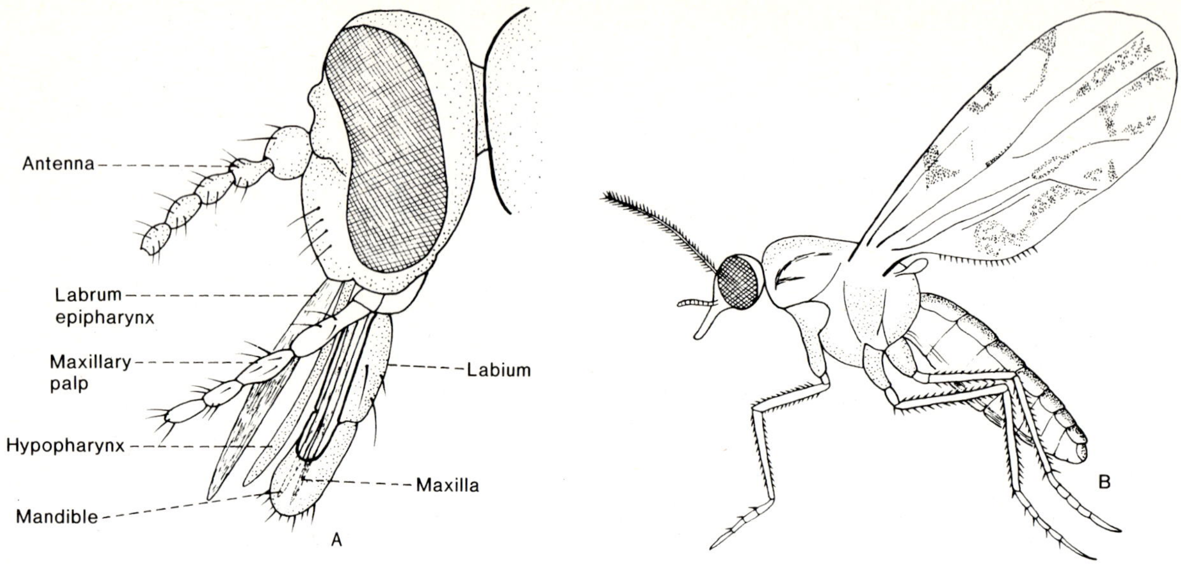

Fig. 11.2. *Culicoides,* a biting midge. (A) Head showing mouthparts; (B) adult female.

naked pupae in about one month. The pupal stage lasts about ten days. Under favorable conditions, the life cycle is completed in about two months. The flies are active only at night and hide during the day in cool, damp places, such as masonry cracks, stone walls, animal burrows, and hollow trees.

Sandflies are vicious biters and carry at least three serious disease agents of humans, although none of the species of *Phlebotomus* occurring in the United States is a vector. The dangerous **verruga** or **Oroya fever** of South America is transmitted principally by *P. verrucarum,* and **sandfly fever** of the Mediterranean region, Near East, southern China, Sri Lanka, and India is transmitted by *P. papatasii.* Kala-azar, a **leishmaniasis** endemic to the Mediterranean region, Near East, southern China, Sri Lanka, India, and clinical variants of this disease in Central and South America, are transmitted by various species of *Phlebotomus* or *Lutzomyia.* Only the females suck blood.

Family Ceratopogonidae

These are minute flies, seldom exceeding 2 to 3 mm in length, that are called **punkies, biting midges,** or **noseeums.** The wings are more or less covered with erect hairs and in many species are marked with pale spots on a dark background.

The eggs of *Culicoides* (fig. 11.2), several hundred in number, are laid in gelatinous masses usually anchored to some underwater object. Larvae, which emerge after a few days, are aquatic or semiaquatic, whitish, with a small brown head and 12 body segments. Slender brown pupae have two short breathing tubes on the thorax. They float nearly motionless in a vertical position, the breathing tubes in contact with the surface film. The life cycle requires from six to 12 months.

The great majority of the species that attack humans belong to *Culicoides.* Only the females suck blood. They become active in the evening and early morning, but if disturbed many will bite in the shade, even on bright days. Their bites can be extremely irritating, and often produce delayed reactions. The adults are aggressive on occasion, crawling into the hair and beneath articles of clothing to obtain a blood meal. However annoying, they are not known to carry human disease agents in this country. Various species of *Culicoides* serve as vectors of three filarial worms of man: *Dipetalonema perstans* in parts of Africa and eastern South America; *D. streptocerca* in Africa; and *Mansonella ozzardi* in the British West Indies. In addition, at least three filariae of animals, *Onchocerca reticulata* of horses, *O. gibsoni* of cattle, and *Icosiella neglecta* of amphibians, are transmitted by species of *Culicoides* (see Nelson, 1964, 1966), as is the virus of blue tongue in ruminants.

Family Simuliidae

These are the small (1 to 5 mm long) dark-colored Diptera, known as **blackflies** (fig. 11.3). The thorax is humped over the head, and the proboscis is short. Antennae are many jointed (10 or 11) but short, with beadlike segments instead of being long and filamentous as in some Nematocera. The wings are broad and unspotted. They have no scales, and they are not hairy, except for bristles on the anterior veins.

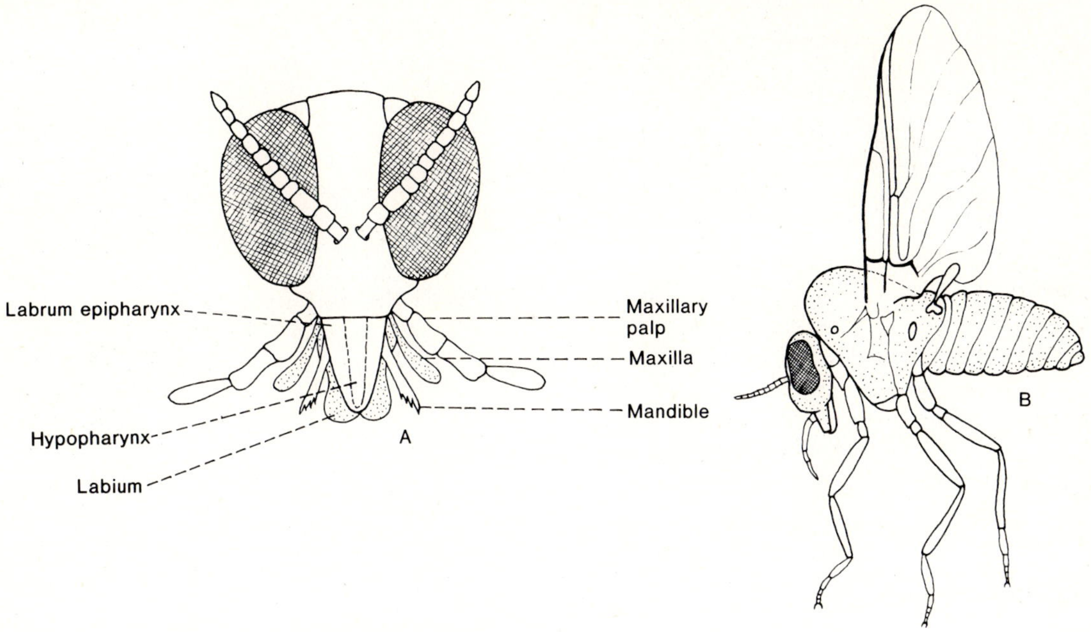

Fig. 11.3. *Simulium,* a blackfly. (A) Head showing mouthparts; (B) adult female.

Shiny eggs are laid in masses of several hundred on stones and plants in flowing streams. They usually hatch in a few days, and the larvae attach to objects in the water, from which they gather food with a set of motile brushes around the mouth. After about a month, the larvae spin weblike cocoons, open at the anterior end for the extrusion of the branching gill filaments used for respiration under the water, and pupate in them. Adults emerge in a few days to a week or more and are carried to the surface by a bubble of air which has been collected inside the pupal case. The life cycle requires from 60 to 105 days or longer, and the number of generations per year, in temperate regions, is from one to six, depending upon the species and climatic conditions. The flies are active mainly during the daylight hours.

Few insects are more capable tormentors of humans and animals than some species of blackflies. One who has never been the victim of their bites cannot fully appreciate the suffering which accompanies and follows them. They often appear suddenly in great clouds, and on such occasions are able, within a few hours, to cause the death of large animals, including humans, if unprotected. Only the females suck blood. Certain species of *Simulium* are vectors of the filarial worm *Onchocerca volvulus* which infects humans in certain parts of Mexico, Central America, and Africa. Blackflies also transmit the blood protozoans, *Leucocytozoon simondi* of young ducks and geese in northern United States and Canada, and *L. smithi* of turkeys in various parts of the United States. The chief vectors are *Simulium rugglesi* and *S. occidentale.*

Family Culicidae

This family includes the **mosquitoes,** Diptera with scales along the wing veins and a prominent fringe of scales along the margin of the wings. The antennae, of 14 or 15 segments, are conspicuous and can usually be used to separate the sexes. In females, they are long and slender with a whorl of a few short hairs at each joint, whereas in the male they have a feathery appearance due to tufts of long and numerous hairs at the joints. Mosquitoes are such a diversified group that it is not possible to write a short general description of the adults that would be generally applicable. Foote and Cook (1959) give excellent descriptions and figures of larvae and adults of mosquitoes of medical importance. The family may be divided into two subfamilies, the Anophelinae (including *Anopheles*) and the Culicinae (fig. 11.4) (including *Culex, Aedes* et al.).

Eggs are usually laid on or in water or in areas subject to flooding at some later time, and each species has its special requirements, which are usually very restricted. *Aedes* and *Psorophora* lay their eggs singly out of water, in places likely to be flooded later; *Anopheles* deposit them singly on the surface of the water; and *Culex, Culiseta, Mansonia,* and *Uranotaenia* lay them in small boat-shaped rafts on the surface of the water, or attach them to submerged objects.

Eggs laid on water hatch in a few days, but those deposited out of water or subjected to desiccation do not hatch until submerged, which may be weeks or months. Even then only a few of them normally hatch at the first flooding; the others remain for subsequent inundations.

Larvae, which are always aquatic, are well known as **wrigglers.** As they feed and grow, they pass through four instars, shedding their skins between each. Since the hairs and other structures that are used in identifying the larvae usually differ between instars, descriptions of mosquito larvae and keys for identifying them usually refer only to the full-grown or fourth-stage larvae. Anophelinae larvae are readily recognized in life by the fact that they have no elongate air tube and lie just under and parallel to the water surface, whereas in Culicinae larvae the air tube is well developed and the body hangs downward at an angle of about 45 degrees, touching the surface film only with the end of the air tube. Species of *Mansonia* absorb air from the air-carrying tissues in the roots of certain aquatic plants, piercing them with the tip of the breathing tube.

After five to 14 days or longer, the fourth-instar larva molts and transforms into a pupa or **tumbler.** The pupal stage is quite short, usually two or three days. Within the pupa, a marked transformation is occurring; new structures are forming that will adapt the mosquito to terrestrial life. Eventually the **adult** within emerges through a slit on the back of the thorax, spreads and dries its wings while clinging to the old pupal case on the surface of the water, and then flies off to mate and feed.

Some mosquitoes, particularly those that breed in relatively permanent water, have several generations each year, the number depending upon water temperatures and the length of the summer season. Others, which breed in temporary rain pools, appear each time their breeding places are flooded. Still others, such as the early spring *Aedes,* produce but one brood each year.

Some species bite at all hours of the day and night; others, such as the Anophelinae, are principally night biters; still others, such as *Aedes aegypti,* usually bite in the daytime. Only females suck blood; males feed on plant nectar or other sweet or fermenting substances. Although females of most species feed on warm-blooded animals, others select cold-blooded animals, such as frogs and snakes. The life span of adult mosquitoes is not well known, but for most species in the southern United States it apparently is only a few weeks during the summer months. In the northern latitudes of this country, the females of *Culex* and *Anopheles* usually hibernate and may live for six months or more.

Mosquitoes probably have had a greater influence on human health throughout the world than any other insect. This is not due wholly to the important human diseases and infections they transmit, but also to the severe annoyance they cause humans and other animals. They are vectors of the causative organisms of **malaria, yellow fever, dengue,** and **filariasis,** four of the most important diseases of the tropical and subtropical regions of the world. *Anopheles, Aedes,* and *Culex* are genera of particular medical importance. *Plasmodium,* the cause of human malaria, is transmitted only by species of *Anopheles,* and the virus causing dengue only by species of *Aedes,* primarily *A. aegypti* and *A. albopictus. Aedes aegypti* also transmits yellow fever virus from person to person under urban conditions, but in tropical rain forests, species of other genera are also involved in transmission of infection from animal to animal, and incidentally to humans. Filarial worms, *Wuchereria bancrofti* and *Brugia malayi,* are transmitted by species of several genera. Today these diseases have been reduced to minor or historical importance in the United States. But epidemics of three types of human **encephalitides** continue to occur in many parts of this country and are the most important mosquito-borne disease in the United States. Species of several genera, among them *Culex tarsalis, C. pipiens,* and *Culiseta melanura,* serve as vectors.

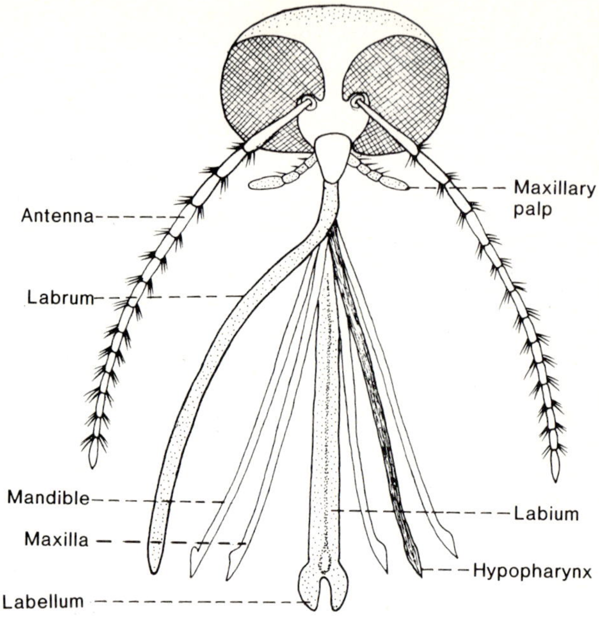

Fig. 11.4. Head of female culicine mosquito, showing typical mouthparts. By comparison, the palpi of anophaline mosquitoes are about as long as the proboscis.

Suborder Brachycera

Antennae shorter than thorax, usually three-segmented, variously formed and usually held horizontally erect. The last segment may be ringed to form several smaller units. **Arista,** when present, terminal or nearly so. The palpi are one- or two-segmented. The only family of parasitological importance is the Tabanidae, which includes the horseflies (*Tabanus*) and the **deerflies** (*Chrysops*).

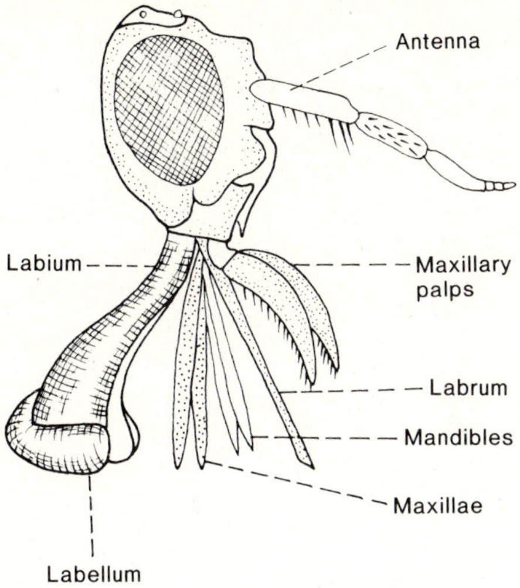

Fig. 11.5. Diagrammatic figure of head of *Chrysops,* showing mouthparts.

Family Tabanidae

The adults are sturdy flies with large heads and eyes. Females are bloodsuckers; their mouthparts are developed for cutting skin and sucking blood that oozes from the wound. Males feed on nectar. Typically, the eggs are deposited in masses on aquatic plants growing over water. Newly hatched larvae drop to the water, or make their way to damp earth, and feed on snails, insect larvae, and other aquatic animals. The larval stage lasts for several months or a year, and the pupal period ranges from one to three weeks.

Chrysops dimidiata (fig. 11.5) and *C. silicea* are important vectors of the filarial worm *Loa loa* in various parts of Africa, particularly in the Belgian Congo, and *C. discalis* is an important transmitter of the etiologic agent of **tularemia** in the western United States. Species of *Tabanus* and *Hybomitra* transmit the filarioid *Elaeophora schneideri* to sheep, deer, and elk in western and southwestern United States. It causes blindness and death in elk and **sorehead** in sheep. Several species of *Tabanus* are known to transmit trypanosomes, causing **surra** in horses, cattle, camels, and dogs.

Suborder Cyclorrhapha

Antennae are very short, generally three-segmented; the first two segments are short, the third is much larger and with an **arista,** a bristle or featherlike structure, usually dorsal in position. Palpi are one-segmented. This suborder, by far the largest of the three suborders of Diptera, is variously divided by taxonomists; the diagnostic characters can be used with certainty only by specialists familiar with the group.

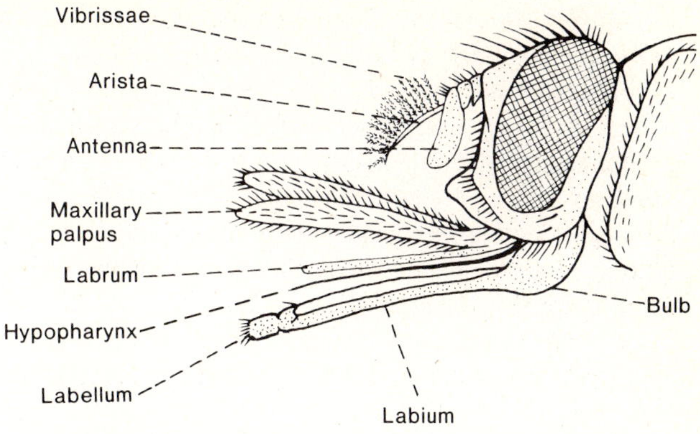

Fig. 11.6. Diagrammatic figure of head of a tsetse, *Glossina,* showing mouthparts.

Family Muscidae

The family is such a large and heterogeneous group that it is not possible to write a description that would be generally applicable to the species of parasitological importance. Because of the great diversity in form, for convenience the family is divided into groups based upon medical importance: (1) the "biters," the most famous of which are the **tsetses** (*Glossina* fig. 11.6); (2) the "nonbiters," of primary importance as agents in the occasional mechanical transmission of certain infectious agents; and (3) the myiasis-producing larvae.

Tsetses are the most important group of bloodsucking muscids directly affecting humans and animals. They are distinguished from other flies by the slender, forward-projecting proboscis; the spiny, long, slender palpi; and the branched dorsal hairs on the dorsal side of the antennal arista. The living fly can be recognized when at rest by the way in which it folds its wings scissorlike above the abdomen.

The female is ovoviviparous and produces one mature larva each ten to 12 days during her lifetime of about 40 days. These larvae pupate almost immediately, the puparia occurring in dry, loose soil in shaded, protected areas. Adult flies are diurnal feeders, and both sexes are bloodsuckers. Tsetses are confined to tropical and subtropical Africa, where they hinder settlement and development. They occur in "fly belts," the type of area infested varies with the species.

Among biting flies, tsetses rank second only to the mosquitoes as vectors of disease-producing agents of humans and animals. They are obligate vectors of the trypanosomes causing sleeping sicknesses. *Glossina palpalis* is the most important vector of *Trypanosoma brucei gambiense,* causing **Gambian sleeping sickness** in people, chiefly in West Africa. The most important vector of *T. brucei rhodesiense,* the organism of **Rhodesian sleeping sickness** in humans and **nagana** in animals (horses, mules, camels, and dogs), chiefly in East

Africa, is *G. morsitans*. Nagana, or African trypanosomiasis of animals, is a disease caused by *T. brucei brucei* and closely related species. In addition, *G. morsitans, G. longipalpis,* and *G. pallidipes* relate to nagana in nearly the same way that other *Glossina* species relate to African human sleeping sickness in that the flies are infective for a day or two after feeding, then become noninfective for about three weeks, when they again become infective and remain so for life.

Family Hippoboscidae

These, the **louse flies,** are degenerate Diptera which have adapted themselves to more or less continuous existence on mammals and birds. The body is broad and flattened dorsoventrally; the abdomen is indistinctly segmented and usually rather leathery in texture. Wings are present in some and absent in others. In winged species, the eyes are large and the antennae exposed, but in the wingless **ked** or **"sheep tick,"** *Melophagus ovinus* (fig. 11.7), the eyes are small and the reduced antennae lie in pits on the forehead. Hippoboscids give birth to mature larvae that are ready to pupate. Usually the larvae are deposited in dry soil or humus, in the nests of birds, or other suitable places, where they transform into black pupae in a few hours. In the case of the ked, however, the females attach their larvae to the wool of the host by means of a gluelike substance.

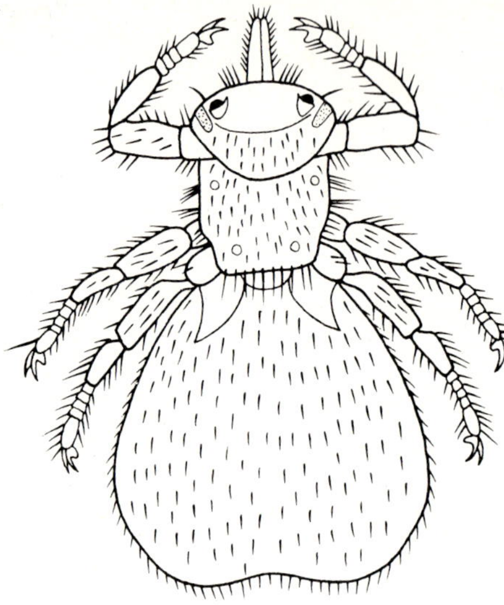

Fig. 11.7. Adult *Melophagus ovinus.*

Two species of great importance in North America are *M. ovinus* and the **pigeon fly,** *Pseudolynchia canariensis.* The ked is especially injurious to young lambs, to which it migrates in large numbers when the older sheep of the flock are sheared. It serves as a vector for *Trypanosoma melophagium,* a nonpathogenic hemoflagellate of sheep. The pigeon fly, which has a pair of transparent wings, in addition to taking blood, especially on squabs, when the feathers begin to form and afford protection, transmits *Haemoproteus columbae,* a blood protozoan of pigeons.

Notes and Sketches

Review Questions

1. What is the name of the remnants of the second pair of wings of flies?
2. Parasitic fly larvae cause a condition called ______ ?
3. Name the three suborders of Diptera.
4. What are the names of the two major genera of sand flies? What diseases do they transmit to humans?
5. To what family does *Culicoides* belong? Are these pests found in the United States?
6. What type of breeding ecotope is necessary for blackflies?
7. *Onchocerca volvulus* is a very important disease of humans. In what countries is it found, and what genus of vector does it need?
8. In terms of human suffering and death, what family of insects is most important?
9. Name all the life-cycle stages of mosquitoes.
10. What genus of flies transmits malaria to humans?
11. What are the two most common genera of horseflies and deerflies? Which transmits tularemia in the United States?
12. Where can larval horseflies be found?
13. What is an aristate antenna?
14. What is the distribution of tsetse flies today?
15. What genus of fly transmits African sleeping sickness?
16. "Louse flies," "keds," and "sheep ticks" refer to which family of insects?

References

Diptera

Adler, S., and Theodor, O. 1957. Transmission of disease agents by phlebotomine sandflies. In *Annual Review of Entomology,* eds. E. A. Steinhaus and R. F. Smith. Annual Reviews, Inc., Palo Alto, Cal., vol. 2, pp. 203–26.

Anthony, D. W. 1962. Tabanidae as disease vectors. In *Biological Transmission of Disease Agents,* ed. K. Maramorosch. Academic Press, New York, pp. 93–107.

Bequaert, J. 1942. A monograph of the Melophaginae, or ked-flies, of sheep, goats, deer, and antelopes (Diptera, Hippoboscidae). *Entomol. Americana* 22:1–64.

Fallis, A. M. 1964. Feeding and related behavior of female Simulidae (Diptera). *Exptl. Parasitol.* 15:439–70.

Foote, R. H., and Cook, D. R. 1959. Mosquitoes of Medical Importance. *U.S. Dept. Agr., Agr. Handbook* No. 152, 158 pp.

Foote, R. H., and Pratt, H. D. 1954. *The Culicoides* of the Eastern United States (Diptera, Heleidae), a Review. *Public Health Monogr.,* No. 18, 53 pp.

Glasgow, J. P. 1963. *The Distribution and Abundance of Tsetse.* Pergamon Press, New York, 241 pp.

———. 1967. Recent fundamental work on tsetse flies. In *Annual Review of Entomology,* eds. R. F. Smith and T. E. Mittler. Annual Reviews, Inc., Palo Alto, Cal., vol. 12, pp. 421–38.

Horsfall, W. R. 1955. *Mosquitoes: Their Bionomics and Relation to Disease.* Ronald Press Company, New York, 723 pp.

James, M. T. 1947. The Flies That Cause Myiasis in Man. *U.S. Dept. Agr. Misc. Publ.* 631, 175 pp.

Kettle, D. S. 1965. Biting ceratopogonids as vectors of human and animal diseases. *Acta Trop.* 22:356–62.

Sources of Study Material for Diptera*

Slides

Aedes aegypti: C, T, Tr, W
Aedes aegypti, head: C, W
Aedes aegypti, eggs: C, W
Aedes aegypti, larva: C, W
Aedes aegypti, pupa: C, W
Anopheles: C, T, Tr, W
Anopheles, head: C, T, W
Anopheles, eggs: C, W
Anopheles, larva: C, T, Tr, W
Anopheles, pupa: C, T, Tr, W
Culex: C, T, Tr, W
Culex, head: C, Tr, W
Culex, eggs: C, T, Tr, W
Culex, larva: C, T, Tr, W
Culex, pupa: C, T, Tr, W
Mosquito life cycle (all stages on one slide): C, T, W
Culicoides: C, W
Cochliomyia, larval spiracles: C, T
Chrysops: W
Chrysops, head: T
Gasterophilus intestinalis, eggs: C
Gasterophilus intestinalis, larval spiracles: C
Glossina: T, W
Glossina, head: T
Hypoderma, larval spiracles: C, T
Melophagus ovinus: C, Tr, W
Musca domestica, head: C, Tr, W
Musca domestica, eggs: C, W
Musca domestica, larva: C, W
Musca domestica, larval spiracles: C, T
Oestrus ovis, larval spiracles: C
Simulium: T, W
Simulium, head: T
Simulium, larva: T
Stomoxys, head: C, W
Tabanus, head: C, Tr, W

*The addresses for the sources are listed on p. 45.

Chapter 12 Order Hemiptera

Members of this order have a jointed fleshy beak attached anteriorly; when not in use it is flexed backwards beneath the head. Mouthparts are of the piercing-sucking type. Winged members of the order have each of the front wings modified into a thickened basal portion, and a membranous distal portion. The membranous portions often overlap. The second pair of wings is membranous and folded beneath the front wings. Metamorphosis is incomplete. Two families are of interest to parasitologists.

Family Reduviidae

Members of the family have a short, three-segmented beak which is attached to the tip of the head. Antennae are four-segmented and the anterior portion of the head is elongate and coneshaped.

Reduviids usually inhabit the burrows or nests of animals, but a few feed on humans and domestic animals. Eggs are glued into cracks and crevices in houses or in the nests of their hosts. They are deposited singly or in small batches, each female ultimately laying about 20. After an incubation period of eight days to a month, first instar nymphs emerge. A blood meal is necessary before each molt, and the time between instars is about 40 to 50 days. The life cycle requires a year or more.

Species of several genera are important vectors of *Trypanosoma cruzi,* the hemoflagellate causing Chagas' disease, which occurs throughout North, South and Central America, especially Brazil, Argentina, and Mexico. In addition to *Panstrongylus megistus* (fig. 12.1), the most important natural vectors are *Triatoma infestans* and *Rhodnius prolixus.* Adults of both sexes attack people.

Family Cimicidae

In this family, the wings are vestigial or lacking. The forewings are reduced to small pads, and the hind ones are absent. The body is broad, the prothorax large; its concave anterior border receives the head. Antennae are four-segmented; the beak and tarsi are three-segmented.

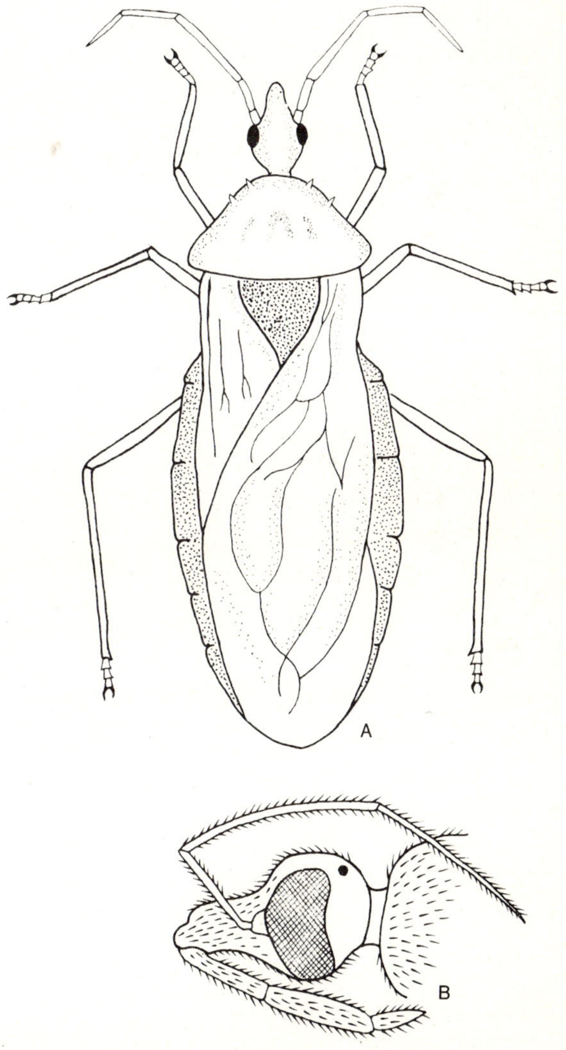

Fig. 12.1. *Panstrongylus megistus,* a reduviid bug. (A) Adult, dorsal view; (B) head, lateral view.

Eggs are laid at the rate of about two to eight each day, glued by secretion in cracks and crevices and under wallpaper. Each female may average 100 to 250 eggs during her lifetime. In warm weather, eggs usually hatch in about eight days; at lower temperatures they may require as much as 30 days. Nymphs, which must

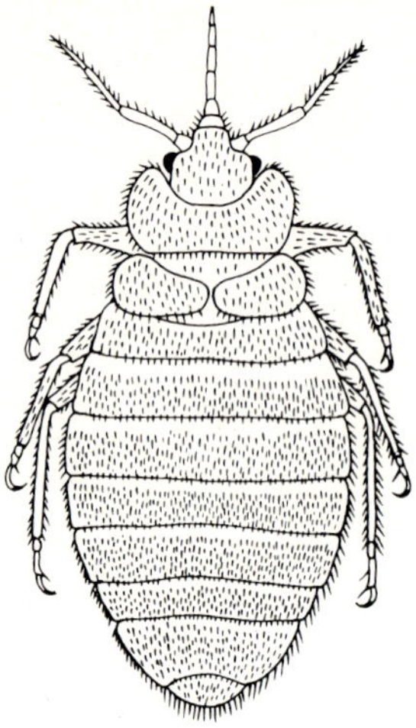

Fig. 12.2. Dorsal view of adult female bedbug, *Cimex lectularius.*

have a blood meal between instars, molt five times before becoming adults. The life cycle requires from seven to ten weeks.

Humans are attacked by the common bedbug *Cimex lectularius* (fig. 12.2), which may become an important pest in living quarters of all kinds. They are most frequently encountered in sleeping quarters, hiding during the day in cracks and crevices of woodwork and furniture and in debris, and emerging at night to suck blood from their victims. Despite the fact that *C. lectularius* experimentally can transmit microbial agents of certain diseases, there is no convincing evidence that it is a vector of any human or animal diseases in nature.

Notes and Sketches

Review Questions

1. What is peculiar about the wings of this order?
2. The mouthparts of Hemiptera are specialized to what function? Are most hemipterans parasitic?
3. Which subfamily is of most medical importance?
4. Explain the epidemiology of Chagas' disease.
5. Are any serious diseases known to be transmitted in nature by *Cimex*?
6. Goodnight, and don't let the ______ bite!

References

Hemiptera

Schmidt, G. D., and Roberts, L. S. 1985. *Foundations of Parasitology.* 3rd ed. The C. V. Mosby Co., St. Louis, pp. 646–54.

Usinger, R. L. 1966. Monograph of Cimicidae (Hemiptera-Heteroptera). Thomas Say Foundation (*Ent. Soc. Amer.*), vol. 7, 585 pp.

Usinger, R. L., Wygodzinsky, P., and Ryckman, R. E. 1966. The biosystematics of Triatominae. *Ann. Rev. Ent.* 11:309–30.

Sources of Study Materials for Hemiptera*

Slides

Cimex lectularius: C, T, Tr, W
Cimex lectularius, nymph: C
Rhodnius prolixus: W
Rhodnius prolixus, eggs: W
Rhodnius prolixus, nymph: W

Preserved

Cimex lectularius: C, W
Cimex lectularius, life cycle: C

*The addresses of the sources are listed on p. 45.

Chapter 13 Phylum Pentastomida

The Pentastomida, commonly referred to as **tongue-worms,** are of uncertain taxonomic position. Some workers consider them as a class of the phylum Arthropoda; others consider them closely related to the phylum Annelida. Still others recognize them as an independent phylum with the belief that they share a common ancestry with the annelids and arthropods (see Self, 1969).

Pentastomes are parasitic, both as larvae and as adults, in the respiratory organs usually of reptiles and carnivorous mammals, while only one species is found in birds. Adults are vermiform, with a short cephalothorax and an elongate, annulate abdomen. On the ventral surface, the cephalothorax bears two pairs of hooklike claws on each side of the mouth.

Life cycles are not well known. Usually an intermediate or paratenic host harbors the immature stage, but in a few species development may be completed in a single host. Late larval stages are referred to as **nymphs.** The pentastome egg contains a fully developed **larva** when oviposited by the female. Eggs are ingested by a vertebrate, and the larvae are liberated in the digestive tract. They bore through the intestinal wall and encapsulate in the viscera and associated mesenteries. In the capsules they undergo metamorphosis and attain the adult form, except that it is not sexually mature.

Family Porocephalidae

Porocephalus crotali

In the United States, adult *P. crotali* (fig. 13.1) occur commonly in the lungs of Crotalidae snakes, especially rattlesnakes and water moccasins; immature stages occur in rodents (muskrats, wood mice). Probably a wide range of wild rodents serve as intermediate hosts, because white mice, rats, and hamsters are satisfactory intermediaries in experimental infections.

Adult females measure up to 7 cm long (fig. 13.1A) with males somewhat smaller. A pair of hooks occurs on each side of the mouth (fig. 13.1B). The genital opening in the male is near the anterior end of the body and in the female near the posterior end.

Eggs containing fully developed four-legged larvae (fig. 13.1C) are deposited in the lungs, pass up the trachea, are swallowed, and voided in the feces. The egg proper consists of a thin, permeable, outer membrane and a thick, inner impermeable one. The space between the two membranes is filled with fluid.

Larvae bear a spearlike anterior penetration apparatus and a short, broad bifurcated tail. There are two pairs of short, stumpy legs that terminate in a pair of sharp, slender claws. Dorsally on the body surface are a pair of stigmata and a median dorsal organ. On the ventral side, between the anterior pair of legs, is a U-shaped, heavily chitinized mouth ring. Several large uninucleate cells, a large circumesophageal ganglion, and a blind gut appear internally.

Upon being swallowed by a rodent, the eggs hatch quickly in the duodenum, and the larvae are in the body cavity within an hour. They wander about in the coelom until the seventh to eighth days, when the first molt occurs. At this time, they become tightly encapsulated in the host tissues.

Once encapsulated, six molts occur, each nymphal stage being followed by a period of growth and differentiation. The sixth-stage nymph (fig. 13.1D) is infective to snakes.

Snakes become infected by ingesting infected rodents. When freed sixth-stage nymphs are fed to snakes, they penetrate the gut wall and are in the body cavity within 24 hours. Since the alimentary canal and lungs are juxtaposed, passage from one to the other is easy and quick. The nymphs attach to the lungs in a few days.

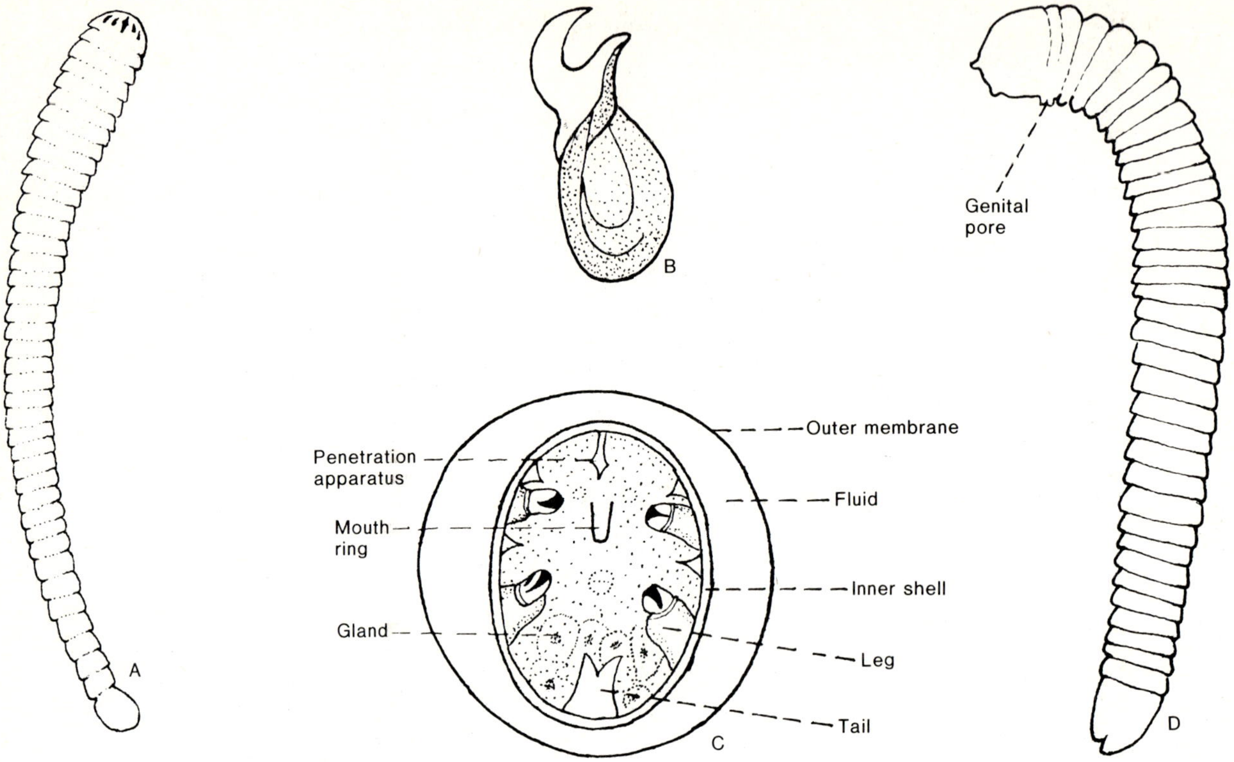

Fig. 13.1. *Porocephalus crotali.* (A) Adult female; (B) hook from anterior end of adult; (C) mature egg, with larva in ventral view; (D) sixth stage of infective male nymph from rodent.

Notes and Sketches

Review Questions

1. What are the most common definitive hosts of tongue worms?
2. Describe the stages in the life cycle of a typical pentastomid.
3. Mammals are the most common intermediate hosts of pentastomids. Can you name any other case of a phylogenetically higher class serving as intermediate host for a lower class?

References

Pentastomida

Esslinger, J. H. 1962. Development of *Porocephalus crotali* (Humboldt, 1808) (Pentastomida) in experimental intermediate hosts. *J. Parasitol.* 48:452–56.

Schmidt, G. D., and Roberts, L. S. 1985. *Foundation of Parasitology.* 3rd ed. The C. V. Mosby Co., St. Louis, pp. 561–68.

Self, J. T. 1969. Biological relationships of the Pentastomida; a bibliography on the Pentastomida. *Exptl. Parasitol.* 24:63–119.

Self, J. T. 1982. Pentastomida. In *Synopsis and Classification of Living Organisms,* ed. S. P. Parker. McGraw-Hill, New York, vol. 2, pp. 726–28.

Sources of Study Material for Pentastomida

At this writing, I do not know of any commercial source of tongueworms. They may be collected in rattlesnakes and water moccasin snakes, if available. But be careful!

SECTION 4 Laboratory Techniques

Chapter 14 Technique Procedures

A Standardization of the Microscope

Calibration of the microscope is required for the measurement of microscopic objects. In reality, however, it is calibration of the ocular scale, used in the microscope, combined with the different objectives. In calibration, two scales are necessary: the ocular micrometer, which is a glass disc bearing an arbitrary scale of usually 100 divisions, and a stage micrometer, which is a slide etched with a known scale of 2 mm usually subdivided into units of 0.1 mm and 0.01 mm. When working with the monocular microscope, both eyes should be kept open. With a little practice, one soon learns to ignore what the unused eye sees. This method avoids eyestrain.

Remove the regular microscope ocular and substitute an ocular equipped with an ocular micrometer disc. If an extra eyepiece with a disc is not available, unscrew the ocular mount and place the ocular micrometer with the engraved side down (figures should appear erect) over the eyepiece diaphragm. Replace the mount and reinsert the ocular in the microscope tube. With the aid of the plane mirror and the iris diaphragm, adjust the microscope so that optimum lighting is obtained, for only then will the scales be sharp and clear. Place the stage or slide micrometer on the microscope stage and adjust the two scales, with the zero points of both coinciding, so that they are parallel (fig. 14.1). Reading is easier if one scale is slightly to one side of the other rather than being superimposed. A similar point where lines coincide should now be found at the extreme right. The farther the matching lines are from the zero point, the less will be the error. Count the number of lines (hundredths of mms) on the stage micrometer (SM) and the number of lines on the ocular micrometer (OM) between the zero point and where the lines coincide exactly. Then divide the number of SM lines by the number of OM lines and you will have the value of one OM unit, with that specific objective. There are 1,000 micrometers (μm) in a mm.

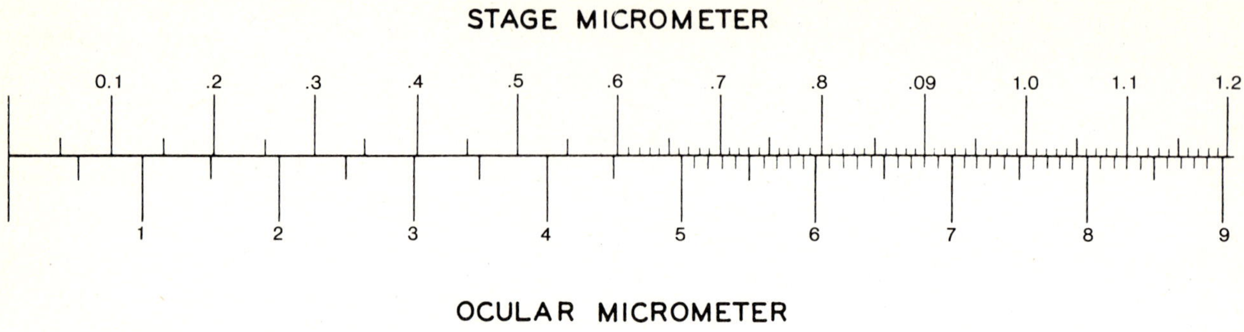

Fig. 14.1. Example of microscope calibration.

In the accompanying example, 83 divisions on the ocular micrometer scale are equal to 110 small divisions or 1.10 mm on the stage micrometer. By dividing the number of SM lines by the number of OM lines, the value of OM unit is obtained, e.g.,

$$\frac{\text{SM}}{\text{OM}} = \text{value for each OM division, or}$$

$$\frac{1.10}{83} = 0.0132 \text{ mm.}$$

Inasmuch as there are 1,000 μm in 1 mm, it follows that one ocular unit equals 13.2 μm (0.0132 $\times$ 1,000). Two ocular units = 26.4 μm, five ocular units = 66.0 μm, etc.

Following the above procedure, calibrate in order the ocular with the low, high dry, and oil immersion objectives. To lessen the error, make five different readings with each objective and take the average. It must be remembered that the calibration thus obtained is accurate only when the same microscope, the same ocular, and the same objectives are used (table 14.1).

Table 14.1 Microscope Calibration.

Microscope No.________ Ocular Micrometer No.________ Oil Imm. Obj. No.________

Objectives		No. OM spaces	No. SM spaces in mm	μm value of one OM space	Average
low 16 mm (10 X)	1				
	2				
	3				
	4				
	5				
high 4 mm (43 X)	1				
	2				
	3				
	4				
	5				
oil 1.8 mm (97 X)	1				
	2				
	3				
	4				
	5				

B Suggestions for Obtaining Fresh Material

Use of fresh material that may be available locally is too frequently neglected in favor of prepared slides and other fixed materials. Nothing will serve to stimulate the student and vitalize the course more than actual contact with living specimens. Fixed and stained material is essential to identify parasites, to understand their structure, and to supplement the study of living material. It seems unwise, however, to use only fixed and stained material to illustrate certain stages that the student will rarely, if ever, see in such condition thereafter. However valuable a stained smear of *Entamoeba histolytica* may be to illustrate the structure of the trophozoite, it is much less important than the living trophozoites of any parasitic amebae in the training of students. Actually the student, in future contacts, may be called upon to recognize the forms in feces, probably without the facilities or opportunity for making stained preparations. One should be mindful that a purely anatomical approach to parasitism is not only sadly stereotyped, often becoming tiresome to the student, but it fails to stimulate students to think in terms of basic parasitological principles and encourage further elucidation of the adaptive properties of parasites and host-parasite relations. Therefore, in conjunction with laboratory studies on the trematodes it is always interesting and advisable to allow students to observe eggs hatching, their miracidia entering snails, and, if possible, recovering sporocysts, rediae, and cercariae. As an alternative project, students may examine local or imported snails for infections (see p. 258).

What is true for *E. histolytica* and the larval stages of trematodes holds with varying degrees for other forms studied. It is emphasized, however, that stained preparations are required because few forms can be correctly identified in the living state. More individual effort is required on the part of the instructor to provide living material, but the reward is worth the extra effort. This is not to say that living materials should supplant prepared mounts; on the contrary, they are necessary to satisfy basic technical needs of students of medical technology, fisheries, and wildlife.

In view of the proven value of fresh material, several generally available hosts that normally harbor a rich parasitic fauna will be used as examples. Some of the parasites included here are listed again along with others under their respective taxa, together with their hosts.

Frogs (*Rana* spp.), generally available locally or from dealers, and routinely used in biology courses, are particularly useful since they harbor a variety of parasites. Following the instructions given below under Internal Examination (p. 243), remove the lungs and tease them open in saline solution. Two helminths are commonly found here. One, a trematode, is a species of *Haematoloechus* (fig. 2.44); the other is the nematode *Rhabdias ranae* (fig. 14.2AA). If removed to a slide in distilled water, both forms will discharge eggs in enormous numbers. In the case of *R. ranae,* the eggs will frequently be seen to hatch, releasing small, writhing larvae. Through manipulating the iris diaphragm to reduce the light of the compound mircoscope, one may readily study the internal anatomy of both larvae and adults. An examination of a blood smear may reveal the hemoflagellate *Trypanosoma rotatorium* (fig. 14.2A), although seldom present in great numbers. The blood from kidney and liver furnishes more specimens than the peripheral blood.

Another source of parasites is the intestine, which should be removed, slit open, and examined in saline solution. If *Rhabdias ranae* is in the lungs, its larvae and eggs will probably be present in the intestine, as will the eggs of flukes. The nematode eggs can be readily distinguished by the color, shape, and the presence of larvae. Moving swiftly about may be seen a small, ovoid protozoan with a sharply pointed end, an undulating membrane, and several flagella that are actively lashing about. This is probably *Tritrichomonas batrachorum* (fig. 14.2B). One or more large ciliates are commonly present: *Opalina obtigonoidea* flattened in cross section; *Cepedea* sp. (fig. 14.2F) circular in cross section; *Balantidium entozoon* (fig. 14.2D) ovoid, with terminal oral groove, smaller than *Opalina* and of interest

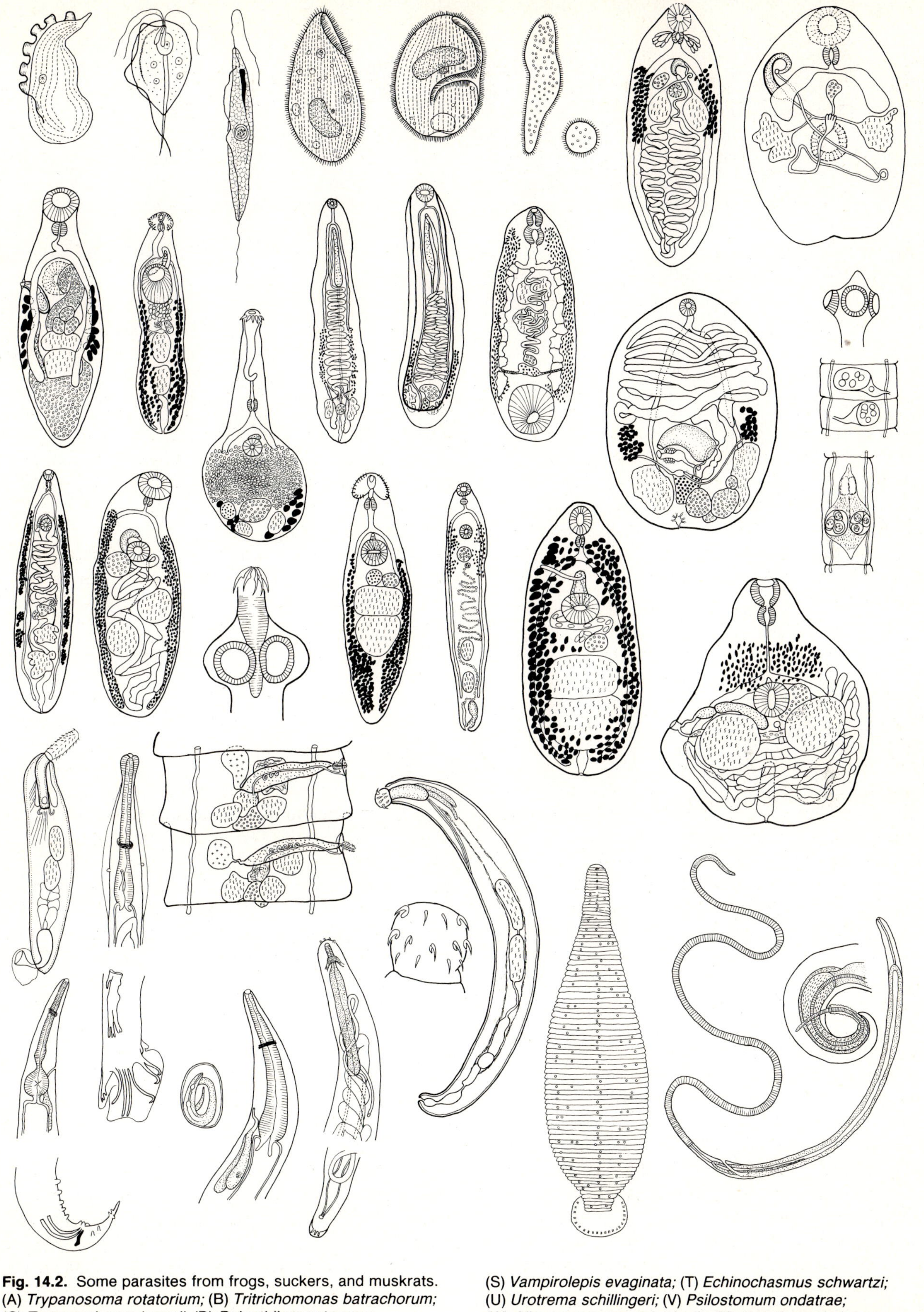

Fig. 14.2. Some parasites from frogs, suckers, and muskrats. (A) *Trypanosoma rotatorium;* (B) *Tritrichomonas batrachorum;* (C) *Trypanoplasma borreli;* (D) *Balantidium entozoon;* (E) *Nyctotherus cordiformis;* (F) *Cepedia* sp., entire specimen and cross section; (G) *Glypthelmins quieta;* (H) *Loxogenes arcanum;* (I) *Triganodistomum simeri;* (J) *Echinopharyphium contiguum;* (K) *Phagicola lageniformis;* (L) *Catatropis filamentis;* (M) *Paramonostomum echinum;* (N) *Wardius zibethicus;* (O) *Nudacotyle novica;* (P) *Cylindrotaenia americana;* (Q) *Opisthorchis tonkae;* (R) *Plagiorchis proximus;* (S) *Vampirolepis evaginata;* (T) *Echinochasmus schwartzi;* (U) *Urotrema schillingeri;* (V) *Psilostomum ondatrae;* (W) *Allasogonoporus marginalis;* (X) *Acanthocephalus ranae;* (Y) *Aplectana americana,* male; (Z) *Oswaldocruzia pipiens,* female; (AA) *Rhabdias ranae,* anterior end and embryonated egg; (BB) *Foleyella americana,* female; (CC) *Octospiniferoides macilentus,* male and proboscis; (DD) *Actinobdella inequiannulata;* (EE) *Trichuris opaca,* female and posterior end of male.

because of its relationship to human *B. coli;* and *Nyctotherus cordiformis* (fig. 14.2E), which resembles *Balantidium* but may be distinguished from it by the prominent oral groove that traverses the body obliquely.

Any nematode found in the subcutaneous connective and muscular tissues is probably *Foleyella americana* (fig. 14.2BB). The following distome trematodes occur in the indicated locations: *Glyphthelmins quieta* (fig. 14.2G), duodenum and adjacent intestine; *Loxogenes arcanum* (fig. 14.2H), bile duct and adjacent region; *Cephalogonimus amphiumae* (fig. 2.69) small intestine; a species of *Halipegus,* the eustachian tube and oral cavity; and *Megalodiscus temperatus* (fig. 2.43), an amphistome, the large intestine. If cysts occur in the skin, they are probably the metacercaria of *M. temperatus.* A tapeworm, *Cylindrotaenia americana* (fig. 14.2P) an acanthocephalan, *Acanthocephalus ranae* (fig. 14.2X); and the nematodes *Oswaldocruzia pipiens* (fig. 14.2Z) and *Aplectana americana* (fig. 14.2Y) also occur in the small intestine.

Two distomes belonging to the genera *Gorgodera* (fig. 2.62) and *Gorgoderina* (fig. 2.68) occur in the urinary bladder. These are characterized by a much-enlarged ventral sucker, with which they may adhere firmly to the bladder wall. A monogenetic fluke, *Polystoma nearcticum* (fig. 2.8), with six muscular suckers and a pair of large hooks on the opisthaptor, occurs in the urinary bladder of the tree frog (*Hyla versicolor*).

Another widely distributed, generally abundant, and fairly easily obtained host that yields a variety of parasitic species is the white sucker (*Catostomus commersoni*). Occurring on the gills is the polystome *Octomacrum lanceatum* (fig. 2.10) and the leech *Actinobdella inequiannulata* (fig. 14.2DD), and attached in the branchial chamber or on the body surface the fish louse *Argulus catostomi* (fig. 7.2). The nematode *Philometroides nodulosa* (fig. 5.32), occurs in the subcutaneous tissue of the cheek and fins. Examination of a scraping made from the outer skin surface may reveal the protozoan, *Myxosoma* sp. The hemoflagellate *Trypanoplasma borreli* (fig. 14.2C) may be present in a blood smear. An examination of the opened intestine is likely to yield one or more species of trematodes belonging to the genus *Triganodistomum* (fig. 14.2I), monozoic cestodes of the genus *Glaridacris,* and acanthocephalans *Pomphorhynchus bulbocolli* (fig. 4.4), *Octospinifera macilentus* (fig. 14.2CC), and a species of *Neoechinorhynchus* (fig. 4.2).

The **muskrat** (*Ondatra zibethicus*), which also has a wide distribution, is unexcelled as a mammalian host for class use. Fresh carcasses, usually available in season from trappers, are small, clean, and yield a wide variety of helminths very suitable for student study and preparation. The following species, among others, in the intestine unless otherwise indicated, from the various groups are commonly recovered: TREMATODES—(1) monostomes, *Catatropis filamentis* (fig. 14.2L), *Nudacotyle novica* (fig. 14.2O), *Paramonostomum echinum* (fig. 14.2M), *Quinqueserialis quinqueserialis*; (2) echinostomes, *Echinoparyphium contiguum* (fig. 14.2J), one or more species of *Echinostoma* (fig. 2.37), *Echinochasmus schwartzi* (fig. 14.2T), *Phagicola lageniformis* (fig. 14.2K); (3) distomes, *Plagiorchis proximus* (fig. 14.2R), *Urotrema schillingeri* (fig. 14.2U), *Psilostomum ondatrae* (fig. 14.2V), *Allasogonoporus marginalis* (fig. 14.2W), *Opisthorchis tonkae* (fig. 14.2Q) in the gallbladder and bile ducts, *Paragonimus kellicotti* (fig. 2.47) in cysts in the lungs; (4) schistosome (blood fluke), *Schistosomatium douthitti* in the mesenteric and portal veins; (5) amphistome, *Wardius zibethicus* (fig. 14.2N); CESTODES—*Vampirolepis evaginata* (fig. 14.2S) and strobilocercus of *Taenia taeniaeformis* (fig. 3.8), and NEMATODES—*Trichuris opaca* (fig. 14.2EE).

C Host Autopsy and Recovery of Parasites

Whether the animal is a piscine, amphibian, reptilian, avian, or mammalian host, the process of autopsy is essentially the same. Animals should be autopsied as soon after death as possible since some species of worms migrate upon death of the host, with the result that their normal location becomes uncertain, and all undergo maceration in the dead host. If autopsy must be delayed, animals should be kept in a refrigerator to retard changes until examined. The instructor will advise as to the disposal of the carcass following autopsy.

External Examination

When examining fish, the entire outer surface should be searched carefully, especially the oral region, the gills and opercula, and the fins. On the gills and often on the fins, one finds Monogenea, larvae (glochidia) of freshwater clams, leeches, and fish lice. On the outer surface, especially the fins, there may be encysted digenetic trematodes (metacercariae), as well as leeches and fish lice. Species of certain metacercariae give the entire surface a salt-and-pepper appearance, known as "black spot." Instructions for the recovery of the smaller, microscopic Monogenea are found on page 258. When examining birds and unskinned mammals, be on the lookout for mites, ticks, fleas, lice, and their eggs. Once collected, small mammals and birds should be promptly placed individually in a paper bag, since some arthropods, especially fleas, leave their host soon after death. The top of the bag is then folded over a few times and secured with paper clips. Upon returning to the laboratory the bag is opened, a small wad of chloroform- or ether-saturated cotton dropped in, and the bag closed again until the vapors have inactivated the parasites. The bag is then torn open and laid out flat, and the host is examined carefully by brushing anteriorly through the hair or feathers with a pair of forceps. Parasites are picked up, using a toothpick or camel's hair brush previously dipped in alcohol, and transferred to a vial containing preserving fluid. Nares and the oral cavity of waterfowl should be dissected open and examined for "duck" leeches and mites.

Internal Examination

Following external examination, make an incision through the midventral body wall, cutting around the anus and urogenital opening. When working with mammals, prior to making the abdominal incision, wet down the ventral surface with waste alcohol as an aid in preventing excessive contamination of the area with hair. Before removing the viscera, look for parasites that may be free in the body cavity, beneath the peritoneum, or in the mesenteries, and examine the liver surface for spots or nodules (cysts) that might be due to parasites. Encysted flukes can be distinguished by their thin hyaline appearance, through which the larva is often obvious. A cyst that is more or less opaque, nonhyaline, can be suspected of being that of a metacestode. These cysts resemble the host tissue, of which they are composed. Furthermore, metacestodes can be distinguished from those of trematodes that contain numerous glasslike granules, known as calcareous bodies. Cysts too small to open with dissecting needles should be pressed between slides and examined under the low power of the compound microscope to determine, if possible, the contents. Myxozoan cysts may be mistaken for those of flukes or tapeworms unless carefully observed. But the presence of polar capsules, which may be expressed by pressure on the glass cover, and minute pyriform spores, identifies them as Myxozoa. When examining fish, in order to facilitate separating the organs, it is suggested that one side be cut away. Now remove the internal organs, one at a time, and place them in separate dishes of saline solution. A temporary label giving relevant data should accompany each organ.

Once the urinary bladder and the segments of the alimentary canal are opened (by making a longitudinal slit with blunt-tipped scissors or scapel) and the other organs teased apart in saline solution, the parasites, by their own movements, usually free themselves from the mucus. If pyloric caeca are present, they should be opened. Tapeworm-infected caeca, due to their whitish, opaque appearance, can be distinguished from worm-free caeca. If the infected caeca are opened at the distal end, by cutting off the tips with scissors, the worms will usually protrude enough so that one can disengage them by placing a dissecting needle or forceps through the exposed loop and pulling gently. Remaining organs should likewise be opened and torn apart by teasing. Not only should every organ and each opened alimentary segment be examined under the dissecting microscope, but the mucosal wall of the intestine should be scraped to loosen small, embedded worms. For the examination of intestinal contents under the dissecting microscope, a dark background reveals the light-colored worms bettter than a light background. When a parasite is recovered, call it to the attention of the instructor.

Acanthocephala occasionally are found free in the intestinal lumen, but usually the proboscis is embedded in the intestinal wall. Unless removed with care, the proboscis will be broken off and left in the intestinal wall, or its hooks, which are important for identification, may be torn loose. A pair of fine-pointed dissecting needles may be inserted into the craterlike opening in the host tissue, and by careful manipulation of the needles the crater may be spread wide enough to free the proboscis of the worm. Or the intestine can be cut into small squares, each holding an embedded proboscis. After these have been preserved, the host tissue should be teased away before the worms are stained.

If pressed for time or when dealing with a large number of animals, the intestine, having first been opened as described above, should be placed together with its contents in saline solution and shaken for a minute or so in a tightly covered jar; then the intestine is removed and discarded. When the contents have settled, the upper, clearer portion of the liquid is poured off. This process should be repeated several times, with the decanted liquid being replaced each time with fresh solution. Parasites, which are heavier than the organic debris, are the first to settle to the bottom and are concentrated in the remaining liquid. By examining only a small portion of the concentrate at a time in a Syracuse watch glass or petri dish under a dissecting microscope in good light, any parasites remaining will be found. These should be transferred with a spatula, bulbed pipette, or tweezers (if caution is exercised so as not to injure the worms), to a Syracuse watch glass filled with saline solution.

D Preparation of Specimens for Study

Fixation

Killing, fixation, and hardening are commonly referred to as separate processes in histology and cytology. When applied to materials to be made into whole mounts, however, nothing is gained by considering them separately, since killing and hardening are usually features of fixation. In its broad sense, fixation consists of arresting the life processes, preserving and hardening the animal in as nearly as possible its natural condition. This is accomplished through the use of certain chemicals known accordingly as fixing reagents. Fixation is probably the most important step in the preparation of materials for microscope slides, for here the elements are set and cannot be changed.

Preparation: Worms should be thoroughly washed in the saline solution and cleansed of mucus prior to fixation. After the mucus has been removed, it often proves advantageous to leave digenetic trematodes and cestodes in distilled water for some minutes prior to killing. This serves to relax the specimens further and in some species of flukes causes them to void most of the eggs. Since an egg-filled uterus usually obscures much of the internal anatomy, voidance of eggs is highly desirable. Trematodes and cestodes give considerable trouble by contracting and thickening upon fixation unless precautions are taken. Some workers prefer to place flukes and tapeworms between two slides, after which the fixing solution is added. But this is objectionable since the pressure distorts the shape of the animal and disturbs the normal position of the internal organs.

Killing and fixing (Table 14.2): Fairly satisfactory results can be obtained if the mucus-free trematodes are dropped into 10% formalin, A-F-A, or Bouin's fluid that is heated until it just starts to bubble (but not boil). Trematodes expand and die instantly upon striking the hot fixing fluid. Cercariae are killed and fixed in a similar manner.

Tapeworms should be plunged into a dish of water heated to about 70 C and agitated with a camel's hair brush or a dissecting needle, or suspended over a hooked needle and dipped repeatedly and rapidly. Since the water is cooled five or more degrees when poured into the dish from the beaker, the original temperature must be adjusted accordingly. In the case of specimens with a thick body wall, e.g., *Ligula, Schistocephalus, Taenia* etc., the water should be heated ten or more degrees higher. Whole worms must be plunged into it quickly so that they come into contact with water simultaneously, thus killing them in a uniformly extended condition. Worms are then promptly transferred to the fixing solution. For plerocercoids and other delicate material, the water temperature should be ten or more degrees lower, to avoid overheating and excessive distention. Using a bulbed pipette, transfer such specimens to the hot water, agitate gently, and transfer promptly to the fixing solution. Whether dealing with larvae or adults, the proper temperature of the water for killing is determined by trial.

Living cestodes can be properly relaxed by placing them in tap water in a refrigerator for a few hours. However, delicate forms may break apart and/or lose their hooks if left in water too long. Only experience can teach the worker how long to keep them immersed.

Living Acanthocephala should be placed in water in a petri dish or Syracuse watch glass. For a time, specimens retain the ability to invert the proboscis, but ultimately the worms take up water and distend so that the proboscis remains fully extended even when stimulated. After the worms have become immobile, the excess water in the dish is poured or pipetted off, leaving only a thin film covering them. The fixing reagent, A-F-A, 10% formalin, or 70% alcohol to which a few drops of glacial acetic acid have been added, is heated until it just begins to bubble, and then poured over the worms. If the specimens are small enough to enter a bulbed pipette, they may be pipetted directly into the steaming fixing reagent.

Large nematodes should be dropped into steaming acetic alcohol (one part glacial acetic acid and three parts 95% alcohol), which results in a little less shrinkage than alcohol alone. The heated solution causes worms to straighten instantly and die in that position, thus avoiding the curled and distorted specimens obtained when using cold fixatives.

Table 14.2. Flow Sheet for Preparing Whole Mounts.

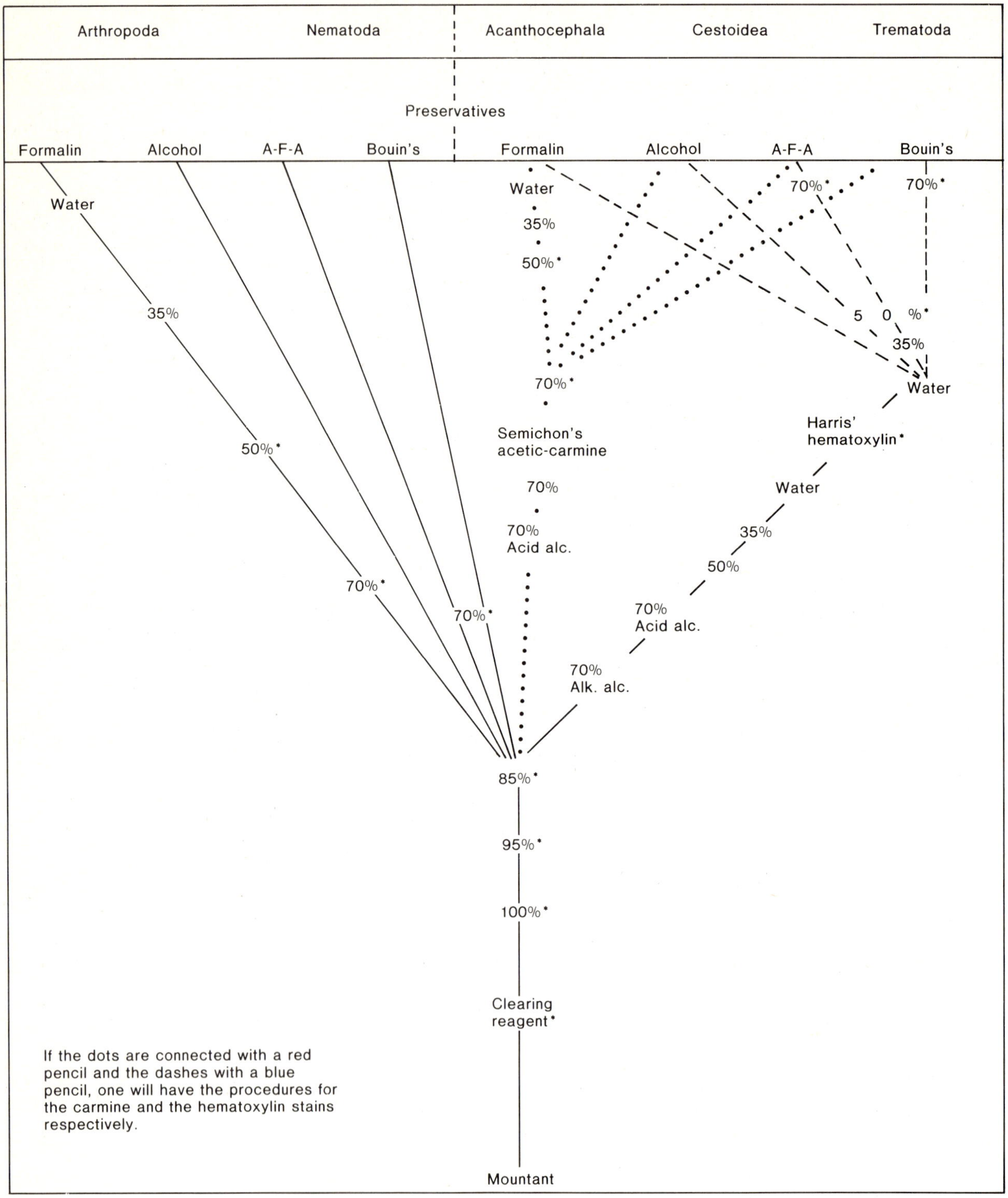

*Solutions in which specimens may be left overnight if necessary.

Small nematodes and larval stages may be studied advantageously while still alive by placing them in a drop of water under a cover glass for examination under the microscope. In applying the cover glass, allow one edge of it to first touch the slide, while it is held at an angle of about 30° to the surface of the slide. Then gently lower the other side of the cover into place. This will to a large extent eliminate the inclusion of air bubbles. Water mounts sealed with Vaseline will last for days. When permanent mounts are desired, the following procedure for fixation produces lifelike specimens. The nematodes are killed by dropping them into 1 to 2 ml of hot (100 C) 0.5% acetic acid in cavity slide and then fixed by transferring to a mixture of 4% formalin and 1 or 0.5% acetic acid for 24 hours. Specimens may be stored in the fixative and studied subsequently as water mounts or processed for permanent mounts in glycerine or semipermanent ones in lactophenol.

It is important that leeches be properly prepared or they are practically worthless for study. They should be moderately extended, reasonably straight, undistorted, and neither macerated nor overhardened. This is accomplished by stupefying them before fixation by the careful use of a good narcotizing agent, such as nembutal, tricaine methanesulphonate, or one of the other recommended reagents. Living specimens are washed, placed in a small amount of water and a little of the anesthetic added. If the solution is too strong they will die quickly, in a contracted state. When completely relaxed they are transferred to a slide and straightened, another slide added, with supports to prevent flattening, and fixative (85% alcohol or 5% formalin) slowly added.

For killing arthropods, Boardman's solution is advised. This has the advantage of relaxing the animals so they die with appendages outstretched, a most important feature to the taxonomist. The solution was originally devised for ticks but works equally well for mites, fleas, and lice. Boardman's is a relaxant only, so the material must be transferred to a preservative. Fly and mosquito larvae are best killed by dropping them for a few seconds, depending upon size, into water heated almost to boiling. Transfer to insect preservative or 70% alcohol plus glycerine for storage.

Preservation and Storage

Smaller worms are usually properly fixed in approximately an hour, but exposure to the fixing reagent for several hours or even days will not impair the value of the specimens for study and insures complete fixation of the larger specimens.

Specimens fixed in Bouin's solution should be washed in several changes of 70% alcohol until the yellow color is lost. This process may be hastened by the addition of a few drops of ammonium hydroxide to the alcohol. Material fixed in 10% formalin should be removed to 5% formalin. For specimens fixed in 70% alcohol plus glacial acetic acid, the solution should be replaced with 70% alcohol. The A-F-A solution should likewise be replaced by 70% alcohol. In each case, worms should be left in the killing solution for some hours (3 to 24) before the solution is replaced by 70% alcohol or 5% formalin (only in case the material has been killed in 10% formalin is it preserved in 5% formalin).

If the material is to be stored for some time (months or longer) before being used, it is recommended that 70% alcohol to which 5% glycerine has been added as a protection against loss from evaporation be used as a preservative instead of straight 70% alcohol. Enough glycerine to make a 5% solution may likewise be added to the formalin preservative.

Important: each vial of material should contain a note bearing: (1) scientific name of host, (2) locality where host was collected, (3) location of parasite within the host, (4) fixing reagent, (5) date (of autopsy), and (6) your name. Specimens without complete accompanying data are relatively worthless.

Staining

The following instructions are based upon methods that have proved satisfactory for laboratory use by experienced workers and students commencing research in parasitology. They are presented as guides in various practical procedures of microtechniques.

When treating specimens with alcohol or other reagents, always use at least four times their bulk of the reagent. Transfer of the specimens from one liquid to another is made by pouring off the first liquid, after which the second is added immediately. If the specimens are very small or delicate, rather than pouring off the liquid, carefully withdraw it with bulbed pipette under a dissecting microscope. Afterwards the second liquid is promptly added. In either case the liquid, whether poured or pipetted off, should be placed temporarily in a preparation or a petri dish. If specimens have inadvertently escaped with the liquid, recovery of them can be made from the dish, which is impossible if the solution with the specimen has been discarded in the sink. While large specimens may be transferred with a spatula from one container to another, it is difficult or impossible with most materials and always involves the risk of damage to the specimen.

Preparation dishes with ground covers are the most satisfactory containers to use in the treatment of specimens. Although less desirable, screw-cap vials of appropriate size may be used in handling small specimens. Containers should always be tightly covered except when changing solutions or when observing specimens. In order that the source of each lot of material may be known, a small pencil-numbered slip of paper should accompany the specimens while in preparation and the same number should be placed in the vial with any remaining material.

Since most structures are relatively transparent in their natural state, their visibility is usually increased by staining specimens with dyes or other color-bearing chemicals. Inasmuch as the various structures absorb certain dyes to different degrees, it is possible to stain some of them conspicuously while others remain less colored, thus giving them various shades of one color; or even to stain certain organs in contrasting colors by combining different colored dyes which have an affinity for different structures.

In the choice of a stain, consideration should be given to its ability to differentiate clearly the anatomy of the specimen. It should (1) be very dependable and easy to use, (2) follow the common fixing agents well, (3) retain optimum staining properties for many years, and (4) not form unsightly precipitates in the material. While a large number of carmine and hematoxylin stains, two of the most widely used classes, have been described, each with its advocates and adversaries, the following few meet practically all needs for whole mounts of helminths. Carmine stains should be used for alcohol-fixed specimens, and hematoxylin for formalin-fixed material. Nematodes and arthropods do not require staining. Important structures show well in nematodes mounted in lactophenol or glycerine, and even in water mounts of small living forms. Specimens well stained with carmine are especially useful, however. Overstaining specimens and then partially decolorizing them is known as regressive staining, in contradistinction to progressive staining, in which the stain once taken by the specimens is not removed. In progressive staining, differentiation is accomplished through the selective affinity of the dye for different organs.

All trace of the fixing reagent should be removed from specimens prior to staining. Wash out formalin in distilled water; A-F-A and Bouin's in 70% alcohol. As the alcohol becomes discolored by the picric acid in Bouin's, it should be replaced with fresh solution. The decoloring process can be hastened by using alkaline 70% alcohol rather than neutral 70% alcohol.

Semichon's Acetic-carmine. Transfer specimens from 70% alcohol to stain (one part stock stain diluted with approximately two parts of 70% alcohol). Stain for 15 to 30 minutes, depending upon the size of the specimens. Destain in 70% acid alcohol, until the cortical layer is nearly free of stain and reproductive organs are a very light pink.

Grenacher's Alcoholic Borax-carmine. Place specimens from 70% alcohol into stain and leave for 24 hours or longer, depending upon the size and permeability of the specimens. After being thoroughly stained they are transferred to the destaining solution and treated as described above.

Lynch's Precipitated Method, Using Grenacher's Alcoholic Borax-carmine. This method gives a much more selective and brilliant stain than that obtained by Grenacher's original method, when the material is treated as follows:

1. Transfer material from 70% alcohol into undiluted Grenacher's borax-carmine and stain for approximately 12 hours.

2. Cautiously add concentrated HCl drop by drop to the dish containing the material and the stain, meanwhile gently agitating the dish, until all the carmine is precipitated as brick red flocculent particles. Generally a drop of HCl is required for approximately each 5 ml of stain. (The volume of a preparation dish with a ground glass cover is about 9 ml, a Syracuse watch glass approximately 20 ml.) Allow the worms to remain in the acidulated solution for six to eight hours, or preferably overnight.

3. Next place the material in 70% acid alcohol and destain until a light pink color has been obtained. If so much of the precipitated carmine remains in suspension that the material cannot be observed satisfactorily under the binocular dissecting microscope, draw off the liquid with a bulbed pipette and replace with fresh acid alcohol. Repeat the process until practically all of the precipitated carmine has been removed. The destaining process usually requires several hours, depending, of course, upon the size of the specimen.

4. When the material is properly destained, draw off the acid alcohol and replace with neutral 85% alcohol. Change the alcohol two or three times, but the total time in 85% should not be less than an hour.

Counterstaining with Fast Green. Fast Green, used progressively as a counterstain for the preparation of whole mounts of specimens previously stained with Semichon's acetic-carmine or Grenacher's alcoholic borax-carmine each used regressively, is especially recommended to bring out the ventral glands of monostomes and spines and/or hooks of appropriate groups.

After the material has been stained, destained, and dehydrated through 95 % alcohol, add a few drops (one or two) of Fast Green stock solution to the dish filled with 95% alcohol. Observe under the dissecting microscope, removing the material immediately to absolute alcohol when properly stained. This step requires constant observation and in all probability will not require more than a minute in the Fast Green solution. Extreme caution should be exercised since Fast Green is not only color fast but fast acting!

Should the specimen become overstained it may be remedied by destaining with a weak 95% alcohol-alkaline solution, which, however, tends to detract from the sharp, bright green contrast otherwise obtained. After counterstaining, dehydrate further in absolute alcohol, clear, and mount in the usual manner.

Harris' Hematoxylin. Although the general procedure is the same as when using the carmines, the hematoxylins differ in one important respect: instead of putting the specimens into the stain from 70% alcohol, as with carmines, they are put into the hematoxylins from water.

Hydrate from 70% alcohol by passing through 50% and 35% alcohols, and leave in distilled water (30 minutes or longer in each), stain in Harris' (stock diluted approximately 1:15 with distilled water) 12 to 24 hours. Rinse in distilled water, to remove the free coloring, and upgrade through 35% to 70% alcohol (30 minutes to an hour in each), preparatory to destaining, as with specimens stained in carmine.

Ehrlich's Hematoxylin. Follow the same procedure outlined for Harris' hematoxylin.

Combination Hematoxylin with a Mordant. The combination hematoxylin is prepared by adding 1 ml each of stock Delafield's hematoxylin and stock Ehrlich's hematoxylin to 40 ml of potassium alum solution. This staining solution should be made as used, following the same procedure outlined for Harris' hematoxylin.

Destaining and Neutralization

Unless otherwise indicated above, destaining is accomplished through the use of acidulated 70% alcohol. While acid alcohol is available for class use, it can be made individually by the addition of a drop or two of concentrated HCl in a preparation dish filled with 70% alcohol.

Since the carmine stains are alcoholic, specimens after being rinsed in 70% can be transferred to the acid alcohol from the stain. It is important that specimens be rinsed only momentarily in 70% alcohol after removal from the stain and before placing them into the destaining solution. Allowing specimens to remain in 70% alcohol for any length of time seems to set the stain, making destaining prolonged and difficult. When hematoxylin has been used, specimens should be rinsed in distilled water and then passed through 35% to 50% alcohol (about 30 minutes in each) before being transferred to the acidulated 70% alcohol. The acid alcohol should be changed occasionally and allowed to act until the worms become a light pink against a white background. The cortical layer should be free of stain, but enough stain should remain to color the internal organs. Since no two worms behave exactly alike in this respect, the process should be observed carefully in good light under a dissecting microscope. Until the point of proper destaining is learned by experience, it is well to have the instructor check your work at this stage. Too much destaining results in a dull colored specimen, minus the brilliance otherwise obtained. When the proper color has been obtained, specimens stained in hematoxylin should be passed through 70% alkaline alcohol, to restore the bluish color and stop the destaining action, before proceeding with dehydration. *Caution:* If too much neutralizing reagent is used, the worms will turn brownish and satisfactory restaining is impossible.

Arranging for Mounting

If specimens are twisted or curved, it is at this point in the process that they must be straightened. This is done by placing them between microscope slides, which are held in position by pressure. A strip of cardboard, of sufficient thickness to prevent too much compression may be inserted at each end of the slide. Keep the specimen on the slide flooded with alcohol, preferably in a petri dish. When the specimen has been oriented and the props are in place, another slide is put flat atop the worm and a clamp(s) or several loops of cotton thread are wound snugly around the slides and the ends tied, while applying some pressure against the faces of the slides with thumb and forefinger. The specimen is then dehydrated through the alcohols in this position in a petri dish or Coplin jar. After an hour or so in 95% alcohol, the worm will be stiffened in the flattened form, after which the slides should be removed and the specimen returned to a preparation dish filled with 95% alcohol. Two changes in absolute alcohol are necessary before clearing. Since only a portion of the specimen, when between the slides, is exposed to the alcohols used in the dehydration, the times normally allowed in each solution should be approximately doubled.

Engorged arthropods, e.g., fleas, sucking lice, ticks, are practically worthless for study unless treated with potassium hydroxide to remove the soft parts so that the chitinized structure, used in making identifications, may be observed more clearly. Such hard parts as the spiracles of fly maggots, mouthparts, etc., should be removed and left in KOH until the chitin is somewhat bleached and fleshy parts destroyed. Instructions for clearing with the alkali solution are on page 262.

Dehydration

This consists of replacing the water in the specimen with an anhydrous solution, such as absolute alcohol. Absolute, often called "100%," alcohol, contains about 1.0–2.0% water. The usual steps of 35%, 50%, 70%, 85%, 95%, and absolute alcohol are generally used in dehydration, allowing about an hour in each grade. Using these as a base, steps may be added or deleted as necessary to dehydrate best the material at hand. Replacing the alcohol with water is known as hydration.

The length of time required in the alcohols for hydration and dehydration depends on the size of the specimen and the permeability of its integument. While approximate times are given, no arbitrary rule can be given on this point; each group of specimens should be judged with respect to its special needs. In general, however, Nematoda and Acanthocephala should be handled very slowly, since their integument, no matter how thin, is easily shrunken by rapid dehydration. To avoid damage while dehydrating nematodes and thorny-headed worms, the body wall should be punctured with a small insect needle at several points along its length. It is essential, however, that material be left in absolute alcohol until every trace of water has been removed, or otherwise clearing will not be possible.

Although it is desirable to have permanent mounts of nematodes for class study, for taxonomy cleared unmounted and/or temporary mounted specimens are often more satisfactory. In the case of the temporary mounts of smaller worms, they may be rolled about under the cover glass, enabling one to view them from all sides. Instructions for making temporary and permanent mounts are on page 250. Leeches and copepods are usually made into permanent mounts for class study, but they should not be mounted for taxonomy.

Clearing

All traces of water having been removed from the specimens, it is next transferred to a clearing reagent, which renders it transparent and miscible with a resinous mountant. Despite the advantages of terpineol (lilacin), benzene, toluene, wintergreen (methyl salicylate), and beechwood creosote for clearing specimens, xylene is the most widely used for this purpose.

The transfer from absolute alcohol to the clearing reagent should be made gradually in order to avoid the formation of violent diffusion currents that tend to distort specimens. First add enough xylene (or other indicated clearing reagent) to have a mixture of approximately ¼ xylene and ¾ alcohol; next a 50–50 mixture; then ¾ xylene and ¼ absolute; and finally pure xylene; after which a second change in xylene is made (about an hour in each).

For Acanthocephala, nematodes, and other delicate material that tend to collapse when subjected to sudden change of solutions, additional caution should be taken when going from absolute alcohol to the clearing reagent. This is effected first by puncturing the body wall, as described above, and second by placing the clearing medium under the alcohol. To a preparation dish about half-filled with absolute alcohol and containing the specimens, a sufficient quantity of clearing reagent to fill the dish is introduced at the bottom of the alcohol with a bulbed pipette. Or, first pour in enough clearing reagent to half fill the dish, after which enough absolute alcohol to fill the dish is carefully added with a bulbed pipette. The specimens to be cleared, being now carefully put into the alcohol, float at the interface of the two fluids; the exchange of the fluids occurs gradually, and the objects slowly sink into the lower layer. When they have sunk to the bottom, the alcohol is pipetted off and the clearing reagent replaced with fresh clearing reagent prior to mounting. Further premounting suggestions to avoid collapsing and subsequent "opacity" are on page 261; they are particularly appropriate for Acanthocephala and nematodes.

Mounting

For permanent mounts, Histoclad, Kleermount, gum damar, or other preferred resinous medium soluble in benzene, toluene, or xylene is satisfactory as a mountant. Canada Balsam is far superior to any of these, however. The use of Permount is strongly discouraged as it crystallizes over time. Small nematodes are commonly mounted in pure glycerine, after which the cover glass is sealed with Thorne's Zut,* or even clear fingernail polish.

Care must be exercised to place the specimen properly on the slide; to select a cover glass of proper size; to add the proper amount of mountant; to prevent the cover glass from tilting to one side; and to prevent the inclusion of air bubbles in the mount. While it is better to use too little rather than too much mountant (in the latter case the specimen will become displaced when the clamp is applied to the cover glass), only by trial and error can one determine the right amount to use.

If the specimen is relatively small, it is best to place it in the center of the slide. Make a guide for this purpose by placing a slide on a piece of white paper, marking around it with a pencil, and then ruling diagonal lines connecting the corners. For mounting, lay the slide on this guide and place the specimen over the intersection of the diagonals.

Arthropods (except fleas) and medium-size worms (those large enough to orient conveniently and not large enough to require a rectangular cover glass) should be oriented perpendicularly to the long axis of the slide. Specimens flattened dorsoventrally should be mounted so that the ventral surface is up, next to the cover glass. Fleas should be mounted on the right side, with the legs pointed away from the technician, so that they will appear in the normal position and facing left when examined under the compound microscope.

*This cover glass ringing cement may be purchased from Bennett Paint and Glass, 65 West First South, Salt Lake City, Utah 84101 or made using Thorne's (1935) original formula.

A large mount, such as several lengths of a tapeworm (scolex + neck, mature and gravid proglottids), or a *Fasciola hepatica* should be placed lengthwise, to one end of the slide, to leave room for the label on the other end. The mount should be placed somewhat away from the end so that approximately 5 mm of the slide will be free of the cover glass.

Differences in the size of various specimens make it impossible to use the same size of cover glass always; the technician's aim should be to use the smallest cover glass that will adequately cover the specimen. Not only does this result in a more attractive mount, but it also means a saving in mountant and cover glasses, both being sold by weight.

When mounting delicate specimens such as larval helminths, pressure should not be applied to the cover glass unless its corners are supported by bits of a glass slide or, depending on the thickness of the specimen, slivers of cover glass. Pieces of fine glass rods of appropriate length and diameter can be made especially for this purpose by heating rods of glass over a Bunsen burner and pulling them to the right diameter.

Use sufficient mountant to spread somewhat beyond the edge of the cover glass. Knowledge of the correct amount of mounting medium for specimens of different volume and cover glasses of different sizes is gained with experience. Air bubbles trapped beneath the cover glass are no cause for serious concern since they will normally disappear after the mount is placed in the drying oven. Any mounting medium naturally shrinks in drying, and if one uses only enough to reach to the edge of the cover, it will eventually shrink back and leave the edge of the cover exposed, or will withdraw unevenly and suck in bubbles of air. If the medium retracts from the edges of the cover while in the drying oven, it should be replaced with new mounting medium so that the entire area between the slide and the cover glass is filled.

As each slide is prepared, the same number that accompanied the specimen in the preparation process is written on it with a wax pencil or an etching pencil before placing it in a drying oven. For the label to stick better, the wax-number is removed before adding the label with the required data. Following mounting, place the slide in the drying oven, perhaps with a weight on the cover glass. Various sized bullets (*not* cartridges!) serve very well. After the mountant has hardened, which will take a week or more, remove the excess medium from the surface of the slide and cover glass by scraping with a single-edged safety razor blade. Next, immerse the slide in a Coplin jar filled with 95% alcohol to remove the powdered mountant, then remove it from the alcohol and touch one end to a paper towel to drain off the excess alcohol, and finally polish it clean with a soft, lint-free cloth. After the mount is cleaned, it is ready for labeling and study. Whole mounts do not develop their maximum transparency until several months after preparation because of the time required for the mountant to diffuse evenly through the specimen and to harden to an increased refractive index with greater clearing power.

Since the image is inverted under the compound microscope, the anterior end of the animal should be directed toward the technician when adding the label. Place the label on the end of the slide indicated by the instructor and add (printed in India ink) the scientific name of the specimen and host, locality, location, date (of mounting), the corresponding number on the "Mount History" card (fig. 14.3), and your name or initials. *Caution:* Always be sure to place the label on the same side of the slide as the cover glass!

It is important, as in all scientific work, that utmost cleanliness be observed throughout the process. Exercise caution in using the right reagent, avoid contamination, and be sure that slide and cover glass are clean. Much chagrin and fruitless explanation can be avoided if these instructions are rigorously followed. Your first attempt may not be cause for much pride, but after a few trials you will succeed and have a mount that will pass inspection.

Storing Slides

Once slides have been placed in a slide box, it should always be placed on end, so that the slides are in a horizontal position, to prevent the drifting of specimens to the margin of the cover glass, thus becoming relatively useless. Any mountant, even when apparently so hard that it will chip, is really a solid fluid (much like molasses in January). If such slides are stored for long periods on edge, particularly in a warm place, specimens will drift toward the bottom edge of the mount. The practical preventive is to store all such slides so that they remain flat.

No.........

MOUNT HISTORY

Solution	Day	Time
1		
2		
3		
4		
5		
6		
7		
8		
9		
10		
11		
12		
13		

Student....... ....Mounting Date.............

HOST (CN).................. (SN)......... ....

Locality.. ....

Autopsy Date..

Parasite...

Location..

Fixed in........... .... Stored in.....................

Autopsied by...

Stained by.................................No...... ....

Fig. 14.3. Mount history card, showing both sides. Actual size 3″ × 5″.

E Special Techniques and Further Notes

Cleaning Slides and Cover Glasses

New slides and cover glasses may be cleaned satisfactorily by dropping them, a few at a time, into 95% alcohol in a Coplin jar. Remove them one at a time after a few minutes, and carefully dry with a soft, lint-free cloth.

Used slides and cover glasses may usually be cleaned by washing in a solution of Bon Ami in a Coplin jar. When dry, wipe clean with a soft, lint-free cloth. If this treatment is insufficient, place them for several hours in equal parts of HCl and 95% alcohol in a Coplin jar, keeping the slides well separated, so that the liquid may act on the entire surface of each. Then rinse in water and place them in 95% alcohol, after which they should be dried.

When cleaning a cover glass, grasp it by the edges in one hand, cover the thumb and the forefinger of the other hand with the cleaning cloth, and rub both surfaces of the glass at the same time. To avoid breaking the cover, keep the thumb and finger opposite each other.

Removing Broken Cover Glasses

Place the damaged slide in the freezing unit of a refrigerator. After about ten minutes the resinous mountant has congealed so that the broken pieces of glass can be easily removed by lifting them off with a dissecting needle or scalpel. After the condensate on the glass disappears, add fresh mountant and another cover glass. This method is much easier and faster than the often-used heating method and the even longer procedure of dissolving off the mountant in xylene or other solvent and mounting anew.

Protozoa: Blood Smears

It is essential that slides used in making smear preparations be unscratched, noncorroded, and meticulously clean, free from grease, dust, acid, or alkali; that slides be handled by their edges; that the blood be taken as it exudes; that the process be done rapidly so as to prevent coagulation; and that smears be left to dry in a horizontal position away from flies and dust. The fingertip or structure to be pricked is cleaned with 70% alcohol, after which a prick is made with a blood lancet or a sterilized needle. The first drop is wiped off with absorbent cotton or gauze. Mark necessary data with wax pencil on the end of each slide. Blood films should be stained as soon as possible after drying to insure proper staining.

Thin Film. On slide "A" place a drop of blood about one-half inch from the end. Take a second slide "B" and place it on the surface of the first slide at about a 45° angle, as indicated in figure 14.4, and move it to the right until contact is made with the drop of blood. The free end of slide "B" may be supported by the third finger. As soon as it touches the blood, the latter will spread. Now push slide "B" toward the left, being careful to keep the edge pressed uniformly against the surface of slide "A."

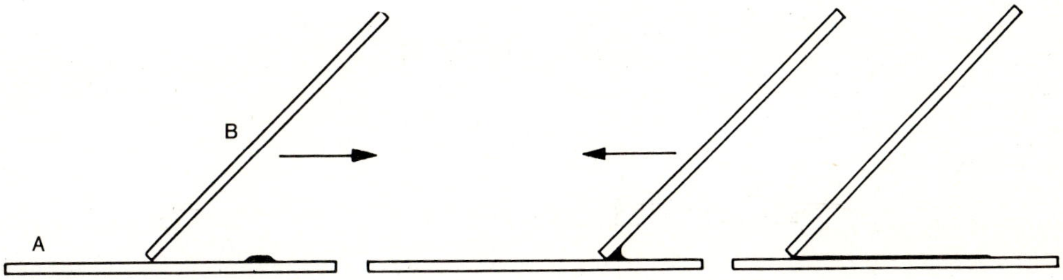

Fig. 14.4. Preparation of thin blood smear.

In this way a thin smear with uninjured host cells and protozoans and/or microfilariae will be obtained. The size of the drop of blood and acuteness of the angle formed between the slides, will determine the thickness of the film, a more acute angle resulting in a thicker film. Allow film to dry thoroughly.

Thick Film. Four or five drops of blood are placed on a slide, spread with a toothpick or the corner of a slide over an area about the size of a dime, and allowed to dry at room temperature. When dry, immerse in distilled water for about 20 minutes to decolorize, then allow it to dry again.

Staining

Giemsa. Place slide in absolute methyl alcohol for five to ten minutes, then rinse in distilled water. Dilute stain with neutral distilled water (one drop of stain to each ml water), add slide and stain for about an hour. Wash in distilled water to remove excess stain, then allow slide to dry.

Wright's. Flood film with undiluted stain and leave for about two minutes. Add an equal amount of distilled water and leave for about two minutes. The stain is then washed in distilled water, after which it is allowed to dry. Determine the time for staining by trial with a series of slides. This is usually about one to four minutes but is variable with every batch of stain. The granules in the neutrophils should stain lilac, the eosinophils bright red, and the basophils deep blue.

Smears of Parasitic Protozoa

For protozoans found in the alimentary tract, gallbladder, urinary bladder, or other organ cavities, a small amount of the material should be spread evenly on a slide or cover glass, preferably the latter, with a toothpick; if necessary, add saline solution. For forms inhabiting tissues, the latter are cut into small pieces in a little saline solution, and smeared uniformly over the cover glass. Do not allow the smear to dry. As the edges of the cover glass start to dry, fix in Schaudinn's fluid at room temperature or warmed to 50 C. Float the cover glass, smeared side down, in a dish of fixative. After about one minute, turn the cover glass over and let it remain on the bottom of the dish for five to ten minutes. Next transfer the cover glass smear to a Columbia staining dish containing 50% alcohol (two changes, ten minutes each); next to 30% alcohol for five minutes, and then to water, which is now placed under gently running tap water for 15 minutes. After rinsing in distilled water, the smear is ready for staining.

While any stains that give good results for cytological and histological work are satisfactory, one of the most commonly used stains is Heidenhain's iron hematoxylin. It requires a mordant, iron alum (ferric ammonium sulphate), followed by the dye. The smear from the distilled water is first placed in the mordant for about three hours, after which it is washed with running water for five minutes and rinsed in distilled water. Place the smear now in a well-ripened 1% solution of hematoxylin for about three hours. After rinsing in distilled water, the smear is destained in a 0.25% iron alum in water. Nuclei should show the stain, and the cytoplasm should be fairly clear. Proper destaining comes with practice. After destaining, wash in running water about 30 minutes. Upgrade the smear through the alcohol series (three minutes in each), clear in xylene (ten minutes), and transfer to a slide in a drop of thin mountant. Make sure that the smear side of the cover glass is down, next to the slide.

Recovery of Helminth Eggs and Protozoan Cysts

Practically all helminth eggs and protozoan cysts are evacuated in the feces, the important exceptions being the kidney worms and *Schistosoma haematobium,* whose eggs are voided in the urine. While various methods of examining feces for helminth ova and their larvae, and/or protozoan cysts, are in vogue, they may be classified as either: (1) the **direct fecal smear,** or (2) **concentration techniques.**

1. To make a direct smear, a small portion of the uncontaminated stool is spread on a slide and mixed, using a wooden applicator, with a few drops of saline solution. Add a cover glass and examine the slide under low power. In the case of doubtful objects and in making the final specific determination, change to high power. It should be remembered that the fecal smear must be thin enough to view all the objects under the cover glass, and there should not be too much solution, so as to float the cover glass.

The direct fecal film should always be made and examined when protozoan and helminthic infections of the intestinal tract and its appendages are suspected. In heavy infections this technique will reveal the presence of the parasite.

2. While the concentration of cysts and eggs may be accomplished in various ways, all methods involve flotation or sedimentation—with or without centrifugation. The methods depend upon mixing the fecal sample with a liquid, the specific gravity of which is different from that of the products being searched for. Although each author advances particular claims for individual methods, some fall short in one or more of the features desired. Any efficient concentration technique should embody: (1) simplicty of operation, (2) be relatively rapid, (3) remove the nonparasitic matter while retaining the parasitic materials, and (4) the parasitic objects obtained should be diagnosable, i.e., not shrunken or so altered in shape so as to be unrecognizable.

Zinc Sulfate Flotation

This technique will concentrate protozoan cysts, most helminth eggs, and larvae. However, most trematode, acanthocephalan and immature *Ascaris* eggs will not float with this technique.

1. Prepare a fecal suspension by mixing one part feces (the size of a pea) with about ten parts of tapwater.

2. Strain suspension through a layer of wet cheesecloth, supported in a small funnel, into a centrifuge tube.

3. Centrifuge for one minute at about 2,500 rpm, decant the supernatant, add 2 or 3 ml of water, shake to break up sediment, fill tube with water, centrifuge and decant as before. Repeat until the supernatant fluid is clear.

4. Decant the last supernatant, add 2 or 3 ml of zinc sulfate solution, shake to resuspend the sediment, and add zinc sulfate solution to within about 5 mm from the rim. (The zinc sulfate solution contains 330 gm of chemical in 1 liter of distilled water, adjusted to a specific gravity of 1.18).

5. The tube is centrifuged for one minute at about 2,500 rpm and is allowed to stop without interference. Do not remove the tube.

6. Using a wire loop, remove several loopfuls of the surface film and place on a slide. Add a drop of Lugol's iodine solution, mix, apply a cover glass, and examine.

Recovery of Sporulation of Coccidian Oocysts by Sugar Flotation

Unsporulated oocysts are easily recovered from the feces of any infected animal. Pulverize freshly passed feces in water to form a paste. *Do not refrigerate* before sporulation is completed. Mix with 2.5% potassium dichromate to retard bacterial growth, and place in a petri dish. Allow five days at room temperature for sporulation to complete. Sieve the material through cheesecloth or a tea strainer to remove coarse material. Suspend the sediment in about ten volumes of sucrose solution, sp. gr. 1.15. Centrifuge at 1,000 to 3,000 rpm for two minutes. Oocysts will be at the surface of the solution in the centrifuge tubes, from which they may be transferred by means of a bacteriological loop to slides for microscopic examination.

Sapero and Lawless' Fixative-stain for Protozoa and Helminth Eggs

In addition to providing good preservation of both protozoan cysts and trophozoites and helminth eggs, this method is simple enough that collections may be taken in vials by unskilled workers in the field or home, for later examination by skilled personnel, and it provides almost immediate staining. MIF technique may be used (1) in making direct fecal smears, or (2) for collection and preservation of bulk stool specimens.

1. For direct smears the ingredients, sufficient for some two dozen preparations, are combined in a Kahn tube. Since the resulting solution is unstable, it must be made up daily; the Lugol's used should not be older than three weeks. To use the stain solution, add a small amount of fecal sample to one drop of MIF solution and a drop of distilled water, mix thoroughly, and add a cover glass. The completed wet preparation should be thin enough to permit the slide to be tipped on edge without shifting the cover glass.

2. For collecting and preservation of fecal samples when not brought directly to the laboratory, a modification of the above MIF stain is advised. The MIF stain-preservation solution consists of (1) a MF stock solution, stored in a brown bottle, and (2) Lugol's solution not over a week old. To use, add 2.35 ml of stock solution into a Kahn tube and stopper with a cork, and in another such tube add 0.15 ml of the fresh Lugol's and close with rubber stopper. Combine the two solutions immediately before the addition of about 0.25 gm (about twice the volume of a medium-size pea) of fecal material, and comminute thoroughly with wooden applicator. If stoppered well to prevent evaporation, specimens will retain a good stain for several months. For examination remove a drop of the supernatant fluid from the top, place on a slide, and add a cover glass.

The above MIF technique has been modified by Blagg et al., to increase its diagnostic yield for helminth eggs, by the addition of ether to dissolve fats and to float fecal detritus. The MIF-preserved specimen is shaken vigorously for five seconds, strained through two layers of wet cheesecloth into a 15 ml centrifuge tube to which 4 ml of refrigerated ether is added. The tube is closed with rubber stopper and shaken vigorously, after which the stopper is removed and the tube left standing for two minutes. It is then centrifuged for one minute at 1,600 rpm, resulting in the formation of four layers; ether at the top, a plug of fecal detritus, a layer of MIF, and a bottom sediment containing helminth eggs. The fecal plug is loosened by an applicator stick, and all but the bottom sediment layer is quickly decanted. Mix the sediment with an applicator stick, and put a drop onto a slide for examination.

Formalin-ether Centrifugation

Eggs of schistosomes are not easy to find in direct fecal smears and are usually destroyed by conventional concentration methods. This technique has been developed for the recovery of helminth eggs, including those of schistosomes, and protozoan cysts.

1. Prepare a fecal suspension by mixing one part feces (the size of a pea) with about 12 ml of saline solution.

2. Strain suspension through two layers of wet cheesecloth, supported in a small funnel, into a centrifuge tube.

3. Centrifuge for two minutes at about 1,800 rpm, decant the supernatant, and add fresh saline solution. Stopper with the gloved finger and shake well.

4. Wash twice in saline solution, centrifuging as above, and decanting the supernatant each time. Repeat until the supernatant fluid is clear.

5. Add 10 ml of 10% formalin to sediment, and leave for five minutes.

6. Add 3 ml of ether, and resuspend the sediment by vigorous shaking (with gloved finger).

7. Centrifuge as above.

8. Loosen the "plug" at the formalin-ether junction with an applicator, and completely decant the supernatant.

9. Using a Pasteur pipette or a wire loop, transfer sufficient sediment to a slide, add a cover glass, and examine. If not enough fluid drains back from the tube wall to enable one to withdraw sediment, add a few drops of saline solution and mix with sediment.

At no time, with either technique, is the oil immersion objective to be used with wet mounts unless a No. 1 cover glass is used and the preparation sealed with Vaseline or other grease.

Diagnosis of Microfilarial Infections

To 1.5 ml of fresh blood, Knott (1939) recommends adding 10 ml of 2% formalin and centrifuging at 1,500 rpm for two to three minutes, after which the supernatant fluid is decanted. Transfer a drop of sediment to a slide, add a cover glass, and examine for microfilariae. A modification involves substituting 1% acetic acid for the formalin and, after transferring to slide followed by drying, stain with Wright's or Giemsa's stain.

Harder and Watson (1964) described useful staining techniques for sheathed and unsheathed bloodborne microfilariae. A simple key to human microfilariae, based upon criteria clearly demonstrated by these stains, is included in their paper.

Papanicolaou-hematoxylin and Eosin Procedure for Sheathed Microfilariae

1. Prepare a thick blood smear on a slide, about 3 cm from one end. Cover a space about the size of a dime with as much blood as will easily spread over this without cracking and peeling when dry. Spread with a toothpick or the corner of a slide so that printing can just be read through the center of a well-made wet film. On the opposite end of the slide put an identifying mark with a wax pencil or other marking device.

2. Lay the slide flat to dry and have it well protected from insects and dust. Air dry for eight hours or more.

3. Dehemoglobinize thoroughly in enough distilled water to cover the smears as the slides stand on end in a Coplin jar, thick film downward. The film should be a milky white, which may take several hours.

4. Air-dry slides, standing on end on absorbent paper or a coin pad.

5. Rinse in 95% alcohol, two changes.

6. Harris' hematoxylin (stock stain diluted equally with a distilled water), five minutes.

7. Rinse in alkaline tap water, two changes.

8. Acidulated 70% alcohol, two quick dips.

9. Running tap water, one minute.

10. Rinse in distilled water.

11. Pass through 70% and 80% alcohols, two minutes in each.

12. Orange G (1% solution in 90% alcohol), one and a half minutes.

13. Rinse in 95% alcohol, two changes.

14. E.A. 50,* two minutes.

15. Rinse in 95% and absolute alcohol, two changes in each.

16. Terpineol, five minutes.

17. Dry. (The remaining steps can be completed later, but within two days for best results.)

18. Harris' hematoxylin, 20 minutes for *Brugia malayi* and *Wuchereria bancrofti;* 45 minutes for *Loa loa,* longer for older slides. (Proper staining time is gained with experience.)

19. Acidulated 70% alcohol, one dip.

20. Running tap water, one minute.

21. Rinse in ammonia water (strong ammonia, 0.5 ml in 1,000 ml distilled water).

22. Stain in eosin (saturated solution of Eosin Y in 80% alcohol), five to ten seconds (according to desired intensity).

23. Rinse in 95% and absolute alcohols, two changes in each.

24. Clear in absolute alcohol-xylene (equal parts) carbol-xylene (melted phenol crystals, 100 ml in 900 ml xylene) and xylene, and mount.

Modified Gomori Trichrome Procedure for Unsheathed Microfilariae

1. Prepare thick blood film and dry, as above.

2. Dehemoglobinize in magnesium sulphate solution (1 gm crystals in 1,000 ml distilled water) for one hour (or longer) until all color disappears from the film.

3. Rinse in distilled water, two changes.

*This polychrome eosin-azure formulation by Papanicolaou, may be purchased from pharmaceutical dealers or from Ortho Diagnostics, Raritan, N.J. 08868.

4. Harris' hematoxylin, five minutes for *Dipetalonema perstans;* 30 minutes for *Mansonella ozzardi.* (Proper staining time is learned with experience.)

5. Distilled water, 10 to 15 dips.

6. Acidulated 70% alcohol, five seconds.

7. Tap water, rinse.

8. Ammonia water, rinse.

9. Distilled water, two changes.

10. Gomori trichrome stain, six minutes.

11. Acetic acid (5.0 ml in 1,000 ml distilled water), two minutes.

12. 95% and absolute alcohols, two changes in each.

13. Clear in xylene and mount.

Improved Method for Pinworm Diagnosis

Since eggs of *Enterobius vermicularis* are not ordinarily recovered in the feces, diagnosis of infection is usually based on the recovery of eggs deposited around the anus by migrating gravid females. For this, a length of transparent, clear, adhesive cellulose tape held adhesive-side-out on the end of a tongue depressor or similar object by the thumb and index finger, is pressed against the right and left perianal folds and then spread flat on a microscope slide for examination. The buttocks should be spread enough to allow pressing the tape firmly against the line of junction between the moist outer part of the anal canal and the somewhat dry, waxy perianal folds. This is where the egg yield is the highest. A strip of paper bearing identification should be placed between the tape and the slide at one end. For microscopic examination, the tape should be grasped at the loose (identification) end and turned back far enough to expose all the surface that had been in contact with the perianal skin. Add one drop of toluene and with a straight wooden applicator smooth the tape back onto the slide and level it. Since the toluene is at once a good solvent for the adhesive material on the tape and a good clearing reagent, air bubbles are allowed to escape as the tape is smoothed out and almost everything except the pinworm eggs becomes cleared, leaving the eggs, as the only conspicuous objects in the preparation. Xylene, benzene, ether, chloroform, and O.1N sodium hydroxide are all unsatisfactory for clearing. Until skill is acquired in handling the toluene and tape, a clean razor blade or scalpel may be used to mark off the portion of the tape that had been in contact with the perianal skin and the remainder of the tape removed. Clear, smooth preparations are best obtained if the tape is only about 20 mm long.

Double Cover Glass Mounting Technique

For making permanent mounts of helminth eggs and other objects of suitable size, which are difficult or impossible to prepare satisfactorily in a resinous medium, an aqueous solution with the addition of a second cover glass, sealed in a resinous mountant, is advised. Number 1 circular cover glasses of two sizes, differing 3 mm or more in diameter, are recommended. The chloral-gum solution used is a modification of Berlese's original formula, but it is believed that other similar aqueous media would prove equally satisfactory.

1. Prepare a series of dilutions of chloral-gum in 10% formalin, starting with a 10% solution and increasing concentrations by 2% for each subsequent step, or prepare each solution as needed.

2. Concentrate fecal suspensions that have been thoroughly fixed in neutral formalin until each drop contains an adequate number of eggs. Each ml of this will yield about ten slides.

3. Pipette 5 ml of the concentrated fecal suspension into a test tube or a vaccine bottle of 15 ml capacity. Next, carefully add an equal volume of 10% chloral-gum-formalin, tilting the receptacle to allow this fluid to run in along the lower side. Each succeeding dilution of the series is heavier than the preceding, and consequently will occupy the lower half of the column of fluid.

4. Cap the containers and place them in a drying oven at about 35 C until the fecal material has completely settled on the bottom. Then remove the clear fluid above the feces with a finely-drawn bulbed pipette, and carefully add an equal volume of the next higher concentration of chloral-gum solution. Continue this process until the eggs are suspended in full-strength chloral-gum medium. When the last sedimentation is complete and the supernatant removed, mix the remaining contents of each container thoroughly, but wait until any bubbles disappear before starting to prepare mounts.

5. Mounts are made by two persons working as a team, one transferring drops of the mixture to the slides, the other adding the smaller cover glasses. Circular cover glasses are advised since they permit an even flow of material to the outer edges. Slides should be laid out in advance on a warming table, if available, since this results in a better spread of the mountant. With some practice, one will come to judge the size of the drop required to spread completely under the cover without excess. Cover glasses should be added quickly, before the medium hardens and spreads poorly. For transferring the material to slides, a finely drawn bulbed pipette or a heavy platinum wire loop, about 4 mm in diameter, is recommended.

6. Place mounts in a horizontal position in a drying oven at about 35 C. Following drying, remove any excess mounting medium with a razor blade and finish cleaning with a clean cloth saturated with water. The margin must be free of mountant if a strong seal is to be obtained.

7. Next add a drop or two of thin resinous mountant atop the previous cover and apply the larger cover glass, being careful to place the second cover so as to have

the same overlapping margin on all sides. Return to drying oven and when dry, clean, if necessary, in the manner suggested for whole mounts. Such preparations will last indefinitely if the cover glass is not cracked, because there is a firm moisture-tight seal at the edge of the larger cover glass.

A modification of the dual mount involves placing the aqueous medium with the material being mounted between the two cover glasses, rather than between the slide and the smaller cover glass as above.

1. Prepare a raised working surface no larger in diameter than the smaller cover glass. A cork of suitable size attached to a base is satisfactory. Place the smaller cover glass on the cork.

2. Transfer a small drop of the warmed concentrated chloral-gum medium containing the eggs to the center of the cover glass. Keep the stock of chloral-gum warm in a water bath. The mounting drop should be of such a size that it just, and only just, covers the cover glass. Judgment, which comes with practice, is needed to gauge the size of the drop, which also depends on the warmth of the mixture.

3. Grasp the larger cover glass with cover glass forceps and quickly place it on the preparation on the cork, keeping the cover glasses as nearly parallel and concentric as possible.

4. When the larger cover glass contacts the chloral-gum, which will spread between the two cover glasses, carefully invert so that the smaller cover glass is uppermost. If necessary, center and flatten out the mount with a dissecting needle. The preparation is placed in a warming oven for a few hours to allow the mountant to set.

5. Place a drop or two of thin resinous mountant on the smaller cover glass. Carefully lower a slide until it touches the mountant on the cover glass, and quickly invert so that the larger cover glass is uppermost. Center the cover glass, apply a spring clamp and return to the drying oven.

Either method may be used for making permanent whole mounts of small hard-to-clear specimens, such as helminth eggs, mites, and small nematodes. The sealing effect of the resinous mountant prevents crushing the specimen, which often results from pulling down the cover when only a single cover glass is used with chloral-gum and similar water-miscible mountants.

Recovery of Monogenea

For recovering small Monogenea, freeze fish gills for six to 24 hours in an appropriately sized jar. Then thaw in tap water, cap the jar, and shake vigorously. The liquid, with the freed parasites, is poured into a Syracuse watch glass and alternately diluted and decanted until clear enough for examination under a dissecting microscope. With the use of a bulbed capillary pipette, the parasites are transferred to clear water.

For recovering *Gyrodactylus,* place fish in a small dish containing a 1:4,000 solution of formalin. After a few minutes the worms fall to the bottom of the dish and can be transferred with a bulbed pipette.

Due to their small size and the tendency of the worms to become entangled in debris and with other specimens, it is practically impossible to handle the small flukes in containers during the ordinary technical processes requisite to mounting. For such forms, the following procedure in making total mounts is advised. With a bulbed pipette transfer specimens from a mixture of water and glycerine to a slide; then remove most of the water-glycerine with pieces of paper toweling or filter paper. Adherence is secured by immersing the slide in a fixative (A-F-A or Bouin's) contained in a Coplin jar. Following fixation, the material is dehydrated (using Coplin jars) through the alcohols, cleared and mounted in the usual way.

Rapid Stain Technique for Showing the Haptoral Bars of *Gyrodactylus*

1. Using a small insect pin inserted in a dowel handle, transfer worms from the formalin fixative to a drop of distilled water on a slide.

2. Draw off water with filter paper and promptly add a drop of Gomori's trichrome solution. Stain until the body of the worm is deep blue (usually one to 15 minutes). Add stain as it evaporates.

3. Orient worm, ventral surface up, add Berlese's chloral-gum or other preferred aqueous mountant, and apply a cover glass.

For permanent mounts, transfer worms directly into 95% alcohol, dehydrate through absolute alcohol, clear and mount in Histoclad or other preferred resinous medium.

Examination of Snails for Cercariae

Large numbers of snails can be dredged up from mud and vegetation with various types of equipment used by limnologists for this kind of collecting. Place the material in pails or large trays with sufficient water to cover the contents. Use pond water for snails from fresh water and seawater for marine species. Cover the surface of the water with paper towels, leaving open spaces. Many of the snails will crawl out on the wet towels and up the sides of the containers, where they can be collected easily. Cover the containers with a cloth or paper to prevent the snails from escaping. It is necessary to search the sandy and muddy bottoms of habitats carefully for some species. Tea strainers and collecting equipment constructed with screen bottoms and designed to strain out sand and mud are necessary to recover the small species of snails.

Wash and separate snails according to species, and transfer several to wide-mouthed glass bottles. Shell vials may be used for the smaller species. Cover the containers. Change water as necessary and feed on fresh lettuce, if kept beyond six days. Examine each container several times daily, in a good light against a dark background. Cercariae will appear as whitish tumbling bodies. When cercariae appear, isolate snails individually, to determine the infected ones. It is necessary to check the containers mornings, afternoons, evenings, and nights because cercariae of many species are not liberated continuously, but are shed only at certain times of the day or night. If snails, especially prosobranch species, are first dried on blotting paper for a few hours and then isolated in water, sometimes cercarial emergence will follow almost immediately.

Cercariae should be studied alive as well as fixed and stained. With a capillary pipette equipped with a rubber bulb, transfer sepcimens to a slide and add a No. 1 cover glass. The addition of a weak solution of a vital stain reveals certain internal structures. Draw sufficient water from under the cover glass by means of absorbent paper to press the cercariae to the point where they are held still but not crushed. In these preparations, flame cells and excretory system show best. Edges of the cover glass should be sealed with Vaseline by running a hot needle around the edge to spread the sealant evenly. Fresh egg albumen, because of its viscosity and physiological properties, is a good medium in which to mount living cercariae for study.

After having studied the cercariae, examine each snail that shed them. For crushing a large number of snails, a small bench vise is recommended. Remove the broken shell and tease the viscera apart in saline solution with dissecting needles, and examine under a dissecting microscope to determine whether cercariae are produced by rediae or daughter sporocysts. The snail's digestive gland and gonad in the apex of the shell are the best sources of trematode larvae. Transfer larval forms in a drop of saline solution to a slide, apply a cover glass and examine under a compound microscope.

Freed cercariae and rediae or daughter sporocysts containing the cercariae should be fixed, stained, and made into permanent mounts. Label each slide, stating the kind of cercaria, whether produced in redia or daughter sporocyst, and the species of snail.

Vital Staining, or Staining in Vivo

Vital (often called supravital) staining, the absorption of coloring by living cells or organisms, with neutral red, Nile blue sulphate, or other suitable stains, is advantageous for distinguishing the number of penetration glands and their secretions, the slender undeveloped caeca, the excretory system, and the presence of spines on those cercariae possessing them. Such stains are also useful for studying pseudophyllidean coracidia. With a fingertip, spread a drop of 0.5% aqueous neutral red on a slide and let dry. Add a drop of water containing cercariae, apply a No. 1 cover glass, and examine. Larvae may be placed directly in the stain if a very dilute solution (0.01% or weaker) is used. Mounts of living larvae last much longer if the edges of the coverglass are sealed with Vaseline.

Artificial Evagination of Tapeworm Cysts

Living cysts are caefully removed from the host and placed in saline solution. Each cysticercus should be liberated from its cystic wall, if necessary, and placed in a bile salts solution. Invaginated specimens, upon being placed in the warmed (37 to 38 C) solution of bile salts, generally show definite signs of movement within 20 seconds. There is a gradual pushing out of the neck region as peristalticlike waves move away from the bladder until there is a complete eversion of the scolex.

Collecting Nematodes from Soil

Larval stages of Strongyloididae, suborders Strongylina, Trichostrongylina, and Metastrongylina occur in the soil and feces, having hatched from eggs passed in the feces of infected hosts. Species of all of these groups occur in domestic and wild animals and some of them in humans. They may be recovered by means of simple apparatus constructed from material readily available in any laboratory.

The Baermann funnel is the apparatus commonly used. Its utility lies in the fact that the larvae are very active and constantly moving. To prepare a Baermann funnel, a short piece of soft rubber tubing is pushed onto the stem of a six to ten inch funnel. The free end of the tubing is closed tightly with a pinch clamp. A tall wide-mouthed jar or a ring stand provides a good support for the funnel, which is filled with water. A piece of loosely woven cloth is spread over the top of the funnel so as to lie loosely on the water and is attached to the rim by means of clothespins, a rubber band, or string. The free margins of the cloth should be arranged over the mouth of the funnel so as not to draw off the water by capillary action. A thin layer of soil from the habitat of infected animals is spread over the slightly submerged cloth. A metal pan similar to a pie tin with a screen bottom, or a piece of hardware cloth, may be used to support the cloth in the funnel.

Larvae and free-living worms wriggling through the soil and cloth gravitate into the rubber tubing. They may be recovered in numbers after a few hours by opening the pinch clamp and drawing a few ml of water into a Syracuse watch glass or petri dish. Fecal material mixed with sand, animal charcoal, or sphagnum moss will permit good aeration and facilitate hatching of eggs. Larvae are recovered after a few days by means of the Baermann funnel.

When baermannizing soil from habitats of parasitized hosts, larvae of the parasitic species and both adults and larvae of nonparasitic soil nematodes will be collected, often with the latter in great numbers. While these can be separated by microscopic examination, the task is tedious and demanding. A quick and easy way to separate them is to add a few drops of concentrated HCl to the dish. The nonparasitic species are killed quickly, whereas the parasitic ones, which are adapted to survive passage through the acidic condition of the stomach, remain alive and very active.

All baermannizing apparatus and glassware used should be scrupulously cleansed with boiling water between each operation. Equipment contaminated with larvae can lead to erroneous conclusions.

Inasmuch as larvae of suborders Strongylina and Trichostrongylina nematodes of horses, cattle, and sheep have a tendency to migrate, they may be collected by another simple method. Moist feces mixed with sterilized sphagnum moss or other diluting material are placed in a small beaker or similar container. The moss should be loose to enable larvae to move about freely, and it should fill the beaker. The beaker is placed in the middle of a large-mouthed bottle that is filled with sufficient water to come to the top of the beaker but not run into it. The large bottle is covered. Larvae migrate up the moist sides of the beaker, over the rim, and into the water, where they sink to the bottom in great numbers. A small petri dish placed inside a larger covered one will work equally well. Once the principle is understood, many applications of it will be apparent.

Artificial Digestion of *Trichinella* Larvae from Muscle Tissue

Rats or other hosts, after being skinned and eviscerated, are ground in a food chopper. Digestion is accomplished by placing the ground meat in a modified Baermann apparatus, to which the digestive fluid is added. A sling of four layers of commercial cheesecloth is suspended in a large glass funnel by means of clip clothespins to retain the ground meat. A bottle is attached by means of a short section of soft rubber hose to the stem of the funnel, which is supported by a ring stand. The mixture should be left for about 24 hours at 37 C. Approximately one liter of solution is used for 30 gms of meat.

If live larvae are present, the action of the digestive fluid, together with their own activity, will free them from muscle tissue. They will excyst and work their way through the cheesecloth sling and fall to the bottom of the bottle. At the end of the digestive period the fluid in the bottle should be poured into a petri dish and examined under a dissecting microscope.

Recovery of Larval Helminths in Host Tissue

A preliminary search for metacercariae and plerocercoids in the viscera and body wall of fresh fish, after both the external and internal examinations have been completed, may be done in the following manner. Fillet the fish and then make incisions to divide the musculature into slices about 1 cm thick, which are carefully inspected. Using small insect needles inserted in dowel handles, carefully dissect out larvae.

Following this cursory examination, the host should be placed in artificial digestive fluid (p. 266) at the rate of about 1 gm of tissue to 10 ml of solution, which frees the larvae alive. Not only are small nematode and acanthocephalan larvae, metacercariae, and plerocercoids obtained in this way, but a qualitative-quantitative study can be made with reasonable accuracy if the different organs are removed and digested separately. For the recovery of larvae other than plerocercoids, grinding of tissue is advised, but grinding is contraindicated when plerocercoids are involved lest they not survive the process intact. Tissue and the digestant are placed in a graduated cylinder or appropriately sized glass jar and left for a few hours at 37 C, with occasional shaking. In the absence of an oven, the necessary temperature can be maintained by placing the glass container in a water bath.

Following digestion, rinse and remove the pieces of host tissue and decant the supernatant. The sedimentation-decantation process is repeated until the material is clear. Pour the sediment into a petri dish and examine in good light under a dissecting mircoscope. In the case of delicate metarcercariae, it is better to use saline solution in the digestant as well as for washing.

Artificial Infection of Laboratory Animals with *Trichinella*

Food should be withheld for 48 to 72 hours from individually caged animals, after which they are offered the infected meat or given the larvae in solution freed by peptic digestion. If they fail to eat the meat within an hour or so, replace it in the refrigerator, and withhold food for an additional 12 to 24 hours. Animals should be given water throughout fasting. Rabbits and guinea pigs should be infected with digested larvae by means of a stomach tube, a method which may also be used with rats and mice.

Each host should be given a dosage in accordance with weight and the degree of infection of the meat. Twenty larvae, either as free larvae by stomach tube or as in infected meat, per gram of host weight is recommended.

Mounts of Nematodes

Nematodes may be prepared in a number of ways for study. The manner of preparation may be determined by the need or the permanency of the mounts.

Water Mounts. Larvae and small adults, both parasitic and nonparasitic forms, should be studied alive in temporary water mounts or dead in fixatives where formalin is used. Such mounts are very valuable and will last for a considerable length of time if the edges of the cover glass are sealed by spreading Vaseline over them with a hot teasing needle.

Temporary Mounts. These may be prepared from fixed specimens cleared in phenol-alcohol solution, lactophenol, or glycerine. Temporary mounts have the advantage of permitting the specimens to be rolled under the cover glass for viewing them from various positions.

Phenol-alcohol solution is useful for observing cuticularized structures. Fixed specimens may be transferred to the solution directly from any fixative. Clearing is rapid and complete; however, the degree of transparency can be controlled by drawing 95% alcohol under the cover glass. The phenol must be removed from specimens to be stored by washing several times in alcohol.

Lactophenol acts rapidly but does not over clear. Put a small amount of 0.01 to 0.0025% aqueous solution of cotton blue (China blue) in lactophenol on a slide and warm to 65 to 70 C on a hot plate. By means of a small needle, transfer worms directly into the lactophenol and leave for two to three minutes. They will be cleared and tinted blue. Mount them in lactophenol tinted with 0.0025% cotton blue. Specimens left in lactophenol for a week or longer tend to macerate and become distorted. If worms are to be preserved for later study, the lactophenol should be washed out through several changes of alcohol.

Glycerine is an excellent mounting medium and is used extensively by nematologists. Nematodes are transferred directly from the fixative into a small amount of the following mixture: 21 ml of 95% alcohol, 1 ml of glycerine, and 79 ml of distilled water.

The dish containing the nematodes is placed on a support in a closed vessel containing 95% alcohol for at least 12 hours at 35 to 40 C, or longer for large nematodes. After this period, the dish is filled with a solution of five parts glycerine in 95 parts of 95% alcohol and placed in a partially closed petri dish kept at 40 C until all of the alcohol has evaporated, leaving the worms in pure glycerine.

Temporary or permanent mounts may be made from specimens cleared in this manner. For permanent mounts, glycerine tinted with 0.0025% cotton blue kept in a desiccator should be used as the mountant.

To make permanent mounts, a drop of the dehydrated glycerine is put on a slide and the worms transferred to it. Small bits of glass or a ring of Zut are used as supports for the cover glass to prevent damage by pressure to the specimens. The amount of glycerine should be sufficient to almost reach the edge of the cover glass when it is in place. The completed mount should be sealed with Zut. When properly prepared, these slides will last indefinitely. Such slides should be stored flat to prevent drifting of specimens.

Resinous Mounts. While time-consuming, they are not difficult to prepare and are especially suitable for nematodes and Acanthocephala stained with carmine dyes and counter-stained with Fast Green. Preparatory to staining and dehydration, the body wall is punctured in a number of places for better movement of the reagents into and out of the worms. For infiltration, the completely dehydrated worms are transferred from absolute alcohol into a mixture of one part xylene and three parts absolute alcohol; equal parts xylene and alcohol; three parts xylene and one part alcohol; several changes of pure xylene; and finally into a very dilute mixture of xylene and the resinous mountant of choice in a Syracuse watch glass containing the mountant. The open dish is covered with a piece of paper to keep out dust and dirt, and the xylene is allowed to evaporate at room temperature or in a warm oven to a point where the resin is the right consistency for mounting the worms on a slide. It is important to use a sufficiently large volume of the dilute xylene-resin mixture to provide adequate mountant in the dish to cover the worms when the solvent has evaporated. Permanent mounts prepared in this way are valuable for study. Fixed worms may be cut in half (so there is free access of the stain, alcohols, clearing agent, and mountant) and processed in the usual manner without difficulty. The halves may be mounted on the same or on different slides.

Preparation of En Face Views of Nematodes

For study of cephalic structures, which are important features in the taxonomy of nematodes, cut off the head end and mount it under a cover glass, orienting it so that the mouth opening is facing up. Such en face mounts may be temporary or permanent, and various methods of preparation have been suggested. However, most of these techniques result in a cover glass exerting pressure on the delicate apical papillae, which tends to obscure them. I prefer to make temporary mounts using a hanging drop slide, as follows.

Clear the specimen in glycerine-alcohol, as described above. Place it on a slide, with the head slightly immersed in glycerine. Place the slide on the stage of a dissection microscope, preferably one with high magnification capabilities if the worm is small. A single-edged razor blade makes a very good guillotine as it can be held in both hands for steadiness. When decapitated, the head should be slightly shorter than wide.

Locate the desired spot for the cut, gently place the edge of the blade on it, slowly raise the blade to vertical, and push down with a steady, firm motion. The detached head is easily lost at this stage, especially if it adheres to the blade. When it is safely located, remove the remainder of the nematode. The head will now be transferred to a cover glass.

Place a very small drop of glycerine on the center of a cover glass. It should not be much larger than the head itself. Using fine needles, pick up the head and place it in the glycerine on the cover glass. Discard the slide. The head must now be oriented with the mouth facing *down,* toward the cover glass. The amount of glycerine on the glass is crucial, for if there is too much or too little the head probably will topple over. Glycerine can be removed by imbibing it with a piece of lens paper, while carefully viewing with the microscope. *Warning:* the greatest care must be taken to avoid the head being sucked into the paper, to be lost forever among the fibers! Once the head is vertically poised, invert the cover glass onto the cavity of a hanging drop slide. The slide should be prepared with four small globs of vaseline or other grease approximately where the corners of the cover glass will fit. The grease will adhere the glass to the slide so that an oil immersion lens can be used for study. If too much glycerine is used, the head may tilt or drift downward out of focus.

Although the above method may sound complicated, it is actually very simple, requiring only a little practice to perfect. Its disadvantage is that a permanent preparation cannot be made. This is more than compensated for by the clear view of papillae and lips, which have no distorting pressure on them.

Revealing Cuticular Structures of Nematodes

If nematodes are left for several hours in aqueous or formalin solutions of cotton blue, particles of stain collect in the depressions, pits, and cavities, revealing the phasmids, cervical papillae, amphids, and excretory pores in relief.

Trisodium Phosphate for Softening Cysts, Helminths, and Insects

Van Cleave and Ross (1947) recommend that hardened acanthocephalans and nematodes be transferred from water into a solution of 0.25% trisodium phosphate. Freshly preserved specimens become soft, pliable, and translucent almost immediately, while specimens that have become hard, brittle, and unyielding after long preservation may require several hours or even days before they attain the proper degree of softness and pliability. If placed in a warming oven, the process is hastened. It is well to keep treated material under observation, since the use of too strong solutions of trisodium phosphate or weaker solutions over too long a time may render specimens too soft and jellylike for easy handling. When the desired degree of softness and translucency has been reached, specimens should be removed to distilled water to check the action.

Specimens thus treated, washed in distilled water and stained, become much more brilliantly stained than untreated specimens. Furthermore, the treated specimens may be dehydrated, cleared, and mounted in resinous medium without developing opacity, except in rare instances.

Another distinct advantage of the softening effect is the ease with which specimens may be straightened prior to dehydration. When softened, specimens may be folded back and forth on a microscope slide along strips of cardboard or toothpicks and may be slightly compressed and held in place by covering with another slide. Spring clamps or several wrappings of thread provide the pressure and hold the slides together so that the stained specimen may be dehydrated in a petri dish or Coplin jar through the grades of alcohol. In 100% alcohol (95% if counterstaining with Fast Green), the specimen becomes firm enough so that it may be removed from between the slides and yet retain its series of folds, ready for clearing and mounting.

Clearing Arthropods with Potassium Hydroxide

Pierce the abdominal region with a small insect needle while the specimen is still in the preservative, and then transfer to water. Replace the water with 10% KOH and leave for 12 to 24 hours, or until the specimen becomes lighter in color. After rinsing in water, replace with acidulated 70% alcohol, dehydrate, clear, and mount in the usual manner. While not advised for class use, the time in KOH can be shortened considerably by heating the solution containing the specimen. About ten minutes in the boiling solution should be adequate. This process should be watched since the chitin is also destroyed with overexposure to KOH.

Cellosolve (ethylene glycol monoethyl ether) can be used for both dehydration and clearing. Following KOH, specimens are passed through water before Cellosolve is added. The advantage of this method is that it reduces the number of times specimens are handled, which minimizes the chance of damaging them.

Collecting and Mounting Parasitic Mites

Parasitic mites such as *Notoedres, Otodectes, Sarcoptes* etc., are usually deeply imbedded in the host tissue and are therefore often lost or damaged when the attempt is made to separate them. For collecting them, a scalpel or knife blade is sterilized by passing it through a flame and then cooled by dipping it in water. Mites are collected more easily if the edge of the scraper is first dipped in acetic-glycerine (1% glacial acetic acid in glycerine). A fold of skin showing lesions is held between the thumb and forefinger, and the crest of the fold scraped until lymph begins to ooze. Care must be taken to avoid drawing blood. Transfer scrapings from scalpel, using a rotary motion, into a drop of acetic-glycerine on a slide. Add a cover glass and additional acetic-glycerine, if necessary, and observe under the low power of a compound microscope.

The material may be stored in the acetic-glycerine indefinitely, and mites may be separated at leisure. Add a small crystal of thymol to discourage growth of molds. Separate parasites with a very fine dissecting needle or finely drawn pipette equipped with a rubber bulb. This is done under a dissecting microscope in good light.

After mites have been separated from the debris, place them on a slide in one of the several well-established chloral-gum preparations. The Double Cover Glass Mounting Technique, described on page 257, is advised for permanent mounts.

The double cover glass mounts containing dorso-ventrally flattened arthropods, as well as trematodes and cestodes, may be mounted between two pieces of thin cardboard cut the same size as a microscope slide. Identical holes are cut in each piece of cardboard, the cover glass mount placed over the hole in one and the two pieces of cardboard glued together by such adhesives as Elmer's glue. With these simple, easily prepared mounts, the specimens between the cover glasses may be viewed from either the dorsal or ventral side. The hole in the cardboard should be large enough to allow either the high dry or oil immersion objectives to swing into position without striking the paper mount.

Microscalpels

Excellent scalpels for minute dissections can be made from double-edged razor blades. Grip the edge of a blade obliquely (over the area desired), between the jaws of needle-nose pliers, Cresson insect pinning forceps, or similar tool. With another pair of pliers bend the blade laterally until it snaps off. Since scalpels with shanks are desirable, first break away a corner of the blade before proceeding down the blade edge. The good pieces, once crudely broken, can be ground to the desired shape of a fine emergy wheel or stone. Do not touch the cutting edge. Insert the shank of the finished scalpel in the cleft of a suitable wooden staff, apply Duco or similar cement, clamp in a vise and leave overnight before using.

Spring Clamps

Fabrication of spring clips is accomplished in five distinct operations: (1) preparation of material, (2) annealing, (3) shaping, (4) hardening, and (5) washing.

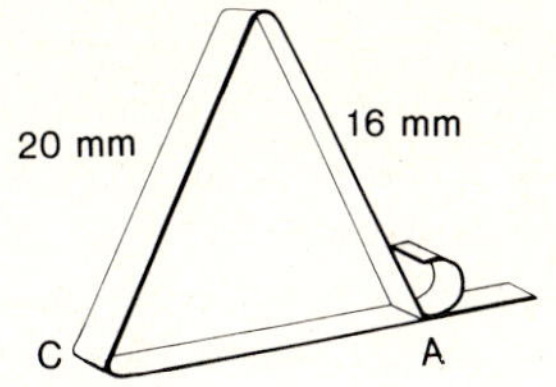

Fig. 14.5. Finished spring clip.

Preparation of Material. Material, consisting of broken watch mainsprings, should first be uncoiled by holding each end, while the spring is pulled back and forth a few times, with the convex surface against a piece of rounded wood. The springs are then cut to the desired length of about 66 mm with a pair of sheet metal shears or cutting pliers.

Annealing. The sections are then annealed by placing them in a piece of nongalvanized iron pipe about 40 mm inner diameter closed at one end, which is placed horizontally on a ring stand, and heated (two Fisher high temperature burners work nicely) until the mass of springs is a bright cherry red. The heat is then turned off and the mass allowed to cool slowly. This slow cooling allows the molecules to align themselves in one plane, thereby making the metal workable.

Shaping. A small circle (approximately 4 mm diameter) is formed at one end with a pair of round-nose pliers and the length of spring then shaped over a triangular piece of wood, planed to the proper shape and size (about 13 × 16 × 20 mm). The second and third bends will have to be finally adjusted by hand to make sure that the arm resulting from the last bend goes beyond that of the first in order to give the spring clip some tension.

Hardening. The hardening process is effectively accomplished by placing the shaped springs inside the segment of pipe (closed at one end as previously), heating the entire mass to a cherry red, and quickly dropping the springs en masse into a container of ordinary No. 10 or 20 motor oil. The purpose of the oil is to cool the springs at the proper rate so that they will be hard but not too brittle. A large pair of pliers should be used in transferring the springs into the oil, and if more than one "batch" of springs is to be hardened, the oil should be cooled again by placing the container in cold water.

Washing. The oil used for cooling should be poured off and the finished spring clips washed in kerosene. Any suitable solvent can be used for washing, but kerosene leaves a thin coat of oil on the metal and thereby prevents rusting.

Appendix 1 Reagents and Solutions

The following list includes reagents and solutions commonly used in parasitology. For convenience, they are listed alphabetically rather than according to use.

Alcohols

Commercial grain alcohol, which is approximately 95% ethyl by volume, is used in making the lower strengths of alcohol. When preparing the different solutions, the procedure given below should be followed. To obtain a given percentage of alcohol through dilution of a higher percentage with distilled water, subtract the percentage required from the percentage of the alcohol to be diluted; the difference is the proportion of distilled water that must be added. Thus, if 70 is the percentage desired, and 95 the percentage to be diluted, then 95 minus 70 equals 25; hence 25 parts of distilled water and 70 parts of 95% alcohol are the proportions used.

This means that in practice one needs only to fill the graduated measuring cylinder to the same number as the percentage required (e.g., 70) with the alcohol being diluted (e.g., 95%) and then fill to the percentage of the latter with distilled water. Thus, one would obtain 95 ml of 70% alcohol, if the measuring cylinder is graduated in milliliters.

Acid Alcohol, Two Percent

Alcohol, 70%	98 ml
Hydrochloric acid, commercial	2 ml

A-F-A (Alcohol-formol-acetic) Fixative

Alcohol, 85%	85 ml
Formalin, commercial	10 ml
Acetic acid, glacial	5 ml

Possibly no other fixing fluid has as many names applied to it. Many modifications in the amount of each constituent have been recommended, and the resulting mixture often bears the name of its sponsor. Evidence indicates, however, that the proportions are not critical, within limits. Water enters into the formula only incidentally through that contained in the formalin and in the diluted alcohol.

Keep a stock solution containing the above proportions of alcohol and formalin. Add the acetic acid to measured quantities of this as needed for use. Thus if one needs 100 ml of the final solution, take 95 ml of stock and add 5 ml of acid; if only 50 ml of this is required, each of the parts should be halved.

Alcohol-glycerine Mixture

Alcohol, 70%	95 ml
Glycerine	5 ml

Alkaline Alcohol

Alcohol, 70%	500 ml
Ammonium hydroxide, concentrated	0.5 ml

Berlese's Chloral-gum Solution, Modified

This aqueous medium is recommended for small, hard-to-clear specimens. Material can be mounted wet or dry, alive or from alcohol or formalin, and the cover glass added. Since clearing takes only a few minutes, it is rapid and efficient. It avoids the necessity of using a fixative; it makes unnecessary the dehydration process of conventional technique, and it avoids the loss of specimens due to their small size and the need for concentrating them after each alcohol change. The mounts harden slowly, but with care they may be used almost immediately.

Water, distilled	50 ml
Gum arabic (acacia), clear lumps or powdered	40 gm
Glycerine	20 ml
Chloral hydrate	50 gm
Acetic acid, glacial	3 ml

Dissolve the gum arabic in the cold water, then dissolve the chloral hydrate with gentle heat in a water bath, add the glycerine and acetic acid, and filter while warm through several layers of cheesecloth.

Bile Salts Solution

Chocolate covered bile salts (available from druggist) .... 1 grain
Physiological solution .... 30 ml

Grind 1 tablet finely in a mortar and place in the saline solution. Filter after a few minutes to remove excess chocolate, thus permitting ready observation of the evagination process by cysticerci placed in the solution.

Bles' Fixative

Alcohol, 70% .... 90 ml
Formalin, commercial .... 1 ml
Acetic acid, glacial .... 3 ml

Boardman's Solution

Alcohol, 20% .... 97 ml
Ether .... 3 ml

Bouin's (picro-formol-acetic) Fixative

Picric acid, saturated aqueous solution .... 75 ml
Formalin, commercial .... 25 ml
Acetic acid, glacial .... 5 ml

This solution keeps indefinitely and is probably the best general purpose fixative. Specimens are usually left in it about 24 hours before being transferred directly to 70% alcohol.

Delafield's Hematoxylin

Hematoxylin crystals .... 1 gm
Alcohol, absolute .... 10 ml
Saturated aqueous ammonia alum
(aluminum ammonium sulphate) .... 100 ml
Glycerine .... 25 ml
Alcohol, methyl .... 25 ml

Dissolve the hematoxylin crystals in the absolute alcohol and add to this solution, a few drops at a time, the ammonia alum. Leave it exposed to the light and air in an unstoppered bottle for several weeks (a month is not too long) to "ripen." Filter, add the glycerine and the methyl alcohol.

Dextrose-salt Solution

Sodium chloride .... 2.25 gm
Calcium chloride .... 0.06 gm
Potassium chloride .... 0.1 gm
Sodium carbonate .... 0.04 gm
Dextrose .... 0.62 gm
Water, distilled .... 1000 ml

Digestive Fluid, Artificial

Powdered pepsin, fresh .... 5 gm
Physiological saline (warmed) .... 1 liter
Hydrochloric acid, commercial .... 7 ml
For fish, increase the pepsin to 7 gm and reduce the HCl to 4 ml/liter of saline solution.

Ehrlich's Hematoxylin

Acetic acid, glacial .... 10 ml
Alcohol, absolute .... 25 ml
Hematoxylin crystals .... 2 gm
Glycerine .... 100 ml
Water, distilled .... 100 ml
Potassium alum (aluminum potassium sulphate) .... 10 gm

Mix the glacial acetic acid with the absolute alcohol and add the hematoxylin crystals. When dissolved, add an additional 75 ml of absolute alcohol and the glycerine. Heat the distilled water and add the potassium alum. When dissolved and still warm, add while stirring to the hematoxylin solution. Let the mixture ripen in the light (with an occasional admission of air) until it acquires a dark red color.

Fast Green, Stock

Fast Green, powdered .... 0.2 gm
Alcohol, 95% .... 100 ml

Formalin

Commercial formalin is approximately a 40% solution of formaldehyde gas in water. This solution is known as 100% formalin and is diluted accordingly when concentrations of formalin are desired. Thus 5% formalin, which is 2% formaldehyde, contains
Water, distilled .... 95 ml
Formalin, commercial .... 5 ml

Giemsa's Stain

The dry powder for preparing the stock solution is available from firms dealing in biological stains. A formula for making a stock solution is:
Giemsa's powder .... 1 gm
Glycerine .... 66 ml
Methyl alcohol, absolute .... 66 ml

To the alcohol-glycerine, add glass beads and the dry powder. Allow the alcohol-glycerine to penetrate the dye for a few minutes and then rotate the flask for about two minutes. This mixture is agitated about every 30 minutes until the procedure has been repeated six times.

When possible, the stain should be made up early in the day so that the final shaking will be completed before the end of the working day. When prepared in this way, the stain is ready for immediate use. Store in a tightly stoppered bottle. The stock solution is diluted just before use with distilled water or dilute buffer solution. A suggested formula is the following:

Giemsa stock solution	1 ml
Water, distilled	20 ml
or	
Phosphate buffer (approx. 0.1 M, pH 6.5)	20 ml

Glycerine Jelly

Soak 7 gm of granulated gelatine in 40 ml of distilled water for 30 minutes. Then melt in a warm water bath and filter through several layers of cheesecloth previously moistened with hot water. Finally, dissolve 1 gm phenol in 50 ml of glycerine and add to the gelatine. Stir until the mixture is homogeneous.

Gomori Trichrome

Chromotrope 2R	0.6 gm
Light Green SF	0.3 gm
Acetic acid, glacial	1.0 ml
Phosphotungstic acid	0.8 gm
Water, distilled	100 ml

Grenacher's Alcoholic Borax-carmine

Carmine	3 gm
Borax, c.p.	4 gm
Water, distilled	100 ml
Alcohol, 70% (preferably methyl)	100 ml

Add the carmine and the borax to the water and boil until carmine is dissolved (30 minutes or more), or, better, allow the mixture to stand for two or three days, with an occasional stirring, until this occurs; then add the alcohol. Allow the solution to stand for a few days and filter.

Harris' Hematoxylin

Hematoxylin crystals	2 gm
Alcohol, absolute	20 ml
Water, distilled	400 ml
Ammonia alum (aluminum ammonium sulphate)	40 gm
Mercuric oxide	1 gm

Dissolve hematoxylin crystals in the absolute alcohol. Heat the distilled water and add ammonia alum. When dissolved and still warm, add, while stirring, the hematoxylin solution. Boil quickly and after a minute of boiling add the mercuric oxide to ripen the solution, which should now be purple. After a minute more of boiling, cool by plunging flask into cold water.

Heidenhain's Iron Hematoxylin

Two solutions are used. They are not to be mixed.

Solution I

Ferric ammonium sulphate (iron alum), use only clear violet crystals	2 gm
Water, distilled	100 ml

This acts as mordant and, when diluted, as differentiator.

Solution II

Hematoxylin crystals	1 gm
Alcohol, absolute	20 ml
Water, distilled	200 ml

First dissolve the hematoxylin in the alcohol, then add the water. Hematoxylin solution must be thoroughly "ripe." This may be accomplished by diluting the stock solution with distilled water until the hematoxylin percentage is approximatey 1%, which is ready for immediate use.

Helix Physiological Solution

Sodium chloride	5.87 gm
Potassium chloride	0.73 gm
Calcium chloride	1.99 gm
Sodium bicarbonate	1.82 gm
Magnesium chloride, hexahydrate	5.62 gm
Potassium bicarbonate	0.22 gm
Water, distilled	1000 ml

Insect Preservative

Alcohol, 95%	53 ml
Ethyl acetate	15 ml
Benzene	5 ml
Water, distilled	27 ml

Kronecker's Solution

Sodium chloride	3 gm
Sodium hydroxide	0.03 gm
Water, distilled	500 ml

Lacto-phenol

Water, distilled	20 ml
Glycerine	40 ml
Lactic acid	20 ml
Phenol, melted crystals	20 ml

This solution should be kept in a brown bottle or in a dark place, because exposure to light causes it to turn yellow.

Lugol's Solution

Lugol's solution consists of 10 gm potassium iodide and 5 gm of iodine crystals in 100 ml of distilled water. Filter into brown bottle, stopper tightly and store away from light. Solution deteriorates rapidly.

Merthiolate-Iodine-Formalin (MIF) Stain-preservative Solution

1. Stock MF Solution (Stored in brown bottle)
   Water, distilled .... 50 ml
   Tincture of Merthiolate, Lilly No. 99 (1:1,000) .... 40 ml
   Formalin (USP) .... 5 ml
   Glycerine .... 1 ml
2. Lugol's Solution, fresh

This may be used (I) in making direct fecal smears and (II) for collecting and preserving fecal specimens.

I. *Direct Fecal Smears.* Place one drop of distilled water and one drop of MIF (MF stock solution plus 5% of freshly prepared Lugol's solution) fixative-stain together on a slide, add a small amount of feces and comminute thoroughly with a toothpick, apply a cover glass and examine.

II. *Collection and Preservation of Bulk Specimens.* For approximately each 0.25 gm of stool to be processed, first introduce 0.15 ml of Lugol's solution followed by 2.35 ml MF solution into a Kahn tube or other suitable glass container, add stool specimen, and comminute thoroughly.

Phenol-alcohol Solution

Phenol .... 80 ml
Alcohol, absolute .... 20 ml

Store in a brown bottle to prevent solution from changing to a dark color.

This solution is useful for clearing specimens quickly, equally well from water or alcohol, or mixtures as occur in various fixatives or preservatives. Its principal use is rapid clearing of nematodes for observing cuticularized structures, such as spicules. Although it overclears quickly, the degree of clearing can be controlled by drawing 95% alcohol under the cover glass, using absorbent paper. Nematodes placed in this solution become round and turgid. If specimens are to be kept, wash several times in 70% alcohol to remove phenol and store in 70% alcohol containing glycerine.

Potassium Alum Solution

Potassium alum (aluminum potassium sulphate) .... 6 gm
Water, distilled .... 100 ml

Potassium Dichromate Solution

Potassium dichromate .... 2.5 gm
Water, distilled .... 100 ml

Ringer's Solution

Sodium chloride .... 8 gm
Sodium bicarbonate .... 0.2 gm
Potassium chloride .... 0.2 gm
Calcium chloride (anhydrous) .... 0.2 gm
Dextrose (optional) .... 1 gm
Water, distilled .... 1000 ml

Saline Solution

Sodium chloride .... 7.5 gm
Water, distilled .... 1000 ml

It is unlikely that any other physiological solution is used more universally for parasitic helminths. Some workers recommend 7 gm NaCl/1000 distilled water for parasites of cold-blooded hosts; 8.5 gm NaCl/1000 distilled water for parasites of warm-blooded hosts.

Schaudinn's Fluid

Mercuric chloride, saturated aqueous solution .... 66 ml
Alcohol, 95% .... 33 ml
Acetic acid, glacial .... 3 ml

The first two can be kept mixed without deterioration, but the acid must be added just before fixation.

Semichon's Acetic-carmine

Acetic acid, glacial .... 100 ml
Water, distilled .... 100 ml
Carmine "in excess" .... about 1.5 gm

Mix distilled water and acetic acid in an Erlenmeyer flask, and add carmine. The objective is to prepare a saturated solution of carmine, but since any excess is wasted, there is no need to add more than will go into solution. Heat in boiling water bath for 15 minutes, then cool the flask in cold water and filter the contents. This stock stain should be diluted with approximately two parts of 70% alcohol before use.

Sheather's Sucrose Solution

Sucrose .... 50 gm
Water, distilled .... 35 ml
Phenole, liquid .... 0.2 ml

Trisodium Phosphate Solution

First, a saturated stock solution is made as follows:
Trisodium phosphate .... 28.3 gm
Water, distilled .... 100 ml

Second, for final use prepare as follows:

Trisodium phosphate, stock	2 ml
Water, distilled	98 ml

At room temperature this gives approximately a 0.25% solution.

Wright's Stain

The preparation of this stain is rather complicated. It is recommended that the solution be purchased ready-made and used as it comes from the bottle.

Zinc Sulfate Solution

Zinc sulfate	330 gm
Warm water	1000 ml

Dissolve zinc sulfate in water and adjust to specific gravity of 1.18.

References

Anderson, R. C. 1958. Methode pour l'examen des nematodes en vue apicale. *Ann. Parasitol. Hum. Comp.* 33:171–72.

Benton, A. H. 1955. A modified technique for preparing whole mounts of Siphonaptera, *J. Parasitol.* 41:322–23.

Blagg, W.; Schloegel, E. L.; Mansour, N. S.; and Khalaf, G. I. 1955. A new concentration technic for the demonstration of protozoa and helminth eggs in feces. *Amer. J. Trop. Med. Hyg.* 4:23–28.

Dubey. J. P.; Swan, G. V.; and Frenkel, J. K. 1972. A simplified method for isolation of *Toxoplasma gondii* from the feces of cats. *J. Parasitol.* 58:1005–6.

Goodey, T. 1963. *Soil and Freshwater Nematodes,* rev. ed. John Wiley & Sons, New York, 544 pp.

Harder, H. I., and Watson, D. 1964. Human filariasis. Identification of species on the basis of staining and other morphologic characteristics of microfilariae. *Amer. J. Clin. Path.* 42:333–39.

Hoffman, G. L. 1967. *Parasites of North American Freshwater Fishes.* Univ. of California Press, Berkeley, 486 pp.

Kritsky, D. C.; Leiby, P. D.; and Kayton, R. J. 1978. A rapid stain technique for the haptoral bars of *Gyrodactylus* species (Monogenea). *J. Parasitol.* 64:172–74.

Mizelle, J. D. 1938. Comparative studies on trematodes (Gyrodactyloidea) from the gills of North American freshwater fishes. *Illinois Biol. Monog.* 17:1–81.

Sapero, J. J., and Lawless, D. K. 1953. The "MIF" stain-preservation technic for the identification of intestinal protozoa. *Amer. J. Trop. Med. Hyg.* 2:613–19.

Thorne, G. 1935. Notes on free-living and plant-parasitic nematodes. *Proc. Helminthol. Soc. Washington* 2:96–98.

———. 1961. *Principles of Nematology.* McGraw-Hill Book Company, Inc., New York, 553 pp.

Van Cleave, H. J., and Ross, J. A. 1947. Use of trisodium phosphate in microscopical technic. *Science* 106:194.

Appendix 2 Some Vertebrate Diseases and Infections, together with the Arthropods Important in Their Transmission

	Hosts	
Diseases/Infections	**Vertebrates**	**Arthropods**
SPIROCHAETES		
Louse-borne relapsing fever	Humans	ANOPLURA
		Pediculus humanus
Tick-borne relapsing fever	Humans	ACARINA
		Ornithodoros
RICKETTSIAE		
Q fever	Humans, cattle, et al.	*Dermacentor* et al.
Rocky Mountain spotted fever	Humans	*Dermacentor* et al.
Rickettsial pox	Humans, mice	*Allodermanyssus*
Scrub typhus	Humans, rodents	*Leptotrombidium*
Epidemic typhus	Humans	ANOPLURA
		Pediculus humanus
Murine typhus	Humans, rodents	SIPHONAPTERA
		Xenopsylla cheopis
BACTERIA		
Plague	Humans, rodents	*X. cheopis* et al.
Tularemia	Humans, rabbits	DIPTERA
		Chrysops
		ACARINA
		Haemaphysalis
		Dermacentor
VIRUSES		
Colorado tick fever	Humans	*Dermacentor andersoni*
		DIPTERA
St. Louis encephalitis (SLE)	Humans, birds	*Culex*
Western encephalitis (WE)	Humans, horses, birds	*Culex*
Eastern encephalitis (EE)	Humans, horses, birds	*Culiseta melanura* et al.
Yellow fever	Humans, monkeys	*Aedes aegypti* et al.
Sandfly fever	Humans	*Phlebotomus*
Blue tongue	Ruminants	*Culicoides parvipennis*
PROTOZOA		
Leishmania spp.	Humans, et al.	*Phlebotomus* et al.
Parahaemoproteus nettionis	Ducks, geese	*Culicoides* sp.
Haemoproteus spp.	Birds	*Lynchia* et al.
Leucocytozoon spp.	Birds	*Simulium*
Plasmodium spp.	Humans, et al.	*Anopheles* et al.
Trypanosoma spp.	Humans, et al.	*Glossina* et al.
		HEMIPTERA
		Triatoma
Babesia spp.	Cattle, et al.	ACARINA
		Haemaphysalis
		Dermacentor
		Boophilus
		Ixodes

Diseases/Infections	Hosts: Vertebrates	Hosts: Arthropods
TREMATODA		
Prosthogonimus macrorchis	Chickens et al.	*ODONATA* *Leucorrhinia* et al.
Haematoloechus medioplexus	Frogs	*Sympetrum*
Paragonimus spp.	Humans, mink	DECAPODA *Cambarus* et al.
Dicrocoelium dendriticum	Sheep et al.	HYMENOPTERA *Formica fusca*
CESTODA		
Anoplocephalidae	Chiefly mammals	ACARINA Oribatid mites
Diphyllobothrium spp.	Humans, et al.	COPEPODA *Diaptomus* *Cyclops*
Spirometra spp.	Cat et al.	*Cyclops*
Proteocephalus ambloplitis	Bass et al.	*Cyclops*
Choanotaenia infundibulum	Chickens et al.	DIPTERA *Musca domestica* et al.
Echinolepis carioca	Domestic fowl et al.	COLEOPTERA Scarabaeidae
Dipylidium caninum	Humans, dog, cat	MALLOPHAGA *Trichodectes* SIPHONAPTERA *Ctenocephalides* et al.
Hymenolepis diminuta	Humans, mice	*Xenopsylla cheopis* et al. MALLOPHAGA *Trichodectes* LEPIDOPTERA *Pyralis* et al. COLEOPTERA *Asopis*
ACANTHOCEPHALA		
Macracanthorhynchus hirudinaceus	Swine	*Cotinus* *Phyllophaga*
Leptorhynchoides thecatus	Fish	AMPHIPODA *Hyalella*
Pomphorhynchus bulbocolli	Sucker et al.	*Hyalella*
Plagiorhynchus cylindraceus	Birds	ISOPODA *Armadillidium vulgare* et al.
Moniliformis moniliformis	Rat	ORTHOPTERA *Periplaneta*
Mediorhynchus grandis	Birds	*Schistocerca americana* et al.
Neoechinorhynchus cylindratus	Fish	OSTRACODA *Cypria*
NEMATODA		
Dracunculus medinensis	Humans et al.	COPEPODA *Cyclops*
Philometroides nodulosa	Suckers	*Cyclops*
Ascarops strongylina	Swine	COLEOPTERA *Aphodius* et al.
Litomosoides carinii	Cotton rat	ACARINA *Ornithonyssus bacoti*
Habronema megastoma	Horse	DIPTERA *Musca*
Wuchereria bancrofti	Humans	*Anopheles* *Aedes* *Culex*
Brugia malayi	Humans et al.	*Anopheles* *Mansonia*
Loa loa	Humans	*Chrysops*
Onchocerca volvulus	Humans	*Simulium*
Mansonella ozzardi	Humans	*Culicoides*
Dirofilaria immitis	Dog	*Anopheles* et al.
D. scapiceps	Hare	*Aedes*
Dipetalonema arbuta	Porcupine	*Aedes*
D. perstans	Humans, monkey	*Culicoides*
D. reconditum	Dog	SIPHONAPTERA *Ctenocephalides*

Appendix 3 Life Cycle Exercises

Protozoa

Trypanosoma lewisi

Development of *T. lewisi* can be followed in white rats by observing stained blood smears made at frequent intervals over a period of 30 days after inoculating the rat with infected blood. Note the dividing rosettes of epimastigotes and the difference in size of the trypanosomes. Observe the rise, peak, and decline of the population in the blood during this period. Do the trypanosomes disappear? What are the time relationships of these events?

In northern rat fleas (*Nosopsyllus fasciatus*), the trypanosomes enter the epithelial cells of the stomach, where binary fission occurs. Upon leaving the cells, they migrate to the rectum and transform to epimastigotes that develop into small trypanosomes known as the metacyclic form. These are voided in the feces and are infective to rats when ingested.

Passage of infective metacyclic forms from the rectum of the flea is known as infection from the posterior station—in contrast to infection from the anterior station by metacyclic forms developing in the anterior part of the alimentary tract and injected by feeding arthropods.

Eimeria tenella

The life cycle of *Eimeria tenella* of chickens can be used as an example of a typical coccidian.

Feed a massive dose of fully sporulated oocysts to eight or ten young chicks. Sacrifice one at each 24 hour interval. Preserve the caeca of each chick in 10% formalin for subsequent sectioning and study. Stain with hematoxylin and eosin. First generation merozoites appear two and a half to three days postinfection and the second during the fourth and fifth days. Make smears of fresh caecal contents at these times to show the merozoites and red blood cells resulting from hemorrhage. Stain the smears with Wright's blood stain.

Gametocytes appear in the epithelial cells between the fifth and seventh days, with the formation of gametes. Oocysts appear in the feces by the eighth day.

From your stained sections and smears, identify all of the stages described in the drawing on the life cycle of *E. tenella* (fig. 1.20).

Trematoda

Cotylurus flabelliformis

Gravid worms may be obtained from the intestines of wild ducks available during the hunting season. An alternative approach to getting them is to find infected snails and feed the snails harboring the tetracotyle metacercariae to baby chicks, ducklings, or goslings. Prepatency takes about a week. Eggs are voided by live, gravid worms in water. They may also be recovered from the feces of experimentally infected birds. Unembryonated eggs hatch in about three weeks at room temperature.

Expose young, preferably laboratory-reared snails to miracidia. They penetrate rapidly and soon transform into mother sporocysts. Mature mother sporocysts up to 5 cm long containing numerous germ balls, embryos of daughter sporocysts, and fully developed metacercariae (tetracotyles) occur in the region of the digestive gland of naturally infected *Lymnaea stagnalis*. About six weeks after infection, daughter sporocysts give birth to cercariae. Daughter sporocysts are large, numerous and entwined in the digestive gland.

These carcariae are among the largest of the pharyngeate, forktailed forms (fig. 2.26). Upon escaping from the snail host, they swim back and forth almost constantly, turning in a spiral manner. They soon lose their tails, sink to the bottom, and creep over the substratum. The furcae are longer than the tail stem. The body is covered by small backward-pointing spines, except for a bare circumoral area on which is a clump of about 18 forward-pointing spines located dorsal to the mouth. The tail stem bears six pairs of long, lateral, hairlike processes. The ventral sucker in the anterior part of the second half of the body is heavily spined.

There are two unpigmented eyespots a short distance anterior to the ventral sucker. The digestive system consists of a short prepharynx, small pharynx, long esophagus reaching almost to the ventral sucker, where it branches to form intestinal caeca that extend to near the caudal end of the body. Two pairs of penetration glands lie near the anterior margin of the ventral sucker. Their ducts pass dorsal to the penetration organ and open lateral to the circular mouth. The reproductive systems are represented by a spherical cluster of cells near the posterior end of the body. The last two cells on each side are in the base of the tail. The execretory bladder sends a long tubule through the tail stem to open at the tip of each furca. The flame cell pattern is

$$2[(2 + 2) + (2 + 2 + 2)].$$

Upon escaping from the snail first intermediate host, the cercariae penetrate other snails and develop into metacercariae of the tetracotyle type. Upon entering certain lymnaeid snails, they develop in the digestive gland and reproductive organs. In physid and planorbid snails, the metacercariae develop as hyperparasites in the rediae and sporocysts of other flukes. They even occur in their own mother sporocysts.

Developing and mature metacercariae occur in the digestive and reproductive organs of the normal snail hosts. About 20 to 30 days at room temperature are required for the cercariae to develop to mature metacercariae and assume the characteristics of the tetracotyles. Young metacercariae are free in the tissue of the snails, but as maturity approaches they always become encysted.

Development of the metacercariae in the snails is one of a striking and complete metamorphosis by which the cercaria transforms into a tetracotyle. The body of a fully developed tetracotyle is pear-shaped with the forebody broad. A large semicircular fold forming the ventral part of the cup-shaped forebody extends from each side of the oral sucker around the posterior margin of the acetabulum. The large holdfast organ is located between the fold and the ventral sucker. Lateral suckers, or cotylae with medial openings, form long clefts extending posterior from the level of the mouth to slightly beyond the acetabulum.

The digestive system consists of a well-developed pharynx, a much shorter esophagus, and very short intestinal caeca that bifurcate at the anterior margin of the acetabulum.

The reproductive system is represented by a large, somewhat bean-shaped cellular mass in the hindbody directly posterior to the holdfast body. The excretory bladder is represented by two vesicles, one at the end of each intestinal caecum. An excretory pore is located at the posterior extremity of the body and opposite an opening in the cyst wall.

While ducks are the natural final hosts, young chicks two weeks old and ducklings become infected when fed mature tetracotyles. Prepatency takes about a week.

Spirorchis parvus

Mature worms must be dissected from the large venules of the intestinal wall. In naturally infected turtles, masses of eggs accumulate in the intestinal wall, embryonate, and eventually break into the lumen of the gut. They are voided unhatched with the feces. Since hatching occurs soon after the eggs reach the water, feces should be collected daily, comminuted and processed by sedimentation and decantation. They are readily recognized, being, along with those of *S. elephantis,* the only embryonated, nonoperculate fluke eggs in painted turtles.

Sediment and decant freshly voided feces to remove fine debris and coloring matter. Pour the cleaned sediment into a small dish of water and change several times daily. At room temperature, hatching occurs between 2 A.M. and 5 A.M. as soon as four days after entering the water.

The miracidia have a large apical gland, flanked by a penetration gland of about the same length, on each side. Immediately behind the apical gland is a pair of crescentic, brown eyespots. The germinal mass lies immediately caudad from the eyespots. A lateral process extends from each side of the body opposite the posterior end of the apical gland. A pair of flame cells on each side of the body opens through excretory pores near the middle of the body. There are four tiers of epidermal plates. Beginning anteriorly, they contain six, six, four, and two plates, respectively.

To assure parasite-free snails for infection, only laboratory-reared *Helisoma* spp. should be used. Young snails between four and 25 days of age are most favorable for infection. Expose them to a large number of newly hatched miracidia. Observe the reaction of the miracidia in the presence of the snails. Watch them attach to and burrow into the snails. Do the latter show evidence of annoyance?

Upon gaining entrance to the snails, miracidia transform to mother sporocysts which are readily found in the free lateral margin of the mantle in five to eight days. By 18 days, they attain a length of 2 mm and contain young daughter sporocysts. In sporocysts five days old, no body cavity or embryos are visible; at seven to eight days germ cells are numerous and a few embryos appear.

Upon leaving the mother sporocyst, daughter sporocysts appear in the lymph spaces of the mantle and in the digestive glands, where they develop.

Dissect sporocysts from the digestive glands, beginning 18 to 20 days after infection of the snails. Continue dissections at intervals of four to five days until cercariae appear.

A birth pore develops at one end. Embryos of cercariae appear when the daughter sporocysts enter the digestive gland. The body wall is thin and the cavity large. It contains germ balls and cercariae.

Study living sporocysts dissected from the liver. Note the birth pore at the anterior end, the germ balls, and cercariae in various stages of development.

The furcocercous cercaria is apharyngeate, brevifurcate, distomate, and bears a cuticular crest along the humped dorsal side (fig. 2.26K). They are very active upon escaping from the snails, swimming and resting intermittently as is characteristic of furcocercous cercariae. They creep inchworm-fashion over the bottom of the dish. Watch for the appearance of cercariae issuing from the experimentally infected snails, beginning about the third week after infection. Hold a vial containing the infected snail so that light falls on it, and view toward a dark background. Swimming cercariae appear as small, white specks moving rapidly in the water.

The protrusible head organ is pyriform. The ventral sucker is smaller than the head organ and located near the union of the second and last thirds of the body. There are six pairs of penetration glands in cercariae that emerge from snails. The ducts open at the anterior end of the head organ. There are seven pairs of sharp penetration spines on the anterior end of the head organ. The mouth is subterminal and ventral, opens into a long slender esophagus that bifurcates into two short caeca just anterior to the ventral sucker. A prominent mass of germ cells located behind the ventral sucker represents the primordium of the reproductive systems.

The excretory system consists of a V-shaped excretory bladder. Two collecting tubes arise from the anterior end of the bladder, and a caudal one from the hind end extends to the end of the tail, bifurcates, and empties at the tips of the furcae. The flame cell formula is

$$2\,[(1 + 1 + 1) + (1 + 1 + 1)].$$

Collect *Helisoma* snails from ponds where painted turtles live and isolate individuals in vials in pond water. Observe them daily for the characteristic and easily identified cercariae of *S. parvus*. Cercariae from naturally infected snails are suitable for infecting turtles.

Young turtles are most suitable for infection experiments. Cercariae attack the soft tissues between the toes, in the flanks, around the eyes, the anus, and in the nasal passages. Turtles being attacked by cercariae react as if trying to escape from them.

After exposing young turtles to infection, keep them in aquaria and examine the feces to determine the prepatent period.

Echinostoma revolutum

A plentiful supply of gravid worms is available to students who have access to wild ducks during the hunting season or muskrats during the trapping season. Flukes from hosts dead for some time are likely to be dead themselves, but the eggs are viable. Eggs are best obtained by dissecting them from the uterus of worms.

In eggs that have been incubated at room temperature for six days, the miracidia are nearly spherical, ciliated, and have two flame cells. During the next few days, the eyespots appear and the cuticular plates become evident. After ten to 12 days, they have the appearance of a hatched individual except for being somewhat smaller. Hatching occurs in 18 to 30 days. Miracidia are covered with unicellular epidermal plates arranged in four rows of six, six, four, and two from anterior to posterior. Eyespots are in the form of dark crescents. There are two flame cells on each side of the body, opening laterally near the junction of the third and last tiers of epidermal plates. An apical gland is present.

Young laboratory-reared snails of the genera *Lymnaea, Physa,* and *Bithynia* are preferable for experimental infections. Place a large number of recently hatched miracidia with the young snails in a small dish of water, leaving them together for several hours.

Upon entering the snails, the miracidia shed the ciliated covering and transform into mother sporocysts. By 40 days postinfection they contain a small number of mother rediae.

There are three generations of rediae. The first can be recognized during its sojourn in the mother sporocyst, the second by rediae within, and the third by the presence of cercariae. Examine the liver of snails 70 to 80 days postinfection for rediae containing rediae and cercariae. The rediae vary greatly in size and, of course, have the pharynx and primitive gut. In addition to rediae or cercariae in the lumen, the rediae contain masses of germ cells.

Watch for cercariae, beginning about 40 days postinfection, when snails are kept at room temperature. Echinostome cercariae are unmistakable because of the head collar and the collar spines on it. They are arranged in the same pattern as on the adult worms (fig. 2.37). Tegumental spines are seen best when cercariae are stained lightly with fast green or methylene blue. A slender prepharynx precedes the oval muscular pharynx. The esophagus bifurcates at the anterior margin of the ventral sucker, and the two caeca extend to the end of the body. The short, broad, saccular excretory bladder sends common collecting tubules to the anterior end of the body, and turn caudad again, sending off six groups of three flame cells each, along the side of the body from the hind end to the front end. From the posterior end of the bladder, a tubule extends caudad into the tail a short distance, divides, and empties laterally on each side. The formula of the flame cell pattern is

$$2 [(3 + 3) + (3 + 3) + 3 + 3)].$$

Infect snails and tadpoles by placing them in a dish with cercariae. Tadpoles and physid snails are killed by too heavy infections, but *Lymnaea* snails tolerate many more cercariae. Cercariae enter the anus of tadpoles and migrate up the ureters to the kidneys, or go through the respiratory duct to the gill chamber and encyst within a few hours. They also enter the anus of small bullheads and encyst in the kidneys. The same specimen of snail that served as the first intermediate host, producing cercariae, may likewise serve as the second intermediary, for the same cercariae may reenter and encyst in it.

Metacercariae become infective within a day after entering the second intermediate hosts. Feed several infected snails to five pigeons, chicks, or rabbits. Examine a host every other day for eight days to obtain stages of developing flukes. Prepatency is about 12 days.

Fasciola hepatica

In endemic areas, such as the southern states, Pacific Coast, parts of the mountain and Great Lakes states, adults can usually be collected in numbers at abattoirs. The complete life cycle can be studied in the snail intermediate host and a mammalian final host such as mice, rats, or rabbits.

Adults can be obtained from the condemned livers of cattle and sheep where these animals are processed. By cutting across infected livers and squeezing them, flukes and eggs often emerge by the handfuls from the open ends of the biliary ducts.

Many eggs can be obtained from the gall bladder or washed from the bile obtained in collecting the flukes. Also, gravid worms discharge eggs when placed in a dish of water. Because of their large size, these eggs can be seen easily with a dissecting microscope.

When incubated in light at room temperature, miracidia develop and begin hatching in about 8 days, with mass emergence by the ninth to tenth days. Eggs kept in darkness do not hatch as readily. When fully embryonated, the miracidium with its semilunar, dark eyespots, and ciliated covering can be seen through the egg shell. The anterior end, with the unciliated papilla-like prolongation, or rostellum, is at the operculated end of the egg. Germ balls appear inside the miracidium. Alongside it in the egg are granules of residual yolk material.

Upon escaping from the eggs, miracidia swim rapidly in an aimless fashion. The anterior end is broader than the posterior and bears the naked rostellum. There are five tiers of epidermal plates, beginning at the anterior end with six, six, three, four, and two plates each. These plates bear the cilia. Several sensory papillae appear between the first and second tiers. There is a large primitive gland at the anterior end of the body, flanked by a pair of small penetration glands. The excretory system consists of large flame cells in the anterior part of the body; each long tubule opens to the outside laterally in the posterior half of the body.

Because of their amphibious nature, the snails (*Lymnaea, Succinea, Fossaria, Practicolla*) require an area of exposed mud. Prepare a large moist chamber dish or small aquarium with the bottom covered with mud so that it will be exposed on one side. Maintain a small pool of water on the low side. Keep the dish or aquarium covered to provide a saturated atmosphere inside. Preferably the mud and water should come from a habitat where the snails thrive. Lettuce or dried maple leaves provide suitable food. An excellent diet is easily prepared by putting chopped grain feed in a pail of water and allowing it to ferment. The sediment after fermentation is readily eaten by the snails and does not foul the water in the aquarium.

Very young laboratory-reared snails or specimens collected from areas not frequented by the vertebrate hosts can be infected by putting the appropriate species in a small container with miracidia. Prevent the snails from crawling out of the water during exposure to the miracidia.

Watch the miracidia attack the snails and the reactions of the latter. Put a Physidae or other kind of snail in the dish and notice how the miracidia shun it. Put some mucus from one of the aceptable snail hosts in a dish containing miracidia and note how they congregate around it. Why do you think they react this way?

Miracidia attack and enter snails over the entire exposed part of the body. Upon entering the epithelium, the miracidia shed their ciliated covering. Eyespots persist but become detached from each other and lose their crescent shape.

After 24 hours, miracidia are present in large numbers in the foot, having undergone very little change. In 48 hours, they are roughly circular. After four days, the sporocysts are elliptical. By the eighth day, sporocysts may attain full development and contain motile rediae. At this time, the sporocysts are numerous in the pulmonary chamber. They are readily recognized as elongate, saclike bodies without a pharynx or primitive gut, and they contain rediae.

There are three generations of rediae. The first two produce rediae and are difficult to differentiate, but the third is readily recognized because it contains cercariae. By the ninth day, the first rediae appear, moving about in the liver. Obviously, these are mother rediae since they are the first to appear. They are sac shaped, with a muscular pharynx at one end followed by a short, thin-walled primitive gut. A pair of lateral outpocketings is present near the posterior end of the body. A collar appears near the anterior end. Development is rapid in the liver. By the 19th day, third-generation rediae with developing cercariae are present. In 24 days, the cercariae have developed tails and exhibit motility. At 31 days, cercariae are well developed, show much pigmentation, and are active.

Cercariae (fig. 2.26D) are shed naturally through the branchial pore of snails around 49 days after infection. As many as 800 or more may be produced in one day by a single heavily infected snail.

The body of the gymnocephalus cercaria is elliptical to almost circular, with a notch where the tail attaches. The suckers are about equal in size. The digestive tract consists of a muscular pharynx, short esophagus, and long simple caeca without a lumen. Cystogenous glands fill the body, tending to obscure all internal organs. The bilobed excretory bladder is thin walled at first but is surrounded later by layers of muscle. The stout tail is up to twice the length of the body. Encystment may occur within a few minutes after escaping from the snail.

Upon coming into contact with a solid object in the water—even the sides and bottom of the dish containing the snails—a swimming cercaria attaches by means of the suckers, rounds up, detaches the tail, and secretes a flow of mucus that completely surrounds it. The mucus hardens and forms a two-layered cyst that adheres to the grass or other objects. Metacercariae are somewhat smaller than the cercariae, having discharged the mucus when forming the cyst. While they still have characters common to the cercariae, they are essentially young flukes.

A plate of glass set upright in the water of a dish containing snails shedding cercaria will accumulate a large number of metacercariae for experimental infection. Scrape them off with a razor blade and store in a vial of water for infection experiments.

Rats, mice, and rabbits serve well as final hosts. Due to their small size and lower cost, mice are considered best for this exercise. Moreover, the flukes grow rapidly in them. However, mice are more susceptible than other laboratory animals to damage by flukes migrating in the liver and may die as a result.

Cysts containing metacercariae may be introduced into the stomach by means of a small flexible tube attached to a syringe or mixed with moist food presented to hungry animals. For short-term experiments, use 100 to 150 cysts per mouse, for longer terms about 50, and for observation on prepatency and longevity of flukes in the liver use two to five cysts. Infect 15 to 20 mice for examination at appropriate times to determine the route of migration and progress of development in them. This number will allow for deaths resulting from liver damage by the flukes.

Metacercariae excyst in the intestine within two to three hours. Within 24 hours after entering the intestine, they have migrated through the intestinal wall and are in the coelom. There is no noticeable change in development of these individuals over the time of excystment.

Necropsy a mouse at about 30 hours after infection. Open the body cavity and flush it and the viscera with saline solution, catching all of the fluid in a tray. Examine the sediment for metacercariea that were migrating through the body cavity to the liver.

By the third day, flukes begin burrowing into the liver. They have increased in size, as growth is rapid in mice. The transverse vitelline ducts are visible and the testicular rudiment is enlarged. They are in the liver parenchyma.

Examine a heavily infected mouse on the fourth day by flushing the coelom with saline solution into a dish. Examine the sediment for young flukes and compare the number found with that recovered on the second day. Macerate the liver in a separate container so as not to destroy the metacercariae, allow the material to stand in saline solution for an hour or two, and then clean it by sedimentation and decantation. Examine the sediment under a dissecting microscope for the small flukes. What percentage of the number administered to the mouse was found in the coelom and in the liver tissue?

By the eighth day, the body has doubled in size and about 13 lateral sacculations appear on each side of the intestinal caeca. The rudiment of the ovary is slightly bulged. Examine the coelom and liver for young flukes. Determine the number of them in each place.

By the 11th day, the anterior body cone characteristic of the adult has appeared. Caecal outgrowths have secondary and tertiary branches. Testicular rudiments are separate and slightly lobed; and the rudimentary ovary is growing out of the shell gland complex. A rudimentary cirrus is visible.

By the 15th day, there is greater branching of the caecal outgrowths. The rudimentary cirrus is prominent inside the cirrus pouch and the testes branch laterally. The posterior end of the body has changed from a rounded to a broadly pointed shape.

By the 21st day, the cephalic cone and body shoulders are well demarcated. Caeca show short outgrowths on the median sides.

By the 28th day, the characteristic body shape of the adult is established. The lateral fields of the body are filled with vitellaria. Some flukes have entered the bile ducts and the others are approaching them. How they gain access to the ducts is unknown. Carefully dissect the liver and look for flukes in the bile ducts.

By the 37th day, flukes have attained sexual maturity and eggs begin to appear in the feces. Demonstrate them by fecal examinations. Prepare the sexually mature flukes from mice as stained permanent mounts and compare them for size with those obtained from cattle and sheep.

Other important Fasciolidae of the liver include *Fasciola gigantica,* common in cattle in Hawaii, and *Fascioloides magna* in deer, elk, and moose, as well as cattle, in North America. Their life cycles are similar to that of *Fasciola hepatica. Fasciolopsis buski* occurs in the intestine of man and swine in the Orient. This species does not have the migratory phase in its cycle as do *F. hepatica* and the other liver-dwelling forms.

Haematoloechus medioplexus

The study will be pursued as a field project rather than a series of controlled laboratory experiments. They should include (1) finding adult worms in the lungs of frogs; (2) obtaining and observing eggs; (3) collecting snail intermediate hosts (*Planorbula*) from the same habitat where infected frogs occur; (4) obtaining the intramolluscan stages of the fluke from naturally infected snails; (5) getting cercariae from them; (6) observing cercariae being drawn through the anus into the respiratory chamber of dragonfly naiads; and (7) collection of metacercariae from naturally infected naiads.

When live gravid worms are placed in water, they void many eggs. Hatching does not normally occur until they are eaten by the snail intermediate host.

From the area where infected frogs occur, collect snails and place several in each of a number of small widemouthed jars containing pond water. Observe the jars each morning by holding them so as to have a dark background and with plenty of light on them for seeing cercariae tumbling in the water. Cercariae are shed mostly during the night. When infections are found, isolate the snails in individual jars in order to find the infected ones.

Cercariae bear a stylet in the dorsal wall of the oral sucker and a finlike structure over the dorsal side, tip, and ventral side of the relatively short tail. The stylet and fin place them in the ornate xiphidiocercaria group (fig. 2.26I).

Dissect an infected snail and find the sporocysts. There are no rediae in the Plagiorchiidae. The sporocysts contain developing and fully formed cercariae. What kind of sporocysts are they? Should you happen to find a sporocyst producing only sporocysts, what would it be? Examine living sporocysts and locate the birth pore near or at the end of the body through which its progeny escape.

Place a dragonfly naiad in a small dish of water with many cercariae. While observing under a dissecting microscope, note how the cercariae are swept through the anus into the branchial basket by the respiratory currents created by the expansion and contraction of the abdomen. Also observe how the naiad reacts to the presence of the cercariae in the branchial basket where they are penetrating the gills. Keep the naiads for several days to allow metacercariae to develop. Open the naiads and observe the metacercariae in delicate cysts in the gills.

If metacercariae are available, feed them to a frog at intervals of two days for a week or thereabouts. Examine the frog one day after the last worms are fed to it. Note the differences in size of the worms. Are they in different locations?

Other commonly encountered Plagiorchiidae with their hosts include: other species of *Haematoloechus,* frogs, and toads; *Glypthelmins quieta,* frogs, *Plagiorchis proximus,* muskrats.

Cestoidea

Proteocephalus ambloplitis

Adults of *P. ambloplitis* are widely distributed and usually present where the host fish (small- and largemouth bass) and copepods occur simultaneously. Place gravid proglottids in biological water, and observe the rupture of the body wall and escape of the eggs. Remove spent proglottids, change the water, add copepods, and watch them eating the eggs. Examine the intestine of the copepods for oncospheres shortly after they have consumed the eggs, and the body cavity at hourly intervals postinfection for development of procercoids. Examination of the copepods is accomplished by concentrating them into a small amount of water, transferring a few in a drawn pipette to a slide, and adding a rectangular cover glass. Within a few minutes, they have quieted sufficiently for examination under the compound microscope, after which the sample is washed back into the culture with a wash bottle. After about three weeks postinfection, allow the appropriate fingerling host fish to feed on the copepods. Examine the viscera of the experimentally infected fish for plerocercoids five, ten, and 15 days postinfection.

Since plerocercoids may occur in the same species of host fish harboring the mature tapeworms, as well as other sunfishes, the plerocercoids can be collected and used to infect the appropriate host, especially when adult worms are not available.

Often other available Proteocephalidae with their hosts include *Proteocephalus parallacticus,* salmonids; *P. pearsei,* yellow perch; *P. pinguis,* chain pickerel; *Corallobothrium fimbriatum,* and *C. parvum,* bullheads.

Taeniidae Life Cycle Exercise

Life cycle exercises may be started with eggs or metacestodes. Adults of *Taenia pisiformis* and *T. serialis* from dogs, and *T. taeniaeformis* from cats are readily available from veterinary hospitals where pets are treated for removal of tapeworms. Eggs dissected from the gravid proglottids should be fed to the appropriate intermediate host, i.e., eggs of *T. pisiformis* to rabbits, and those of *T. taeniaeformis* to rats or mice. Examination of sacrificed experimental intermediate hosts over spaced intervals will show the migration through the circulation and muscles and development of the metacestodes. About a week after ingesting eggs, the metacestodes appear as pearly bodies in the liver and other parts of the body, depending on the species; they are infective to the final host about 60 days postinfection. Dogs and cats may be infected by feeding them mature metacestodes.

As a likely source of cysticerci the following generally available hosts are suggested: rabbits (*Sylvilagus* spp.) and hares (*Lepus* spp.) for *Cysticercus pisiformis,* the metacestode stage of a dog tapeworm *Taenia pisiformis;* muskrats and rats (*Rattus* spp.) for *C. fasciolaris* (as a strobilocercus), the metacestode stage of a cat tapeworm *T. taeniaeformis.*

Whether starting with eggs or metacestodes, care should be taken to see that the experimental hosts are clean. If eggs or proglottids are in the feces, the intended final hosts should be treated with an anthelmintic before feeding them metacestodes. Only laboratory reared and unexposed hosts should be used in egg feeding experiments.

Dipylidium caninum

Flea eggs can be obtained by placing a pan beneath a infested caged dog or cat. Eggs placed in petri dishes with moist sand at room temperature hatch within a few days. Feed the larvae dried blood or dog biscuits. Larvae intended for infection should be isolated and not allowed to eat for about a day. Tease gravid proglottids apart in saline solution to release the egg-capsules. Add sufficient powdered blood to make a paste and feed it to hungry larvae. After feeding, the larvae must be kept on moist sand until they become adults. Larvae, pupae, and adult fleas should be dissected, in a small amount of saline solution, to observe the development of the cysticercoids. If final hosts are to be fed experimentally infected fleas, they should first be given an anthelmintic, such as Praziquantel, to remove any tapeworms that may already be present.

Another commonly available Dilepididae is *Choanotaenia infundibulum* of chickens. Eggs develop into cysticercoids when fed to houseflies (*Musca domestica*), grasshoppers (*Melanoplus differentialis*), and a wide variety of beetles.

Other often available Dilepididae with their hosts include *Amoebotaenia cuneata,* chickens; and *Metroliasthes lucida,* turkeys.

Hymenolepis diminuta

Either *Tribolium* or *Tenebrio* beetles will serve as experimental hosts for this tapeworm. Both are easily maintained in the laboratory. About a week before feeding them eggs of *H. diminuta,* sift the beetles from the stock colony. Transfer them to petri dishes or other shallow glass containers lined on the bottom with filter paper. Cover the dishes to prevent escape of the insects. Label the dishes by number and date. Starve the beetles for about seven days.

After removing a gravid *H. diminuta* from a rat, cut off the last few proglottids, not exceeding 10 mm, and mince them together with a sliver of apple or a raisin. Place some of the mixture into each of the dishes, together with several beetles. The number of beetles depends upon the size of cysticercoid yield desired. Add the date of feeding to the label and secure the lid once again. Allow the beetles to feed overnight or up to 24 hours. Then transfer them to clean dishes with rolled oats or flour and keep them at room temperature for at least 15 days. A beetle can be dissected after five days, seven days, and so on to record progress of development. Fully developed cysticercoids are easily seen with a dissection microscope.

To infect rats, pipette cysticercoids into the back of the buccal cavity of a restrained rat, using a blunt pipette, such as a medicine dropper. One can expect a yield of about one adult tapeworm for each five cysticercoids administered. Worms should be shedding eggs in about two weeks.

Vampirolepis nana

Tapeworms for experimental purposes may be obtained from wild mice and rats. Feed gravid proglottids to flour beetles, from which food has been withheld for a day or so. At room temperature cysticercoids develop in two to three weeks. Examine the intestinal contents of beetles four to five hours postinfection for the presence of hatched oncospheres and again one or two days thereafter to observe the development of the cysticercoids in the hemocoel. For further details consult Voge and Heyneman (1957).

Infect mice or rats with mouse-line or rat-line cysticercoids. Also try cross-infection of rats with larvae originating from mice and vice versa to determine whether there are differences of susceptibility with the use of different physiological stains. Determine the prepatent period.

Also try the direct life cycle, by feeding gravid proglottids directly to mice. Examine the intestinal contents of one mouse four to five hours postinfection for the presence of hatched oncospheres. At daily intervals examine the intestinal mucosa to observe: (1) the penetration of the hexacanths into the villi and the development of the cysticercoids. (2) when the cysticercoids escape into the lumen of the intestine and attach to the mucosa, and (3) the rate of development to adults. Determine prepatency. Serial sections are required for the first two steps.

Davainea proglottina

Worms for experimental purposes can be obtained from farm-reared chickens. Eggs, teased from the gravid proglottids, should be placed in a petri dish with the appropriate molluscan intermediate hosts. To insure against using naturally infected intermediate hosts, they should be collected from localities where chickens have not been kept recently or, better still, reared in the laboratory. Dissection of the intermediary gastropods under a dissecting microscope and examination of the intestinal contents a few hours postinfection will show oncospheres and empty egg shells. Later examinations over spaced intervals will show the passage of the oncospheres through the gut wall and the developing cysticercoids. After about 14 days during the summer, the cysticercoids are infective to chicks.

Other often available Davaineidae include *Davainea meleagridis,* turkeys; *Raillietina* (*Raillietina*) *echinobothrida,* chickens; and *Raillietina* (*Skrjabinia*) *cesticillus,* chickens.

Acanthocephala

Leptorhynchoides thecatus

Eggs, obtained by dissecting gravid females in a dish of pond water, should be placed in a small jar and the amphipods added. After exposure for about an hour, wash the amphipods to remove any eggs that may be adhering and place them in a clean container. If left much longer, overinfection results, which kills many of the amphipods or causes abnormal and retarded development of the parasites. In order to insure that clean *Hyalella* are used, they should be collected from a habitat lacking fish. Amphipods may be kept in containers with a few willow rootlets and duckweed. Aeration and the addition of a bit of yeast to the culture each week are recommended.

Dissection of the intermediary crustaceans under a dissecting microscope and examination of the intestinal contents a few hours after exposure will show acanthors and empty egg shells. Later examination over spaced intervals will show the passage of the acanthors through the gut wall and subsequent larval development.

When cystacanths mature, feed them to small rock bass or other sunfish. Infect fish by feeding them infected amphipods or by dissecting out the cystacanths and feeding them with a bulbed pipette. After infecting fish, place them in individual containers and observe whether the cystacanths are regurgitated. To insure having clean fish, they should be held for about two months during which the feces are examined for parasite eggs. For further life cycle details, consult De Giusti (1949).

Macracanthorhynchus hirudinaceus

Adult worms are available from abattoirs where swine, which have had access to pasture, are processed. Eggs, teased from gravid females in saline solution, should be mixed with slightly moist soil. Add the larval beetles and allow them to feed for several days. Grubs for experimental purposes should be collected from soil to which swine have not had access for several years. To observe the hatching of the eggs, the passage of the acanthors through the intestinal wall, and subsequent development of the larval stages in the body cavity, examine the experimentally infected grubs, kept at room temperature, at spaced intervals varying from a few hours to about 90 days postinfection.

Moniliformis moniliformis

Adults can be obtained from rats in habitats where cockroaches are present. Eggs, teased from gravid females, in saline solution, should be mixed with a bit of pabulum in a small glass jar. Introduce clean cockroaches, from which food has been withheld for 24 hours, to the jar with the egg-contaminated pabulum. After being fed, the roaches should be removed to battery jars or similar containers with moist sand, over which are placed wet paper towels, to maintain the moisture and provide shelter for the roaches. Cover the jars to retain the moisture. Roaches thrive on pabulum and dry dog food and are fond of apples.

Infect a sufficient number of roaches so that dissections can be made over spaced intervals from two days through two months, to observe the hatching of the eggs, the passage of the acanthors through the intestinal wall, and subsequent development of the larval stages in the body cavity. To insure against using naturally infected roaches, they should be collected from rat-free localities or reared in the laboratory. Dissection of an adequate sample of a wild population will determine if they are suitable for experimental purposes.

White rats can be used as experimental final hosts. Cystacanths can be fed orally by means of a bulbed pipette. Experimental rats should be kept in a roach-proof enclosure to prevent further infections. Beginning about the fifth week postinfection, the prepatency period can be determined by examining the rats' feces.

Neoechinorhynchus sp. in Turtles

Adult worms can be obtained from the appropriate turtle hosts. Eggs, teased from gravid females, should be placed in a small dish of water to which ostracods are added. In order to insure that clean *Cypria* are used, they should be collected from a habitat lacking turtles.

Dissection of the ostracods under a dissecting microscope and examination of the intestinal contents a few hours after exposure will show the acanthors and empty egg shells. Later examinations over spaced intervals will show the passage of the acanthors through the gut wall and further larval development.

Nematoda

Ancylostoma caninum

The life cycle of *A. caninum* (fig. 5.13) is used as the representative of all hookworms, including those of humans, because important parts of it can be studied in the laboratory.

In warm regions where *A. caninum* commonly occurs in dogs, eggs for experimental studies may be obtained from feces of infected animals or gravid female worms expelled by dogs treated in small animal clinics. Fecal material should be mixed with sand or sphagnum moss to provide good aeration and reduce putrefaction, kept moist, and incubated at room temperature. Eggs dissected from gravid females may be incubated in petri dishes in shallow water. Larvae are easily recovered from the culture with the Baermann funnel. Place a thin layer of fecal material on the submerged screen in the funnel containing water about 40 to 42 C. Larvae will soon forsake the feces and appear in the rubber tubing on the stem of the funnel. By frequent examinations of the material, find the first- and second-stage rhabditiform larvae, and the third-stage filariform, infective larvae. When the buccal capsule of the filariform is viewed in optical section, it appears to consist of a pair of spears, extending from the mouth caudad for a short distance. These are actually the walls of the buccal capsule seen in optical section.

For experimental infection of mice, place 4,000 to 5,000 third-stage larvae in a small volume of water on a clean, dry, shaved area of skin of each of three restrained mice. A desk lamp placed near them will hasten evaporation of the water and facilitate the penetration of the larvae.

The first mouse should be killed and examined 12 hours postinfection, by grinding the skin finely in a food chopper and baermannizing for recovering the larvae. Each of the remaining mice should be examined on consecutive days after exposure for the recovery of migrating larvae, according to the following procedure.

Heart Blood. Ligate the hepatic portal vein near the intestine and the pulmonary artery near the lungs. Remove the heart and the two tied blood vessels intact, wash out the blood, and examine for larvae after sedimenting and decanting until clear. Allow 30 minutes for sedimentation so larvae will reach the bottom of the container.

Lungs. Examine lungs for hemorrhages due to blood diffusing from wounds produced by larvae migrating from blood vessels into the air sacs. Heavy infections will cause pneumonia. Grind the lungs and place them in Baermann funnels to recover migrating larvae.

Trachea and Esophagus. Slit open the trachea and esophagus, wash, and scrape them in separate dishes. Examine the sediment for migrating larvae.

Stomach and Intestine. Examine separately the contents and scrapings of the wall of each for larvae, using the high power of a dissecting microscope.

Body Muscles and Organs. Grind and baermannize separately voluntary muscles and various major organs for larvae.

Central Nervous System. Examine the brain for larvae by macerating and baermannizing it.

In comparison with the free-living third-stage larvae, do the ones recovered from the tissue of mice appear similar or have they developed beyond the third stage? How do you account for larvae that have entered the mice via the skin being in the body muscles and central nervous system?

Other common Ancylostomatidae are *Ancylostoma tubaeforme* of cats and *A. brazilense* from both cats and dogs.

Nematospiroides dubius

This species is a good illustration of a Trichostrongylina life cycle, in which the parasite enters its host passively through the mouth and has a limited migration into the mucosal lining of the intestine for only a short time before returning to the lumen. An additional advantage is its ready availability.

To collect eggs, allow feces from infected mice to accumulate for about 12 hours on moist paper towels placed beneath the cages in trays containing a shallow layer of water. Transfer the pellets to a flask or jar containing 0.1N NaOH and leave in the refrigerator overnight to comminute. Next morning, add water, shake vigorously, and remove the coarse material by straining through several layers of cheesecloth. The strained portion contains fine particulate material, coloring matter, and eggs.

Next remove the fine material and coloring matter, and concentrate the eggs. This is done by repeated sedimentation, by gravity or centrifugation, and decantation. Divide the strained materal into several tall cylinders, fill each with water, and let stand for 30 minutes. At the end of this time, decant the supernatant fluid and repeat until the water is clear. Concentrate the sediment that contains the eggs by a short period of centrifugation, transfer to a vial, and store in a refrigerator until ready to study them.

The centrifugation method of cleaning is more rapid. Fill two or four centrifuge tubes, preferably those of 50 ml capacity, spin briefly to precipitate the eggs but not the fine material, decant, and repeat until the water is clear. The trick is to spin the tubes only enough to throw the eggs down but leave the maximum amount of particulate and coloring matter in suspension for removal with the supernate. When the water is clear, concentrate the sediment from the various tubes and store in a vial in a refrigerator for subsequent observation on the development of the larvae.

Eggs incubated in clean water in a small dish with a cover are good for observing embryonation, hatching, and development of the three stages of larvae. Examine eggs and larvae at frequent intervals by placing them on glass slides in water during the first three to four days. Use a finely tipped, bulbed pipette to transfer them. Sealing the edges of the cover glass with Vaseline will maintain the water mounts, enabling one to observe the development of individual eggs and larvae over a long period. Hatching occurs in 20 to 24 hours after the feces are passed, and third-stage larvae are present in three to four days.

Large numbers of third-stage larvae for infection of experimental hosts can be obtained from eggs collected as described above. It is easier, however, to prepare fecal cultures from infected mice in which the eggs have hatched and the larvae developed to the third stage in three to four days at room temperature.

To obtain larvae, the fecal pellets may be handled in several ways, among which are: (1) in the petri dish-filter paper method, feces pulverized in a small amount of water are spread in a thin layer over moist discs of filter paper, placed in a petri dish, covered, and kept at room temperature; (2) in charcoal-fecal cultures, comminuted pellets are mixed with an equal amount of powdered charcoal spread on moist discs of filter paper, placed in a petri dish, covered, and kept at room temperature; (3) in charcoal-fecal cultures, comminuted pellets are mixed with an equal amount of moist, powdered charcoal and incubated (sand or sphagnum moss may be used in lieu of charcoal); and (4) fecal pellets may be incubated in petri dishes. In each case, larvae appear in large numbers.

Upon hatching, the active rhabditiform first-stage larvae are up to 0.4 mm long. The intestine is composed of indistinct cells. Ten hours after hatching, the larvae become lethargic for an equally long time in preparation for the first molt, which begins about 48 hours after hatching. It is indicated by the loosened cuticle appearing at the ends of the body.

Third-stage larvae are enclosed in the shed cuticle of the second-stage larvae. They are present between 48 and 56 hours after the beginning of the second molt, and measure up to 0.6 mm long. The buccal capsule is much shorter than in the preceding stages, and the esophagus has lost its rhabditiform character and becomes strongyliform, i.e., long and slender with a slight swelling at the posterior end. The tail is shorter, and blunter. The strongyliform esophagus, short blunt tail, and ensheathing loose cuticle are characteristics of this stage. Further development takes place only in the body of the host.

Third-stage ensheathed larvae may be exsheathed chemically by placing them in a fresh solution of one part commercial Clorox and three parts of water for ten to 15 minutes. Transfer the sheathed larvae from water into the exsheathing solution on a glass slide, cover, and watch the process under a microscope. Normal molting of third-stage larvae takes place in the mucosal lining under chemical conditions very different from those on the microscope slide.

The route of infection is by mouth only. Infect ten mice three to four weeks old by placing about 200 third-stage larvae, as determined by aliquot counts, in 0.2 ml of water on a small piece of bread and feeding it to the hungry animals.

The objective of examining mice after they have swallowed the third-stage larvae is to: (1) see third-stage larvae free in the lumen on the small intestine; (2) see them in the mucosal lining; (3) observe the third molt and subsequent rapid growth in the mucosa; (4) see them return to the lumen; (5) observe the fourth and final molt; and (6) note the development to sexual maturity.

Remove the small intestine, slit it open, and wash out the contents to obtain the three stages of worms at appropriate times. Press sections of it between two glass plates to see the worms in the mucosa.

According to the following protocol, examine the mice in search of the developing worms. Examine one mouse 24 hours after infection and look for larvae still free in the lumen of the intestine and those that have entered the mucosa. On the fourth day, examine another mouse and look for larvae in the lumen and mucosa. What stage are they, and is there any indication of molting? Examine a mouse on the sixth day and another on the eighth day. Where are the larvae, and what stage are they? Do any show evidence of the final molt approaching? Note the stage of development of the reproductive organs in both sexes. Are the sexes distinguishable?

Larvae may be removed from the intestinal wall by digesting the gut in a solution of 1% HCl and 1% pepsin in tap water for eight to ten hours at 37 C, with periodic shaking. When the digestive process is completed, wash and decant several times, and finally add sufficient formalin to make a solution of 10% (10 ml of commercial formalin in 90 ml of water) for fixing and preserving the worms.

Counting the worms is accomplished best after staining for 30 minutes in bulk in the formalin solution with iodine (30 gm I, 40 gm KI in 100 ml of hot water). Excessive coloring of the debris may be removed by washing in 5% aqueous sodium thiosulphate solution. Count all the worms in the intestinal wall and those found in the lumen to get the percentage of those swallowed that infected the mouse and those that failed to do so.

After determining the prepatent period, sacrifice the remaining mice to recover the adult worms, which should be prepared for study, using one of the standard methods.

Ascaris suum

Living gravid females placed in Kronecker's solution ovulate for six to 12 days. Eggs may also be obtained quickly by removing them from the uteri of gravid females, but many undeveloped ones will be present. Eggs kept in 1% to 2% formalin at 20 to 33 C develop to the second-stage larvae in three to four weeks but do not hatch. **CAUTION:** Care must be taken in handling embryonated eggs because they are able to hatch and the larvae migrate throughout the body in humans.

First-stage larvae are easily liberated mechanically by gently rolling the eggs back and forth under a cover glass, or tapping the cover glass with the handle of a teasing needle. Watch for developing larvae, which appear about the eighth day, and liberate them mechanically from the eggs, beginning when they first appear and continue at frequent intervals until satisfied that you have second-stage larvae. They will be the largest ones enclosed in the loose cuticle of the first molt.

Third-stage larvae do not appear until eggs containing fully developed second-stage larvae have hatched in the gut of the host and the larvae have reached the lungs about five days after ingestion.

Young guinea pigs are good experimental hosts, but before infecting them, some eggs embryonated for four weeks should be tested for infectivity by giving them to white mice. Examine the contents of the small intestine six to eight hours later for empty egg shells and newly hatched larvae. If there are no larvae in the intestine, the eggs are not sufficiently developed.

Infect ten guinea pigs weighing 140 to 150 gm each with 40,000 to 50,000 fully embryonated eggs by placing them back of the tongue in a minimum volume of water in a bulbed pipette. Make estimates of the number of eggs by aliquot counts.

Examine each animal in the same manner and look for the following: (1) intestinal contents for eggs, shells, and newly hatched larvae; (2) wall of small intestine for hemorrhages, congestion, and second-stage larvae; (3) liver for hemorrhages, white spots, pale color, swelling, congestion, and second-stage larvae; (4) heart for swelling; (5) lungs for hemorrhages, congestion, and second-, third-, and fourth-stage larvae; (6) congestion and pale color of kidneys, and (7) note whether labored breathing occurs.

To examine experimentally infected hosts, put finely diced intestine, liver, and lungs in separate Baermann funnels kept at about 37 C for several hours. Press bits of intestinal wall, liver, and lungs between two pieces of glass about four inches square and examine each under a microscope for migrating larvae. Wash intestinal contents from gut and clean by repeated sedimentation and decantation in water, using a hand centrifuge to precipitate the material. Put a drop of cleaned sediment on a microscope slide and examine under the low power of a compound microscope for eggs, shells, and larvae.

Examine each of the guinea pigs according to the protocol given below and search for the migrating larvae.

Two Hours Postinfection. A few macroscopic lesions in the small intestine. Many unhatched eggs and empty shells, and larvae in the stomach and small intestine. A few larvae in the mucosa; none in the liver and lungs.

Four Hours. Minute hemorrhages in intestine; newly hatched larvae but none in lymph glands, liver, or lungs.

Eighteen Hours. Numerous small hemorrhages in the intestine; liver pale, with numerous hemorrhages; few larvae in the liver and lungs.

One Day. No unhatched eggs but many shells in the gut; lymph nodes, liver, and lungs with many larvae; liver pale and with white spots.

Three Days. Many intestinal hemorrhages, congestion of intestine and liver. Many second-stage larvae in liver and lungs. Lungs with many hemorrhages; kidneys pale.

Five Days. Many larvae in lungs, few in liver. Liver enlarged and with white spots. Second molt begins in lungs and third-stage larvae appear. Many second-stage larvae still present.

Eight Days. Lungs with excessive damage. Heart and spleen enlarged. Many third-stage larvae in lungs and some in stomach and intestine. Larvae in trachea. Labored breathing.

Ten Days. Third molt begins in lungs. Intestine with hemorrhages, liver congested, lungs with hemorrhages. Labored breathing.

Eleven Days. Liver enlarged, congested, and dark. Lungs with many hemorrhages and area of consolidation. Kidneys pale. Small intestine with hemorrhages and congestion. Are larvae still in the intestine?

Sixteen Days. Host recovering. Liver normal in size but pale and with white spots. Lungs with small area of hemorrhages. Are larvae in intestine? Development of the adult does not occur in guinea pigs.

Some other common Ascaridae with their hosts include: *Baylisascaris columnaris,* skunks and racoons; *B. devosi,* fisher and martens; *Parascaris equorum,* horses; *Toxascaris leonina,* carnivores.

Toxocara canis

The objective of this experiment is to follow the migration of the larvae in white mice, determining the course taken and the time required for them to complete the somatic migration to the various tissues of the body. From the information obtained, it is possible to extrapolate what happens in children when infective eggs are swallowed.

Obtain gravid female *T. canis* from a small animal clinic where young dogs are treated to remove the worms. Dissect eggs from the terminal part of the uteri and incubate them in moist charcoal at room temperature. They will be infective when the second-stage larvae appear in them.

Infect each of 16 to 20 white mice of equal size and five to six weeks old with 2,500 embryonated eggs. Counts of eggs are made from dilution aliquots. Mix the eggs with a small amount of meal and serve in a shallow dish to hungry mice. Kill one mouse each day for 14 consecutive days and at 21 and 28 days and examine for larvae. From the results, it is possible to construct the route of migration by the larvae, the time required for each phase, final destination of the larvae, and their ultimate fate in the mouse.

Kill the mice by decapitation or cervical dislocation; remove and discard the skin, feet, and tail. Save the liver, lungs, alimentary tract, kidneys, carcass, and brain in separate containers for individual examination and enumeration of larvae. Grind the parts individually in physiological saline containing pepsin or trypsin in a blender. Use trypsin, pH adjusted to 7.0 with 0.1 N sodium hydroxide, for larvae less than eight days old and pepsin, pH adjusted to 1.0 with hydrochloric acid, for larvae eight days or older.

Grind carcasses in blender for 45 seconds in 50 ml of saline solution containing one gm of pepsin or trypsin and digest at 37 C for two hours. Grind liver and lungs for 45 or 30 seconds, respectively, in 25 ml of saline solution, pour into a pointed 50 ml centrifuge tube containing 0.05 gm trypsin in 5 ml saline, and digest for three to four hours. Stomach, intestine, and caecum are freed of contents, which are saved, and all three parts are ground together for 45 seconds in 20 ml of saline and treated similarly to the liver and lungs. Treat the kidneys and brain in a similar manner for counts of larvae in them. Before grinding the brain, note the hemorrhagic spots on the surface caused by the migrating larvae.

Each digested organ is carefully screened to remove tissue debris, sedimented and decanted several times by suction, and the remainder fixed by adding an equal volume of boiling 6% formalin. The volume of material from each organ is reduced after final sedimentation to 10 ml and stored in screw-top vials.

For counting the larvae, reduce the completely sedimented contents of each vial to 5 ml by suction. Shake the vial vigorously and quickly remove 1 ml of the contents with a graduated, bulbed pipette to a shallow, lined, glass dish, and count the larvae in one-half of it. If the larvae are too numerous to count, or the mixture too opaque to see them, reduce the volume in the vial to 3 ml and use 0.1 ml on a microscope slide, covering the fluid with a cover glass. To arrive at the approximate number of larvae per organ, multiply the number in 1 ml aliquots by 10 and the average of three counts of 0.1 ml by 30.

From the data acquired, prepare a report in which the migratory route, time of appearance and disappearance of larvae in each organ, and their final destination will be made. Begin the account with larvae in the intestine. For fuller details, consult Sprent (1952).

Syphacia muris or *Syphacia obvelata*

Eggs dissected from females do not develop in vitro, but those laid by them do. To obtain egg donors, place ten to 12 parasite-free mice three to four weeks old on contaminated fecal trays with infected mice for a period of 24 hours. The maximum number of gravid females appear in the colon and caecum on the twelfth day postinfection. When the gravid females are transferred quickly from the intestine to saline solution, they promptly lay about 300 eggs each. Such worms are in the procss of leaving the host to oviposit, after which they die. Eggs begin to appear in the perianal region on the eleventh day, as shown by the transparent adhesive tape examination method (see p. 257), indicating that the worms are in the colon and rectum.

When kept at room temperature, these eggs contain first-stage infective larvae in about 20 hours and at 37 C in about three to five hours. Infect ten to 12 young mice three to four weeks old with 300 to 400 embryonated eggs by means of a stomach tube or by putting eggs on a small piece of bread for individual hungry mice to eat.

Since the eggs hatch in the presence of trypsin rather than pepsin, newly hatched first-stage larvae appear in the small intestine. Inasmuch as development can proceed only in the environment provided by the caecum, the larvae quickly enter it, where subsequent development takes place.

Having selected an infected animal, dispatch it with deep anesthesia by ether, chloroform, or natural gas. Separate the parts of the small intestine from the caecum and large intestine and put in individual dishes of saline solution. For examination, each segment is slit, spread open in a dish, and scrutinized for worms. After examination, each segment of gut is put in a jar of saline solution, shaken vigorously, allowed to stand sufficiently long for the worms to settle to the bottom, and the supernatant fluid decanted. Clean the samples by repeated sedimentation and decantation until they are clear. Examine the sediment, in a petri dish in good light and under a dissecting microscope, for worms. Examine mice according to the protocol given below.

Two Hours Postinfection. Empty egg shells and first-stage larvae are present. Larvae up to 0.2 mm long are in the intestine and caecum. Where are most of the larvae? Why are there more in one place than the other? Do you see any larvae with loose cuticular sheaths, indicating commencement of the first molt?

Twenty-four Hours. The majority of the larvae are in the caecum but they are undifferentiated sexually. Enveloping cuticular sheaths are present. This is probably the first molt and the larvae are second-stage. They are up to 0.4 mm long.

At the Third Day. Sexual differentiation is apparent together with early development of the genital system. Males show the cuticular mamelons and females a column of cells leading to the vulva, which is not open. Some larvae enveloped in sheaths are still present. These may be in the process of undergoing the second molt. Males up to 0.5 mm long, females 0.6 mm. Note the relative proportion of males to females on this and succeeding days.

By the Fourth Day. Males show distinct mamelons; the spicule, preanal, and postanal papillae are forming. The female reproductive system shows little change over the previous day. These are probably third-stage larvae. Males are up to 0.9 mm long.

By the Fifth Day. About a third of the females have brown plugs closing the vulva, indicating that fertilization has occurred. The female reproductive system shows differentiation. These are probably fourth-stage larvae. Males measure up to 1.1 mm long; females are up to 2.1 mm long.

At the Sixth Day. The majority of the females are inseminated. The males have attained adult size of up to 1.5 mm long. Females are up to 2.6 mm long.

By the Seventh Day. The reproductive system of the female shows distinct uteri, oviducts, and ovaries, and the males are sexually mature. Males are up to 1.5 mm long by 150 μm in diameter, with the esophagus 229 μm long and the tail 122 μm. Females measure up to 3.8 mm by 173 μm in diameter, with the esophagus 312 μm long.

At the Eighth Day. Eggs appear in the uteri for the first time. Having completed their function, the males begin to disappear. This also occurs in *Enterobius vermicularis.* Females are up to 4.0 mm long.

By the Ninth Day. Nearly all of the females are gravid and the eggs are in the early stages of cleavage. They are still in the caecum.

By the Tenth Day. Practically all females are gravid and the eggs show further development. Females are up to 4.7 mm long. They are still in the caecum. Check the perianal region with transparent adhesive tape for eggs. What results did you get? How do you interpret this?

By the Eleventh Day. Larvae enclosed in eggs show further development, some with the gut formed. Where are the females? What results do you get with the transparent tape? What is your interpretation of this?

On the Twelfth Day. Adult female worms migrate from the rectum, as revealed by the presence of eggs on the tape. Sometimes one may see worms emerging from the anus. Many of them die in the perianal folds after oviposition. The manner of migration is similar to that of *Enterobius vermicularis.*

By the Thirteenth Day. Larval stages appear in the intestine and caecum. These are the result of the availability of eggs from the experimental infections two weeks earlier. The mice are becoming reinfected.

Other common species of Syphaciidae include *Wellcomia evaginata* of porcupines, and *Syphacia muris,* normally of white rats but capable of infecting white mice. In *S. muris,* the anterior mamelon is near the midbody, and the eggs are much smaller (75 $\times$ 29 μm). For further descriptions of *S. muris* and *S. obvelata,* consult Hussey (1957).

Litomosoides carinii

Animals required are infected cotton rats that may be live-trapped in the southeastern United States or purchased from Florida, where the parasite is enzootic. White rats may be used as nurse animals for developing colonies of mites. Cotton rats, wild brown rats, and often white rats are naturally infested with tropical rat mites.

To obtain mites to start a colony, place one or two infested rats in a cage provided with a tray beneath containing wood shavings. Engorged females drop off the host animals to lay their eggs. Twenty-four hours after putting the infested animals in the cages, examine the shavings by scattering and moving them over a large sheet of white paper to disclose the engorged female mites. Collect 50 to 100 mites by means of an aspirator bottle or a moist camel's hair brush to start a colony of uninfected mites.

A large number of mites can be reared for this experiment in an artificial nest constructed from a small, tight wooden box and homemade cages. Cover the bottom of the box with several layers of newspapers and then with a half-inch layer of loam soil (not sand) over which straw (not grass) is placed and bunched around the sides. The soil absorbs the urine voided by the rats, and the straw provides a place for the mites to hide and oviposit and for the young to develop.

Construct small but comfortable cages for individual rats from one-half-inch mesh hardware cloth bent and fitted appropriately. Put them on the straw and pack more about them. Put an uninfected white rat in each small cage, provide it with food and water, and place mites on them as culture animals. A thick ring of Vaseline or soft grease around the inside of the box above the cages and straw will discourage mites from crawling out. If the box is supported over a tray containing old lubricating oil, those mites negotiating the grease ring will be unable to escape into the room.

Within ten to 14 days, the colony will contain hundreds of mites. It will consist of: (1) bloodsucking females and nonbloodsucking males, (2) eggs, (3) hexapod nonfeeding larvae, (4) octopod bloodsucking protonymphs, and (5) nonfeeding deutonymphs that molt and develop into adults.

Engorged females are reddish and have a white Y-shaped mark on the dorsal side. Eggs are laid in two to three days after each engorgement and hatch in about 30 hours at room temperature. The whitish hexapod larvae about the size of the eggs do not feed, but they molt when about 24 hours old to form eight-legged protonymphs. They must engorge before being able to molt into deutonymphs 24 hours later. Deutonymphs do not feed but molt when 24 to 36 hours old into adults. The total time required to complete the life cycle of mites is 11.5 days at room temperature.

Uninfected mites are essential as a starting point for experimental studies. They can be obtained easily by taking advantage of the developmental stages of mites. Place 50 or more adult females on a rat whose cage stands over a tray of water. Upon engorgement, the mites drop onto the surface of the water and are transferred to brood vials kept at room temperature. Oviposition takes place in them and the eggs hatch. Hexapod larvae soon molt to protonymphs, which must have a blood meal. Use uninfected white rats for feeding the mites to continue their development. After the protonymphs have molted to the deutonymphs, they transform, without feeding, into the adult stage. Hungry and uninfected female mites are ready for infection.

To infect the mites, put adult females (protonymphs are not easily infected) on an infected cotton rat known to have microfilariae in the peripheral blood as demonstrated by stained blood smears. Allow the mites to engorge for 24 hours, at which time around 50% of them become infected; collect them from the surface of the water, and put in incubation vials. The mites must be allowed to feed on parasite-free rats on the fifth and tenth days following the initial engorgement. By the 15th day, the larvae have developed to the third stage. They are infective to rats and may be transmitted through the bites of the mites.

Dissect the mites in saline solution, beginning two hours after they drop off the rat and every four hours thereafter during the first day to determine the sequence of events in the development of the larvae. Dissect other mites daily for the next 14 days and examine them for larvae, molts, and growth.

Larval worms in the mites fall into three separate and distinct stages. They are: (1) first-stage larvae, which include the microfilariae and finally the thick, stumpy, sausage forms which appear between the ninth and thirteenth days; (2) somewhat slender second-stage worms appearing between the ninth and thirteenth days; and (3) third-stage larvae, which are slender and longer than the preceding stages.

First-stage microfilariae in the stomach at the end of the blood meal are similar to those in the peripheral blood. After six hours they are distributed randomly throughout the stomach. During a 24-hour residence in the stomach, the embryonic sheath is lost, and by the end of this time most of the microfilariae have migrated through the stomach wall into the hemocoel.

During the first week in the hemocoel, microfilariae gradually transform from the long, slender shape to the short, thick sausage form. At first they become thickened equatorially. On the sixth day the slender sickle-shaped tail appears along with other characteristics of the sausage form, which is attained on the seventh day. Molting begins on the eighth day, as indicated by the appearance of the loose cuticle at the ends of the body.

The first molt is completed by the ninth day while the body is still stumpy. By the end of the thirteenth day, nearly half of the larvae have molted and are in the third stage.

Third-stage larvae are heralded by loosening of the cuticle as early as the ninth day, and the second molt in some individuals is completed by the tenth day, producing third-stage larvae. The majority of the larvae have completed the molt by the thirteenth day. The resultant infective larvae are up to 1 mm long.

Female mites that have fed on parasitized rats with microfilarae circulating in the peripheral blood harbor infective third-stage larvae by the end of two weeks. Infection of parasite-free cotton rats can be accomplished easily and with a high degree of confidence by allowing these mites to feed on them. In order to follow the basic aspects of development in the vertebrate host, infect five clean, young cotton rats by exposing each of them to 50 or more infected mites.

Four rats are to be examined postmortem for the developing stages of the worms in the pleural cavity. Keep the fifth one to determine the prepatent period.

After killing each infected rat, the skin is removed and the thoracic cavity opened. Carefully remove the lungs and heart to a large dish of saline solution and wash them to remove any larvae from the serosal surfaces. Next wash out the thoracic cavity with the solution, catching all of the fluid. Sediment and decant the fluid until clear of blood. Any worms that have reached the pleural spaces from the site of inoculation by the mites can be recovered in this manner.

Within 18 hours after inoculation by the feeding mites, third-stage larvae are nearly 1 mm long and begin arriving in the thoracic cavity. Examine one rat on the second day after infection.

During the first week, the body attains a length of 1.5 mm. A well-defined group of cells in the anterior half of the body represents the ovary and vulva. The cloaca is distinct and the spicules show development. The buccal capsule, which is characteristic of this stage, is long and narrow with the outer surface of the wall bearing faint encircling thickened rings. By the end of the seventh day, evidence of the beginning of the third molt appears when the cuticle loosens and the third-stage buccal capsule is hanging free. Examine another rat on the seventh day.

The third molt is accomplished as clealy as the eighth day after infection. Recently molted fourth-stage larvae are about 1.2 mm long. Growth continues for 24 days, during which time the worms develop sexually and grow in size. Males become 6.4 mm long and females 8.8 mm. The reproductive organs assume recognizable forms. Examine a rat 20 and 22 days postinfection.

The final molt occurs about 23 to 24 days postinfection and the worms enter the preadult stage. They increase in length 1 to 2 mm during the molt and grow rapidly. The buccal capsule is thick-walled, cylindrical, and the outer surface of the wall has well-defined encircling rings. Examine a rat on the twenty-fifth to twenty-sixth days.

Examine a drop of blood prepared as a smear from the tail of the remaining rat at weekly intervals to determine the prepatent period. Microfilariae may appear in the blood between the fiftieth and ninety-third days. When were they first seen in your rat?

Trichinella spiralis

Infective larvae in muscle tissues, a few white mice, some cages, and a bit of ingenuity are necessary for demonstrating the life cycle. Larval stages from the intestine, body fluid, blood, serosa, and muscles should be studied alive in saline solution, using vital stains (see p. 259), as well as fixed and stained permanent mounts. There are three general developmental periods in the life cycle. They occur sequentially and in different parts of the body. The purpose of this study is to learn where to find them and how to recognize them.

The first period begins in the uterus of the adults living in the small intestine, where larviposition occurs. The newly born larvae may enter the intestinal circulation and be carried to the heart, from which they are carried by the arterial system to the skeletal muscles, or they may reach the muscles via the body cavities and serous membranes.

The second period pertains to the development of the different parts of the reproductive system and other organs and takes place in the muscles of the host. Encapsulation of the larvae terminates this period.

The third period is in the intestine of the host, beginning when the encapsulated larvae are ingested with meat and liberated from the capsule. The period ends with the appearance of sexually mature worms.

Wild rats obtained on the garbage dumps where organic refuse is deposited generally can be depended on to provide a source of larvae for beginning the study. Often, however, the department will have infected white rats on hand or be able to obtain one.

Because of their small size and low cost, white mice are preferred as experimental hosts in this study. Times given below for features in the life cycle are those that occur when mice are used. If white rats, which work equally well, are substituted, consult Villella in Gould (1970). In rats, the timing of events, particularly molting, apparently differs from that in mice. Feed each of 30 hungry mice about 300 encapsulated larvae in small pieces of heavily infected muscle tissue from the donor rats.

Beginning four hours postinfection and continuing through the next 40 hours, examine a mouse every four hours. Remove intestine, cut into short lengths, slit open longitudinally, and place in saline solution at 37 C. Occasionally agitate the intestinal segments with a dissecting needle to separate the emerging worms. After about an hour, remove the intestine and clean the solution by sedimentation and decantation preparatory to examination for larvae under the compound microscope. Compress the opened intestine between two glass plates and examine it with transmitted light under the high power of a dissecting microscope (or low power of a compound microscope) for worms that have entered the mucosa. Adult females will be present in the intestinal mucosa.

Some larvae fail to enter the intestinal mucosa and are passed alive in bits of undigested material or free in the feces. Feed fresh fecal pellets passed by mice during the first two days after a meal of trichinous meat to three trichina-free mice. About ten days later, examine the diaphragm, tongue, masseter muscles, and intercostal muscles by the compression method. Presence of larvae in these muscles is evidence that they were in the feces.

Between five and 14 days postinfection, examine the body fluid, blood, and serous membranes for larvae to determine the time and route of migration. Beginning with the fifth day, examine a mouse every 24 hours for ten consecutive days. Kill the animal, place it on its back, make a midventral and two lateral incisions, free the diaphragm, and fold back the body wall to expose the thoracic and abdominal cavities. With a pipette, transfer any body fluid to a slide and add a cover glass for examination under a compound microscope for young larvae. To collect the blood, ligate the hepatic portal vein near the intestine and the aortic arch above the heart. Cut the blood vessels so that the ligatures will keep the blood within the limits bounded by them and remove the lungs, heart, and liver as a unit to a dish of saline solution, where they should be opened to remove the blood. Sediment and decant until the solution is clear and examine for the larvae. Remove some serosa, transfer to a slide, smooth out in a drop of saline solution, add a cover glass, and examine under a compound microscope.

Larvae begin entering the skeletal muscle bundles on about the sixth day postinfection and undergo a long period of development, which for convenience is divided into an early and a late one.

The early larval stages extends from the sixth to the twelfth days. On the seventh day, they are shorter than the blood forms, there are 18 to 24 cuboidal cells in the esophagus, a syncytial cord extends to the posterior tip of the body, and a separate rectum has formed. Larvae are coiled on the ninth day, the intestine is a thin-walled tube, and the rectum differentiated. By the twelfth day, larvae are 270 μm long and 21 μm in diameter.

The late larval stage includes the period from the fourteenth day through the twenty-ninth day, during which time development has advanced markedly and encapsulation has been completed. By the fourteenth day, the stichosome is 146 μm long; on the sixteenth day, the sexes are differentiated, with the rectum of the females 21 μm long and the males 28 μm; by the seventeenth day, larvae are U-shaped or coiled. Encapsulation begins about the twenty-first day and is completed by the twenty-ninth. No molt occurs while larvae are in the muscles, either free or encapsulated. On each of the eight days indicated, kill a mouse, and examine one or more of the heavily infected muscles, compressed between slides, under a compound microscope, for the different developmental stages.

Four molts occur in the intestine of the host, following ingestion of infective larvae, as judged by the presence of a loose cuticular sheath and distinct anatomical changes.

First Molt. Males molt in nine to ten hours after entering the intestine and females in 12 to 13 hours. Are they in the lumen or intestinal wall? The body increased in length prior to the molt. The female genital primordium appears as a long chain of cells ventral to the stichosome and the male primordium is hook shaped.

Second Molt. This function occurs on about the seventeenth hour for males and the nineteenth for females. By this time, the cone-shaped copulatory papillae of the males are evident. The cuticular lining of the muscular part of the esophagus and the rectum is loose. Where are the larvae?

Third Molt. This molt occurs about the twenty-fourth hour for males and the twenty-sixth for females. The sex organs are well developed, the copulatory papillae curved, and the vaginal plate present. Where are the larvae at this stage?

Fourth Molt. Males molt about the twenty-ninth hour and females the thirty-sixth hour. After molting, the worms are mature.

Examine mice at appropriate times in an attempt to locate and identify the different larval stages.

Index

D

E

F

Q

R

S